Chloroplast Biogenesis and Plastid Interconversions

Vijay Kumar Dalal
Amarendra Narayan Misra
Editors

Chloroplast Biogenesis and Plastid Interconversions

Springer

Editors
Vijay Kumar Dalal
Department of Botany
Dayalbagh Educational Institute
Agra, Uttar Pradesh, India

Amarendra Narayan Misra
Jagadguru Kripalu University
Cuttak, Odisha, India

ISBN 978-981-95-0771-9 ISBN 978-981-95-0772-6 (eBook)
https://doi.org/10.1007/978-981-95-0772-6

This Springer imprint is published by the registered company Springer Nature Singapore Pte Ltd.
The registered company address is: 152 Beach Road, #21-01/04 Gateway East, Singapore 189721, Singapore

...Ra Dha Sva Aa Mi...
Dedicated to
Prof B C Tripathy, My PhD supervisor
and my other mentors
Prof Zach Adam,
Prof Ziv Reich
Prof Agepati S Raghavendra
Prof Subramanyam Rajagopal
Prof Govindjee

Vijay Kumar Dalal

Foreword

I am highly delighted to write a foreword to the new book *Chloroplast Biogenesis and Plastid Interconversions*, edited by Vijay Kumar Dalal and Amarendra Narayan Misra. Basic knowledge and exploitation of plastids is a key for the commercial utilization of plants in the form of starch grain production by amyloplasts, oil by leucoplasts, and the color of fruits and flowers in the chromoplasts. The book is a wonderful contribution in *plastid biology* providing the readers in-depth information on the generation of green plastids from the proplastids and etioplasts, as well as on the formation and interconversions of other types of plastids.

Scientific enquiry into the plastid interconversion has revealed a wealth of knowledge about the genetic and biochemical processes that enable these organelles to interconvert and adapt to changing environments, developmental stages, and metabolic demands. *Chloroplast Biogenesis and Plastid Interconversions* highlights the dynamic interplay between the chloroplast and molecular mechanisms of other plastid types, their functioning, conditions, and factors significant for interconversions, gene expression, and the orchestration of various signaling pathways that ensure their functionality and survival.

There are 12 chapters in the book, and Chap. 1, by Vijay Kumar Dalal, Fahima Dilnawaj, and Amarendra N. Misra, justifies the title of the book by introducing as well as describing key details of all types of plastids, their functioning as well as their regulation. This is followed by five chapters that clearly explain the functioning of chloroplasts, factors involved in their biogenesis, as well as the latest developments in the field. Thus, these five chapters serve the goal of unlocking the secrets of chloroplast biology. The following Chap. 2, coauthored by Shina Sasi, Tanveer A. Khan, Mayank A. Gururani, tells us, in depth, about the intricacies of the genetic systems of chloroplasts. In Chap. 3, Tulika, V. Agarwal, Neha Singh, Vijetna Singh, and Bhumi N. Tripathi focus on the retrograde signaling, which is an important process governing the stoichiometry of proteins in developing chloroplasts. How the chloroplasts import all the required (~3000) proteins, although the chloroplast DNA codes for only ~5% of the total proteins, is the topic of Chap. 4 by Siddhartha Dutta which describes the fascinating protein import apparatus. Chapter 5 (by Joanna Grzyb and Beata Myśliwa-Kurdziel) includes a through discussion about the various components of the thylakoid membranes, their assembly through numerous assembly factors, all the elements that are essential for their proper functioning, as well as the degradation of the thylakoids. Chapter 6, by Marija Vidović, takes us through the wonderful journey of variegated leaf-plants—when, why and how they are formed? Since the biggest obstacle in the achievement of United Nations Sustainable Development Goals is the changing climate and the various environmental stresses, Kaushik K. Panigrahi and coauthors (Jaspreet Kaur; Kishan Kumar, Sudhan S. Kaintura, Sourav R. Mohapatra, Prabhasmita Shatpatthy, Ayesha Mohanty, and Kaushik K. Panigrah), in Chap. 7, have wonderfully explored the mechanism of the various stresses and their responses by chloroplasts.

The rest of the book is dedicated to various types of plastids—their origin, function, propagation, and interconversion mechanisms. In Chap. 8, Akhilesh K. Singh, Santendra Soni, and Vijay Kumar Dalal discuss the current concepts about etioplasts, their significance, composition, and transition to chloroplasts. In Chaps. 9, 10, and 11, the readers would be pleased to read the importance of plastids and their interconversions—essential from the economic point of view. In Chap. 9, Nur A. Rejab and Najiah M. Sadali, from Malaysia, discuss this topic from the point of view of chromoplast biogenesis; whereas, in Chap. 10, Vijay Kumar Dalal discusses about the amyloplasts, the starch-synthesizing plastids which are very significant for the food processing industry for determining the dough characteristics. In a multi-authored Chap. 11, Najiah M. Sadali, with five coauthors (Mohammad A. Loh, Siddi A.M. Roszelin, Zul I.Z. Lubes. Mohammed N. Husri, and Nur A. Rehjab) present an in-depth chapter that deals with the transition steps of chloroplasts becoming elaioplasts, which synthesize and store oils. These chapters not only give us tremendous information about the current literature and the know-how about the commercially important plastids but also shed light on the molecular events responsible for their function, potentially paving the way for their manipulation to get commercial gains and to improve food availability as well as security for us all. Chapter 12 (by Jitendra Pandey and Gunjan Dubey) that discusses the senescing plastids-Gerontoplasts, is also critical for increasing the food production by delaying the senescence which is mediated through gerontoplasts formation. Understanding chloroplast to gerontoplast transition is important as the food synthesized in the plants is transformed providing us grains, tubers, fleshy roots, fruits, vegetables, and other beneficial structures.

Considering all of the above, I highly recommend this book to all plant biologists, plant physiologists, and scientists working to improve plant productivity; further, in my judgment, this book will prove to be a milestone for not only Masters, PhDs, and Postdoc students but also research scientists working in areas including chloroplast biogenesis and photosynthesis. All the contributors to this book are reputed scientists in their respective areas, and their collective insight is expected to deepen our understanding of this fascinating area.

In the end, I hope that by reading this book, we all can appreciate how even the smallest organelle such as the plastid, can influence the tapestry of the plant life in the territory of ecological, physiological, and evolutionary events.

Govindjee Govindjee
Department of Plant Biology, Department of Biochemistry,
and the Center of Biophysics & Quantitative Bioplogy
University of Illinois at Urbana- Champaign,
Urbana, IL, USA
(e-mail: gov@illinois.edu;
urls: https://www.life.illinois.edu/govindjee/;
https://www.life.illinois.edu/govindjee/recent_papers.html)

Preface

Rapidly expanding population has exacerbated the demand for food and plant-based products, for the generation of which photosynthesis has a great role to play. Photosynthesis is the fundamental process supporting life on Earth. Photosynthesis is achieved through a type of plastid called chloroplast. Therefore, after the first chapter in which we have described various types of plastids and set a framework of the book, the initial four chapters describe the genetic machinery responsible for chloroplast homeostasis, assembly, and disintegration of thylakoid membranes during biogenesis/degradation of chloroplasts, translocation of proteins into chloroplast and retrograde signaling essential for chloroplast biogenesis. Understanding the effects of climate change, particularly water and temperature stress on plant photosynthesis and carbon dynamics, is critical for plant yield, thus, seventh chapter is dedicated to stress responses of chloroplasts. Beside this, many later chapters dedicated to particular type of plastids, also briefly discuss the stress responses of that plastid, e.g. etioplasts chapter, beside discussing various aspects of etioplasts, also covers their responses to stress. Furthermore, every budding photosynthesis scientist would have been curious about the variegated leaves once in their lifetime; therefore, sixth chapter is dedicated to variegated leaves.

The yield of photosynthesis is derived in the form of fruits, grains, biomass production, flowers and other storage organs. This storage is materialized through other types of plastids such as amyloplasts, elaioplasts, proteinoplasts, chromoplasts, etc. Therefore, understanding the formation and functions of these plastids is of utmost importance to enhance commercial gains through better quality and yield of the produce.

The different types of plastids perform various processes such as food synthesis, storage, pollination, stress tolerance and help carry the plastid-gene-repertoire to the next generation. Although the genome of varied plastid types is same and quite small, still they achieve such diverse type of functions through importing different sets of proteins. This is very fascinating and presents an example of another layer of regulation for achieving diverse functions in an organism. The understanding of the functions of plastids is important for achieving better quantity and quality of the produce in terms of protein, oil, starch, and flavour and aroma content.

We have chosen the authors with their bona fide firmly established, and are in their prime of lives. There had been a distinct lack of books on the topic "plastids", although this subject being in the forefront of crop improvement strategies such as creating higher and desired profiles of sugar and starch in grains and fruits through amyloplasts manipulation, food quality improvement by enhancing the contents of antioxidants (e.g. carotenoids) through chromoplasts, oils through elaioplasts and stress tolerance (xanthophylls) through the manipulation of dessicoplasts and gerontoplasts. Understanding the chloroplast to chromoplast transition can help in invoking synchronous maturation which is essential for mechanical harvesting of fruits, e.g. tomatoes, chillies, apples, citrus etc and flower colour manipulations; chloroplast to gerontoplast for delayed senescence, increasing fruit shelf life, increased source to sink translocation of food; and etioplast to chloroplast for producing healthy seedlings that are more robust to stresses.

This book is suitable for undergraduate, postgraduate, doctorate students, early career researchers and established scientists in this area. This book genuinely reflects the facts and knowledge about plastids, written passionately in plain yet persuasive language, and starts with a fascinating introductory chapter on various types of plants, which tees up to the complex developmental, physiological, functional and regulatory aspects of varied plastids. Although the less abundant and newly discovered plastids are briefly introduced, they are not thoroughly discussed. The number of such plastids is so high that a separate book can be written about these.

The book serves as a fundamental knowledge as well as covers recent findings on plastids and their interconversions. Although proteinoplasts have not been covered separately, these topics have been dealt with in the introductory Chap. 1. The knowledge in the field of biology is growing at a fast face, more so with the integration of tools and technologies from the other fields such as bioinformatics, instrumentations, physics, maths, chemistry, material science and computing power, which has made it necessary to put forth a compendium of knowledge at regular intervals. It is our understanding that at least every 3–5 years a book covering a specialized topic such as this, with the latest information in the area, should be on shelves although there are only few precedences, e.g. *Structure and Function of Plastids* by Robert Wise and Kenneth Hoober (Editors; 2006th edition), *Annual Plant Reviews, Plastids* by Simon Geir Møller (Editor; 2004), *Cell and Molecular Biology of Plastids* by Bock, R. (Editor, 2007) and *Plastid Development in Leaves During Growth and Senescence* by Basanti Biswal, Karin Krupinska and Udaya C. Biswal (Editors, 2013). Therefore, we think, our book will potentially fill this gap. We hope an established scientist working in the area of plastids as well as a newcomer to the field will learn a lot about plastid types and intricacies of plastid interconversions by reading this book.

Agra, Uttar Pradesh, India — Vijay Kumar Dalal
Cuttak, Odisha, India — Amarendra Narayan Mishra

Acknowledgment

This book is visualized as a foundation for the basic understanding of the development of the photosynthetic process in developing plastids in higher plants. The project would not have been possible without the unwavering support and dedication of our authors and the reviewers who had given their best efforts to bring out the chapters in a decent form. It is of significance to present the current knowledge on the topic of plastid development from pro-plastids to chloroplasts and chromoplasts in an easily comprehendible and meaningful manner. We thank our authors who have brought out new insights to a topic of interest to Plant Biologists, Developmental Biologists, Molecular Biologist, and Biotechnologists, looking forward to a more colorful and beautiful new world for sustaining human life on earth. The theme of the book was cherry-picked based on the input of many of the colleagues who underlined the importance of this subject—plastid interconversions.

The book covers the following UN Sustainable Development Goals (SDGs) directly or indirectly for a livable, healthy, clean, and green earth. Although the book deals with plastid development and interconversions, it certainly focuses on a broader theme of photosynthesis, and carbohydrate, lipid, and protein storing bodies like amyloplasts, elaioplasts, and proteinoplasts. The energy harvesting mechanism and subsequent carbon dioxide fixation in chloroplasts, metabolic conversion of the fixed carbon compounds to other storage forms in plastids, and several plastid-dependent processes are vital for sustaining life on earth. Therefore, the following SDGs are related to this book.

SDG 1. No poverty
SDG 2. Zero hunger
SDG 3. Good health and well-being
SDG 7. Affordable and clean energy
SDG 11. Sustainable cities and communities
SDG 13. Climate action
SDG 15. Life on land
SDG 17. Partnership for the goals

To get a firsthand feedback of the book, foreword is an important feature. Therefore, we are thankful to honorable Prof Govindjee who agreed to write the foreword for this book, sparing his precious time. In addition, we would like to thank the publishing team who quickly responded to our queries and clarified doubts, and extended the submission deadlines whenever requested.

Last but not least, we thank our families who had been always with us to bear our "not so perceptive mind" during the tenure of book-writing. It is their happiness that adds to our joy of completing the book.

Vijay Kumar Dalal
Amarendra Narayan Misra

Contents

Editors and Contributors

About the Editors

Vijay Kumar Dalal earned his undergraduate and postgraduate degrees from the University of Delhi and IIT Kharagpur, respectively. He completed his Ph.D. from MDU, Rohtak, India in 2012 while working at the School of Life Sciences, Jawaharlal Nehru University, India, where he studied the effects of drought on photosynthesis and chlorophyll biosynthesis. His research interests are diverse and include chlorophyll biosynthesis, chloroplast biogenesis, abiotic stress responses, osmolyte protection, plant tissue culture, production of secondary metabolites, biofertilizers, biopesticides, protected cultivation, and hydroponics. During his postdoctoral research in Israel, he focused on chloroplast biogenesis in the tomato shoot apical meristem using RNA sequencing of laser-microdissected samples. He later conducted research on the protective effects of proline against oxidative stress on enzymes at the University of Hyderabad. Dr. Dalal also has industry experience, having worked at Bharat Biotech, Hyderabad. Currently, he is serving as an Assistant Professor in the Department of Botany at Dayalbagh Educational Institute (DEI), Agra, where he joined in 2019. He has authored several research papers in prominent plant science journals and contributed chapters to books published by reputed publishers. He has reviewed numerous manuscripts for international journals and has delivered talks at various national and international conferences and webinars. Dr. Dalal is a recipient of fellowships and awards from DST, DBT, CSIR, and PBC, and has secured research grants from NABARD, DEI, and AADEI.

Amarendra Narayan Misra brings over 48 years of research experience and 40 years of postgraduate teaching in Plant Biology, Molecular Biology, and Biotechnology. He currently serves as the Vice Chancellor of Jagadguru Kripalu University (JKU), Cuttack, Odisha, India. Following his superannuation in 2021 from Central University of Jharkhand, he held the position of Distinguished Professor at Centurion University of Technology and Management, Jatni, Bhubaneswar. He also served as Senior Professor and Dean at Sri Sri University, Cuttack. He has been a Guest Researcher at the Institute of Frontier Sciences, University of Tokyo, Japan. Prof. Misra has contributed significantly to academic leadership, having served as Vice Chancellor of three Indian universities and held various roles in university governance and administration. He has authored over 135 publications in the areas of Plant Biology, Biotechnology, and Bioinformatics. His research interests include the structure and function of photosynthetic membranes, luminescence, biosensors, plant tissue culture, and mechanisms of stress adaptation in tropical crop plants. He is a recipient of several prestigious fellowships, scholarships, viz. Fulbright (USA), JSPS (Japan), and the Commission of European Community. He has also received research support and recognition from agencies such as DBT, DST, CSIR, INSA, UGC, CEC, and FEBS. Prof. Misra has actively collaborated with scientists across more than ten countries, including Japan, Poland, Switzerland, Bulgaria, Czech Republic, Ukraine, the UK, and the USA.

Contributors

Variyata Agrawal Department of Biotechnology, Indira Gandhi National Tribal University, Amarkantak, Madhya Pradesh, India

Vijay Kumar Dalal Department of Botany, Dayalbagh Educational Institute, Agra, Uttar Pradesh, India

Fahima Dilnawaz School of Biotechnology, Centurion University of Technology and Management, Bhubaneswar, Odisha, India

Gunjan Dubey Department of Botany, University of Rajasthan, Jaipur, Rajasthan, India

Siddhartha Dutta Department of Biotechnology, School of Health Science and Translational Research, Sister Nivedita University, Kolkata, West Bengal, India

Joanna Grzyb Department of Biophysics, Faculty of Biotechnology, University of Wroclaw, Wroclaw, Poland

Mayank Anand Gururani Biology Department, College of Science, UAE University, Al Ain, United Arab Emirates

Mohd Naqiuddin Husri Malaysian Palm Oil Board, Kajang, Selangor, Malaysia

Zul Ilham Institute of Biological Sciences, Faculty of Science, Universiti Malaya, Kuala Lumpur, Malaysia

Anamika Jangra Division of Genetics & Tree Improvement, Forest Research Institute, Dehradun, Uttarakhand, India

Sudhan S. Kaintura Division of Genetics & Tree Improvement, Forest Research Institute, Dehradun, Uttarakhand, India

Jaspreet Kaur Division of Genetics & Tree Improvement, Forest Research Institute, Dehradun, Uttarakhand, India

Tanveer Alam Khan Khalifa Center for Genetic Engineering and Biotechnology, Al Ain, United Arab Emirates

Kishan Kumar Division of Genetics & Tree Improvement, Forest Research Institute, Dehradun, Uttarakhand, India

Mohd Azrai Loh Institute of Biological Sciences, Faculty of Science, Universiti Malaya, Kuala Lumpur, Malaysia

Purabi Mazumdar Centre for Research in Biotechnology for Agriculture (CEBAR), Universiti Malaya, Kuala Lumpur, Malaysia

Amarendra Narayan Misra Jagadguru Kripalu University, Cuttack, Odisha, India

Faculty of Sciences, Sri Sri University, Cuttack, Odisha, India

Ayesha Mohanty AINP on Soil Biodiversity Biofertilizers, Odisha University of Agriculture and Technology, Bhubaneswar, Odisha, India

Sourav Ranjan Mohapatra Department of Forest Biology and Tree Improvement, Odisha University of Agriculture and Technology, Bhubaneswar, Odisha, India

Beata Myśliwa-Kurdziel Department of Plant Physiology and Biochemistry, Faculty of Biochemistry, Biophysics and Biotechnology, Jagiellonian University, Krakow, Poland

Jitendra Kumar Pandey Department of Botany, Government Post Graduate College, Tikamgarh, India

Maharaja Chhatrasal Bundelkhand University, Chhatarpur, India

Kaushik Kumar Panigrahi AICRN on Potential crops, Odisha University of Agriculture and Technology, Bhubaneswar, Odisha, India

Nur Ardiyana Rejab Centre for Research in Biotechnology for Agriculture (CEBAR), Universiti Malaya, Kuala Lumpur, Malaysia

Institute of Biological Sciences, Faculty of Science, Universiti Malaya, Kuala Lumpur, Malaysia

Sitti' Aisyah Mohd Roszelin Institute of Biological Sciences, Faculty of Science, Universiti Malaya, Kuala Lumpur, Malaysia

Najiah Mohd Sadali Centre for Research in Biotechnology for Agriculture (CEBAR), Universiti Malaya, Kuala Lumpur, Malaysia

Shina Sasi Khalifa Center for Genetic Engineering and Biotechnology, Al Ain, United Arab Emirates

Prabhasmita Shatpathy Department of Plant Physiology, College of Agriculture, Odisha University of Agriculture and Technology, Bhubaneswar, Odisha, India

Akhilesh K. Singh Department of Biotechnology, School of Life Sciences, Mahatma Gandhi Central University, Motihari, Bihar, India

Neha Singh Department of Biotechnology, Indira Gandhi National Tribal University, Amarkantak, Madhya Pradesh, India

Vijetna Singh Department of Biotechnology, Indira Gandhi National Tribal University, Amarkantak, Madhya Pradesh, India

Santendra K. Soni Department of Botany, Dayalbagh Educational Institute, Agra, Uttar Pradesh, India

Bhumi Nath Tripathi Department of Biotechnology, Indira Gandhi National Tribal University, Amarkantak, Madhya Pradesh, India

Tulika Department of Biotechnology, Indira Gandhi National Tribal University, Amarkantak, Madhya Pradesh, India

Marija Vidović Group for Plant Molecular Biology, Institute of Molecular Genetics and Genetic Engineering, University of Belgrade, Belgrade, Serbia

Plastids and Their Interconversions

1

Vijay Kumar Dalal, Fahima Dilnawaz, and Amarendra Narayan Misra

Abstract

Plastids are crucial subcellular organelles developed to carry out specific functions, such as photosynthesis and the synthesis and storage of metabolites. Plastids exist in various forms with distinct attributes, permitting them to operate in a wide range of specialized developmental processes in plants, tissues, and cells, as well as responding to different environmental cues. The diverse functions of plastids and their transitional states are based on their characteristics. Moreover, the current knowledge of their regulatory mechanisms is emphasized at both the level of gene expression and protein modification. With a specific emphasis on the processes involved in the development of green and nongreen traits and their interconversion, the gene-specific molecular network provides substantial support for enhancing various features associated with plastid interconversion.

Keywords

Plastids · Proplastids · Chloroplasts · Chromoplast · Gerontoplasts · Plastid interconversion · Green traits · Nongreen traits

V. K. Dalal
Department of Botany, Dayalbagh Educational Institute, Agra, Uttar Pradesh, India

F. Dilnawaz
School of Biotechnology, Centurion University of Technology and Management, Bhubaneswar, Odisha, India

A. N. Misra (✉)
Jagadguru Kripalu University, Cuttack, Odisha, India

Faculty of Sciences, Sri Sri University, Cuttack, Odisha, India

V. K. Dalal, A. N. Misra (eds.), *Chloroplast Biogenesis and Plastid Interconversions*, https://doi.org/10.1007/978-981-95-0772-6_1

Abbreviations

AFM	atomic force microscopy
ARF	auxin response factor
ARR	auxin response regulator
Aux/IAA proteins	auxin/indole-3-acetic acid family proteins
CAS	calcium-sensing receptor
CCD1	carotenoid cleavage dioxygenase 1
Chl	chlorophyll
CHUP1	chloroplast unusual positioning 1
CGA1	cytokinin-responsive GATA factor 1
CUE1	CAB underexpressed
CV	chloroplast vesiculation
CRTI	bacterial phytoene desaturase
DXS	deoxy-D-Xylulose 5-phosphate synthase
EC	epidermal chloroplasts
EDS5	enhanced disease susceptibility 5
FCC	fluorescent chlorophyll catabolites
GNC	GATA-nitrate-inducible carbon metabolism-involved
GNL	GNC-LIKE
G6P	glucose 6-phosphate
GPT	G6P translocator
GS-GOGAT cycle	glutamine synthetase (GS) and glutamate synthase (GOGAT) cycle
GSH1	γ-glutamylcysteine synthetase
GST	glutathione S-transferase
HY5	elongated hypocotyl 5
LCYB	lycopene cyclase-beta
MSH1	MutS HOMOLOG 1
MVA	mevalonate pathway
NAC TFs	NAM, ATAF1/2, and CUC2 transcription factors
NCED	9-cis-epoxycarotenoid dioxygenase
NOR	non-ripening
OR	orange
PCCs	pavement cell chloroplasts
Pchlide	protochlorophyllide
PDS	phytoene desaturase
PFTFs	plastid fusions and/or translocation factors
PG	plastoglobules
PhANGs	photosynthesis activated nuclear genes
PIN3	PIN formed 3
pINV	plastidic invertase
PLBs	prolamellar Bodies
POR	protochlorophyllide oxidoreductase
PPT1	phospho*enol*pyruvate (PEP)/ Phosphate Transporter 1

PSII	Photosynthesis complex II
*psb*A	gene encoding for PSII D1 protein
pSuT	sucrose transporter
PSY	phytoene synthase Y
RIN	ripening inhibitor
ROS	reactive oxygen species
rrn	gene encoding for ribosomal RNA
SAG	senescence-associated genes
SAM	shoot apical meristem
SAV	senescence-associated vacuoles
SEPs	sensory epidermal plastids
TAGL1	tomato agamous like1
TCP14	teosinte branched, cycloidea, and PCF 14
TIC236	translocon of inner membrane complex 236
WRKY	TFs containing sequence WRKYGQK at the N-terminus of DNA-binding domain

1.1 Introduction

Plastids originated from a process called endosymbiosis, which was an evolutionary process of embedding photosynthetic prokaryotes into the eukaryotic cells. The process ensued co-evolution of the enveloped plastid and eukaryote cells, with significant alterations in the roles of the plastids, which were involved in the establishment of gene regulatory networks (Inaba et al. 2011; Choi et al. 2021). The gene-specific molecular networks and their participating genes can help us understand plastid interconversion-related characteristics. A plastid is a specialized organelle that is typically found in plant cells, algae, and certain parasitic organisms. Most known plastids are partially independent organelles having the genes coding for photosystem-associated polypeptides (Miyagishima et al. 2011). Plastids are well-known for their capacity to convert solar energy to chemical energy in metabolites. Chloroplast in photosynthetic cells is the archetypical plastid (Kato and Sakamoto 2010). Although new plastids are interesting since they are replicated or transitioned from existing plastids, they cannot be produced or born; therefore, the plastid interconversion is crucial and of particular interest to cells, tissues, and species (Sierra et al. 2023). The unique characteristic of every type of plastid is briefly discussed in this review to aid in an understanding of their overall nature.

1.2 Types of Plastids and Their Roles

Plastid color, form, stored metabolite or product, and ultrastructure can be used to classify them into a variety of groups (Wise 2007). Each plastid type has a unique role that is closely linked to its characteristics. The "proplastids" are undifferentiated plastids that are commonly seen in meristematic and reproductive tissues. They

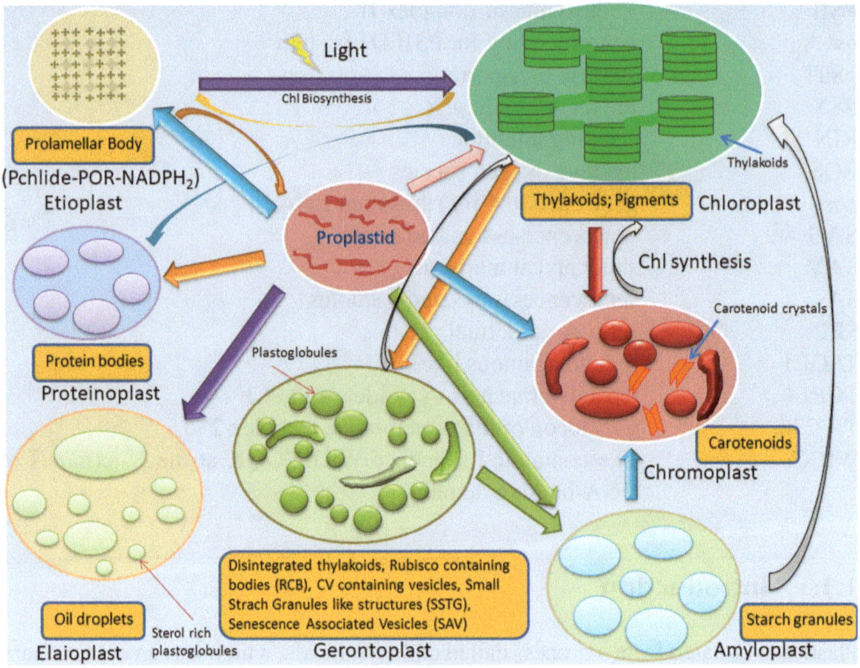

Fig. 1.1 Different types of plastids in plants and their interconversion. Broader and straight arrows denote the prevalent processes; curved arrows denote lesser prevalent processes. Chl: chlorophyll; Pchlide: protochlorophyllide; POR: protochlorophyllide oxidoreductase; NADPH: reduced nicotinamide adenine dinucleotide phosphate; CV: chloroplast vesiculation.

can be transformed to "chromoplasts" (any color except green, e.g., yellow, orange, blue, or red), "leucoplasts" (white), and "chloroplasts" (green color). Leucoplasts can be further divided into starch-storing "amyloplasts," protein-storing "proteinoplasts," and lipid-storing "elaioplasts." "Etioplasts" are intermediate forms, and "gerontoplasts" are senescent forms of the chloroplasts (Fig. 1.1) (Jarvis and López-Juez 2013). Usually, the whole leaf in dicots or a segment of the monocotyledon cereal leaf has been used to characterize the development of plastids. This hampers the evaluation, where distinct plastids are present but their contribution cannot be distinguished due to the availability of lower sample and the presence of rare cell types (Sierra et al. 2023). This is also the case of other organs, such as embryos, floral organs, gametophytes, or roots (Chu and Li 2015). Thus, plastids of epidermal cell pavement and vascular cells of the *Arabidopsis* were earlier defined as leucoplasts, with the exception of stomatal cells (Lawson et al. 2014; Brunkard et al. 2015). However, with the advent of varied novel techniques, e.g., Chl a fluorescence (Mishra 2018), AFM, super resolution microscopy, and single-cell analysis techniques, e.g., single-cell transcriptomics (scTranscriptomics), scProteomics, and scMetabolomics, and pertinent mutant's isolation and characterization, it has been possible to re-designate the previous categorization of plastid types (Barton et al.

2018). In the L1 layer, the cells were found to contain chloroplasts above the SAM, but disintegrated away from SAM, toward the leaf (Charuvi et al. 2012; Dalal et al. 2018). These cells were earlier designated as leucoplasts (Brunkard et al. 2015). Recent evidence suggested the involvement of chloroplasts in the sensing and perception of environmental signals and coordinating the defense response, although evidence is still needed to demarcate them from the mesophyll cells, as this property may also be possessed by specialized mesophyll cells. The details about different plastids are given below.

1.2.1 Proplastids

Proplastids are undifferentiated plastids that preserve a modest plastid structure in order to facilitate intergenerational organelle transmission. Proplastids are minuscule in size, lacking any discernible morphological characteristics, and colorless in comparison to other plastid forms (Jarvis and López-Juez 2013; Liebers et al. 2017). They are mostly found in the meristematic and egg cells of plants, and in some species, like *Pelargonium* and barley (*Hordeum vulgare*), they are also occasionally present during the formation of pollen (Hagemann 2004; Sakamoto et al. 2008; Gajecka et al. 2020). *Arabidopsis* TIC236, a translocon component in the inner membrane of chloroplast, is critical for importing chloroplast-targeted preproteins in proplastids and chloroplast division (Choi et al. 2021; Liu et al. 2024). The proplastids can directly give rise to several types of plastids, e.g., chloroplasts, etioplasts, chromoplasts, amyloplasts, and leucoplasts, which are described below.

1.2.2 Etioplasts

Etioplasts are formed mostly when the seeds germinate under the soil and when they have not been exposed to light (Bykowski et al. 2020). In some fully developed green plants, shoot parts often remain in the soil after germination and can develop special plastids that contain Pchlide or Pchlide esters, called etioplasts (Kakuszi et al. 2017). Although in some plants, e.g., *Phaseolus vulgaris*, a little amount of Chl develops in the hypocotyl inside the soil due to the light percolation effect (Kakuszi et al. 2016). Etioplasts are of yellow in color due to the accumulation of carotenoid pigments. Besides developing in the shoot parts of seeds germinating below the soil, these plastids are also found in the darkened/covered parts of the light exposed aerial tissues of the plants, e.g., in the inner layers of the cabbage, lettuce leaves, covered leaf base/sheaths of some monocots and sometime even during the dark periods of diurnal cycles. Etioplasts contain paracrystalline prolamellar bodies (PLBs) that are tubular crystalline structures and contain the oligomeric complexes of Pchlide-POR-NADPH (protochlorophyllide-protochlorophyllide oxidoreductase-nicotinamide adenine dinucleotide phosphate reduced) (Wójtowicz et al. 1996). The Chl biosynthesis is arrested at the Pchlide stage in the etioplasts due to the absolute requirement of light by the enzyme POR, which is also called

LPOR in angiosperms. Gymnospermic POR does not require light, so it is also called dark POR (DPOR). The PLB lattices are hollow tubules. Flattened porous disc-like structures extend beyond the lattice and are called prothylakoids (PTs). When the light falls on the etioplasts, the Chl biosynthesis resumes; Pchlide is converted to Chlide, which, through further steps and the Chl cycle, generates Chl a and Chl b pigments that are integrated into the proteins to form functional photosynthetic complexes involved in photosynthetic electron transport chain and other processes. The size and number of etioplasts are usually smaller than the chloroplasts. Etioplasts are the intermediate stage in the development of chloroplasts from proplastids. They prepare the plants for the formation of thylakoid upon exposure to light. In the absence of etioplasts, plants would have to prepare pigments and lipid components of thylakoids from scratch; however, etioplasts prepare the intermediary stages of these components, so that plants save time in making thylakoids and active photosynthetic complexes form as soon as 14 hours after light exposure (Pipitone et al. 2021). Moreover, etioplasts can synthesize certain metabolites and proteins that help plants survive under dark conditions. They produce plastoglobules mostly filled with carotenoids, violaxanthin, and lutein (Choi et al. 2021). They produce signaling molecules. e.g., phytochromes and their downstream components, e.g., PIFs, that help light reception and response of plants upon exposure to light. There are several aberrations that arise in etioplasts and PLBs due to stress conditions (Dalal and Tripathy 2012, 2018; Tripathy and Dalal 2013). For example, PLBs change from the closed structure to open type structures as a result of chilling stress (Skupień et al. 2017). Furthermore, carotenoid and galactolipid composition are modified in such PLBs (Skupień et al. 2017). In an elegant cryoelectron and confocal microscopy study, Kowalewska et al. (2016) showed that the tubular structure of the PLB transforms directly to linear slates without dispersion to the vesicles.

1.2.3 Leucoplasts and Derivatives

Leucoplasts are characterized by their white structures, which are typically present in non-photosynthetic tissues and serve as storage organelles (Carde 1984; Altamura et al. 2024). Leucoplasts are further classified into three subtypes: amyloplasts, elaioplasts, and proteinoplasts (Sadali et al. 2019).

1.2.4 Amyloplasts

Amyloplasts are distinguished by their high-density starch content, which is stored as starch granules. During the synthesis of amyloplast membranes, lipids such as free fatty acids, lysophospholipids, lysophosphatidylcholine, and lysophosphatidylethanolamine are added to the starch granules (Gayral et al. 2019). Amyloplasts are primarily found in sink tissues such as seeds, fruits, tubers, and roots for carbon storage, however, they are also frequently present in diverse tissues such as leaves, stems, and roots for temporal storage (Jarvis and López-Juez 2013). Interestingly, amyloplasts exhibit instability in certain instances, such as in *Arabidopsis* leaves. The storage and depletion of starch in these leaves are very dynamic, fluctuating on a daily basis in response to the presence or absence of photosynthetic activity (Fernandez et al. 2017). In addition to their role in storing substances, the amyloplasts found in the roots of *Arabidopsis* plants also play a part in transmitting signals related to gravitropism (Nakamura et al. 2019). Amyloplasts, in contrast to other plastid forms, frequently coexist in the same cell with other plastid types. Nevertheless, combinatory forms of plastid known as amylo-chromoplasts are found in the tissues of species, like winter squash, peach palm fruit, and sweet potato tuber. These plastids contain starch granules alongside carotenoid crystals (Hempel et al. 2014). Additional starch granules can be found inside many plastid types, including chloroplasts.

Amyloplasts are characteristic of the parenchymatic tissues of storage organs (tubers, rhizomes, roots, etc.) but can be also found in root cap cells, where they are associated with geotropism and with root gravisensing in the columella region of the calyptra (Thomson and Whatley 1980). In addition, amyloplasts of some species contain key enzymes of the GS-GOGAT cycle and are involved in nitrogen assimilation similarly to root nodule proplastids (Wise 2007). The tubers and stolons of potatoes are representative organs that accumulate amyloplasts (Naeem et al. 1997; Sagisaka 2008). Starch mainly accumulates during the "maturation" period, such as with banana (Solis-Badillo et al. 2020) and Italian arum (Bonora et al. 2000). For apples, in vitro experiments showed that the formation of amyloplasts occurs in the callus and endosperm (Sagisaka 2008). For rice and wheat, the accumulation of amyloplasts is mainly found in the endosperm (Wurtzel et al. 2012). In amyloplasts, starch is produced in the matrix space (stroma) and forms starch grains, which exhibit different morphologies depending on the plant species and have been extensively studied in various staple crops. The diameter of starch grains in corn and in sorghum is about 10–20 μm, in rice 3–10 μm, while in barley and wheat, two types of granules are present: large granules of 15–35 μm and 20–35 μm and small granules of 1–10 μm and 2–10 μm, respectively (Matsushima et al. 2014; Matsushima and Hisano 2019).

Transcriptional and translational activities are very low in amyloplasts as compared with chloroplasts (Brosch et al. 2007). Ribosomal operon *rrn* and the *psb*A gene are some of the more transcribed genes in amyloplasts (Brosch et al. 2007). Transcription of *rrn* is mainly performed by PEP in amyloplasts. In tomato (*Solanum tuberosum*), many TFs are known to regulate starch formation in amyloplasts (reviewed in Li et al. 2021; Liu et al. 2022; Li et al. 2024; Mathura et al. 2024). Certain TFs are known to induce amyloplast formation from proplastids, e.g. bZIP58 (Mingo-Castel et al. 1991; Wang et al. 2013a; Brosch et al. 2007; Yang et al. 2024). StbHLH93 TF also regulates the proplastid to amyloplast transition (Yang et al. 2024). Lower number of proplastids differentiated into amyloplasts in StbHLH93-KD lines in contrast to more amyloplast differentiation in StbHLH93-OX lines (Yang et al. 2024).

Besides the transcriptome, proteomic studies have also been conducted to assess the proteome profile of amyloplasts or the chloroplast-amyloplast transition. Based on the proteomics study, Balmer et al. (2006) concluded that amyloplasts possess high metabolic activity, although others had reported the opposite for amyloplasts. Liu et al. (2022) compared the chloroplast proteome of the potato with the amyloplast proteome. As per their investigation, these two organelles showed more differences than similarities. The starch metabolic pathway was present in both organelles. Fructose and mannose metabolism pathways were expressed in both organelles but their biosynthesis/degradation pathway is distinct in the chloroplasts and amyloplasts. Plastidic invertase (pINV) and sucrose transporter (pSuT) proteins were only present in chloroplasts. In contrast, two isoforms of starch synthases (SS5 and SS6) and glucose-6-phosphate (G6P) translocator 1.1 (GPT1.1) were explicitly detected in amyloplasts (Liu et al. 2022). Although earlier considered inactive, the potato was found to form starch granules in leaves in a recent study (Cui et al. 2020). Li et al. (2024) determined that the starch synthesizing/breakdown proteins, Chl breakdown proteins, and plastoglobule proteins are increased in amyloplasts during kiwifruit ripening. Additionally, there is an increase in the proteins responsible for the ATP synthesis and protein import in the amyloplasts.

1.2.5 Elaioplasts

Elaioplasts have ultrastructure that is packed with hydrophobic substances, including lipids and terpenoids. Although they have various purposes in particular tissues, they are well-suited for lipid storage and production. Elaioplasts are exported into secretory pockets in citrus fruits and may significantly affect taste and scent (Zhu et al. 2018). Lipoxygenases have a role for the generation of volatiles chemicals and aromas. Different types of lipoxygenases accumulate in the elaioplasts of citrus (Zhu et al. 2018). Exine synthesis in pollen was found to be substantially dependent on elaioplasts (Quilichini et al. 2014; Kobayashi et al. 2019). In *Brassica*, elaioplasts of the tapetum in the anther, supply nutrition and structural components to the developing pollen coat (Suzuki et al. 2013; Kobayashi et al. 2019)). In *Arabidopsis*, major lipids of the elaioplasts are sterol and wax esters, derived from the cytosolic isoprenoid mevalonate pathway (MVA). These isoprenoids are transported through the endoplasmic reticulum to elaioplasts and stored there (Hernández-Pinzón et al. 1999). Besides the tapetum, elaioplasts have been reported in ovaries and ovary epidermis in flowers (Chen et al. 1988; Ciampolini et al. 1993; Kwiatkowska et al. 2010). Elaioplasts were also reported in secretory ducts of the stem and leaf epidermis from *Centaurea cyanus* and *Haemanthus albiflos* (Kwiatkowska et al. 2010; Pacek et al. 2012), young leaves of *Vanilla planifolia*, roots, hypocotyls of *Althaea rosea* (Kwiatkowska et al. 2010), seeds of canola and sunflower (Howitt and Pogson 2006; Lersten et al. 2006), mesocarp of the fruit in avocados (Scott et al. 1963), and green pericarps of citrus fruits (Zhu et al. 2018). The formation of elaioplasts has been reported to occur by diverse mechanisms that vary by species.

1.2.6 Proteinoplasts

Protein bodies, also known as proteinoplasts (or proteoplast, aleuroplast) (Dashek and Miglani 2017), can be found in plastid structures, mainly in the cytosolic area. Proteinoplasts are usually found in various kinds of cells at several phases of plastid development. Proteinoplasts are supposed to be involved in protein storage based on their location and contents. Moreover, the significant oxidase activity of the proteinoplasts from tobacco roots could indicate a specialized use (Vigil and Ruddat 1985). Protein-containing plastids are mainly observed in root caps and vacuolate cells of roots and slender tubules of the plastids. With the division of cells, tubules expand, proteins accumulate, and protein bodies of dense spheroidal structures appear. Proteinoplasts were observed in the roots of tobacco (Vigil and Ruddat 1985), and leaves and seeds of *Helleborus corsicus* and *Zea maize*, respectively (Härtel and Thaler 1953; Denic et al. 1971). In the leaves of mung bean, proteinoplasts were observed that contained a large amount of protein in their granular matrix (Dashek and Miglani 2017). Proteinoplasts contain a tubular network that originates from the invagination of the plastid inner membrane and accumulate large amounts of proteins in protein bodies (PBs). In tobacco roots, PBs contain heme proteins with strong oxidase activity. In tobacco root cap cells, proteinoplasts contain a large PB, but their ultrastructure is variable (Vigil and Ruddat 1985). The PBs are often reported in plastids of a variety of tissues, such as tapetal cells, anther, and pollen in *Miconia*, and apical meristem cells of potato (Platonova et al. 2010; Cortez et al. 2015), indicating that this particular structure might have specialized functions in different plastid types.

1.2.7 Chloroplasts

Chloroplasts are ubiquitous in photosynthetic species and have been extensively investigated for their content, biogenesis, and functions (Rottet et al. 2015; Rudowska et al. 2012; Ostermeier et al. 2024). Photosynthesis uses pigment-protein complexes to convert light energy into chemical energy. Chloroplasts contain thylakoids or lamellae that are hollow structures performing photosynthesis. Thylakoids stacks of lipid-protein layers that resemble disks form grana in chloroplasts. Photosynthetic protein complexes are attached to these lipid surface layers. Disk-like thylakoids create plastoglobules, hydrophobic pocket structures that increase the interior area of the lipid bilayer (Rottet et al. 2015). Plastoglobules degrade apocarotenoids and accumulate non-endogenous carotenoids (Mortimer et al. 2017). The primary component of photosynthesis, chlorophyll, gives chloroplasts its green color. The lutein, β-carotene, violaxanthin, and neoxanthin are few of the many hydrophobic terpenes that are abundant in thylakoids and assist photosynthesis (Ruiz-Sola and Rodríguez-Concepción 2012). The carotenoids in the chloroplast not only convert light in the UV-blue range to electro-energy for photosynthesis (Dall'Osto et al. 2014) but also modulate the non-radiative dissipation of excess excitation energy, which is important for photoprotection (Dall'Osto et al. 2014). In particular, hydroxylated carotenoids, also known as xanthophylls, contribute to photoprotection by either scavenging the reactive oxygen species (ROS) produced during photosynthesis or mediating the direct quenching of triplet chlorophylls (^{3}Chl) (Dall'Osto et al. 2007); thus, keeping a balance between photosynthesis, photoprotection, and ROS scavenging, which is one of the fundamental functions of chloroplasts (Stefanov et al. 2024). Chloroplasts are frequently used to produce lipid bodies, protein bodies, amino acids, secondary metabolites, and starch granules for short-term storage as well as to help meet the demands of environmental and developmental fluctuations (Apostolova and Misra 2014). Chloroplasts mainly arise from proplastids or etioplasts (Dalal and Tripathy 2018; Liebers et al. 2017); however, other type of plastids, e.g., amyloplasts in potato tubers, roots, and callus; chromoplasts in tomato and papaya; and gerontoplasts under mild stress conditions also give rise to chloroplasts (Choi et al. 2021).

The chloroplast shows light sensitivity and movement based on light perception. Upon exposure to high light, chloroplasts move toward the periphery/on the side walls of the cells to avoid light exposure (Abhijita et al. 2023; Schramma et al. 2023). High as well as low light causes the remodeling of the photosynthetic complexes, higher triacylglycerol content and flux of fatty acid through the endoplasmic reticulum pathway, and a lower rate of fatty acid synthesis. Moreover, thylakoid membrane unstacking is increased (Yu et al. 2021; Abhijita et al. 2023). High light causes inhibition of the photosynthetic reaction centers, which is mediated primarily through the degradation of the PSII reaction center D1 protein, affected by the generation of singlet oxygen species (Tripathy and Oelmüller 2012). Under high light and stress conditions, plants tend to decrease their antenna size and vice versa under high light. Interestingly, increased LHCII oligomerization and monogalactosyl diacylglycerol (MGDG) amount and decreased anionic lipids made LHCII less sensitive to high light in pea plants (Dankov et al. 2011). Low light causes the opposite effect of increased antenna size of photosynthetic complexes, adjustment of the antennae through state transitions, and accumulation and spreading of chloroplasts in a layer on the bottom of the cell (Shi et al. 2022; Schramma et al. 2023).

1.2.8 Chromoplasts

1.2.8.1 Structure

Chromoplasts are colorful due to the accumulation of carotenoids, with unique colors defined by the type of carotenoids. They accumulate high amounts of carotenoids, which are of specific types and display certain colors. The carotenoids can be stored in crystals (tomato fruits), membranes (*Narcissus pseudonarcissus* petals), plastoglobules (lipid droplets in carotenoid bearing petals), and tubules/fibrils (Capsicum fruit) (Camara et al. 1995). The carotenoids that mature into globular, spherical, or coiled-shaped crystals are created during the chromoplast formation and retained in their hydrophobic structures known as plastoglobules (PG) (Schweiggert et al. 2011). The half-lipid bilayer that connects the lipoprotein particles/plastoglobules to thylakoids allows them to perform two functions: lipid production and storage. In reproductive tissues, colored plastids with developed plastoglobules store carotenoids, also attract pollinators, and act as seed dispersers (Rottet et al. 2016). An example of chromoplast differentiation from proplastids is the orange curd mutant of *Brassica oleracea* L. var. *botrytis* DC. In the apical meristems of the shoots, the mutants accumulate significant quantities of carotenoids. The sole plastid-type seen in the apex of this mutant, the chromoplast, harbors these carotenoids. In contrast, several proplastids are present in the apices of wild-type shoots, indicating that chromoplasts originated from proplastids (Paolillo et al. 2004). Proplastids can be transformed into chromoplasts in *Arabidopsis thaliana* (L.) Heynh. callus (Choi et al. 2021). Chromoplast differentiation and carotenoid accumulation take place in *Citrullus lanatus* (Thunb.) Matsum. & Nakai during fruit development that gives rise to a range of colors to the fruits. Fruit color at maturity is influenced by the number of plastoglobules in the chromoplasts. According to Fang et al. (2020), the red-fleshed accession of *Citrullus lanatus* (Thunb.) Matsum. & Nakai has a higher PG number than the yellow and orange accessions (Fang et al. 2020). The latter two also exhibit a higher number of plastoglobules than the white-fleshed accession. This type of color range is also shown by the callus of *Daucus carota* L. Proplastids are present, and the carotenoid level is minimal in pale yellow calli (Oleszkiewicz et al. 2018). On the other hand, proplastid's quantity is greatly decreased in dark orange calli, where chromoplasts with high carotenoid concentrations are predominantly found. During fruit development, the proplastid to chromoplast transition can also be observed in *Carica papaya* L. (Schweiggert et al. 2011).

Chromoplasts are the only plastids found in the colorless basal region of the saffron style. They evolve from amyloplasts in red stigmas of saffron (*Crocus sativus* L.); amylo-chromoplasts, which are transitional forms of plastids, may also be found in saffron (Grilli Caiola and Canini 2004). Chromatoplasts immediately separate from amyloplasts when the color of *Nicotiana tabacum* L. floral nectaries turns orange. This conversion process is accompanied by the breakdown of starch and the creation of nectar sugars (Horner et al. 2007). Amyloplasts constitute the endosperm of *Oryza sativa* L. caryopses, which does not produce a considerable quantity of carotenoids. Nonetheless, even in caryopses, overexpression of carotenogenic genes, including ORANGE (OR), deoxy-D-xylulose 5-phosphate synthase (DXS), PSY, and bacterial phytoene desaturase (CRTI), leads to chromoplast formation (You et al. 2020). During fruit and root development, chromoplasts can also originate from leucoplasts and amyloplasts, such as that seen in stigmas and floral nectaries.

1.2.8.2 Regulations of Chromoplast Differentiation (Genetics, ROS, Metabolites)

The differentiation of chromoplasts necessitates an alteration in the organelle's developmental program, which is dictated by endogenous and environmental cues. Apart from the genes responsible for disassembling the photosynthetic apparatus, multiple other elements govern the differentiation of chromoplasts. A functioning endogenous ATP production mechanism is present in chromoplasts. Consequently, increased ATP production is linked to chromoplast differentiation during the transition from chloroplast to chromoplast, and among the most prevalent proteins in the chromoplast proteome, across a range of crops, are ATP synthase (Barsan et al. 2012; Wang et al. 2013b) and adenine nucleotide translocator (Wang et al. 2013b). Thus ATP synthesis process is actively performed by chromoplasts (Angaman et al. 2012). Chromoplasts of a range of species contain similar functional classes of proteins, i.e., carotenoid metabolism, antioxidant system, and ATP synthesis (Wang et al. 2013b).

Carotenoid-derived volatile metabolites, the apocarotenoids, greatly augment during fruit ripening and contribute significantly to the flavor and aroma of fruits (Simkin et al. 2004). Carotenoid cleavage dioxygenase (CCD1) was found in a proteomics study of chromoplasts of six crops (Wang et al. 2013b). Similarly, 9-cis-epoxycarotenoid dioxygenase (NCED), key enzyme in the ABA biosynthetic pathway, was also determined in chromoplast proteomes of several crop species (Wang et al. 2013b; Zeng et al. 2011). Chromoplasts have a highly active redox/antioxidant system; superoxide dismutase enzyme is enhanced during chromoplast development in pepper (Marti et al. 2009) and glutathione S-transferase (GST) during papaya fruit ripening (Nogueira et al. 2012). Besides the common proteins, species-specific proteins are also found in chromoplasts, such as CCD1, which synthesize capsanthin and capsorubin, responsible for the typical color of red pepper and fibrillin in red bell pepper (Siddique et al. 2006).

Apart from the genes responsible for disassembling the photosynthetic apparatus, multiple other elements govern the differentiation of chromoplasts. For instance, it has been discovered that plastid fusions and/or translocation factors (PFTFs) are involved in vesicle fusion and internal plastid remodeling in *Capsicum annuum* L. during the chloroplast-to-chromoplast transition (Hugueney et al. 1995). Furthermore, plastome gene expression is active throughout chromoplast differentiation, even if it is inhibited in mature chromoplasts. The translational level of photosynthesis-related protein expression is downregulated, but FA biosynthesis gene expression is maintained (Egea et al. 2010), because it is essential for the chromoplast's internal membrane reorganization prior to carotenoid synthesis (Barsan et al. 2012). One of the environmental factors influencing the production of chromoplasts is temperature; ideal temperature conditions are necessary for the best possible differentiation of organelles. This is especially important to maintain color consistency in fruits that are sensitive to temperature, like *Solanum lycopersicum* L. and *Carica papaya* L. (Sadali et al. 2019). A study conducted by Karlova et al. (2011) revealed that the ethylene-induced Apetala2a (AP2a) gene plays an important role in the initiation of chromoplast differentiation. AP2a gene disruption imposes interference in the carotenoid production, resulting in yellow rather than red fruits, as in *Solanum lycopersicum* L. (Karlova et al. 2011). This suggests that the gene plays a role in the initiation of chromoplast differentiation. Heat-shock protein 70 (Hsp70), which functions as a chaperone for plastid protein import and facilitates the import modifications required for chromoplast transition, has been demonstrated to be upregulated by AP2a (Karlova et al. 2011). Membrane proliferation depends on the fatty acid production and lipid metabolism, and the lipids are necessary for plastoglobules in chromoplasts as well as for the genesis of carotenoid-lipoprotein sequestration substructures. Accordingly, the chromoplast proteome of numerous crop species contains a substantial number of proteins involved in the lipid metabolism, including essential enzymes for the synthesis of FAs, sulfolipids, and glycolipids as well as for the lipid catabolism and homeostasis (Wang et al. 2013b).

1.2.9 Gerontoplasts

Chloroplast-derived plastids, known as gerontoplasts, are designed mainly for plastid recycling. They are primarily observed during senescence processes or in stressful environments (Biswal et al. 2012; Joshi et al. 2013). Plant survival depends on the breakdown of chloroplasts and the recycling of their nutrients, because they contain up to 80% of the leaf nitrogen pool (Mu and Chen 2021). Furthermore, three distinct pathways through which chloroplast proteins are degraded are autophagy, senescence-associated vacuoles (SAV), and chloroplast vesiculation (CV) (Xie et al. 2015). Besides this, other pathways operate, which are discussed in another chapter of this book. Plastids show different ultrastructural stages during the senescence process, which are less prominent when senescence begins. Initially, starch granules disappear due to the inhibition of photosynthesis. Secondly, chlorophyll (Chl) and thylakoids degrade. Lipid structures, lipophilic materials, and broken lipids accumulate due to damage to thylakoids. Plastoglobules grow in size and quantity (Biswal et al. 2012; Joshi et al. 2013). Mostly, NAC and WRKY TF families help in these catabolic processes by regulating the expression of senescence-associated genes (SAGs) and other genes in the regulatory network of transition from chloroplast to a gerontoplast (Biswal et al. 2012; Joshi et al. 2013; Woo et al. 2019). The gerontoplast formation happens through the steps: grana unstacking, thylakoid flattening, intra-thylakoid space enlargement, and stroma lamellae destruction (Mulisch and Krupinska 2013). Grana unstacking happens through the dissolution of PSII–LHCII (photosystem II–light harvesting complex of photosystem II) supercomplexes (Albanese et al. 2020), followed by the degradation of LHCII and Chl (Domínguez and Cejudo 2021). The multistep process of Chl degradation includes the conversion of Chl *b* to Chl *a*, the removal of the central magnesium atom, the formation of pheophytin *a* molecules (Shimoda et al. 2016), the removal of the phytol chain of pheophytin *a* (Schelbert et al. 2009), and finally the generation of non-phototoxic red and blue-fluorescent chlorophyll catabolites (FCC) (Pruzinská et al. 2007; Kuai et al. 2018). Photosystem I (PSI) complex and ATPase activities appear to be intact until the last stages of leaf senescence, in contrast to PSII, whose activity is inhibited in the early stages of senescence (Krupinska et al. 2012). Gerontoplasts formation involves additional rearrangements in the macromolecular structure to generate alternative electron transport pathways to continue the production of ATP, although most of the chloroplasts' functions and the linear electron transport are vanished (Biswal et al. 2012; Joshi et al. 2013; Krieger-Liszkay et al. 2019).

1.2.10 Specialized Types: Desiccoplasts, Phenyloplasts, and Xyloplasts

1.2.10.1 Desiccoplasts

Plastids known as desiccoplasts can be produced from chloroplasts and proplastids in plants that can withstand desiccation, i.e. resurrection plants (Solymosi et al. 2013; Challabathula et al. 2016). They contain several large plastoglobules and few thylakoids arranged as vesicles (Altamura et al. 2024). The thylakoids are overlapping and highly stacked in desiccoplasts. During dehydration, the thylakoids of the chloroplasts are dismantled—at this stage, they are called desiccoplasts—and upon rehydration, desiccoplasts are transformed into functionally active chloroplasts (Altamura et al. 2024). In contrast to homoiochlorophyllous plants that retain their Chl upon dehydration, the poikilochlorophyllous resurrection plants degrade their Chl and turn brownish and dry. Upon rehydration, they readily absorb moisture up to levels almost equal to green plastids within 48 h, and their leaves turn pale green. Granal stacking starts loosening in these leaves within 12 h of rehydration. The green leaves have around twofold higher fatty acids (FA) than the dried resurrection plants. The greening process in desiccoplasts is accompanied by an increase in the FAs. The linoleic acid increased threefold in pale green leaves of resurrection plant *Xerophyta scabrida* during regreening after dehydration (Georgieva et al. 2011). Desiccation tolerance has been studied through transcriptome, proteome, and metabolome analysis (Dinakar and Bartels 2013)—transcriptomics studies revealed that signaling molecules, kinases, TFs, LEA proteins, ELIPs, cell wall modifications, osmolytes, carbohydrates, and antioxidant machinery were induced, along with several unknown function proteins and those involved in general metabolism. Proteomics showed overlapping data with transcriptomics with abundance of LEA proteins, TFs, and antioxidant enzymes increasing and decreased photosynthetic complex proteins during dehydration (reviewed in Dinakar and Bartels 2013). A large chunk of altered proteins are unknown proteins, e.g., in *Selaginella tamariscina* (Wang et al. 2010). Upon rehydration of poikilochlorophyllous *Xerophyta humilis* in the dark for 2 weeks, leaves showed the development of pregranal plastids with newly synthesized PTs and accumulated non-photoactive Pchlide. During longer dark incubation of 3 weeks, basal regions of the leaves formed etioplasts containing PLBs and phototransformable 655 nm emitting Pchlide complexes (Solymosi et al. 2013).

1.2.10.2 Phenyloplasts

Phenyloplasts are identified as a new plastid type due to phenol-enriched colorful plastids compared to chromoplasts because of their different storage contents (Brillouet et al. 2014).

Phenyloplasts are found in the vascular tissue of the mesocarp of vanilla fruit. These are similar to chloroplasts in size but lack internal membranes and differentiate from chloroplasts. Phenyloplasts accumulate high levels of secondary metabolites, such as vanillin glucoside and phenylpropanoid-derived glucosides in solid amorphous deposits. Differentiation of phenyloplasts starts with the formation of loculi from the thylakoid membranes that generate osmiophilic plastoglobuli filled with glucosides (Brillouet et al. 2014; Gallage et al. 2018).

1.2.10.3 Xyloplasts

Xyloplasts are specialized plastids that are devoted to synthesizing precursors for the manufacture of monolignols in the secondary vascular tissues (Pinard and Mizrachi 2018). These precursors are generated from proplastids or, more frequently, amyloplasts (Pinard and Mizrachi 2018).

1.2.10.4 Tannoplasts

Tannoplasts are derived from chloroplasts and present in Tracheophyta, e.g., *Dryopteris sp*, *Cycus revoluta*, *Ginkgo biloba* and *Vitis vinifera*. They contain membranous vesicular spheroid structures called tannosomes. They produce and accumulate tannins (proanthocyanidins), essential for defense in vascular plants. Tannoplast differentiation starts with chloroplast swelling and grana unstacking; deposits of osmiophilic material appear in thylakoidal lamellae that extend into the stroma. In these organelles, both plastid envelope membranes are apparently replaced by a diffuse mass, where abundant cargo vesicles filled with tannosomes and osmiophilic materials with high tannin content are expelled from the tannoplast and transported to the vacuole (Brillouet et al. 2013; Kumachova et al. 2018).

1.2.10.5 Iridoplasts

When under low light conditions, strong blue iridescence falls on the leaves, under the canopy of densely populated forests, plants adapt to capture green-red light, thus enhancing quantum yield (Jacobs et al. 2016). This is achieved by producing regularly spaced thylakoid lamellae (TLs) with photonic properties that can produce structural color. It leads to the formation of iridoplasts, which are modified chloroplasts, in the adaxial epidermal cells of iridescent leaves (Pao et al. 2018), in species such as *Begonia*, *Phyllagathis*, *Trichomanes*, and *Selaginella* (Masters et al. 2018). Iridoplasts show a strong blue peak in reflectance observed at a spectral width of 435–500 nm (Jacobs et al. 2016). Ultrastructural studies of iridoplasts in taxa with no visible iridescence, combined with shading experiments, show that the iridescence is affected by the spacing between the thylakoid lamellae. In *Selaginella erythropus*, the iridoplasts (also referred to as bisonoplasts) show two distinct regions of thylakoid membranes. The lower region shows a standard arrangement of grana, and the upper region (facing the adaxial surface of the leaf) has highly ordered thylakoid lamellae similar to those observed in Begonia's iridoplasts (Masters et al. 2018). Similar to Begonia's iridoplasts, *Selaginella* plants grown under high light conditions display limited structural color compared with those grown in low light, and this correlates with the presence of orderly arranged thylakoid lamellae (Masters et al. 2018). The photonic properties of the lamellae structures of iridoplasts constitute an excellent model to study light absorption optimization through morphological adjustments in leaves during changing light conditions (Castillo et al. 2021).

1.3 Sensory Epidermal Plastids (SEPs)

SEPs are formed from chloroplasts. They physiologically and functionally responded to environmental stimuli and coordinate the defense responses of plants (Barton et al. 2018; Beltrán et al. 2018; Virdi et al. 2016; Gorecka et al. 2014; Irieda and Takano 2021). Earlier, plastids present in the pavement cells, except stomata were called leucoplasts, but later studies found them to act as sensors, and thus, they were proposed to be named as pavement cell chloroplasts (PCCs), epidermal chloroplasts (ECs), or most recently sensory epidermal plastids (Barton et al. 2018; Mackenzie and Mullineaux 2022). With the exception of stomata cells, which contain chloroplasts (Lawson et al. 2014), plastids present in pavement cells are SEPs. These cells accumulate chlorophyll and display light avoidance responses as observed in chloroplasts. Despite these morphological similarities, these are different from chloroplasts due to their smaller size, fewer plastids per cell, low Chl content, and impaired grana development (Barton et al. 2018). SEPs show distinct morphology, proteome composition, and plastid–nuclear interactions to facilitate environment and pathogen sensing and signaling similar to the vascular parenchyma plastids (vascular sensory plastids) (Virdi et al. 2016). They also have sensory similarities with mesophyll chloroplasts, which produce organelle extensions (stromules) in response to redox signals, while SEPs make them in response to sugar levels (Brunkard et al. 2015). SEPs display specific relocation movements in *Arabidopsis*, called the epidermal chloroplast response (ECR) as a result of fungal infection (Irieda and Takano 2021). Furthermore, since the fungi secrete cell wall-degrading enzymes for causing infection, these ECR movements are regulated by the chloroplast photo-relocation regulator protein CHLOROPLAST UNUSUAL POSITIONING 1 (CHUP1) and are stimulated by cell wall-degrading enzymes (Irieda and Takano 2021). The ECR movements are independent of blue light receptors. SEPs stromules from fungus infected plants, accumulate immune response proteins ENHANCED DISEASE SUSCEPTIBILITY 5 (EDS5), CAS (calcium-sensing receptor), and γ-glutamylcysteine synthetase 1 (GSH1), suggesting their roles in defense response to fungus infection (Irieda and Takano 2021). DNA-binding MutS HOMOLOG 1 (MSH1) protein with homology to the bacterial mismatch repair MutS protein is exclusive to the plants (Virdi et al. 2016). It accumulates in tissues, including the epidermis, vascular parenchyma, and meristems. The phenotypes observed in mutants of the plastid and mitochondrial-targeted MSH1 protein also support the role of SEPs as key environmental sensors. *msh1* mutants display heritable nongenetic developmental defects, including leaf variegation, dwarfism, delay in juvenility to maturity transition, late flowering, and altered stress responses (Shedge et al. 2010); however, hemi-complementation experiments demonstrate that the pleiotropic responses associated with MSH1 suppression derive from SEPs (Xu et al. 2011; Virdi et al. 2016).

Another important protein involved in defense responses signaling in PCC, CAB underexpressed (CUE1) encodes one of the PPT1 (a phospho*enol*pyruvate (PEP)/phosphate transporter) of the inner plastid membrane, which imports PEP to the

PCCs and is specifically expressed in the epidermal and vascular plastids (Streatfield et al. 1999; Beltrán et al. 2018). Proteome analysis demonstrated that SEPs have stress-associated protein enrichment related to oxidation–reduction, fatty acid and amino acid biosynthesis, and shikimate pathway (Shende et al. 2024). Transcriptomic analyses of *msh1* and other mutants defective in epidermal and vascular parenchyma also showed altered expression of genes for ABA biosynthesis, cold, oxidative, osmotic, and salt stresses, calcium signaling, and auxin responses (Xu et al. 2011; Gorecka et al. 2014; Beltrán et al. 2018).

Based on available data, the sensory plastids from the epidermis and vascular tissues appear to be central for sensing environmental conditions and triggering diverse stress-response mechanisms. It is still unclear whether the SEPs from the epidermis and those from the vascular tissue are the same or different plastid subtypes. Resolving this remains as an important topic for future investigations.

1.4 Chloroplasts to Albino Phenotype

The absence of chloroplasts in tissues that are typically green is known as albinism. Albinism has been reported in several plant species and also under stress conditions, e.g., in *Delonix regia*. The chloroplasts and thylakoid membrane system of albino plants are poorly developed, and numerous plastoglobules are present (Silva et al. 2020). Decrease expression of photosynthesis genes, lower expression of chlorophyll and carotenoid biosynthetic genes is observed under albinism, e.g., in *Agave angustifolia* (Us-Camas et al. 2017). In some plants, e.g. in tea mutants, albinism is a periodic event, where tea leaves exhibit albinism at very low temperatures but regreen when temperatures become high. This phenomenon is controlled by a regulatory network of miRNAs (Xu et al. 2023). Albinism is related to auxin hormone. An *Arabidopsis* albino mutant, *shikimate kinase-like1* (*skl1*) is hampered in tryptophan biosynthesis that results in low auxin biosynthesis (Xu et al. 2018). Transcript expression of auxin receptors proteins *IAA1*, *IAA2*, *IAA19*, and *IAA29*, and transporters, *PIN Formed 3* (*PIN3*), and *PIN4* were also reduced in this mutant (Xu et al. 2018).

Another atypical plastid, unnamed so far, is present in phloem-associated cells. This contains a single starch grain and a few internal membranes, but no further description of this organelle exists (Evert and Deshpande 1971; Wise 2007). Besides these plastids, muroplasts, rhodoplasts, kleptoplasts, sieve cell plastids, nodular plastids, and apicoplasts have also been proposed.

1.5 Transition of Plastids

Plastids developed various forms and activities concurrently with the growing morphological, developmental, and genetic variety of plant tissues. Transition dynamics exist between the various plastid types' morpho-functional characteristics which are mentioned below.

1.5.1 Development of Chloroplast: Greening Phenotype

Almost all biotic processes derive their organic mass and energy from photosynthesis. It is responsible for producing all oxygen in the atmosphere, supporting life on Earth. Chl aids the conversion of light into chemical energy. Apart from chloroplasts, various cell structures that serve different purposes also can evolve into or be derived from chloroplasts (Dilnawaz et al. 2001; Misra et al. 2001).

1.5.1.1 Proplastid to Chloroplast Transition

The proplastid to chloroplast transition mostly takes place in the shoot apical meristem and during embryogenesis. The process of chloroplast differentiation in *Arabidopsis* begins at the shoot apical meristem of the young leaf and progresses into leaf development. Both the central subtending cell layers and the top layer undergo differentiation, with a requirement of light duration of 5–10 h (Yadav et al. 2019). The development of chloroplast from undifferentiated proplastids was identified at the globular stage of embryogenesis in *Arabidopsis*. In in vitro experiments, calluses/calli developed in the dark only included proplastids, but those grown in the light showed short thylakoids and chloroplasts with an immature membrane (Ladygin et al. 2008). In monocot leaves, which exhibit cell proliferation and differentiation along a linear developmental gradient, the leaf primordia develop on the sides of the stem cell population located in the shoot apical meristem. The cellular differentiation takes place in the leaf primordia in a downward direction, resulting in a gradient of distinct cellular morphologies. The distal differentiated cells at the tip of the leaf have easily recognizable characteristics, while the basal progenitor proliferating cells next to the shoot apical meristem display different stages of development (Dilnawaz et al. 2001).

1.5.1.2 Root Greening

When the roots of *Arabidopsis* are cut, they show greening. In these roots, plastids synthesize Chl and differentiate a functional photosynthetic apparatus, which is based on the cytokinin (CK) hormone. In severed roots, increased CK (mediated through upregulated auxin response regulators (ARRs) and deficiency of auxin) triggers the increased expression of B-GATA TFs, GNC, and GNL, which induce the expression of photosynthesis-activated nuclear genes (PhANGs). This usually happens in the root cortical parenchyma and pericycle (Kobayashi et al. 2017). It has been proposed that HY5 and CGA1 may be the interactors between CK and IAA in the regulation of root greening. The GNC- and GNL-mediated root plastid differentiation is controlled by HY5. The inhibition of root greening by auxin occurs by the IAA signaling protein IAA14 and the auxin responsive transcription factors ARF7 and ARF19, which repress *HY5* and *GNC/CGA1* expression (Kobayashi et al. 2017).

1.5.1.3 Etioplast to Chloroplast Transition

Etioplast, when exposed to light, light-dependent NADPH-Pchlide oxidoreductase converts Pchlide, the precursor of Chl in prolamellar bodies, to Chlide. Chlide is converted to Chl by several enzymes (Fujii et al. 2019). This transition is commonly found in darkened leaf tissues and in some tissues of light-grown plants, e.g., inner leaves of white cabbage, lettuce, and cucumber cotyledons (Solymosi et al. 2004; Kanamoto et al. 2006; Solymosi et al. 2012), and results in greening or de-etiolation. De-etiolation events in tobacco leaves revealed that exposure to light caused immediate changes in the physical structures of the etioplast PLBs, including reduced size and regularity (Armarego-Marriott et al. 2019). In *Arabidopsis*, chloroplast biogenesis is divided into two phases: the "structure establishment phase," which involves disassembling the prolamellar body, which forms the thylakoid membrane, and increases galactolipids and proteins linked to the photosynthesis, and the "chloroplast proliferation phase," which entails cell division, a linear rise in prokaryotic and eukaryotic galactolipids and proteins related to the photosynthesis (Pipitone et al. 2021).

1.5.1.4 Leucoplast to Chloroplast Transition

Non-photosynthetic leucoplasts can transform into photosynthetic chloroplasts in plants. In the cortical parenchyma tissues of potato tubers, directly underneath the periderm, amyloplasts (starch-enriched leucoplasts) undergo a transformation into chloroplasts upon exposure to light. This process results in the accumulation of Chl (Tanios et al. 2018). Seasonally, huge levels of starch accumulate in the fall and winter, which causes the appearance of amyloplasts that again convert into chloroplasts in the spring and summer. Greening occurs in arum (Italian fruits) after the fruits are developed. The thylakoid membrane develops, converting the amyloplasts into chloroplasts when the green color appears, and the size and quantity of plastoglobules decrease (Bonora et al. 2000).

1.5.1.5 Chromoplast to Chloroplast Transition

Regreening is defined as a reversible transition from chromoplast to chloroplast. It is present in citrus fruits (Mayfield and Huff 1986), pumpkin (Devidé and Ljubešíc 1974), and cucumber fruits (Rodríguez-Villalón et al. 2009). Cucumber thylakoids disintegrate as the fruit matures and are recreated by regreening. Regreened chloroplasts and ripe fruit plastids share morphological similarities, indicating redifferentiation. Thylakoids reconstitute from membrane-bound entities, resulting in surface expansion and fragmentation. Tubules and double membrane sheets are then created. Plastoglobules stay in the plastid even after rebuilding thylakoids. Prebeg et al. (2008) found that during chloroplast redifferentiation, several membrane structures are connected with the plastid envelope. Light plays a significant role in regreening. Irradiation of Valencia orange citrus fruit, using blue LED light resulted in a gradual greening of the tissue- after 4 weeks, the Chl concentration was nearly double that of the unirradiated tissue (Ma et al. 2020). Light irradiation transformed carrot roots from beta-carotene-rich to lutein-containing chloroplasts (Rodriguez-Concepcion and Stange 2013).

1.5.2 Development of Chromoplasts: Non-greening Phenotype

Chromoplasts, which produce and store carotenoids, are mostly present not only in reproductive organs, such as petals and fruits, but also in leaves and roots. Chromoplast formation begins with the breakdown of thylakoids and Chl. Concomitantly, plastoglobuli grow in size, and carotenoids are synthesized (Wrischer et al. 2009). There are examples of plants that exhibit irreversible processes from chloroplast to chromoplast, such as the tomato (D'Andrea et al. 2014), red pepper (Jeong et al. 2020), Italian arum (Bonora et al. 2000), and lily (Juneau et al. 2002). Inner envelope and thylakoid membranes disappear during the transition to chromoplasts, resulting in the formation of structures such as plastoglobules and crystal remnants via membrane fusion and vesicle budding (Wang et al. 2020).

1.5.2.1 Proplastid to Chromoplast Transition

During fruit ripening, the chloroplasts convert to orange- or red-colored chromoplasts due to the thylakoid disintegration, and synthesis and the accumulation of high levels of carotenoids. Watermelon, papaya, and carrot calluses demonstrate the shift from proplastid to chromoplast.

In papaya, spherical and undifferentiated proplastids predominate during the early stages of white-pulp maturation. Before chromoplasts appear, intermediate plastids like chloroplasts and amyloplasts are absent (Schweiggert et al. 2011). Watermelon plastid differentiation patterns showed increased accumulation of carotenoids and chromoplasts as the fruit matures. Yellow and orange watermelons have much more plastoglobuli in their chromoplasts than white watermelons. The red watermelon chromoplasts become more elongated and irregular, with a rise in plastoglobuli (Fang et al. 2020). In case of pale-yellow carrot calluses, the carotenoid content was low while most of the plastids were proplastids; nevertheless, in the case of dark-orange carrot calluses, there was a substantial quantity of chromoplasts with high carotenoid contents, while the number of proplastids was significantly reduced (Oleszkiewicz et al. 2018).

1.5.2.2 Leucoplast to Chromoplast Transition

A shift usually happens throughout the maturation of fruits, flowers, and roots. Carrots, orange, and mutant orange cauliflower are common examples of this shift (Sun et al. 2018). Carrot roots exhibit distinct plastid types based on their hue. Orange carrot roots had more chromoplasts with crystal-shaped structures due to the carotene, while white carrots had fewer chromoplasts and no crystal-shaped structure and had amyloplasts loaded with starch grains (Paolillo et al. 2004). In the case of orange cauliflower, the chromoplasts had accumulated beta-carotene, whereas the white-wild-type accumulated leucoplast (Paolillo et al. 2004). Chromoplast transitions have also been observed in rice and corn endosperms, which are primarily composed of amyloplasts. Rice endosperms do not generate carotenoids, rather a mixture of carotenogenic genes leads to chromoplast development (You et al. 2020).

1.5.3 Development from Chloroplasts: De-greening Phenotype

1.5.3.1 Chloroplast to Leucoplast Transition

Arabidopsis flower development is a common example of leucoplast de-greening. Newly blooming petals have green chloroplasts, but as they grow, they lose Chl and turn white. During this process, plastids degrade the Chl and do not produce carotenoids (Pyke 2013).

1.5.3.2 Chloroplast to Chromoplast Transition

The chloroplast-to-chromoplast transition occurs during fruit maturation, due to the changes in biochemical composition (e.g., the elevation of carotenoid and quinone contents) and structural organization (e.g., fibril assembly and degradation of thylakoids) (Ljubičić et al. 1998). This happens in the fruits, e.g., tomato, papaya, apple, etc. During the development of chromoplasts, due to the accumulation of carotenoid esters, the size and number of plastoglobules increase. Some chromoplasts accumulate carotenoid crystals, e.g., carrot roots, in their stroma (Oleszkiewicz et al. 2018), tomato fruits contain lycopene, and crystalloids in membranous sacs (Harris and Spurr 1969). Carotenoid biosynthesis is enhanced by key enzymes, like PSY, lycopene beta-cyclase (LCYB,) and phytoene desaturase (PDS). Chromoplast carotenoids act as antioxidants and protect cells from oxidative damage from reactive oxygen species (ROS) under stress conditions (Muñoz and Munné-Bosch 2018). A recently discovered protein, orange (OR), increases the carotenoid biosynthesis by directly interacting with PSY, which is the rate-limiting enzyme in the carotenoid biosynthesis and stabilizes its structure (Zhou et al. 2015). OR is responsible for the generation of chromoplasts in floral organs of cauliflower (*Brassica oleracea* var. *botrytis*) (Lu et al. 2006; Kim et al. 2018; Welsch et al. 2020; Li et al. 2022) and chloroplast development through binding with *Arabidopsis* TEOSINTE BRANCHED, CYCLOIDEA, AND PCF (TCP14) to increase the stability and transcription of early light-inducible proteins (ELIP1 and ELIP2) (Sun et al. 2019). Proteins like OR in chromoplasts have a role in improving heat tolerance by acting as chaperones and stabilizing the photosystem II complex (Sun and Li 2020). Another protein STAY-GREEN 1 (SlSGR1) regulates the chromoplast transformation in tomato by interacting and stabilizing PSY1, thus enhancing carotenoid formation (Luo et al. 2013).

Other factors and TFs have also been reported to increase chromoplast development through upregulated carotenoid synthesis. A MADS box TF, RIPENING INHIBITOR (RIN) was found to act as a core positive regulator of carotenoid pathways, regulating the activities of rate-limiting enzymes (Martel et al. 2011). TOMATO AGAMOUS LIKE1 (TAGL1) forms a complex with RIN, and its overexpression enhanced the lycopene content in tomato (Itkin et al. 2009; Lü et al. 2018). Another MADS-box protein called SlCMB1 also enhances the chromoplast development in tomato. The SlCMB1-RNAi tomato plants showed decreased PSY1 and PDS expression and decreased ethylene production and related signal pathway genes (Zhang et al. 2018). A mutant of NON-RIPENING (NOR), a NAC TF, also showed a phenotype similar to the *rin* mutants by inhibition of the plastid transition to chromoplasts (Giovannoni 2004, 2007). TF HY5 increases the carotenoid synthesis through enhanced expression of PSY and phytoene desaturase (PDS) enzymes (Toledo-Ortiz et al. 2014). A special chapter is dedicated to chromoplasts in this book.

1.5.3.3 Chloroplast to Gerontoplast Transition

As chloroplasts mature, they develop plastoglobuli and lipophilic compounds, and structural alterations occur in the thylakoid membranes while the outer shell stays intact. However, another type of plastid, called gerontoplasts, is different from the aged chloroplasts. In *Arabidopsis*, gerontoplasts have defective outer shells and thylakoid membranes, as well as larger plastoglobuli and grana, which gradually expand with age (Evans et al. 2010). The structural properties of gerontoplasts, during the development of *Jatropha curcas* seeds, showed that the plastoglobuli gradually broke down after the inner membrane system—the thylakoid membrane and the plastid outer envelope membrane—decomposed (Shah et al. 2016). Ultrastructural changes, such as disintegration of chloroplasts, happen in three stages, i.e., degradation of thylakoid membrane system, increased accumulation and size of plastoglobules, and degradation of thylakoid envelopes. Besides these anatomical changes, molecular changes, such as degradation of proteins, pigments, lipids, and starch, also take place (Mulisch and Krupinska 2013; Domínguez and Cejudo 2021). Degradation sequence and extent of stacked grana thylakoids and unstacked stroma thylakoids are variable in different plant species and depend on both external and internal conditions. Early degradation of stroma thylakoids is reported frequently as compared to that of grana thylakoids. Grana stacks without the interconnecting stroma thylakoids are also reported. Generally, thylakoid degradation starts with the unstacking of grana thylakoids, followed by the formation of elongated and loose lamellae. Simultaneous swelling and degradation of all thylakoids is also reported (Eckardt 2007). Since most of the Chls are attached to the light-harvesting complexes, a decrease in the Chls content or a change in Chl *a*/*b* ratio indicates degradation or other changes in the thylakoid membranes. Grana unstacking, flattening of thylakoids, swelling of intrathylakoid space, and degradation of stroma lamellae are the characteristic features of thylakoid disorganization (Biswal and Biswal 1988; Joshi et al. 2013; Domínguez and Cejudo 2021). Cup-shaped, stacked membranes, thylakoid coiling, and formation of dilations at the thylakoid ends have also been reported (Wrischer et al. 2009). Chl is disintegrates through the formation of red/yellow fluorescent Chl catabolites (Lichtenthaler 2013).

1.6 Conclusions

There are different kinds of plastids in higher plants that play different roles based on their structure, function, and localization. Many of these plastids can interconvert (plasticity) depending on the developmental stage of the plant and their localization in the plant tissue. The specialized roles of plastids and their interconversion in land plants are strictly interconnected and represent the dynamics of plant development.

Plastid plasticity and the dynamics of plant development are closely related. Certain alterations in the proteome and morphology of organelles are caused by internal positional signals; in this regard, understanding the underlying causes of extreme morpho-functional and metabolic heterogeneity in various plastid types may aid in identifying processes that will help in the development of crops adaptable to climate change. The plastid plasticity determines the survival of plants and is also of significance due to their role as commercial gain determinants. When plastid interconversion is thoroughly understood, it will be possible to manipulate it to get higher starch, protein or oil in the fruits/seeds and get color variations in flowers/fruits as well as an increase the carotenoids that would tremendously benefit human and assist the plants in stress tolerance.

References

Abhijita S, Dalal VK, Misra AN (2023) The effect of light quality and quantity on photosynthesis. In: Dalal VK, Misra AN (eds) A closer look at photosynthesis. Nova Science Publications, NY, pp 46–81

Albanese P, Tamara S, Saracco G, Scheltema RA, Pagliano C (2020) How paired PSII-LHCII supercomplexes mediate the stacking of plant thylakoid membranes unveiled by structural mass-spectrometry. Nat Commun 11(1):1361. https://doi.org/10.1038/s41467-020-15184-1

Altamura MM, Piacentini D, Della Rovere F, Fattorini L, Valletta A, Falasca G (2024) Plastid dynamism integrates development and environment. Plant Physiol Biochem 213:108813. https://doi.org/10.1016/j.plaphy.2024.108813

Angaman DM, Petrizzo R, Hernandez-Gras F, Romero-Segura C, Pateraki I, Busquets M, Boronat A (2012) Precursor uptake assays and metabolic analyses in isolated tomato fruit chromoplasts. Plant Methods 8:1

Apostolova EL, Misra AN (2014) Alterations in structural organization affect the functional ability of photosynthetic apparatus. In: Handbook of plant and crop physiology, pp 103–120

Armarego-Marriott T, Kowalewska Ł, Burgos A, Fischer A, Thiele W, Erban A et al (2019) Highly resolved systems biology to dissect the etioplast-tochloroplast transition in tobacco leaves. Plant Physiol 180:654–681. https://doi.org/10.1104/pp.18.01432

Balmer Y, Vensel WH, DuPont FM, Buchanan BB, Hurkman WJ (2006) Proteome of amyloplasts isolated from developing wheat endosperm presents evidence of broad metabolic capability. J Exp Bot 57:1591–1602

Brosch M, Krause K, Falk J, Krupinska K (2007) Analysis of gene expression in amyloplasts of potato tubers. Planta 227:91–99

Barsan C, Zouine M, Maza E, Bian W, Egea I, Rossignol M et al (2012) Proteomic analysis of chloroplast-to-chromoplast transition in tomato reveals metabolic shifts coupled with disrupted thylakoid biogenesis machinery and elevated energy-production components. Plant Physiol 160(2):708–725. https://doi.org/10.1104/pp.112.203679

Barton KA, Wozny MR, Mathur N, Jaipargas EA, Mathur J (2018) Chloroplast behaviour and interactions with other organelles in Arabidopsis thaliana pavement cells. J Cell Sci 131:jcs202275. https://doi.org/10.1242/jcs.202275

Beltrán J, Wamboldt Y, Sanchez R, LaBrant EW, Kundariya H, Virdi KS et al (2018) Specialized plastids trigger tissue-specific signaling for systemic stress response in plants. Plant Physiol Biochem 178:672–683. https://doi.org/10.1104/pp.18.00804

Biswal B, Biswal U (1988) Ultrastructural modifications and biochemical changes during senescence of chloroplasts. Int Rev Cytol 113:271–321. https://doi.org/10.1016/S0074-7696(08)60851-7

Biswal B, Mohapatra PK, Biswal UC, Raval MK (2012) Leaf senescence and transformation of chloroplasts to gerontoplasts. In: Eaton-Rye J, Tripathy B, Sharkey T (eds) Photosynthesis advances in photosynthesis and respiration, vol 34, pp 217–230. https://doi.org/10.1007/978-94-007-1579-0_10

Bonora A, Pancaldi S, Gualandri R, Fasulo MP (2000) Carotenoid and ultrastructure variations in plastids of *Arum italicum* Miller fruit during maturation and ripening. J Exp Bot 51(346):873–884. https://doi.org/10.1093/jexbot/51.346.873

Brillouet JM, Romieu C, Schoefs B, Solymosi K, Cheynier V, Fulcrand H, Verdeil JL, Conéjéro G (2013) The tannosome is an organelle forming condensed tannins in the chlorophyllous organs of Tracheophyta. Ann Bot 112:1003–1014. https://doi.org/10.1093/aob/mct168

Brillouet JM, Verdeil JL, Odoux E, Lartaud M, Grisoni M, Conéjéro G (2014) Phenol homeostasis is ensured in vanilla fruit by storage under solid form in a new chloroplast-derived organelle, the phenyloplast. J Exp Bot 65:2427–2435. https://doi.org/10.1093/jxb/eru126

Brunkard JO, Runkel AM, Zambryski PC (2015) Chloroplasts extend stromules independently and in response to internal redox signals. Proc Natl Acad Sci U S A 112:10044–10049. https://doi.org/10.1073/pnas.1511570112

Bykowski M, Mazur R, Buszewicz D, Szach J, Mostowska A, Kowalewska Ł (2020) Spatial nanomorphology of the prolamellar body in etiolated *Arabidopsis thaliana* plants with disturbed pigment and polyprenol composition. Front Cell Dev Biol 8:586628

Camara B, Hugueney P, Bouvier F, Kuntz M, Monéger R (1995) Biochemistry and molecular biology of chromoplast development. Int Rev Cytol 163:175–247

Carde J (1984) Leucoplasts: a distinct kind of organelles lacking typical 70S ribosomes and free thylakoids. Eur J Cell Biol 34:18–26

Castillo MA, Wardley WP, Lopez-Garcia M (2021) Light-dependent morphological changes can tune light absorption in iridescent plant chloroplasts: a numerical study using biologically realistic data. ACS Photonics 8(4):1058–1068. https://doi.org/10.1021/acsphotonics.0c01600

Challabathula D, Puthur JT, Bartels D (2016) Surviving metabolic arrest: photosynthesis during desiccation and rehydration in resurrection plants. Ann N Y Acad Sci 1365(1):89–99

Charuvi D, Kiss V, Nevo R, Shimoni E, Adam Z, Reich Z (2012) Gain and loss of photosynthetic membranes during plastid differentiation in the shoot apex of Arabidopsis. Plant Cell 24(3):1143–1157. https://doi.org/10.1105/tpc.111.094458

Chen ZK, Wang FH, Zhou F (1988) On the origin, development and ultrastructure of the orbicules and pollenkitt in the tapetum of Anemarrhena asphodeloides (Liliaceae). Grana 27:273–282. https://doi.org/10.1080/00173138809429949

Choi H, Yi T, Ha S-H (2021) Diversity of plastid types and their interconversions. Front Plant Sci 12:692024. https://doi.org/10.3389/fpls.2021.692024

Chu CC, Li H (2015) Protein import into isolated pea root leucoplasts. Front Plant Sci 6:690. https://doi.org/10.3389/fpls.2015.00690

Ciampolini F, Nepi M, Pacini E (1993) Tapetum development in *Cucurbita pepo* (Cucurbitaceae). In: Hesse M, Pacini E, Willemse M (eds) The Tapetum plant systematics and evolution (Supplementum 7), vol 7. Springer, Vienna, pp 13–22. https://doi.org/10.1007/978-3-7091-6661-1_2

Cortez PA, Caetano APS, Carmello-Guerreiro SM, Teixeira SP (2015) Anther wall and pollen development in Neotropical species-rich Miconia (Melastomataceae). Plant Syst Evol 301:217–230. https://doi.org/10.1007/s00606-014-1067-z

Cui P, Li Y, Cui C, Huo Y, Lu G, Yang H (2020) Proteomic and metabolic profile analysis of low-temperature storage responses in *Ipomoea batata* Lam. tuberous roots. BMC Plant Biol 20:1–15

D'Andrea L, Amenós M, Rodríguez-Concepción M (2014) Confocal laser scanning microscopy detection of chlorophylls and carotenoids in chloroplasts and chromoplasts of tomato fruit. Methods Mol Biol 1153:227–232. https://doi.org/10.1007/978-1-4939-0606-2_16

Dalal VK, Tripathy BC (2012) Modulation of chlorophyll biosynthesis by water stress in rice seedlings during chloroplast biogenesis. Plant Cell Environ 35(9):1685–1703

Dalal V, Dagan S, Friedlander G, Aviv E, Bock R, Charuvi D et al (2018) Transcriptome analysis highlights nuclear control of chloroplast development in the shoot apex. Sci Rep 8:881. https://doi.org/10.1038/s41598-018-27305-4

Dalal VK, Tripathy BC (2018) Water-stress induced downsizing of light-harvesting antenna complex protects developing rice seedlings from photo-oxidative damage. Sci Rep 8(1):5955

Dall'Osto L, Bassi R, Ruban A (2014) Photoprotective mechanisms: carotenoids. In: Thegnand S, Wollman FA (eds) Plastid biology advances in plant biology, vol 5. Springer, New York, NY, pp 393–435. https://doi.org/10.1007/978-1-4939-1136-3_15

Dall'Osto L, Cazzaniga S, North H, Marion-Poll A, Bassi R (2007) The Arabidopsis aba4-1 mutant reveals a specific function for neoxanthin in protection against photooxidative stress. Plant Cell Environ 19:1048–1064. https://doi.org/10.1105/tpc.106.049114

Dankov KG, Dobrikova AG, Ughy B, Bogos B, Gombos Z, Apostolova EL (2011) LHCII organization and thylakoid lipids affect the sensitivity of the photosynthetic apparatus to high-light treatment. Plant Physiol Biochem 49(6):629–635

Dashek WV, Miglani GS (2017) Plant cells and their organelles. In: Vacuoles and protein bodies, vol 2. Wiley, Chichester, pp 351–370. https://doi.org/10.1002/9781118924846

Denic M, Dzamic M, Dumanovic J (1971) On the role of proteinoplasts in genetic regulation of amino acid composition of maize endosperm proteins. Agrochimica 15:564–568

Devidé Z, Ljubešíc N (1974) The reversion of chromoplasts to chloroplasts in pumpkin fruits. Z Pflanzenphysiol 73:296–306. https://doi.org/10.1016/s0044-328x(74)80130-3

Dinakar C, Bartels D (2013) Desiccation tolerance in resurrection plants: new insights from transcriptome, proteome and metabolome analysis. Front Plant Sci 4:482

Dilnawaz F, Mohapatra P, Misra M, Ramaswamy NK, Misra AN (2001) The distinctive pattern of photosystem 2 activity, photosynthetic pigment accumulation, and ribulose-1, 5-bisphosphate carboxylase/oxygenase content of chloroplasts along the axis of primary wheat leaf lamina. Photosynthetica 39:557–563. https://doi.org/10.1023/A:1015656112903

Domínguez F, Cejudo F (2021) Chloroplast dismantling in leaf senescence. J Exp Bot 72(16):5905–5918. https://doi.org/10.1093/jxb/erab200

Eckardt NA (2007) Thylakoid development from biogenesis to senescence, and ruminations on regulation. Plant Cell 19(4):1135–1138. https://doi.org/10.1105/tpc.107.052779

Egea I, Barsan C, Bian W, Purgatto E, Latché A, Chervin C et al (2010) Chromoplast differentiation: current status and perspectives. Plant Cell Physiol 51(10):1601–1611. https://doi.org/10.1093/pcp/pcq136

Evans I, Rus A, Belanger E, Kimoto M, Brusslan JA (2010) Dismantling of Arabidopsis thaliana mesophyll cell chloroplasts during natural leaf senescence. Plant Biol 12:1–12. https://doi.org/10.1111/j.1438-8677.2009.00206.x

Evert RF, Deshpande BP (1971) Plastids in sieve elements and companion cells in Tilia americana. Planta (Berl) 96:97–100. https://doi.org/10.1007/BF00397909

Fang X, Liu S, Gao P, Liu H, Wang X, Luan F et al (2020) Expression of ClPAP and ClPSY1 in watermelon correlates with chromoplast differentiation, carotenoid accumulation, and flesh color formation. Sci Hortic 270. https://doi.org/10.1016/j.scienta.2020.109437

Fernandez O, Ishihara H, George GM, Mengin V, Flis A, Sumner D et al (2017) Leaf starch turnover occurs in long days and in falling light at the end of the day. Plant Physiol 174(4):2199–2212. https://doi.org/10.1104/pp.17.00601

Fujii S, Nagata N, Masuda T, Wada H, Kobayashi K (2019) Galactolipids are essential for internal membrane transformation during etioplast-tochloroplast differentiation. Plant Cell Physiol 60(6):1224–1238. https://doi.org/10.1093/pcp/pcz1041

Gajecka M, Marzec M, Chmielewska B, Jelonek J, Zbieszczyk J, Szarejko I (2020) Plastid differentiation during microgametogenesis determines green plant regeneration in barley microspore culture. Plant Sci 291:110321. https://doi.org/10.1016/j.plantsci.2019.110321

Gallage NJ, Jørgensen K, Janfelt C, Nielsen AJZ, Naake T, Duński E et al (2018) The intracellular localization of the vanillin biosynthetic machinery in pods of Vanilla planifolia. Plant Cell Physiol 59:304–318. https://doi.org/10.1093/pcp/pcx185

Gayral M, Fanuel M, Rogniaux H, Dalgalarrondo M, Elmorjani K, Bakan B et al (2019) The spatiotemporal deposition of lysophosphatidylcholine within starch granules of maize endosperm and its relationships to the expression of genes involved in endoplasmic reticulum–amyloplast lipid trafficking and galactolipid synthesis. Plant Cell Physiol 60:139–151. https://doi.org/10.1093/pcp/pcy198

Georgieva K, Ivanova A, Doncheva S, Petkova S, Stefanov D, Péli E, Tuba Z (2011) Fatty acid content during reconstitution of the photosynthetic apparatus in the air-dried leaves of *Xerophyta scabrida* after rehydration. Biol Plant 55:581–585

Gorecka M, Alvarez-Fernandez R, Slattery K, McAusland L, Davey PA, Karpinski S et al (2014) Abscisic acid signalling determines susceptibility of bundle sheath cells to photoinhibition in high light-exposed Arabidopsis leaves. Philos Trans R Soc Lond Ser B Biol Sci 369:20130234. https://doi.org/10.1098/rstb.2013.0234

Grilli Caiola M, Canini A (2004) Ultrastructure of chromoplasts and other plastids in Crocus sativus L. (Iridaceae). Plant Biosyst 138:43–52. https://doi.org/10.1080/11263500410001684116

Harris WM, Spurr AR (1969) Chromoplasts of tomato fruits. II. The red tomato. Am J Bot 56(4):380–389

Härtel O, Thaler I (1953) Die Proteinoplasten vonHelleborus corsicus Willd. Protoplasma 42:417–426. https://doi.org/10.1007/BF01247925

Hempel J, Amrehn E, Quesada S, Esquivel P, Jiménez VM, Heller A et al (2014) Lipid-dissolved g-carotene, b-carotene, and lycopene in globular chromoplasts of peach palm (Bactris gasipaes Kunth) fruits. Planta 240:1037–1050. https://doi.org/10.1007/s00425-014-2121-3

Hernández-Pinzón I, Ross JHE, Barnes KA, Damant AP, Murphy DJ (1999) Composition and role of tapetal lipid bodies in the biogenesis of the pollen coat of Brassica napus. Planta 208:588–598. https://doi.org/10.1007/s004250050597

Horner HT, Healy RA, Ren G, Fritz D, Klyne A, Seames C et al (2007) Amyloplast to chromoplast conversion in developing ornamental tobacco floral nectaries provides sugar for nectar and antioxidants for protection. Am J Bot 94(1):12–24. https://doi.org/10.3732/ajb.94.1.12

Howitt CA, Pogson BJ (2006) Carotenoid accumulation and function in seeds and non-green tissues. Plant Cell Environ 29:435–445. https://doi.org/10.1111/j.1365-3040.2005.01492.x

Hugueney P, Bouvier F, Badillo A, d'Harlingue A, Kuntz M, Camara B (1995) Identification of a plastid protein involved in vesicle fusion and/or membrane protein translocation. Proc Natl Acad Sci USA 92(12):5630–5634. https://doi.org/10.1073/pnas.92.12.5630

Inaba T, Yazu F, Ito-Inaba Y, Kakizaki T, Nakayama K (2011) Retrograde signaling pathway from plastid to nucleus. Int Rev Cell Mol Biol 290:167–204. https://doi.org/10.1016/B978-0-12-386037-8.00002-8

Irieda H, Takano Y (2021) Epidermal chloroplasts are defense-related motile organelles equipped with plant immune components. Nat Commun 12:2739. https://doi.org/10.1038/s41467-021-22977-5

Itkin M, Seybold H, Breitel D, Rogachev I, Meir S, Aharoni A (2009) TOMATO AGAMOUS-LIKE 1 is a component of the fruit ripening regulatory network. Plant J 60:1081–1095. https://doi.org/10.1111/j.1365-313X.2009.04064.x

Giovannoni JJ (2004) Genetic regulation of fruit development and ripening. Plant Cell 16:S170–S180. https://doi.org/10.1105/tpc.019158

Giovannoni JJ (2007) Fruit ripening mutants yield insights into ripening control. Curr Opin Plant Biol 10(3):283–289. https://doi.org/10.1016/j.pbi.2007.04.008

Jacobs M, Lopez-Garcia M, Phrathep OP, Lawson T, Oulton R, Whitney HM (2016) Photonic multilayer structure of Begonia chloroplasts enhances photosynthetic efficiency. Nat Plants 2:16162. https://doi.org/10.1038/nplants.2016.162

Jarvis P, López-Juez E (2013) Biogenesis and homeostasis of chloroplasts and other plastids. Nat Rev Mol Cell Biol 14:787–802. https://doi.org/10.1038/nrm3702

Jeong HB, Jang SJ, Kang MY, Kim S, Kwon JK, Kang BC (2020) Candidate gene analysis reveals that the fruit color locus C1 corresponds to PRR2 in pepper (Capsicum frutescens). Front Plant Sci 11:399. https://doi.org/10.3389/fpls.2020.00399

Joshi P, Nayak L, Misra AN, Biswal B (2013) Response of mature, developing and senescing chloroplasts to environmental stress. In: Biswal B, Krupinska K, Biswal U (eds) Plastid development in leaves during growth and senescence, Advances in photosynthesis and respiration, vol 36. Springer, Dordrecht. https://doi.org/10.1007/978-94-007-5724-0_28

Juneau P, Le Lay P, Böddi B, Samson G, Popovic R (2002) Relationship between the structural and functional changes of the photosynthetic apparatus during chloroplast–chromoplast transition in flower bud of *Lilium longiflorum*. Photochem Photobiol 75(4):377–381. https://doi.org/10.1562/0031-8655(2002)0750377RBTSAF2.0.CO2

Kakuszi A, Solymosi K, Böddi B (2017) Transformation of plastids in soil-shaded lowermost hypocotyl segments of bean (*Phaseolus vulgaris*) during a 60-day cultivation period. Physiol Plant 159(4):483–491

Kakuszi A, Sárvári É, Solti Á, Czégény G, Hideg É, Hunyadi-Gulyás É, Böddi B (2016) Light piping driven photosynthesis in the soil: Low-light adapted active photosynthetic apparatus in the under-soil hypocotyl segments of bean (Phaseolus vulgaris). J Photochem Photobiol B: Biology 161:422–429

Kanamoto H, Yamashita A, Asao H, Okumura S, Takase H, Hattori M et al (2006) Efficient and stable transformation of Lactuca sativa L. cv. Cisco (lettuce) plastids. Transgenic Res:205–217. https://doi.org/10.1007/s11248-005-3997-2

Karlova R, Rosin FM, Busscher-Lange J, Parapunova V, Do PT, Fernie AR et al (2011) Transcriptome and metabolite profiling show that APETALA2a is a major regulator of tomato fruit ripening. Plant Cell 23(3):923–941. https://doi.org/10.1105/tpc.110.081273

Kato Y, Sakamoto W (2010) New insights into the types and function of proteases in plastids. Int Rev Cell Mol Biol 280:185–218. https://doi.org/10.1016/S1937-6448(10)80004-8

Kim HS, Ji CY, Lee CJ, Kim SE, Park SC, Kwak SS (2018) Orange: a target gene for regulating carotenoid homeostasis and increasing plant tolerance to environmental stress in marginal lands. J Exp Bot 69:3393–3400. https://doi.org/10.1093/jxb/ery023

Kobayashi K, Ohnishi A, Sasaki D, Fujii S, Iwase A, Sugimoto K et al (2017) Shoot removal induces chloroplast development in roots via cytokinin signaling. Plant Physiol 173(4):2340–2355. https://doi.org/10.1104/pp.16.01368

Kobayashi K, Suzuki M, Nagata N (2019) Wax-deficient Arabidopsis mutant cer1-1 shows abnormal tapetosomes, elaioplasts, and pollen coat. Cytologia 84:221–225. https://doi.org/10.1508/cytologia.84.221

Kowalewska Ł, Mazur R, Suski S, Garstka M, Mostowska A (2016) Three-dimensional visualization of the tubular-lamellar transformation of the internal plastid membrane network during runner bean chloroplast biogenesis. Plant Cell 28(4):875–891

Krieger-Liszkay A, Krupinska K, Shimakawa G (2019) The impact of photosynthesis on initiation of leaf senescence. Physiol Plant 166(1):148–164. https://doi.org/10.1111/ppl.12921

Krupinska K, Mulisch M, Hollmann J, Tokarz K, Zschiesche W, Kage H et al (2012) An alternative strategy of dismantling of the chloroplasts during leaf senescence observed in a high-yield variety of barley. Physiol Plant 144(2):189–200. https://doi.org/10.1111/j.1399-3054.2011.01545.x

Kumachova TK, Voronkov AS, Orlova YV, Zhirov VK (2018) Tannosomes in the pericarp cells of Maloideae (Rosaceae). Dokl Biol Sci 482:214–218. https://doi.org/10.1134/S0012496618050149

Kwiatkowska M, Stępiński D, Popłońska K, Wojtczak A, Polit J (2010) "Elaioplasts" of Haemanthus albiflos are true lipotubuloids: cytoplasmic domains rich in lipid bodies entwined by microtubules. Acta Physiol Plant 32:1189–1196. https://doi.org/10.1007/s11738-010-0514-x

Ladygin VG, Bondarev NI, Semenova GA, Smolov AA, Reshetnyak OV, Nosov AM (2008) Chloroplast ultrastructure, photosynthetic apparatus activities and production of steviol glycosides in Stevia rebaudiana in vivo and in vitro. Biol Plant 52:9–16. https://doi.org/10.1007/s10535-008-0002-y

Lawson T, Simkin AJ, Kelly G, Granot D (2014) Mesophyll photosynthesis and guard cell metabolism impacts on stomatal behaviour. New Phytol 203:1064–1081. https://doi.org/10.1111/nph.12945

Lersten NR, Czlapinski AR, Curtis JD, Freckmann R, Horner HT (2006) Oil bodies in leaf mesophyll cells of angiosperms: overview and a selected survey. Am J Bot 93:1731–1739. https://doi.org/10.3732/ajb.93.12.1731

Li A, Lin J, Zeng Z, Deng Z, Tan J, Chen X et al (2024) The kiwifruit amyloplast proteome (kfALP): a resource to better understand the mechanisms underlying amyloplast biogenesis and differentiation. Plant J 118(2):565–583

Li R, Tan Y, Zhang H (2021) Regulators of starch biosynthesis in cereal crops. Molecules 26(23):7092

Li Y, Jian Y, Mao Y, Meng F, Shao Z, Wang T et al (2022) "Omics" insights into plastid behavior toward improved carotenoid accumulation. Front Plant Sci 13:1001756. https://doi.org/10.3389/fpls.2022.1001756

Lichtenthaler HK (2013) Plastoglobuli, thylakoids, chloroplast structure and development of plastids. In: Biswal B, Krupinska K, Biswal U (eds) Plastid development in leaves during growth and senescence, Advances in photosynthesis and respiration, vol 36. Springer, Dordrecht, pp 337–361. https://doi.org/10.1007/978-94-007-5724-0_15

Liebers M, Grübler B, Chevalier F, Lerbs-Mache S, Merendino L, Blanvillain R et al (2017) Regulatory shifts in plastid transcription play a key role in morphological conversions of plastids during plant development. Front Plant Sci Rep 8(23):10.3389/fpls.2017.00023

Liu S, Liu T, Wang E, Cheng Y, Liu T, Chen G et al (2022) Dissecting the chloroplast proteome of the potato (*Solanum Tuberosum L.*) and its comparison with the tuber amyloplast proteome. Plants 11:1915

Liu M, Yu J, Yang M, Cao L, Chen C (2024) Adaptive evolution of chloroplast division mechanisms during plant terrestrialization. Cell Rep 43:113950. https://doi.org/10.1016/j.celrep.2024.113950

Ljubičić JM, Wrischer M, Ljubešić N (1998) Formation of the photosynthetic apparatus in plastids during greening of potato microtubers. Plant Physiol Biochem 36(10):747–752. https://doi.org/10.1016/S0981-9428(98)80025-9

Lü P, Yu S, Zhu N, Chen YR, Zhou B, Pan Y et al (2018) Genome encode analyses reveal the basis of convergent evolution of fleshy fruit ripening. Nat Plants 4:784–791. https://doi.org/10.1038/s41477-018-0249-z

Lu S, Van Eck J, Zhou X, Lopez AB, O'Halloran DM, Cosman KM et al (2006) The cauliflower Or gene encodes a DnaJ cysteine-rich domain-containing protein that mediates high levels of β-carotene accumulation. Plant Cell 18:3594–3605. https://doi.org/10.1105/tpc.106.046417

Luo Z, Zhang J, Li J, Yang C, Wang T, Ouyang B et al (2013) A STAY−GREEN protein S l SGR 1 regulates lycopene and β−carotene accumulation by interacting directly with S l PSY 1 during ripening processes in tomato. New Phytol 198:442–452. https://doi.org/10.1111/nph.12175

Ma G, Zhang L, Kitaya Y, Seoka M, Kudaka R, Yahata M et al (2020) Blue LED light induces regreening in the flavedo of Valencia orange in vitro. Food Chem 335:127621. https://doi.org/10.1016/j.foodchem.2020.127621

Mackenzie SA, Mullineaux P (2022) Plant environmental sensing relies on specialized plastids. J Exp Bot 73:7155–7164. https://doi.org/10.1093/jxb/erac334

Martel C, Vrebalov J, Tafelmeyer P, Giovannoni JJ (2011) The tomato MADS-box transcription factor RIPENING INHIBITOR interacts with promoters involved in numerous ripening processes in a COLORLESS NONRIPENING-dependent manner. Plant Physiol 157:1568–1579

Marti MC, Camejo D, Olmos E, Sandalio LM, Fernandez-Garcia N, Jimenez A, Sevilla F (2009) Characterisation and changes in the antioxidant system of chloroplasts and chromoplasts isolated from green and mature pepper fruits. Plant Biol 11:613–624

Masters NJ, Lopez-Garcia M, Oulton R, Whitney HM (2018) Characterization of chloroplast iridescence in *Selaginella erythropus*. J R Soc Interface 15:20180559. https://doi.org/10.1098/rsif.2018.0559

Mathura SR, Sutton F, Rouse-Miller J, Bowrin V (2024) The molecular coordination of tuberization: current status and future directions. Curr Opin Plant Biol 82:102655

Matsushima R, Hisano H (2019) Imaging amyloplasts in the developing endosperm of barley and rice. Sci Rep 9:1–10. https://doi.org/10.1038/s41598-019-40424-w

Mayfield SP, Huff A (1986) Accumulation of chlorophyll, chloroplastic proteins, and thylakoid membranes during reversion of chromoplasts to chloroplasts in Citrus sinensis epicarp. Plant Physiol 81:30–35. https://doi.org/10.1104/pp.81.1.30
Mingo-Castel AM, Pelacho AM, De Felipe MR (1991) Amyloplast division in kinetin induced potato tubers. Plant Sci 73(2):211–217
Mishra AN (2018) Chlorophyll fluorescence: a practical approach to study ecophysiology of green plants. In: Advances in plant ecophysiology techniques, pp 77–97
Misra AN, Dilnawaz F, Misra M, Biswal AK (2001) Thermoluminescence in chloroplasts as an indicator of alterations in photosystem 2 reaction centre by biotic and abiotic stresses. Photosynthetica 39:1–9. https://doi.org/10.1023/A:1012480114479
Miyagishima S, Nakanishi H, Kabeya Y (2011) Structure, regulation, and evolution of the plastid division machinery. Int Rev Cell Mol Biol 291:115–153. https://doi.org/10.1016/B978-0-12-386035-4.00004-5
Mortimer CL, Misawa N, Perez-Fons L, Robertson FP, Harada H, Bramley PM et al (2017) The formation and sequestration of nonendogenous ketocarotenoids in transgenic Nicotiana glauca. Plant Physiol 173:1617–1635. https://doi.org/10.1104/pp.16.01297
Mu X, Chen Y (2021) The physiological response of photosynthesis to nitrogen deficiency. Plant Physiol Biochem 158:76–82. https://doi.org/10.1016/j.plaphy.2020.11.019
Mulisch M, Krupinska K (2013) Ultrastructural analyses of senescence associated dismantling of chloroplasts revisited. In: Biswal B, Krupinska K, Biswal UC (eds) Plastid development in leaves during growth and senescence, Advances in photosynthesis and respiration, vol 36. Springer, Dordrecht, pp 307–335. https://doi.org/10.1007/978-94-007-5724-0_14
Muñoz P, Munné-Bosch S (2018) Photo-oxidative stress during leaf, flower and fruit development. Plant Physiol Biochem 176(2):1004–1014. https://doi.org/10.1104/pp.17.01127
Naeem M, Tetlow I, Emes MJ (1997) Starch synthesis in amyloplasts purified from developing potato tubers. Plant J 11:1095–1103. https://doi.org/10.1046/j.1365-313X.1997.11051095.x
Nakamura M, Nishimura T, Morita MT (2019) Gravity sensing and signal conversion in plant gravitropism. J Exp Bot 70(3495–3506). https://doi.org/10.1093/jxb/erz1158
Nogueira SB, Labate CA, Gozzo FC, Pilau EJ, Lajolo FM, Oliveira do Nascimento JR (2012) Proteomic analysis of papaya fruit ripening using 2DE-DIGE. J Proteome 75:1428–1439
Oleszkiewicz T, Klimek-Chodacka M, Milewska-Hendel A, Zubko M, Stróz D, Kurczýnska E et al (2018) Unique chromoplast organisation and carotenoid gene expression in carotenoid-rich carrot callus. Planta 248:1455–1471. https://doi.org/10.1007/s00425-018-2988-5
Ostermeier M, Garibay-Hernández A, Holzer VJ, Schroda M, Nickelsen J (2024) Structure, biogenesis, and evolution of thylakoid membranes. Plant Cell 36(10):4014–4035
Pacek A, Stpiczyńska M, Davies KL, Szymczak G (2012) Floral elaiophore structure in four representatives of the Ornithocephalus clade (Orchidaceae: Oncidiinae). Ann Bot 110:809–820. https://doi.org/10.1093/aob/mcs158
Pao SH, Tsai PY, Peng CI, Chen PJ, Tsai CC, Yang EC et al (2018) Lamelloplasts and minichloroplasts in Begoniaceae: iridescence and photosynthetic functioning. J Plant Res 131:655–670. https://doi.org/10.1007/s10265-018-1020-2
Paolillo D, Garvin D, Parthasarathy MV (2004) The chromoplasts of or mutants of cauliflower (Brassica oleracea L. var. botrytis). Protoplasma 224:245–253. https://doi.org/10.1007/s00709-004-0059-1
Pinard D, Mizrachi E (2018) Unsung and understudied: plastids involved in secondary growth. Curr Opin Plant Biol 42(30–36). https://doi.org/10.1016/j.pbi.2018.1001
Pipitone R, Eicke S, Pfister B, Glauser G, Falconet D, Uwizeye C et al (2021) A multifaceted analysis reveals two distinct phases of chloroplast biogenesis during de-etiolation in Arabidopsis. eLife 10:e62709. https://doi.org/10.7554/eLife.62709
Platonova TA, Evsyunina AS, Korableva NP (2010) Changes in the plastid apparatus of apical meristem cells of potato tubers upon growth regulation with jasmonic acid. Appl Biochem Microbiol 46:352–358. https://doi.org/10.1134/S0003683810030178

Prebeg T, Wrischer M, Fulgosi H, Ljubešiæ N (2008) Ultrastructural characterization of the reversible differentiation of chloroplasts in cucumber fruit. J Plant Biol 51:122–131. https://doi.org/10.1007/BF03030721

Pruzinská A, Anders I, Aubry S, Schenk N, Tapernoux-Lüthi E, Müller T et al (2007) In vivo participation of red chlorophyll catabolite reductase in chlorophyll breakdown. Plant Cell 19(1):369–387. https://doi.org/10.1105/tpc.106.044404

Pyke KA (2013) Divide and shape: an endosymbiont in action. Planta 237(2):381–387. https://doi.org/10.1007/s00425-012-1739-2

Quilichini TD, Douglas CJ, Samuels AL (2014) New views of tapetum ultrastructure and pollen exine development in Arabidopsis thaliana. Ann Bot 114:1189–1201

Hagemann R (2004) The sexual inheritance of plant organelles. In: Daniell H, Chase C (eds) Molecular biology and biotechnology of plant organelles, vol 1. Springer, Dordrecht, pp 93–113. https://doi.org/10.1007/978-1-4020-3166-3_4

Rodriguez-Concepcion M, Stange C (2013) Biosynthesis of carotenoids in carrot: an underground story comes to light. Arch Biochem Biophys 539:110–116. https://doi.org/10.1016/j.abb.2013.07.009

Rodríguez-Villalón A, Gas E, Rodríguez-Concepción M (2009) Phytoene synthase activity controls the biosynthesis of carotenoids and the supply of their metabolic precursors in dark-grown Arabidopsis seedlings. Plant J 60:424–435. https://doi.org/10.1111/j.1365-313X.2009.03966.x

Rottet S, Besagni C, Kessler F (2015) The role of plastoglobules in thylakoid lipid remodeling during plant development. Biochim Biophys Acta Bioenerg 1847:889–899. https://doi.org/10.1016/j.bbabio.2015.02.002

Rottet S, Devillers J, Glauser G, Douet V, Besagni C, Kessler F (2016) Identification of plastoglobules as a site of carotenoid cleavage. Front Plant Sci 7:1855. https://doi.org/10.3389/fpls.2016.01855

Rudowska Ł, Gieczewska K, Mazur R, Garstka M, Mostowska A (2012) Chloroplast biogenesis—correlation between structure and function. Biochim Biophys Acta Bioenerg 1817(8):1380–1387

Ruiz-Sola MÁ, Rodríguez-Concepción M (2012) Carotenoid biosynthesis in Arabidopsis: a colorful pathway. Arabidopsis Book 10:e0158. https://doi.org/10.1199/tab.0158

Sagisaka S (2008) The proliferation of amyloplasts in meristematic cells of developing stolons of potato and apple callus: progenitors of proplastids. J Plant Physiol 165:1678–1690. https://doi.org/10.1016/j.jplph.2008.02.003

Sadali NM, Sowden RG, Ling Q, Jarvis RP (2019) Differentiation of chromoplasts and other plastids in plants. Plant Cell Rep 38:803–818. https://doi.org/10.1007/s00299-019-02420-2

Sakamoto W, Miyagishima SY, Jarvis P (2008) Chloroplast biogenesis: control of plastid development, protein import, division and inheritance. Arabidopsis Book 6:e0110. https://doi.org/10.1199/tab.0110

Schelbert S, Aubury S, Burla B, Agne B, Kessler F, Krupinska K et al (2009) Pheophytin pheophorbide hydrolase (pheophytinase) is involved in chlorophyll breakdown during leaf senescence in Arabidopsis. Plant Cell 21(3):767–785. https://doi.org/10.1105/tpc.108.064089

Schramma N, Perugachi Israëls C, Jalaal M (2023) Chloroplasts in plant cells show active glassy behavior under low-light conditions. Proc Natl Acad Sci 120(3):e2216497120

Schweiggert RM, Steingass CB, Heller A, Esquivel P, Carle R (2011) Characterization of chromoplasts and carotenoids of red-and yellow-fleshed papaya (Carica papaya L.). Planta 234(5):1031–1044. https://doi.org/10.1007/s00425-011-1457-1

Scott FM, Bystrom BG, Bowler E (1963) Persea americana, mesocarp cell structure, light and electron microscope study. Bot Gaz 124:423–428. https://doi.org/10.1086/336230

Shah M, Soares EL, Lima MLB, Pinheiro CB, Soares AA, Domont GB et al (2016) Deep proteome analysis of gerontoplasts from the inner integument of developing seeds of Jatropha curcas. J Proteome 143:346–352. https://doi.org/10.1016/j.jprot.2016.02.025

Shedge V, Davila J, Arrieta-Montiel MP, Mohammed S, Mackenzie SA (2010) Extensive rearrangement of the Arabidopsis mitochondrial genome elicits cellular conditions for thermotolerance. Plant Physiol 152:1960–1970. https://doi.org/10.1104/pp.109.152827

Shende VV, Bauman KD, Moore BS (2024) The shikimate pathway: gateway to metabolic diversity. Nat Prod Rep 41(4):604–648. https://doi.org/10.1039/d3np00037k

Shi Y, Ke X, Yang X, Liu Y, Hou X (2022) Plants response to light stress. J Genet Genomics 49(8):735–747

Shimoda Y, Ito H, Tanaka A (2016) Arabidopsis STAY-GREEN, Mendel's green cotyledon gene, encodes magnesium-dechelatase. Plant Cell 28(9):2147–2160. https://doi.org/10.1105/tpc.16.00428

Siddique MA, Grossmann J, Gruissem W, Baginsky S (2006) Proteome analysis of bell pepper (*Capsicum annuum* L.) chromoplasts. Plant Cell Physiol 47:1663–1673

Sierra J, Escobar-Tovar L, Leon P (2023) Plastids: diving into their diversity, their functions, and their role in plant development. J Exp Bot 74(8):2508–2526. https://doi.org/10.1093/jxb/erad044

Silva LAS, Sampaio VF, Barbosa LCS, Machado M, Flores-Borges DNA, Sales JF et al (2020) Albinism in plants – far beyond the loss of chlorophyll: structural and physiological aspects of wild-type and albino royal poinciana (Delonix regia) seedlings. Plant Biol (Stuttg) 22(5):761–768. https://doi.org/10.1111/plb.13146

Simkin AJ, Schwartz SH, Auldridge M, Taylor MG, Klee HJ (2004) The tomato carotenoid cleavage dioxygenase 1 genes contribute to the formation of the flavor volatiles beta-ionone, pseudoionone, and geranylacetone. Plant J 40:882–892

Skupień J, Wójtowicz J, Kowalewska Ł, Mazur R, Garstka M, Gieczewska K, Mostowska A (2017) Dark-chilling induces substantial structural changes and modifies galactolipid and carotenoid composition during chloroplast biogenesis in cucumber (*Cucumis sativus* L.) cotyledons. Plant Physiol Biochem 111:107–118

Solis-Badillo E, Agama-Acevedo E, Tiessen A, Lopez Valenzuela JA, Bello-Perez LA (2020) ADP-glucose pyrophosphorylase is located in the plastid and cytosol in the pulp of tropical banana fruit (Musa acuminata). Plant Foods Hum Nutr 75:76–82. https://doi.org/10.1007/s11130-019-00788-w

Solymosi K, Martinez K, Kristóf Z, Sundqvist C, Böddi B (2004) Plastid differentiation and chlorophyll biosynthesis in different leaf layers of white cabbage (Brassica oleracea cv. capitata). Physiol Plant 121:520–529. https://doi.org/10.1111/j.0031-9317.2004.00349.x

Solymosi K, Morandi D, Bóka K, Böddi B, Schoefs B (2012) High biological variability of plastids, photosynthetic pigments and pigment forms of leaf primordia in buds. Planta 235:1035–1049. https://doi.org/10.1007/s00425-011-1559-9

Solymosi K, Tuba Z, Böddi B (2013) Desiccoplast–etioplast–chloroplast transformation under rehydration of desiccated poikilochlorophyllous Xerophyta humilis leaves in the dark and upon subsequent illumination. J Plant Physiol 170:583–590

Stefanov MA, Rashkov GD, Borisova PB, Apostolova EL (2024) Changes in photosystem II complex and physiological activities in pea and maize plants in response to salt stress. Plants 13(7):1025

Streatfield SJ, Weber A, Kinsman EA, Häusler RE, Li J, Post-Beittenmiller D et al (1999) The phosphoenolpyruvate/phosphate translocator is required for phenolic metabolism, palisade cell development, and plastid-dependent nuclear gene expression. Plant Cell 11(9):1609–1622. https://doi.org/10.1105/tpc.11.9.1609

Sun T, Li L (2020) Toward the 'golden' era: the status in uncovering the regulatory control of carotenoid accumulation in plants. Plant Sci 290:110331. https://doi.org/10.1016/j.plantsci.2019.110331

Sun T, Zhou F, Huang XQ, Chen WC, Kong MJ, Zhou CF et al (2019) ORANGE represses chloroplast biogenesis in etiolated Arabidopsis cotyledons via interaction with TCP14. Plant Cell 31(12):2996–3014. https://doi.org/10.1016/j.plantsci.2019.110331

Sun T, Yuan H, Cao H, Yazdani M, Tadmor Y, Li L (2018) Carotenoid metabolism in plants: the role of plastids. Molecular plant 11(1):58–74

Suzuki T, Tsunekawa S, Koizuka C, Yamamoto K, Imamura J, Nakamura K et al (2013) Development and disintegration of tapetum-specific lipid-accumulating organelles, elaioplasts and tapetosomes, in Arabidopsis thaliana and Brassica napus. Plant Sci 207:25–36. https://doi.org/10.1016/j.plantsci.2013.02.008

Tanios S, Eyles A, Tegg R, Wilson C (2018) Potato tuber greening: a review of predisposing factors, management and future challenges. Am J Potato Res 95:248–257. https://doi.org/10.1007/s12230-018-9648-y

Thomson WW, Whatley JM (1980) Development of nongreen plastids. Annu Rev Plant Physiol 31:375–394. https://doi.org/10.1146/annurev.pp.31.060180.002111

Toledo-Ortiz G, Johansson H, Lee KP, Bou-Torrent J, Stewart K, Steel et al (2014) The HY5-PIF regulatory module coordinates light and temperature control of photosynthetic gene transcription. PLoS Genet 10(6):e1004416. https://doi.org/10.1371/journal.pgen.1004416

Tripathy BC, Oelmüller R (2012) Reactive oxygen species generation and signaling in plants. Plant Signal Behav 7(12):1621–1633

Us-Camas R, Castillo-Castro E, Aguilar-Espinosa M, Limones-Briones V, Rivera-Madrid R, Robert-Díaz ML et al (2017) Assessment of molecular and epigenetic changes in the albinism of Agave angustifolia Haw. Plant Sci 263:156–167. https://doi.org/10.1016/j.plantsci.2017.07.010

Vigil E, Ruddat M (1985) Development and enzyme activity of protein bodies in proteinoplasts of tobacco root cells. Histochemistry 83:17–27. https://doi.org/10.1007/BF00495295

Virdi KS, Wamboldt Y, Kundariya H, Laurie JD, Keren I, Sunil Kumar KR et al (2016) MSH1 is a plant organellar DNA binding and thylakoid protein under precise spatial regulation to alter development. Mol Plant 9:245–260. https://doi.org/10.1016/j.molp.2015.10.011

Wang X, Chen S, Zhang H, Shi L, Cao F, Guo L, Dai S (2010) Desiccation tolerance mechanism in resurrection fern-ally Selaginella tamariscina revealed by physiological and proteomic analysis. J Proteome Res 9(12):6561–6577

Wang X, Sun F, Han C, Chen Q, Yang X, Cheng J et al (2020) The E-2- hexenal play critical role in carotenoids activated inner membrane structures transformation in the chloroplasts. Authorea:1–17. https://doi.org/10.22541/au.158316318.85083277

Wang Y, Yu B, Zhao J, Guo J, Li Y, Han S et al (2013a) Autophagy contributes to leaf starch degradation. Plant Cell Physiol 25(4):1383–1399. https://doi.org/10.1105/tpc.112.108993

Wang YQ, Yang Y, Fei Z, Yuan H, Fish T, Thannhauser TW et al (2013b) Proteomic analysis of chromoplasts from six crop species reveals insights into chromoplast function and development. J Exp Bot 64(4):949–961. https://doi.org/10.1093/jxb/ers375

Welsch R, Zhou X, Koschmieder J, Schlossarek T, Yuan H, Sun T et al (2020) Characterization of cauliflower OR mutant variants. Front Plant Sci 10:1716. https://doi.org/10.3389/fpls.2019.01716

Wise RR (2007) The diversity of plastid form and function. In: Wise RR, Hoober JK (eds) The structure and function of plastids, Advances in photosynthesis and respiration, vol 23. Springer, Dordrecht, pp 3–26. https://doi.org/10.1007/978-1-4020-4061-0_1

Woo HR, Kim HJ, Lim PO, Nam HG (2019) Leaf senescence: systems and dynamics aspects. Annu Rev Plant Biol 70(1):347–376. https://doi.org/10.1146/annurev-arplant-050718-095859

Wójtowicz J, Węgrzyn A, Bykowski M, Bukat A, Mostowska A, Kowalewska Ł (1996) 3 Prolamellar body—getting ready for photosynthesis. In: Handbook of photosynthesis. CRC Press, pp 3–23

Wrischer M, Prebeg T, Magnus V, Ljubesic N (2009) Unusual thylakoid structures appearing during degradation of the photosynthetic apparatus in chloroplasts. Acta Bot Croat 68(1):1–9. https://hrcak.srce.hr/36153

Wurtzel ET, Cuttriss A, Vallabhaneni R (2012) Maize provitamin a carotenoids, current resources, and future metabolic engineering challenges. Front Plant Sci 3(29). https://doi.org/10.3389/fpls.2012.00029

Xie QJ, Michaeli S, Peled-Zehavi N, Galili G (2015) Chloroplast degradation: one organelle, multiple degradation pathways. Trends Plant Sci 20(5):264–265. https://doi.org/10.1016/j.tplants.2015.03.013

Xu C, Li J, Wang H, Liu H, Yu Z, Zhao Z (2023) Whole-transcriptome sequencing reveals a ceRNA regulatory network associated with the process of periodic albinism under low temperature in Baiye No. 1 (*Camellia sinensis*). Int J Mol Sci 24(8):7162. https://doi.org/10.3390/ijms24087162
Xu H, Zhang L, Li R, Wang X, Liu S, Liu X et al (2018) SKL1 is essential for chloroplast development in Arabidopsis. Front Plant Sci 9:1–12. https://doi.org/10.3389/fpls.2018.00179
Xu YZ, Arrieta-Montiel MP, Virdi KS, de Paula WB, Widhalm JR, Basset GJ et al (2011) MutS HOMOLOG1 is a nucleoid protein that alters mitochondrial and plastid properties and plant response to high light. Plant Cell 23:3428–3441. https://doi.org/10.1105/tpc.111.089136
Yadav D, Zemach H, Belausov E, Charuvi D (2019) Initial proplastid-tochloroplast differentiation in the developing vegetative shoot apical meristem of Arabidopsis. Biochem Biophys Res Commun 519:391–395. https://doi.org/10.1016/jbbrc.2019.09.019
Yang R, Sun Y, Zhu X, Jiao B, Sun S, Chen Y et al (2024) The tuber-specific StbHLH93 gene regulates proplastid-to-amyloplast development during stolon swelling in potato. New Phytol 241(4):1676–1689
You MK, Lee YJ, Kim JK, Baek SA, Jeon YA, Lim SH et al (2020) The organ-specific differential roles of rice DXS and DXR, the first two enzymes of the MEP pathway, in carotenoid metabolism in Oryza sativa leaves and seeds. BMC Plant Biol 20(1):167. https://doi.org/10.1186/s12870-020-02357-9
Yu L, Fan J, Zhou C, Xu C (2021) Chloroplast lipid biosynthesis is fine-tuned to thylakoid membrane remodeling during light acclimation. Plant Physiol 185(1):94–107
Zhang J, Hu Z, Yao Q, Guo X, Nguyen V, Li F et al (2018) A tomato MADS-box protein, SlCMB1, regulates ethylene biosynthesis and carotenoid accumulation during fruit ripening. Sci Rep 8(1–15):3558–3563. https://doi.org/10.1038/s41598-018-21672-8
Zeng Y, Pan Z, Ding Y, Zhu A, Cao H, Xu Q, Deng X (2011) A proteomic analysis of the chromoplasts isolated from sweet orange fruits [*Citrus sinensis* (L.) Osbeck]. J Exp Bot 62:5297–5309
Zhou X, Welsch R, Yang Y, Álvarez D, Riediger M, Yuan H, Fish T, Liu J, Thannhauser TW, Li L (2015) Arabidopsis OR proteins are the major posttranscriptional regulators of phytoene synthase in controlling carotenoid biosynthesis. Proc Natl Acad Sci 112(11):3558–3563. https://doi.org/10.1073/pnas.1420831112
Zhu M, Lin J, Ye J, Wang R, Yang C, Gong J et al (2018) A comprehensive proteomic analysis of elaioplasts from citrus fruits reveals insights into elaioplast biogenesis and function. Hortic Res 5(1–11). https://doi.org/10.1038/s41438-017-0014-x

2 Chloroplasts: Genetic Machinery and Regulatory Mechanisms in Plant Function and Adaptation

Shina Sasi, Tanveer Alam Khan, and Mayank Anand Gururani

Abstract

Plants rely on chloroplasts for photosynthesis and metabolic functions as their genomes developed through endosymbiotic events with cyanobacteria. Next-generation sequencing breakthroughs have exposed structural differences in chloroplast genomes among plant species, which demonstrate the essential function of these genomes in environmental stress adaptation. Chloroplast transcription is governed by two RNA polymerases: Photosynthetic genes are transcribed by PEP, while non-photosynthetic genes are transcribed by NEP, with nuclear factors controlling both processes. The stability and functionality of RNA depend on posttranscriptional modifications, including RNA editing and splicing, which pentatricopeptide-repeat (PPR) proteins and RNA-binding enzymes drive forward. Light-regulated translation within chloroplasts happens through a bacterial-like 70S ribosome that incorporates components from both plastid and nuclear origins. The process of ribosomal pausing plays a critical role in achieving correct protein folding as well as the assembly of photosynthetic complexes. The complex regulatory molecular mechanisms associated with chloroplast gene expression and translation that promote plant growth and stress adaptations are discussed in this chapter.

Keywords

Chloroplast · Genome · Photosynthesis · Ribosomes · RNA polymerase · Transcription

S. Sasi · T. A. Khan
Khalifa Center for Genetic Engineering and Biotechnology, Al Ain, United Arab Emirates

M. A. Gururani (✉)
Biology Department, College of Science, UAE University, Al Ain, United Arab Emirates
e-mail: gururani@uaeu.ac.ae

V. K. Dalal, A. N. Misra (eds.), *Chloroplast Biogenesis and Plastid Interconversions*,
https://doi.org/10.1007/978-981-95-0772-6_2

2.1 Introduction

Chloroplasts are the organelles found in plant and green algal cells that carry out photosynthesis and other fundamental metabolic pathways. By transforming light energy into chemical energy through photosynthesis, chloroplasts are vital for the growth and survival of plants. Chloroplasts have been known to originate from an ancient symbiotic relationship between a eukaryotic host cell and a photosynthetic cyanobacterium (De et al. 2024). The plant chloroplast that we know today is the result of this association, which developed over the years. While modern chloroplasts contain remnants of the ancient cyanobacterial genome, a large portion of this genome has been transferred to the nuclear genome of the plant. This gene transfer led to the continuous coevolution between the host plant and its endosymbiotic partner. The expression of genes in these two compartments is tightly coordinated to form complex multimeric protein complexes, in which individual subunits are encoded by both the nuclear and plastid genomes (Jarvis and López-Juez 2013; Kleine and Leister 2016).

Different plant species show vast differences in their chloroplast genome, containing up to 315 genes, encoding more than 3000 proteins. A significant number of these proteins are necessary for the survival of the plant, since they are implicated in vital processes (i.e., photosynthesis) (Gururani et al. 2015). Chloroplast genes encode crucial components of the photosynthetic machinery, including the large subunit of Rubisco, some subunits of the thylakoidal protein complexes, such as photosystem I and II (PSI and PSII), cytochrome b6f complex, and ATP synthase. Along with this, it also contributes to the plant gene expression system (including rRNAs, tRNAs, and a set of RNA polymerase core subunits of bacterial type). Retention of chloroplast gene expression is still a matter of scientific curiosity (Allen et al. 2011; Green 2011).

Aside from photosynthesis, chloroplasts take part in several significant biochemical processes. These metabolic processes are relevant for the synthesis of fatty acids, phytohormones, vitamins, sulfur, and nitrogen, among others (Daniell et al. 2016). Chloroplasts are also involved in the generation of color pigments that impact plant development and morphology. Chloroplasts provide metabolites that help in plant development as well as enable the plant to respond to various environmental stresses, like light, temperature, and nutrient availability (Daniell et al. 2016).

Chloroplasts are known to contain their own DNA; however, most of their proteins in these organelles are encoded by the nuclear genome. These proteins are synthesized in the cytoplasm and required to be imported into the chloroplast, which depends on certain specialized protein pores on the chloroplast membrane called TOC (translocon of the outer chloroplast membrane) and TIC (translocon of the inner chloroplast membrane). This recognition is mediated by short amino acid sequences at the N-terminal of the protein called transit peptides, to guide it through the chloroplast membranes (Sun and Zerges 2015).

SP1 stands for SUPPRESSOR OF PPI1 LOCUS1, which encodes a group of proteins regulating protein import into the chloroplast. SP1 encodes an E3 ligase that shuttles unnecessary proteins for degradation, thus maintaining essential proteins in the chloroplast. SP1 has two homologs: SPL1 and SPL2. While both proteins share some sequence similarity with SP1, their exact function remains unclear (Mohd Ali et al. 2023). SP2 regulates the degradation of TOC proteins. SP2 acts in conjunction with SP1 to create a novel ubiquitin-proteasome system (UPS) pathway referred to as the chloroplast-associated protein degradation (CHLORAD) pathway. In this way, SP2 also mediates the retrotranslocation of substrates, allowing import of only the correctly assembled proteins into the chloroplasts. This system enables chloroplasts to make functional changes during development and in response to environmental stresses (Mohd Ali et al. 2023). In fact, SPL1 and SPL2 are also implicated in leaf senescence but play opposite roles. This implies that these proteins function within an extensive regulatory network to provide a counterbalance to protein import into the chloroplast versus degradation to maintain chloroplast homeostasis, which is critical for normal chloroplast function and plant health.

2.2 Chloroplast Genomic Architecture

Chloroplast DNA (cpDNA) was once believed to exist as a circular molecule inside plant cells. Recent advances have revealed that cpDNA can also develop multi-branched linear structures in numerous angiosperms. The shape of cpDNA can differ, depending on tissue type, developmental stage, and experimental conditions, and varies within plant species (Mower and Vickrey 2018). The first cpDNA sequencing was successful with tobacco (*Nicotiana tabacum*) and liverwort (*Marchantia polymorpha*) (Ohyama et al. 1986; Shinozaki et al. 1986). The sequencing of *Synechocystis* sp. marked a major milestone in this direction, PCC6803, a *Cyanobacterium* (Sato et al. 1999), provided vital insights into the chloroplast origin. This work supported the endosymbiotic theory, which states that chloroplasts evolved from photosynthetic cyanobacteria. Extensive phylogenetic studies have shown that most plant proteins have a cyanobacterial origin, with more than 4500 genes in *Arabidopsis thaliana* showing footprints of this ancestry (Zoschke and Bock 2018).

The cpDNA size varies remarkably between plant species, ranging from only 15,553 base pairs (bp) in *Asarum minus* to 521,168 bp in *Floydiella terrestris* (Dobrogojski et al. 2020). But this variation is uncorrelated with plant nuclear genome size. The general features, including the structure, content, and gene arrangement of cpDNA, are highly conserved among terrestrial plants despite these differences. cpDNA is organized into dense structures termed nucleoids, and the number of cpDNA copies per plant cell ranges from 1000 to 1700. For instance, in *Arabidopsis thaliana*, the chloroplast carries ~20–35 cpDNA copies of cpDNA (Zoschke et al. 2007), and this number changes according to the plant growth process (Oldenburg and Bendich 2015). The structure of cpDNA comprises two identical inverted repeat (IR) regions ranging between about 20,000 and 25,000 bp in size, separated by a long single copy (LSC) region and a short single copy (SSC) region. These IR regions are extremely conserved among plant species (Morley et al. 2019).

Besides, the intron sequences of cpDNA are usually highly conserved with few exceptions, such as chickpeas (*Cicer arietinum*), barley (*Hordeum vulgare*), bamboo (Bambusoideae), and cassava (*Manihot esculenta*), and intron loss is observed in genes encoding for specific proteins (Dobrogojski et al. 2020). Notably, this loss of introns in genes has been observed for proteins, such as ATP synthase (atpF), Clp protease (clpP), RNA polymerase (rpoC2), and ribosomal proteins (rpl2, rps12, rps16) (Kaila et al. 2017). Extensive intron loss occurs in certain plant groups, like gymnosperms (*Pinus*), eudicots (*Onagraceae* and *Oleaceae*), and monocots (*Poaceae*), but intron loss appears to be restricted to certain plant families or species.

2.3 Chloroplast DNA Structure

The entire cpDNA sequence of *Arabidopsis thaliana* has been published and reveals the gene organization pattern of the chloroplast genome (Sato et al. 1999). *A. thaliana* is one of the most well-studied model plants and has brought many important discoveries in various areas of plant biology, mostly related to plant genomics, and its cpDNA sequence has constructed a foundation for the comprehensive understanding of plant biology.

Arabidopsis cpDNA is about 154,478 bp in length and contains a canonical quadrupole structure. It contains two inverted repeat (IR) regions (IRa and IRb) of approximately 26,264 base pairs. These are separated by a large single copy (LSC) region (84,197 bp) and a small single copy (SSC) region (17,780 bp). The overall A + T content of the chloroplast genome is 63.7%, which is analogous to that of the cpDNA reported for tobacco (62.2%), rice (61.1%), maize (61.5%), and Thunberg pine (61.5%) (Maier et al. 1995; Wakasugi et al. 1994). A + T content is not uniform among the regions; inverted repeats have 57.7% A + T, while the corresponding LSC and SSC have higher values: 66.0% and 70.7%, respectively (Sato et al. 1999). There are 87 open reading frames (ORFs), with a duplicated ORF77 (YCF15), and this can be considered 85 ORFs when ORFs other than the duplicated ORFs are counted. Among these, 79 encode proteins, and eight genes are duplicated in the IR regions. In addition to this, three ribosomal RNA (rRNA) genes, including 16S, 23S, and 4.5S rRNA, were located in the IR regions, while a 5S rRNA gene was found in the LSC region. A cluster of rRNA genes is interspaced with two tRNA genes (TRN1 and TRNA), and 37 tRNA-encoding genes are dispersed between the LSC and SSC regions are observed (Sato et al. 1999).

The *A. thaliana* chloroplast genome (cpDNA) encodes important proteins needed for gene expression and photosynthesis. These genes include:

2.3.1 Photosynthetic Genes

A total of 45 genes are involved in encoding proteins that are responsible for photosynthesis and are required for energy production in the plant cells (Dobrogojski et al. 2020).

The cytochrome b6f complex subunits are encoded by six genes (PETA, PETB, PETD, PETG, PETL, and PETN), a crucial component of the electron transport chain. It carries electrons between PSII and PSI and also helps in creating a proton gradient across the thylakoid membrane, which is essential for ATP production. Photosystem I (PSI), encoded by five genes (PSAA, PSAB, PSAC, PSAI, and PSAJ), is essential to the light reactions of photosynthesis. PSI captures light energy and uses energy to reduce NADP+ to NADPH, which is then later used to make sugars in the Calvin cycle. Photosystem II (PSII) components are encoded by 15 genes (PSBA, PSBB, PSBC, PSBD, PSBE, PSBF, PSBH, PSBI, PSBJ, PSBK, PSNL, PSBM, PSBN, PSBT, and PSBZ), and they are responsible for carrying out the initial phases of photosynthesis, that is, splitting water molecules to release oxygen and electrons (Dobrogojski et al. 2020). PSII is important in energizing the electron transport chain. ATP synthase (ATPase) complex is encoded by six genes (ATPA, ATPB, ATPE, ATPF, ATPH, and ATPI) for synthesizing ATP in the light reactions. It uses the proton gradient to synthesize ATP from ADP, thereby providing energy to the plant. NADH Dehydrogenase subunits are encoded by 11 genes (NDHA, NDHB, NDHC, NDHD, NDHE, NDHF, NDHG, NDHH, NDHI, NDHJ, NDHK, or PSBG) that are involved in the transferring electrons from NADH to plastoquinone in the electron transport chain and is responsible for establishing the proton gradient, needed for ATP synthesis. RuBisCO is encoded by a single gene (RBCL), catalysis the conversion of atmospheric carbon dioxide into organic molecules, which serve as the building blocks of plant sugars.

2.3.2 Transcription and Translation Associated Genes

Other than photosynthetic genes, the *Arabidopsis* chloroplast genome encodes several other genes involved in gene expression and protein synthesis:

RNA polymerase is encoded by four genes (RPOA, RPOB, RPOC1 and RPOC2). This enzyme is responsible for the transcription of the chloroplast genome into RNA (Börner et al. 2015). Ribosomal RNA (rRNA) is encoded by 22 genes ((RPL2, RPL14, RPL16, RPL20, RPL22, RPL23, RPL32, RPL33, RPL36 for large ribosomal proteins) (RPS2, RPS3, RPS4, RPS7, RPS7, RPS8, RPS11, RPS12, RPS14, RPS15, RPS16, RPS18, and RPS19 for small ribosomal proteins) and is responsible for forming ribosomes (Dobrogojski et al. 2020). These genes guarantee that the chloroplast operates correctly, allowing it to perform photosynthesis and produce the proteins required for the growth and development of the plant.

2.4 The Inverted Repeat-Lacking Clade

The IRLC (inverted repeat-lacking clade) provides an interesting case, being a clade of *Leguminosae* species that have undergone a major evolutionary transition in their chloroplast genomes. In these plants, one of the IR regions (~25 kb) has been lost (Mehmetoğlu et al. 2023). This long-term evolutionary modification has now been found in almost 4000 species belonging to 56 genera in the IRLC, revealing variations in chloroplast genome organization within this group.

This kind of evolutionary IRLC trend has been extensively studied in the wild relative of the common chickpea, *Cicer reticulatum* cpDNA (Mehmetoğlu et al. 2023). The cpDNA of *Cicer reticulatum* is 125,794 bp in length, which is consistent with that of other IRLC members. In contrast to the majority of plants, the *Cicer reticulatum* genome is composed of a single inverted repeat (IR) region (29,949 bp), a LSC (82,531 bp) and a SSC (13,314 bp) (Fig. 2.1). This loss of one IR region is particularly reflective of IRLC plants, distinguishing it from most of the plants with the standard two-IR structure (Mehmetoğlu et al. 2023). The cpDNA of *Cicer reticulatum* contains 117 genes: 4 rRNA genes, 34 tRNA genes, and 79 protein-coding genes. Two protein-coding genes (clpP1 and ycf3) have two introns, and 23 other genes (ndhA, ndhB, rps12, atpF, and others) have one intron (Mehmetoğlu et al. 2023). The first and most critical thing that makes these introns essential is that they play a part in gene regulation and expression, which can affect how well these genes are functioning. These introns contribute to the genome complexity of the chloroplast in this species and exemplify the evolutionary modifications within the IRLC.

Cicer reticulatum possesses an extraordinarily large, 2488 bp intron, the trnK-UUU (trnTERM-UUA) gene (Mehmetoğlu et al. 2023). This intron is the largest intron in the plant genome. Conversely, the smallest intron is in gene rrn23 and is only 198 bp. Intron size in these genes is variable and therefore could affect the evolutionary flexibility of the chloroplast genome. Intron losses and gains can affect patterns of gene expression, and in *Cicer reticulatum* as well, intron structure seems to play a role in its adaptability to diverse environmental and evolutionary pressures.

Studies by He et al. (2017) revealed that introns play a vital role in alternative splicing and can also serve to stabilize transcripts in certain eukaryotic lineages. About 16 genes containing introns were identified in *L. japonica*, 14 of which contained 1 intron while rps18 and ycf3 contained two introns. Rps18 is highly important for plastid translation, and ycf3 is required for the stable accumulation of the photosystem I complex (He et al. 2017). Both genes from the polycistronic unit are present in *Chlamydomonas reinhardtii* but are otherwise found isolated in land plants (Lee et al. 2007). Therefore, a gain of two introns in the rps18 and ycf3 genes of *L. japonica*'s plastid genome is rare and not seen in any other plastid genomes, adding an intriguing dimension to the investigation of the evolution of photosynthesis. On the other side, in *L. japonica* cp genome, introns were lost in the clpP, petB, petD, and rpl16 (He et al. 2017). For example, in most angiosperms, the rpl16 intron is a stable component but is lost in a couple of angiosperm families, including *Geraniaceae*, *Goodeniaceae*, and *Plumbaginaceae*. The loss of introns in petB, petD, and rpl16 first occurred in Asterid lineages. Such introns play a key role in

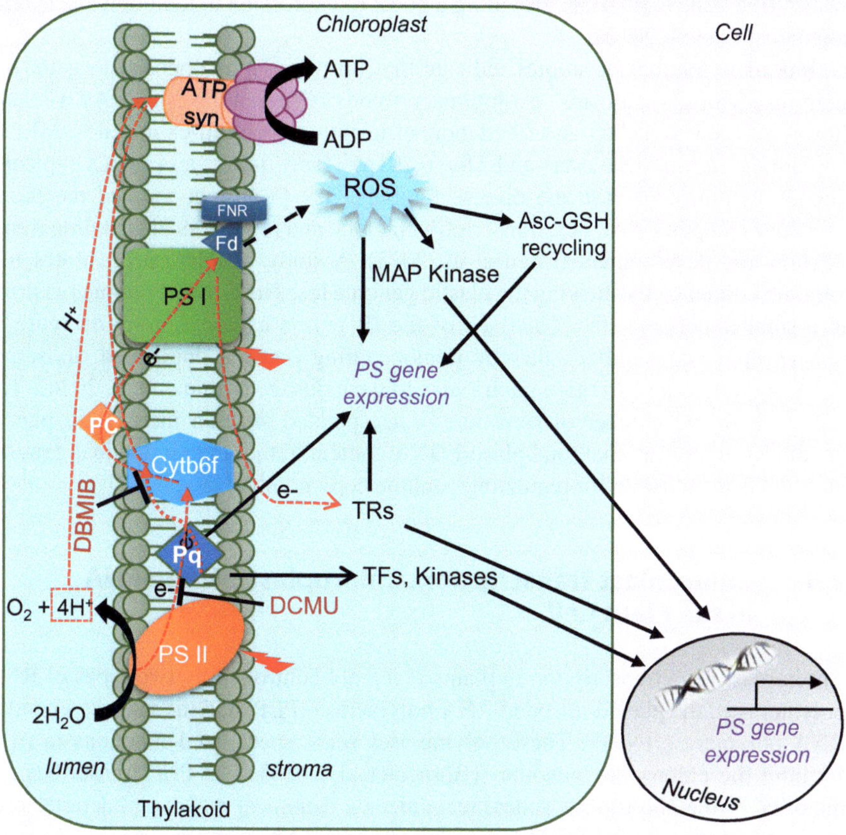

Fig. 2.1 Activation of photosynthetic gene expression by light-driven redox chemistry. The electron transport chain in the thylakoid membrane involves the electron flow (depicted as red arrows). In photosystem II (PSII), light-induced charge separation splits water into protons (H^+), electrons (e^-), and oxygen (O_2). Protons are then transferred across the membrane to ATP synthase (ATP syn), while electrons are transferred to PSI through the cytochrome b6f (Cytb6f) protein complex. Plastoquinone (Pq) accepts electrons from PSII and transfers them to PSI via Cytb6f and the plastocyanin (PC) pool. Various functional and metabolic processes, including light acclimation and photosynthetic (PS) gene expression, depend on the redox state of the Pq pool. This redox state is sensitive to electron transport inhibitors 3-(3,4-dichlorophenyl)-1,1-dimethylurea (DCMU) and 2,5-dibromo-3-methyl-6-isopropyl-p-benzoquinone (DBMIB). DCMU inhibits the reduction of Pq, whereas DBMIB inhibits the oxidation of Pq. Pq pools also induce the expression of certain protein kinases and transcription factors (TFs), which can activate transcriptional signal transduction in the nucleus. After receiving electrons from the Cytb6f complex, PSI transfers them to the electron end acceptor NADP+ via ferredoxin (Fd). Additionally, Fd reduces thioredoxins (TRs) that control the activity of key enzymes involved in the Calvin cycle. TRs are reduced by PSI via ferredoxin and ferredoxin NADP+ reductase (FNR). Abiotic stresses, such as high light, drought, low temperature, etc., can result in a transfer of electrons from PSI to oxygen, which in turn generates reactive oxygen species (ROS). Ascorbate (Asc) and glutathione (GSH) recycling then detoxify ROS molecules, preserving the redox state of cellular proteins. Incomplete detoxification of ROS can cause activation of mitogen-activated protein kinases, which then trigger nuclear gene expression

controlling gene regulation, providing a boost of expression in temporal and spatial manners (Yi et al. 2012).

Indeed, a number of studies indicate that intron loss, as opposed to gain, has dominated most organisms' evolutionary trends (Chen et al. 2023). As a result, intron-less genes can be transcribed more efficiently than multi-exon genes (MEGs) containing introns (Savisaar and Hurst 2016). Truly intron-less genes typically encode members of vast and diverse families, like G protein-coupled receptors, olfactory receptors, histones, transcription factors, and regulators of signaling transduction and development (Chen et al. 2023). A study of tobacco illustrates this genome efficiency by showing the plastid genome loses its large IR region, resulting in a smaller genome size. This lack of the IR region caused a small decrease in plastid ribosome numbers, thereby demonstrating that the duplicated ribosomal RNA operon in the IR region contributes to gene dosage (Krämer et al. 2024). The reduction in the number of ribosomes corresponded with an increase in plastid genomes, suggesting that total plastid DNA content, rather than individual genome number, is responsible for regulating genome copy number.

2.5 Chloroplast Transcription: Chloroplast DNA to RNA in the Plant Cell

Chloroplast gene transcription in plants is mainly controlled by two types of RNA polymerases: the plastid-encoded RNA polymerase (PEP) and the nucleus-encoded RNA polymerase (NEP). These polymerases serve specialized functions in transcribing the chloroplast genomes (Börner et al. 2015). The PEP transcribes the majority of the chloroplast genes, recognizes a canonical σ70-type bacterial promoter. This promoter of PEP has two core consensus sequences, TATAAT at position −10 and TTGACT at position −35, from the transcription initiation site. These sequences serve as signals to direct the polymerase to the appropriate location to initiate transcription (Ortelt and Link 2021). But the specific location of such sequences varies from plant to plant as a result of plant genetic diversity. PEP is essential for transcribing key genes involved in photosynthesis, as it produces proteins critical for the light-dependent reactions. It consists of multiple subunits. Although the majority of the genes for these subunits have been relocated to the nuclear genome, some crucial genes are still retained in the cpDNA. These genes encode the α, β, β′ and β″ subunits of the polymerase core (Börner et al. 2015). The molecular weight of the various subunits is α subunit (38 kDa), β subunit (120 kDa), β′ subunit (85 kDa), β″ subunit (185 kDa). Notably, the genes encoding the α subunit (RPOA) and ribosomal proteins are co-located in an operon and transcribed by a single same promoter. Conversely, the genes coding for the β, β′, and β″ subunits (RPOB, RPOC1, and RPOC2) constitute their own operon, RPOBC (Börner et al. 2015). The PEP function is additionally regulated by sigma factors (SIGs) and polymerase-associated proteins (PAPs), where both are encoded by the nuclear genome. These SIGs function as transcription initiation factors, directing the polymerases to particular promoters (Puthiyaveetil et al. 2021). In *Arabidopsis*, there are around six sigma factors that participate in the process of transcription. These SIGs are conserved at their C-terminal ends, and non-conserved regions determine specific functions and regulation (Chi et al. 2015). PAPs are proteins that associate with the Rpo catalytic core of PEP, and there are 12 identified PAPs in *Arabidopsis*, which play important roles in various aspects of transcription. Mutation in these PAPs caused disruption of chloroplast transcription and typically resulted in albino phenotypes, producing effects similar to PEP-related gene knockouts (Zhang et al. 2023).

Some of the notable PAPs are:

(i) PAP5 (pTAC12 or HEMERA): This PAP binds to ssRNA and ssDNA, indicating its role at active sites of the PEP complex in positioning DNA and RNA (Chang et al. 2017).
(ii) PAP8 (pTAC6): It contains an RNA-binding motif, suggesting its involvement in nascent RNA management (Chambon et al. 2022).
(iii) PAP10 (THIOREDOXIN Z) and PAP6 (FRUCTOKINASE-LIKE 1): Participate in the redox regulation of transcription for maintaining transcription stability (Yang et al. 2021).
(iv) PAP4 and PAP9: These PAPs, enriched with superoxide dismutase domains, protect the PEP complex against oxidative stress induced by reactive oxygen species (Favier et al. 2021).

Overall, PAP seems to play a role in almost all stages of transcription (Pfannschmidt et al. 2015). These proteins respond with different expression levels to ensure transcription of specific chloroplast genes is initiated at the correct time and in response to environmental signals (Zhang et al. 2023) by interacting with PEP. It also shields the transcription machinery against stressors such as oxidative damage.

The NEP enzyme carries out the entire transcription cycle in the chloroplast from promoter recognition until the end of transcription. It was derived from a duplication event of a gene that initially encoded a mitochondrial RNA polymerase and has similarity to phage-type RNA polymerases, which also have a single subunit (Börner et al. 2015).

There are three classes of NEP, encoded by genes referred to as RPOT (RNA polymerase of phage type T3/T7). This includes:

(i) RpoTp: chloroplast-targeted.
(ii) RpoTm: mitochondrion-targeted.
(iii) RpoTmp: dual-targeted (in both, in chloroplasts and mitochondria).

In dicots such as *Arabidopsis* and tobacco, RpoTmp is active in young, dividing cells and in non-photosynthetically active tissues, whereas RpoTp has higher activity in green, photosynthetically active tissues (Puthiyaveetil et al. 2021). Notably, in monocots, only RpoTp has been detected in chloroplasts, and RpoTm is detected in mitochondria (Dobrogojski et al. 2020). In chloroplasts, NEP transcribes a defined set of genes that are not related to or involved in photosynthesis. Specifically, it transcribes:

(i) RPL23 (ribosomal protein L23).
(ii) ACC D (acetyl-CoA carboxylase, vital for some dicots).
(iii) CLPP (subunit encoding gene of the ClpP protease, important for most monocots).
(iv) RPOB (the β subunit of the PEP core required for all higher plants) (Liere and Börner 2007).

These genes perform multiple cellular functions, like synthetic proteins and metabolic functions, but do not participate in the photosynthetic processes.

Chloroplast genes and their promoters can be classified into three groups, depending on which polymerase they are transcribed by:

(i) NEP-transcribed genes (the rpoB operon and accD gene).
(ii) PEP-transcribed genes (the majority of chloroplast genes).
(iii) Genes transcribed by both NEP and PEP (Hajdukiewicz et al. 1997).

More than 215 transcription start sites have been discovered using Terminome-seq technology in *Arabidopsis* in mature chloroplasts for both PEP and NEP (Castandet et al. 2019). NEP and PEP also recognize particular promoter sequences but bind to different motifs. NEP just recognizes three types of promoters (Ia, Ib, and II). The Ia and Ib promoters contain the conserved motifs YRTA and GAA box (Morley et al. 2019). In contrast, PEP binds to bacterial-type σ70 promoters consisting of a TATAAT sequence at position −10, and a TTGACT sequence at −35 as mentioned above.

NEP and PEP are engaged in various developmental stages of a plant, also in non-photosynthetic organs (e.g., fruits, seeds, and roots). They also consistently maintain activity in genes essential for basic cellular functions, including those encoding tRNAs. (Börner et al. 2015). PEP is a more dominant polymerase in mature chloroplasts, especially in photosynthetic tissues, and produces >80% of chloroplast primary transcripts (Puthiyaveetil et al. 2021).

Transcriptional pausing is where the RNA polymerase stalls at specific sequences of the DNA. The pause facilitates gene expression regulation since it allows time for the cell to decide on how to process the genetic instruction (Core and Adelman 2019; Wissink et al. 2019). Within bacteria and animals, pausing/stalling at well-defined locations proximal to the start of a gene is a fundamental aspect of controlling gene activity and modulating the structure of the genome (Core and Adelman 2019). Pause is also involved in cellular processes, such as connecting transcription with translation, splicing RNA and folding RNA, and stabilizing the genome (Akhtar et al. 2019).

In plants like *Arabidopsis*, RNA polymerase II is paused at 20–40% of the protein-coding genes, influencing transcription efficiency (Zhu et al. 2018). Approximately 30% of genes in chloroplasts exhibit evidence of transcriptional pausing. PEP frequently stalls early in these genes, particularly within the associated promoter regions (Zhu et al. 2018; Palomar et al. 2022). This would indicate that PEP does not force the transcription process forward immediately but instead pauses for a second, probably to enable proper regulation of gene expression. As pauses aid in transcription (Chi et al. 2015; Palomar et al. 2022)—if this pausing is interrupted—transcription can be disrupted. Of note, there are protein factors that regulate pausing within chloroplasts. One of these protein groups, mitochondrial transcription termination factors (mTERFs), was first found in mitochondria but has recently also been identified in chloroplasts. The *Arabidopsis* mTERF5 helps to pause the PEP complex at unique sites within the gene psbEFLJ, which not only ensures correct transcription but even stabilizes the RNA being produced (Ding et al. 2019). Hence, the RNA stability was significantly decreased in mutants devoid of mTERF5, indicating that pausing is a significant factor in not only transcription but also the preservation of RNA integrity (Méteignier et al. 2020).

2.6 PEP Complex Activation Through Light-Induced Redox Changes in Chloroplasts

In chloroplasts, light-driven photosynthetic redox reactions are essential; various environmental changes impact the electron transport efficiency of the components of the electron transport chain as well as the amount of the redox-active compounds involved in photosynthesis (Fig. 2.1). In the nucleus, genes, the coding of PAPs, are expressed through light signals that induce histone H4 acetylation as a form of modification. This occurs early within the light signaling pathway, paving the way for subsequent events (Barrero-Gil et al. 2022). When light strikes the plant, it also triggers the synthesis of the complex of proteins called PEP, which is essential to transcription in the chloroplast. Light accomplishes this by shutting down certain proteins known as phytochrome-interacting factors (PIFs). Putting all this together, PIFs facilitate the communication of signals generated in the nucleus, and they do so to regulate the assembling of the PEP complex (Yang et al. 2019; Hwang et al. 2022).

Once formed, PEP assembles as a complete machinery with the gene promoters to initiate transcription (Finster et al. 2013). PRIN2, a protein that senses chloroplast redox status, is a central component of this light-driven process (Kindgren et al. 2012). In the dark, PRIN2 adopts a dimeric formation connected via a disulfide bond. In contrast, when light strikes the plant, thioredoxins are activated in the chloroplast, switching PRIN2 to its active monomeric form. The active monomers subsequently mediate PEP complex activation, potentially via interactions with the protein PAP10/THIOREDOXIN Z (Díaz et al. 2018). This communication connects the light-induced (chloroplast) electron flow to PEP complex activation. Additionally, the activity of the PEP complex and the sigma factors are phosphorylated in response to the redox state of the plastoquinone pool by plastid transcription kinases. As an example, SIG1 phosphorylation, an important sigma factor, regulates the transcription of photosystem genes, allowing the plant to adapt to light (Shimizu et al. 2010).

In this way, the light induces a transcriptional cascade of events in the chloroplast as well as in the nucleus. In addition to redox changes in the chloroplast, it achieves this by signaling the nucleus to check the proper chloroplast function and the turnover of the PEP complex. This highly synchronized process enables the plant to adapt to light fluctuations throughout the day and maximize photosynthetic efficiency.

2.7 Posttranscriptional Modifications of RNA in Chloroplasts

The primary chloroplast transcripts generated first as long polycistronic chains require multiple important posttranscriptional processing events for its maturation into mRNA. These include RNA editing, splicing, trimming of the 5′ and 3′ ends, and intercistronic cleavage between genes (Zhang et al. 2023). These modifications are crucial for the stability of the newly formed chloroplast transcripts and ensuring to form the secondary structures correctly, thereby allowing efficient translation by the 70S ribosomes present within the chloroplast (Zoschke and Bock 2018; Zhang et al. 2020).

2.8 Chloroplast Intercistronic Cleavage and RNA End Processing

In land plants, chloroplast gene expression is regulated by an mRNA maturation process, where long primary transcripts are processed into various mature RNA isoforms. This is done by ribonucleases, a wide class of RNA-modifying enzymes. There are mainly two general subclasses of ribonucleases found in chloroplasts, exonucleases and endonucleases (Zhang et al. 2023). The same type of enzymes found in bacteria, called exonucleases, degrade RNA by removing nucleotides from the 5′ or 3′ end of RNA molecules. In contrast, endonucleases cleave the RNA internally, generating shorter RNA subfragments. Many chloroplast exonucleases have been identified in *Arabidopsis*, including RNase J (5′ → 3′), RNase R, and polynucleotide phosphorylase (3′ → 5′). Internal RNA cleavage is performed by endonucleases, predominantly RNases E and J (Manavski et al. 2018). Other key ribonucleases are involved in specialized processes such as tRNA maturation. These include RNase P and RNase Z (TRZ2), which process tRNAs. Two distinct RNA-binding endonucleases termed chloroplast stem-loop binding proteins (CSP41a and CSP41b) serve to stabilize non-translated mRNAs and precursor rRNAs, especially in low-light environments (Leister 2014). Finally, members of the miniribonuclease III family, RNC3 and RNC4, process the 23S–4.5S rRNA precursor into 23S RNA and 4.5S RNA (Hotto et al. 2015). In conclusion, these ribonucleases target common structural motifs found in chloroplast RNA molecules instead of their sequences and facilitate the appropriate processing required for functional chloroplast RNAs.

The process of primary chloroplast transcript processing begins with endonucleolytic cleavage of RNA by endonucleases (CSP41; RNases E and J), cutting RNA at unstructured portions. After these cleavages, exonucleases will subsequently cut the 5′ and the 3′ ends to refine the RNA. The 5′ ends of transcripts are normally protected both by RNA protectors against attacks by 5′ to 3′ exonucleases, and the 3′ ends are additionally protected by secondary structures in the 3′ untranslated region (UTR) generated by short inverted repeat sequences in chloroplasts (Zhang et al. 2023). Pentatricopeptide repeat (PPR) proteins comprise a major class of RNA protectors in chloroplasts. They also protect RNA from exonuclease degradation by binding to its terminal regions. PPR proteins consist of repeated 35-amino-acid motifs that fold into a helical structure that can bind specific nucleotides in the RNA. This "one-repeat:one-nucleotide" binding model promotes PPR proteins to stabilize the 5′ ends of transcripts and functions as a physical barrier against the activity of endogenous exonucleases. These can also compete with other RNA structures, such as stem-loop structures, that allow ribosomal binding sites (e.g., Shine-Dalgarno sequence) to be exposed. In *Arabidopsis*, the PPR protein SUPPRESSOR OF THYLAKOID FORMATION 1 (STF1) and the maize protein PPR53 help stabilize the 5′ ends of the 23S precursor RNA and the ndhA transcript. Some PPR proteins also safeguard and define transcript 3′ ends without short inverted repeat sequences (Zoschke et al. 2016). PPR10 preferentially associates with intergenic regions of the atpI–atpH and psaJ–rpl33 transcripts and prevents RNA degradation from either end. Another instance involves the PPR protein CHLOROPLAST RNA PROCESSING 1 (CRP1), which interacts with the S1-domain-containing RNA chaperone PETB/PETD STABILIZING FACTOR to maintain the integrity of the 3′ end of the petB transcript and facilitate translation initiation of petD (Jiang et al. 2023). Another PPR protein, PGR3, stabilizes RNA of the PETL operon. PGR3 thus promotes the correct translation of the genes which function as components of the chloroplast's energy production pathways, such as PETL and NDHA (Shikanai and Fujii 2013). Overall, these results emphasize that PPR proteins are critical to preserving RNA integrity and enabling efficient translation in the chloroplast (Shikanai and Fujii 2013).

2.9 Posttranscriptional RNA Editing in Chloroplasts

RNA editing is an important posttranscriptional event in which the nucleotide sequence of RNA is modified, potentially generating a protein-coding signal that is different from the one encoded in the genomic DNA. In chloroplasts, this is done predominantly by converting cytosine (C) to uracil (U) in an event referred to as C-to-U RNA editing (Small et al. 2020). This editing can alter the coding sequences of chloroplast mRNAs, thereby affecting their stability, secondary structures, splicing, and even the generation of translational start sites (AUG). Chloroplast RNA editing is mediated by a class of specific proteins, leading to 20–60 revealed editing positions in flowering plants. These proteins interact with the 5′ end of the RNA, generally 20–25 nucleotides upstream of the edited nucleotide. This association results in an RNA–protein complex that is postulated to associate with an enzyme with deaminase activity that is important in the editing process (Takenaka et al. 2013). Notably, in chloroplasts, the RNA editing process is mediated by factors coded in the nuclear genome, rather than on the plastid genome. Studies, including those in barley mutants lacking plastid ribosomes and in the presence of translation inhibitors, have shown that RNA editing can still occur independently of active chloroplast translation, further highlighting its dependence on nuclear-encoded factors (Dobrogojski et al. 2020).

PLS- class PPR proteins are part of a distinct structural class of PPR proteins and are central to RNA editing in the chloroplast (Wang et al. 2021). They possess repetitive motifs of P, L, and S, and typically have a C-terminal DYW domain, which is homologous to cytidine deaminases (Dedow and Bailey-Serres 2019). The signature HxE(x)nCxxX is conserved in the DYW domain, which suggests that it is the deaminase in the RNA editing pathway. An example of a PPR-DYW protein capable of C-to-U editing on its mRNA target transcript, ccmFC, is PPR56 from *Physcomitrium patens* and highlights the DYW domain contribution to uniqueness within the RNA editing process (Hayes and Santibanez 2020). The crystal structure of the DYW domain of the OTP86DYW protein in *Arabidopsis* showed a unique autoinhibition mechanism for the domain. A gating domain occludes the DYW domain deaminase fold and prevents cytidine from accessing the active site (Takenaka et al. 2021). When the gating domain is activated, it undergoes a conformational change, permitting the active site to surround and attract a water molecule necessary for catalyzing the C-to-U deamination (Takenaka et al. 2021). This mechanism guarantees complete generation precisely at the goal sites, precluding off-target impacts in the cytosol. Interestingly, reverse U-to-C editing in plants has also been reported, which has its origins associated with a subclass of DYW domains called DYW:KP. This set features a conserved Lys-Pro dipeptide, implying that slightly different DYW domain architectures could allow the reversal of the C-to-U editing mechanism (Small et al. 2020). In addition to the PPR-DYW proteins, flowering plants evolved a number of other essential factors involved in RNA editing, including MORF, RNA helicase ISEL2, and organelle zinc finger proteins (Wang et al. 2021). When combined with PPR proteins, these factors assemble to form editosomes, large multi-protein complexes that mediate RNA editing. For example, MORF9 binds to the L motif of PLS-type PPR proteins, converting them to active forms with higher-binding affinity to their cognate RNA. However, the functional importance of other components of the editosome that do not interact directly with PPRs remains poorly defined (Yan et al. 2017; Royan et al. 2021).

The ORRM1 gene, which encodes a principal plastid RNA editing factor, has been shown to be mutated in *Arabidopsis*, leading to a severe decrease in RNA editing efficiency (around 62% reduction) at chloroplast editing sites (Searing et al. 2020). It is interesting that orrm1 mutants do not display any apparent symptoms at normal temperatures but are extremely sensitive to cold. The young leaves of these mutants turn yellow when exposed to low temperatures (4 °C), indicating that chloroplast RNA editing plays a role in helping plants cope with cold stress (Wang et al. 2016). But the specific RNA editing site that conferred this cold tolerance remains elusive. A similar phenomenon has been reported in rice. The defects in RNA editing of the rps8 gene, which is necessary for ribosome function, render the Dular rice cultivar (dua1) cold-sensitive (Cui et al. 2019). This editing is essential in keeping the RPS8 protein stable and allows the plant to thrive at lower temperatures. These results indicate that RNA editing in chloroplasts has probably developed as a mechanism to improve protein stability, particularly in the face of stressful cold environments (Cui et al. 2019).

2.10 The In-depth Molecular Mechanism of Chloroplast Transcription and RNA Splicing

2.10.1 RNA Splicing

Introns are removed from the primary transcripts in a process termed RNA splicing so that a mature mRNA, whose pattern is appropriately represented for genetic information, can be created.

2.10.2 Different Types of Introns

The chloroplast genes have two kinds of introns—type I introns and type II introns. Both are classified as self-splicing introns, though the self-splicing assay has only been validated in the lab for the group I introns (Inaba-Hasegawa et al. 2021). Nuclear genes have no direct effects on the splicing of group I introns in chloroplasts, surprisingly enough. For group II introns, genetic studies of genomes in plants have identified factors regulating their splicing (Germain et al. 2013).

However, most chloroplast introns have lost the ability to self-splice, a consequence of the degradation of RNA structures and the loss of intron-encoded maturases during evolution. Consequently, new proteins are required to efficiently splice these introns (Zhang et al. 2023). The trnK intron that encodes the maturase MatK is an example. Its role is particularly important in splicing a number of group IIA introns (Zhang et al. 2023).

Meanwhile, other nuclear-encoded splicing factors facilitate this process, for example:

(i) The RNase III-domain protein RNC1.
(ii) The plant organelle RNA recognition protein (WTF1) in maize.
(iii) ACCUMULATION OF PHOTOSYSTEM ONE 1 (known from maize and Arabidopsis) (Watkins et al. 2011).

In *Arabidopsis*, the DEAD-box RNA helicase RH3 plays a key role in regulating the splicing of several chloroplast genes. RH3-deficient mutants are embryo lethal, whereas mutants containing a weaker rh3–4 allele have slow growth, pale-green leaves, and poor splicing of trnA, trnI, rpl2, and rps12 genes (Asakura et al. 2012).

The DEAD-box RNA helicase RH3 has been shown to influence the production of the hormone abscisic acid (ABA), which plays a role in stress responses. In rh3–4 seedlings, ABA levels are approximately 50% lower because the genes for the biosynthesis of ABA are downregulated, resulting in these plants demonstrating higher susceptibility toward abiotic stresses, such as salinity, cold, and dehydration (Gu et al. 2014). Therefore, a decrease in photosynthetic efficiency would be seen by interfering with chloroplast RNA splicing, which would make the plants more vulnerable to environmental stresses. On the other hand, accurate splicing may trigger retrograde signaling from chloroplasts to the nucleus, which helps regulate stress responses in plants.

2.11 Chloroplast Ribosome Maturation (CRM) Proteins

Chloroplast ribosome maturation (CRM) proteins, CAF1, CAF2, CRS1, CFM2, and CFM3, are essential for splicing group IIB introns of chloroplast RNA (Wang et al. 2022). CAF1 and CAF2 associate with CRS2 to form complexes that enable the efficient splicing of certain introns. The CRS2–CAF1 complexes splice seven introns, and the CRS2–CAF2 complexes splice five, with the ycf3 intron 1 being spliced by both complexes (Wang et al. 2020). These complexes are guided in specificity by interactions with PPR proteins, like PHOTOSYSTEM I BIOGENESIS FACTOR 2 (associating with CAF1 and CAF2) and EMBRYO DEFECTIVE 1270 (binding with CFM2) (Wang et al. 2020; Zhang et al. 2021). RNA-binding PPR proteins bind RNA in a sequence-specific way to guide the splicing of the correct intron sites. Although chloroplasts have a small genome, the process of RNA splicing is quite complex and requires a variety of factors. Unlike self-splicing introns that are abundant in prokaryotes, chloroplast introns are spliced with great dependence on these nuclear-encoded splicing factors (Wang et al. 2022). This dependency potentially evolved via the coevolution of the nucleus and the chloroplasts. This indicates the importance of these splicing factors in plant growth and development and their potential mirroring of environmental stressors, further emphasizing their importance for the ability of the plant to adapt to shifting conditions (Zhang et al. 2021).

2.12 Involvement of Plastid Noncoding RNAs in Their Regulation and Function

Chloroplasts contain a wide variety of abundant (~45% of all accumulated RNA) plastid noncoding RNAs (pncRNAs) ranging in length from 15 nucleotides to over 3 kilobases (Hotto et al. 2011). The majority of these pncRNAs are produced from intergenic, intronic, or antisense regions of the chloroplast genome, commonly due to read-through transcription from neighboring genes. All pncRNAs are not though; some are primary transcripts regulated by their own unique promoters (Anand and Pandi 2021).

Processing of these molecules requires ribonucleases, such as the 3′ to 5′ direction exonuclease RNase R, to mature these pncRNAs. If RNase R is not present, these pncRNAs do not form and thus are deleted. Other ribonucleases are also considered to contribute to the stabilization or accumulation of pncRNAs (Hotto et al. 2011), including RNase J and RNase E. Among them, PPR proteins, including PPR10 and PPR53, generate small pncRNAs, commonly called PPR footprints (Zoschke et al. 2016).

pncRNAs are important mediators of posttranscriptional regulation of chloroplast gene expression. As an example, antisense RNA (asRNA) such as AS5, which is complementary to the rrn5-trnR region, might regulate the maturation of the 5S rRNA (Hotto et al. 2011). NdhB transcript stability may be influenced or RNA editing may be affected by long asRNAs targeting the ndhB gene. Likewise, other asRNAs targeting psbT can block its translation and also stabilize the psbT transcript. In addition, chloroplast sRNAs also have the potential to compete with mRNAs for binding to PPR proteins, thereby affecting RNA metabolism and subsequent regulation of gene expression (Ruwe et al. 2016).

2.13 Chloroplast Translation

Chloroplasts contain a 70S ribosome (bacterial-like), which, together with transfer RNAs (tRNAs) and ribosomal RNAs (rRNAs), is crucial for protein synthesis within the organelle (Pulido et al. 2018). This ribosome consists of a large (50S) and a small (30S) subunit, each one containing rRNAs and proteins encoded by the plastid genome and the nucleus. Although chloroplasts and bacteria have a common ancestor and an incredibly similar ribosomal structure to each other, their ribosomes have diverged in different fashions. In addition, chloroplast translation is linked to the thylakoid membrane system and is light-regulated to ensure efficient photosynthesis (Zhang et al. 2023).

Only a full spectrum of bacterial-type rRNAs is used by the chloroplast ribosome, including the 23S, 16S, and 5S rRNAs (Zoschke and Bock 2018). It has been known for a very long time that the 23S rRNA works as a peptidyl transferase responsible for elongation during protein synthesis and that the 16S rRNA, in combination with other factors, helps decode the genetic messages and assists in the assembly of ribosomes (Moore and Steitz 2011). Moreover, the chloroplast has an additional 4.5S rRNA that is absent in bacteria but is believed to have originated from a piece of the prokaryotic 23S rRNA (Chi et al. 2012). The pre-23S rRNA undergoes posttranscriptional maturation through two cleavages, leading to three products (0.5, 1.2, and 1.1 kb). These cleavages, performed by specialized enzymes including RNA helicase RH39 (Bieri et al. 2017), are responsible for chloroplast translation. It has, however, severely changed during its evolution, most notably at the level of its protein content. For instance, in chloroplasts, some bacterial ribosomal proteins have been lost, such as Rpl25 and Rpl30 (Ahmed et al. 2016). Concurrently, plastid-specific ribosomal proteins (PSRPs) 1–6 have also appeared, including PSRP5 and PSRP6 in the large subunit (50S) and PSRP2 and PSRP3 in the small subunit (30S) (Bieri et al. 2017). These PSRPs are essential components of the chloroplast ribosome (Zhang et al. 2020). Some, such as PSRP1, serve specific functions, like binding to the mRNA channel of the small subunit, possibly regulating translation by obstructing tRNA access. Other PSRPs aid in compensating for evolutionary modifications of the ribosomal RNA structure (Perez Boerema et al. 2018).

Mitochondrial ribosomes, or mitoribosomes essential for the synthesis of proteins involved in oxidative phosphorylation, drive energy generation in plant cells. These ribosomes are composed of three distinct classes of RNA, namely, 5S, 18S, and 26S, all synthesized by the mitochondrial genome itself (Petersen et al. 2015). Although the nuclear and mitochondrial genomes primarily encode the proteins that make up these ribosomes, species-specific variations in the number and types of these ribosomal proteins make it more difficult to regulate this vital plant function (Harper et al. 2023). The effect of limiting plastid transcription or translation on plant growth and organelle function was studied in *Arabidopsis* seedlings (Sajib et al. 2023). In this particular study, the etiolated *Arabidopsis* seedlings were treated with two antibiotics, spectinomycin and rifampicin. The antibiotic-treated seedlings displayed increased apical hook bending (a typical reaction to loss of plastid functions). A similar kind of twisting effect was also exhibited by RPOTp RNA polymerase mutants, suggesting that the phenotype was not just an off-target effect of the antibiotics but rather was directly linked to a restriction of plastid gene expression. They also found that rifampicin treatment in germinating seedlings significantly altered the mitochondrial metabolism, which was characterized by an increase in reactive oxygen species and increased activity of alternative oxidases (Turrens 2003; Sajib et al. 2023). These characteristic changes suggested a potential connection between plastids and mitochondria, where plastid dysfunction initiates changes in mitochondrial metabolism. They discovered that rifampicin-treated seedlings were enriched with intermediates derived from the TCA cycle and glycolysis, as well as showing marked alterations in amino acid metabolism (Sajib et al. 2023). This indicated that constraining plastid transcription could have far-reaching effects, which can also affect mitochondrial activity and metabolism (Sajib et al. 2023).

These changes in mitochondrial metabolism appeared to send signals to the nucleus. Among the genes that were highly upregulated in the rifampicin-treated seedlings were those linked to mitochondrial stress, such as AT12CYS and AOX1A (Garmash et al. 2021; Sajib et al. 2023). It implies that in instances of limited gene expression of plastids, mitochondria can relay these stress signals to the nucleus—perhaps via physical or signaling pathways between the organelles. These studies reveal inter-organellar communication between plastids and mitochondria. When one organelle is stressed—by problems in plastid gene expression or mitochondrial dysfunction—the other appears to compensate. This suggests that a much deeper level of coordination exists between the nucleus, the plastids, and the mitochondria to enable responses to cellular stress.

2.13.1 Chloroplast Translation and Ribosomal Pausing Regulation

Chloroplast translation is an intricate process that entails four main phases: initiation, elongation, termination, and ribosome recycling. The initiation of translation begins with the recruitment of the 30S ribosomal subunit to the ribosome-binding site, mediated by initiation factors (IF1, IF2, and IF3) (Zheng et al. 2016). The vast majority of chloroplast genes exhibit a bacterial-style translation initiation process reliant on Shine-Dalgarno (SD) sequences. The SD sequence on the mRNA is complementary to the anti-SD sequence in the 16S rRNA of the ribosome, so that the ribosome is correctly positioned to begin translation (Bieri et al. 2017). In contrast, translation initiation for genes that do not contain SD sequences is primarily mediated by the lack of structural folding of the 5′ untranslated region (UTR), probably by a separate mechanism. After formation of the preinitiation complex, the 50S ribosomal subunit binds, and a complete 70S ribosome is generated, and the process transitions to the elongation phase (Masuda et al. 2012). Prokaryotic elongation factors EF-Tu, EF-G, and EF-Ts are needed for the addition of amino acids to the growing protein during elongation.

Chloroplasts also use the standard genetic code, but their sets of tRNAs have been reduced, and they utilize special base-pairing mechanisms (wobble and super-wobble) to decode mRNA with a smaller set of tRNAs (Zoschke and Bock 2018). Termination occurs when the ribosome encounters a stop codon (UAA, UAG, or UGA). Release factors RF1/PrfA and RF2/PrfB1 promote the release of the full-length protein, while ribosome recycling factors recycle the 70S ribosome to make it available for another round of translation (Zoschke and Bock 2018). Chloroplast ribosomes routinely pause in specific RNA regions in seed plants (Gawronski et al. 2018), including *Arabidopsis* and maize. These pauses are governed by RNA structure, internal SD sequences, and positively charged amino acids, etc. These pauses are necessary to allow proper transmembrane domains insertion, cofactors integration, and protein folding, all of which are essential processes that must occur to assemble photosynthetic complexes correctly. This mechanism is conserved in photosynthetically dependent plants, but appears varied in parasitic plants with reduced or absent photosynthesis (Gawronski et al. 2018). This indicates that ribosome pausing is particularly relevant for photosynthetic complex formation, reinforcing its importance for the biogenesis of these fundamental macromolecular structures (Gawronski et al. 2018).

2.13.2 Posttranslational Modifications in the Chloroplast: Ensuring Correct Protein Function and Stability

It begins in the chloroplasts, where the proteins in their newly produced state go through a few processes to become functional. The first steps involve the removal of the N-formylmethionine (fMet) from the protein in a deformylation process enzymatically mediated by methionine aminopeptidase (Meinnel and Giglione 2022).

The D1 precursor undergoes cleavage by a protease called carboxyl terminal peptidase at the C-terminal tail to ensure that the mature protein is properly formed (Gao et al. 2023).

On the other hand, the proteases in the chloroplast are also important players in protein quality control. They assist proteins to be folded correctly and get rid of the damaged proteins (Gao et al. 2023). D1 (one of the most important light-absorbing proteins in photosynthesis) is among those proteins in need of constant degradation and binding-compartment repair, as it is constantly damaged by light. Several proteases are involved in the degradation of damaged D1, including Deg1, Deg5, Deg7, and Deg8, from *Arabidopsis* (Sun et al. 2010). Deg7 specifically has a role in the degradation of other PSII subunits. Another protease suggested in relation to the PSII complex is ftsH2, which is involved in D1 degradation during PSII complex formation; however, the nature of ftsH2 and the specific cut sites in the D1 protein remain unclear and require further exploration (Yi et al. 2022).

2.13.3 Controlling Translation and Repair of PSII by Light-Induced pH Changes in the Chloroplast

In land plants, chloroplasts modulate translation rates in response to light linked to energy-consuming protein biosynthesis, with energy-producing photosynthesis. Light-induced translation facilitates the repair of the light-damaged PSII reaction center protein D1 (Chotewutmontri and Barkan 2020). Rapid activation associated with translation elongation in chloroplasts is mainly triggered by the transitional change from dark to light. (Zhang et al. 2023). An increase in pH levels in the thylakoid lumen is the main cause of this instant activation, which is also probably impacted by the cytochrome b6f complex. This pH shift favors translation elongation; however, the mechanism is unclear, though proteins such as PSRP1 and EF-Tu might play a role (Chotewutmontri and Barkan 2020; Zoschke and Bock 2018). This is particularly true for the translation of psbA RNA, which encodes D1 and is upregulated by light. This adds to the complexity of HCF244 function, as it is also involved in regulating psbA translation together with the OHP1/OHP2 proteins, as nascent D1 interacts with this protein complex and is recruited to PSII repair. In the absence of D1, the complex activates translation to generate sufficient amounts of D1 for PSII repair. This autoregulatory mechanism allows plants to coordinate the repair of light-induced damage with the translation of D1 (Chotewutmontri and Barkan 2020).

2.13.4 Environmental Factors That Influence Chloroplast Genome Expression and Stability

One of these environmental factors that affects chloroplasts is temperature, and other factors are light and cold, which greatly influence the behavior of the chloroplast genomes of the plants. Given that cpDNA is responsible for photosynthesis and stress responses, and that the plants have developed mechanisms to regulate chloroplast gene expression under stress conditions. Genes like psbA, which encodes D1 protein linked to PSII repair, show higher expression during high light or temperature stress (Danilova et al. 2018). In tomatoes, a protein named SlWHY1 increases the transcription of psbA when challenged with cold to repair photodamage (dealing with chilling stress) (Zhuang et al. 2019). The psbD gene, coding for PSII repair, is also under the control of a light-responsive promoter that modulates under certain stress conditions such as salinity or low temperature (Noordally et al. 2013).

The processing of RNA is another essential part of chloroplast regulation, especially under stressful environmental. RNA editing in plant chloroplasts enables them to adapt as well as survive in cold-temperature conditions (Adamiec and Luciński 2024). Defective RNA editing of rps8 is shown to reduce the stability of the protein it encoded, thereby severely affecting tolerance of rice towards low temperature (Zhang et al. 2020). Moreover, splicing and RNA stability are both crucial for stress responses. Plant mutants with the RH3 gene (involved in RNA splicing) had reduced ABA hormone production and reduced ability to respond to drought and cold stress (Cui et al. 2019). In addition, chloroplast translation is important for plants to adapt to stressful conditions or exhibit stress tolerance. In high light or cold, plants need to rapidly repair damaged proteins to perform photosynthesis smoothly, and this process involves ribosomal proteins such as Rps15 and Rpl33. Mutations in these proteins have seem to increase the sensitivity to cold in plants (Rogalski et al. 2008). Mutants of tobacco Rpl33 were severely damaged at low temperatures, indicating that maintaining efficient chloroplast translation is essential for cold tolerance (Yao et al. 1989; Rogalski et al. 2008). Likewise, chloroplast translation is affected by heat stress. In *Arabidopsis*, RPS1 knockdown showed reduced capacity to activate heat stress responses, linking chloroplast translation to plant thermotolerance (Yu et al. 2012). Environmental stressors regulate chloroplast gene expression at different levels of transcription, RNA processing, and translation. In short, the different pathways form a multi-faceted defense system enabling plants to respond and thrive in suboptimal environments, thereby ensuring photosynthesis in stressful conditions.

2.14 Conclusion

Chloroplast is considered as a key player in photosynthesis and plant metabolism. The genomic structure and function of the chloroplast is tightly tied to the evolutionary adaptation as well as environmental stress tolerance. Chloroplast transcription, translation, and its posttranslational modification are highly regulated to maintain the function and stability of these organelles. Chloroplasts control and regulate protein synthesis and repair, particularly in the case of light-damaged proteins, through RNA editing, RNA splicing, RNA maturation, ribosomal pausing, and protein quality control. There is still much to uncover, for example, surrounding the specific mechanisms that control genetic machinery in chloroplasts, the underlying macular mechanism, and the factors involved, but it is clear that this system is critical to maintaining the health and productivity of plants. Understanding how chloroplasts balance their genetic system could also have implications for helping to improve the yields and resilience of crop plants exposed to environmental change.

References

Adamiec M, Luciński R (2024) The roles of RNA modifications in regulating chloroplast performance and photosynthesis efficiency. Int J Mol Sci 25(22):11912. https://doi.org/10.3390/ijms252211912

Ahmed T, Yin Z, Bhushan S (2016) Cryo-EM structure of the large subunit of the spinach chloroplast ribosome. Sci Rep 6:35793

Akhtar J, Kreim N, Marini F, Mohana G, Brüne D, Binder H, Roignant JY (2019) Promoter-proximal pausing mediated by the exon junction complex regulates splicing. Nat Commun 10:521

Allen JF, de Paula WB, Puthiyaveetil S, Nield J (2011) A structural phylogenetic map for chloroplast photosynthesis. Trends Plant Sci 16:645–655

Anand A, Pandi G (2021) Noncoding RNA: an insight into chloroplast and mitochondrial gene expressions. Life 11:49

Asakura Y, Galarneau E, Watkins KP, Barkan A, van Wijk KJ (2012) Chloroplast RH3 DEAD box RNA helicases in maize and Arabidopsis function in splicing of specific group II introns and affect chloroplast ribosome biogenesis. Plant Physiol 159(3):961–974. https://doi.org/10.1104/pp.112.197525

Barrero-Gil J, Bouza-Morcillo L, Espinosa-Cores L, Piñeiro M, Jarillo JA (2022) H4 acetylation by the NuA4 complex is required for plastid transcription and chloroplast biogenesis. Nat Plants 8:1052–1063

Bieri P, Leibundgut M, Saurer M, Boehringer D, Ban N (2017) The complete structure of the chloroplast 70S ribosome in complex with translation factor pY. EMBO J 36:475–486

Börner T, Aleynikova AY, Zuboa YO, Kusnetsovb VV (2015) Chloroplast RNA polymerases: role in chloroplast biogenesis. Biochim Biophys Acta 9:761–769

Castandet B, Germain A, Hotto AM, Stern DB (2019) Systematic sequencing of chloroplast transcript termini from *Arabidopsis thaliana* reveals >200 transcription initiation sites and the extensive imprints of RNA-binding proteins and secondary structures. Nucleic Acids Res 47:11889–11905

Chambon L, Gillet FX, Chieb M, Cobessi D, Pfannschmidt T, Blanvillain R (2022) PAP8/pTAC6 is part of a nuclear protein complex and displays RNA recognition motifs of viral origin. Int J Mol Sci 23:3059

Chang SH, Lee S, Um TY, Kim JK, Do Choi Y, Jang G (2017) pTAC10, a key subunit of plastid-encoded RNA polymerase, promotes chloroplast development. Plant Physiol 174:435–449

Chen Y, Ma T, Zhang T, Ma L (2023) Trends in the evolution of intronless genes in Poaceae. Front Plant Sci 14:1065631

Chi W, He B, Mao J, Jiang J, Zhang L (2015) Plastid sigma factors: their individual functions and regulation in transcription. Biochim Biophys Acta 9:770–778

Chi W, He B, Mao J, Li Q, Ma J, Ji D, Zou M, Zhang L (2012) The function of RH22, a DEAD RNA helicase, in the biogenesis of the 50S ribosomal subunits of Arabidopsis chloroplasts. Plant Physiol 158(2):693–707. https://doi.org/10.1104/pp.111.186775

Chotewutmontri P, Barkan A (2020) Light-induced *psbA* translation in plants is triggered by photosystem II damage via an assembly-linked autoregulatory circuit. Proc Natl Acad Sci USA 117:21775–21784

Core L, Adelman K (2019) Promoter-proximal pausing of RNA polymerase II: a nexus of gene regulation. Genes Dev 33:960–982

Cui X, Wang Y, Wu J, Han X, Gu X, Lu T, Zhang Z (2019) The RNA editing factor DUA 1 is crucial to chloroplast development at low temperature in rice. New Phytol 221(2):834–849
Díaz MG, Hernández-Verdeja T, Kremnev D, Crawford T, Dubreuil C, Strand Å (2018) Redox regulation of PEP activity during seedling establishment in *Arabidopsis thaliana*. Nat Commun 9:50
Daniell H, Lin CS, Yu M, Chang WJ (2016) Chloroplast genomes: diversity, evolution and applications in genetic engineering. Genome Biol 17:134. https://doi.org/10.1186/s13059-016-1004-2
Danilova MN, Kudryakova NV, Andreeva AA, Doroshenko AS, Pojidaeva ES, Kusnetsov VV (2018) Differential impact of heat stress on the expression of chloroplast-encoded genes. Plant Physiol Biochem 129:90–100
De BC, Cournoyer J, Gao Y, Wallace CL, Bram S, Mehta AP (2024) Photosynthetic-directed endosymbiosis to investigate the role of bioenergetics in chloroplast function and evolution. Nat Commun 15:10622. https://doi.org/10.1038/s41467-024-54051-1
Dedow LK, Bailey-Serres J (2019) Searching for a match: structure, function and application of sequence-specific RNA-binding proteins. Plant Cell Physiol 60:1927–1938. https://doi.org/10.1093/pcp/pcz072
Ding S, Zhang Y, Hu Z, Huang X, Zhang B, Lu Q, Wen X, Wang Y, Lu C (2019) mTERF5 acts as a transcriptional pausing factor to positively regulate transcription of chloroplast *psbEFLJ*. Mol Plant 12:1259–1277
Dobrogojski J, Adamiec M, Luciński R (2020) The chloroplast genome: a review. Acta Physiol Plant 42:98. https://doi.org/10.1007/s11738-020-03089-x
Favier A, Gans P, Boeri Erba E, Signor L, Muthukumar SS, Pfannschmidt T, Blanvillain R, Cobessi D (2021) The plastid-encoded RNA polymerase-associated protein PAP9 is a superoxide dismutase with unusual structural features. Front Plant Sci 12:668897
Finster S, Eggert E, Zoschke R, Weihe A, Schmitz-Linneweber C (2013) Light-dependent, plastome-wide association of the plastid-encoded RNA polymerase with chloroplast DNA. Plant J 76:849–860
Gao LL, Hong ZH, Wang Y, Wu GZ (2023) Chloroplast proteostasis: a story of birth, life, and death. Plant Commun 4:100424
Garmash EV, Belykh ES, Velegzhaninov IO (2021) The gene expression profiles of mitochondrial respiratory components in Arabidopsis plants with differing amounts of *ALTERNATIVE OXIDASE1a* under high intensity light. Plant Signal Behav 16(3):1864962. https://doi.org/10.1080/15592324.2020.1864962
Gawronski P, Jensen PE, Karpinski S, Leister D, Scharff LB (2018) Pausing of chloroplast ribosomes is induced by multiple features and is linked to the assembly of photosynthetic complexes. Plant Physiol 176:2557–2569
Germain A, Hotto AM, Barkan A, Stern DB (2013) RNA processing and decay in plastids. Wiley Interdiscip Rev RNA 4:295–316
Green BR (2011) Chloroplast genomes of photosynthetic eukaryotes. Plant J 66:34–44
Gu L, Xu T, Lee K, Lee KH, Kang H (2014) A chloroplast-localized DEAD-box RNA helicaseAtRH3 is essential for intron splicing and plays an important role in the growth and stress response in Arabidopsis thaliana. Plant Physiol Biochem 82:309–318
Gururani MA, Venkatesh J, Tran LSP (2015) Regulation of photosynthesis during abiotic stress-induced photoinhibition. Mol Plant 8:1304–1320
Hajdukiewicz PT, Allison LA, Maliga P (1997) The two RNA polymerases encoded by the nuclear and the plastid compartments transcribe distinct groups of genes in tobacco plastids. EMBO J 16:4041
Harper NJ, Burnside C, Klinge S (2023) Principles of mitoribosomal small subunit assembly in eukaryotes. Nature 614:175–181. https://doi.org/10.1038/s41586-022-05621-0
Hayes ML, Santibanez PI (2020) A plant pentatricopeptide repeat protein with a DYW-deaminase domain is sufficient for catalyzing C-to-U RNA editing *in vitro*. J Biol Chem 295:3497–3505
He L, Qian J, Li X, Sun Z, Xu X, Chen S (2017) Complete chloroplast genome of medicinal plant Lonicera japonica: genome rearrangement, intron gain and loss, and implications for phylogenetic studies. Molecules 22(2):249

Hotto AM, Castandet B, Gilet L, Higdon A, Condon C, Stern DB (2015) Arabidopsis chloroplast mini-ribonuclease III participates in rRNA maturation and intron recycling. Plant Cell 27:724–740

Hotto AM, Schmitz RJ, Fei Z, Ecker JR, Stern DB (2011) Unexpected diversity of chloroplast noncoding RNAs as revealed by deep sequencing of the *Arabidopsis* transcriptome. G3 (Bethesda), 1:559–570

Hwang Y, Han S, Yoo CY, Hong L, You C, Le BH, Shi H, Zhong S, Hoecker U, Chen X, Chen M (2022) Anterograde signaling controls plastid transcription via sigma factors separately from nuclear photosynthesis genes. Nat Commun 13:7440

Inaba-Hasegawa K, Ohmura A, Nomura M, Sugiura M (2021) Development of an in vitro chloroplast splicing system: sequences required for correct pre-mRNA splicing. Plant Cell Physiol 62(8):1311–1320. https://doi.org/10.1093/pcp/pcab095

Jarvis P, López-Juez E (2013) Biogenesis and homeostasis of chloroplasts and other plastids. Nat Rev Mol Cell Biol 14:787–802

Jiang D, Cai X, Gong M, Xia M, Xing H, Dong S, Li HL (2023) Complete chloroplast genomes provide insights into evolution and phylogeny of Zingiber (Zingiberaceae). BMC genomics 24(1):30

Kaila T, Chaduvla PK, Rawal HC, Saxena S, Tyagi A, Mithra SVA, Solanke AU, Kalia P, Sharma TR, Singh NK, Gaikwad K (2017) Chloroplast genome sequence of clusterbean (Cyamopsis tetragonoloba L.): genome structure and comparative analysis. Genes (Basel).19:212

Kleine T, Leister D (2016) Retrograde signaling: organelles go networking. Biochim Biophys Acta 1857:1313–1325

Krämer C, Boehm CR, Liu J, Ting MKY, Hertle AP, Forner J, Ruf S, Schöttler MA, Zoschke R, Bock R (2024) Removal of the large inverted repeat from the plastid genome reveals gene dosage effects and leads to increased genome copy number. Nat Plants 10:1–13

Kindgren P, Kremnev D, Blanco NE, de Dios Barajas López J, Fernández AP, Tellgren-Roth C, Strand Å (2012) The plastid redox insensitive 2 mutant of Arabidopsis is impaired in PEP activity and high lightdependent plastid redox signalling to the nucleus. The Plant Journal 70(2):279–291

Lee HL, Jansen RK, Chumley TW, Kim KJ (2007) Gene relocations within chloroplast genomes of Jasminum and Menodora (Oleaceae) are due to multiple, overlapping inversions. Mol Biol Evol 24(5):1161–1180

Leister D (2014) Complex(iti)es of the ubiquitous RNA-binding CSP41 proteins. Front Plant Sci 5:255

Liere K, Börner T (2007) Transcription and transcriptional regulation in plastids. In: Cell and molecular biology of plastids. Springer, Heidelberg, Berlin, pp 121–174

Maier RM, Neckermann K, Igloi GL, Kössel H (1995) Complete sequence of the maize chloroplast genome: gene content, hotspots of divergence and fine tuning of genetic information by transcript editing. J Mol Biol 251:614–628

Manavski N, Schmid LM, Meurer J (2018) RNA-stabilization factors in chloroplasts of vascular plants. Essays Biochem 62:51–64

Masuda T, Petrov AN, Iizuka R, Funatsu T, Puglisi JD, Uemura S (2012) Initiation factor 2, tRNA, and 50S subunits cooperatively stabilize mRNAs on the ribosome during initiation. Proc Natl Acad Sci USA 109(13):4881–4885. https://doi.org/10.1073/pnas.1118452109

Méteignier LV, Ghandour R, Meierhoff K, Zimmerman A, Chicher J, Baumberger N, Alioua A, Meurer J, Zoschke R, Hammani K (2020) The Arabidopsis mTERF-repeat MDA1 protein plays a dual function in transcription and stabilization of specific chloroplast transcripts within the *psbE* and *ndhH* operons. New Phytol 227:1376–1391

Mehmetoğlu E, Kaymaz Y, Ateş D, Kahraman A, Tanyolaç MB (2023) The complete chloroplast genome of Cicer reticulatum and comparative analysis against relative Cicer species. Sci Rep 13(1):17871

Meinnel T, Giglione C (2022) N-terminal modifications, the associated processing machinery, and their evolution in plastid- containing organisms. J Exp Bot 73:6013–6033

Mohd Ali S, Li N, Soufi Z, Yao J, Johnson E, Ling Q, Jarvis RP (2023) Multiple ubiquitin E3 ligase genes antagonistically regulate chloroplast-associated protein degradation. Curr Biol 33(6):1138–1146

Moore PB, Steitz TA (2011) The roles of RNA in the synthesis of protein. Cold Spring Harb Perspect Biol 3(11):a003780. https://doi.org/10.1101/cshperspect.a003780

Morley SA, Peralta-Castro A, Brieba LG, Miller J, Ong KL, Ridge PG, Nielsen BL (2019) *Arabidopsis thaliana* organelles mimic the T7 phage DNA replisome with specific interactions between Twinkle protein and DNA polymerases Pol1A and Pol1B. BMC Plant Biol 19:241

Mower JP, Vickrey TL (2018) Chapter nine—structural diversity among plastid genomes of land plants. Adv Bot Res 85:263–292

Noordally ZB, Ishii K, Atkins KA, Wetherill SJ, Kusakina J, Walton EJ, Kato M, Azuma M, Tanaka K, Hanaoka M, Dodd AN (2013) Circadian control of chloroplast transcription by a nuclear-encoded timing signal. Science 339(6125):1316–1319

Ohyama K, Fukuzawa H, Kohchi T, Shirai H, Sano T, Sano S, Umesono K, Shiki Y, Takeuchi M, Chang Z, Aota S, Inokuchi H, Ozeki H (1986) Chloroplast gene organization deduced from complete sequence of liverwort *Marchantia polymorpha* chloroplast DNA. Nature 322:572–574

Oldenburg DJ, Bendich AJ (2015) DNA maintenance in plastids and mitochondria of plants. Front Plant Sci 6:1–15

Ortelt J, Link G (2021) Plastid gene transcription: an update on promoters and RNA polymerases. In: Chloroplast biotechnology: methods and protocols. New York, NY, Springer, pp 49–76

Palomar VM, Jaksich S, Fujii S, Kucinski J, Wierzbicki AT (2022) High-resolution map of plastid-encoded RNA polymerase binding patterns demonstrates a major role of transcription in chloroplast gene expression. Plant J 111:1139–1151

Perez Boerema A, Aibara S, Paul B, Tobiasson V, Kimanius D, Forsberg BO, Wallden K, Lindahl E, Amunts A (2018) Structure of the chloroplast ribosome with chl-RRF and hibernation- promoting factor. Nat Plants 4:212–217

Petersen G, Cuenca A, Møller IM, Seberg O (2015) Massive gene loss in mistletoe (Viscum, Viscaceae) mitochondria. Sci Rep 5(1):17588

Pfannschmidt T, Blanvillain R, Merendino L, Courtois F, Chevalier F, Liebers M et al (2015) Plastid RNA polymerases: orchestration of enzymes with different evolutionary origins controls chloroplast biogenesis during the plant life cycle. J Exp Bot 66(22):6957–6973

Pulido P, Zagari N, Manavski N, Gawronski P, Matthes A, Scharff LB, Meurer J, Leister D (2018) Chloroplast Ribosome Associated supports translation under stress and interacts with the ribosomal 30S subunit. Plant Physiol 177(4):1539–1554. https://doi.org/10.1104/pp.18.00602

Puthiyaveetil S, McKenzie SD, Kayanja GE, Ibrahim IM (2021) Transcription initiation as a control point in plastid gene expression. Biochim Biophys Acta Gene Regul Mech 1864:194689

Rogalski M, Schottler MA, Thiele W, Schulze WX, Bock R (2008) Rpl33, a nonessential plastid-encoded ribosomal protein in tobacco, is required under cold stress conditions. Plant Cell 20(8):2221–2237

Royan S, Gutmann B, Colas des Francs-Small C, Honkanen S, Schmidberger J, Soet A, Sun YK, Vincis Pereira Sanglard L, Bond CS, Small I (2021) A synthetic RNA editing factor edits its target site in chloroplasts and bacteria. Commun Biol 4:545

Ruwe H, Wang G, Gusewski S, Schmitz-Linneweber C (2016) Systematic analysis of plant mitochondrial and chloroplast small RNAs suggests organelle-specific mRNA stabilization mechanisms. Nucleic Acids Res 44:7406–7417

Sajib SA, Grübler B, Oukacine C, Delannoy E, Courtois F, Mauve C, Lurin C, Gakière B, Pfannschmidt T, Merendino L (2023) Limiting etioplast gene expression induces apical hook twisting during skotomorphogenesis of Arabidopsis seedlings. Plant J 114(2):293–309

Sato S, Nakamura Y, Kaneko T, Asamizu E, Tabata S (1999) Complete structure of the chloroplast genome of *Arabidopsis thaliana*. DNA Res 6:283–290

Searing AM, Satyanarayan MB, Donnell JP, Lu Y (2020) Two organelle RNA recognition motif proteins affect distinct sets of RNA editing sites in the *Arabidopsis thaliana* plastid. Plant Direct 4(4):e00213. https://doi.org/10.1002/pld3.213

Shikanai T, Fujii S (2013) Function of PPR proteins in plastid gene expression. RNA Biol 9:1446–1456

Shimizu M, Kato H, Ogawa T, Kurachi A, Nakagawa Y, Kobayashi H (2010) Sigma factor phosphorylation in the photosynthetic control of photosystem stoichiometry. Proc Natl Acad Sci USA 107:10760–10764

Savisaar R, Hurst LD (2016) Purifying selection on exonic splice enhancers in intronless genes. molecular biology and evolution 33(6):1396–1418

Shinozaki K, Ohme M, Tanaka M, Wakasugi T, Hayashida N, Matsubayashi T, Zaita N, Chunwongse J, Obokata J, Yamaguchi-Shinozaki K, Ohto C, Torazawa K, Meng BY, Sugita M, Deno H, Kamogashira T, Yamada K, Kusuda J, Takaiwa F, Kato A, Tohdoh N, Shimada H, Sugiura M (1986) The complete nucleotide sequence of the tobacco chloroplast genome: its gene organization and expression. EMBO J 5:2043–2049

Small ID, Schallenberg-Rüdinger M, Takenaka M, Mireau H, Ostersetzer-Biran O (2020) Plant organellar RNA editing: what 30 years of research has revealed. Plant J 101:1040–1056

Sun Y, Zerges W (2015) Translational regulation in chloroplasts for development and homeostasis. Biochim Biophys Acta 9:809–820

Sun X, Fu T, Chen N, Guo J, Ma J, Zou M, Lu C, Zhang L (2010) The stromal chloroplast Deg7 protease participates in the repair of photosystem II after photoinhibition in Arabidopsis. Plant Physiol 152:1263–1273

Takenaka M, Takenaka S, Barthel T, Frink B, Haag S, Verbitskiy D, Oldenkott B, Schallenberg-Rüdinger M, Feiler CG, Weiss MS, Palm GJ, Weber G (2021) DYW domain structures imply an unusual regulation principle in plant organellar RNA editing catalysis. Nat Catal 4(6):510–522. https://doi.org/10.1038/s41929-021-00633-x

Takenaka M, Zehrmann A, Verbitskiy D, Härtel B, Brennicke A (2013) RNA editing in plants and its evolution. Annu Rev Genet 47(1):335–352

Turrens JF (2003) Mitochondrial formation of reactive oxygen species. J Physiol 552:335–344. https://doi.org/10.1113/jphysiol.2003.049478

Wakasugi T, Tsudzuki J, Ito S, Nakashima K, Tsudzuki T, Sugiura M (1994) Loss of all ndh genes as determined by sequencing the entire chloroplast genome of the black pine *Pinus thunbergii*. Proc Natl Acad Sci U S A 91:9794–9798

Wang S, Bai G, Wang S, Yang L, Yang F, Wang Y et al (2016) Chloroplast RNA-binding protein RBD1 promotes chilling tolerance through 23S rRNA processing in Arabidopsis. PLoS Genet 12(5):e1006027

Wang X, An Y, Xu P, Xiao J (2021) Functioning of PPR proteins in organelle RNA metabolism and chloroplast biogenesis. Front Plant Sci 12:627501. https://doi.org/10.3389/fpls.2021.627501

Wang X, Wang J, Li S, Lu C, Sui N (2022) An overview of RNA splicing and functioning of splicing factors in land plant chloroplasts. RNA Biol 19(1):897–907. https://doi.org/10.1080/15476286.2022.2096801

Wang X, Yang Z, Zhang Y, Zhou W, Zhang A, Lu C (2020) Pentatricopeptide repeat protein Photosystem I Biogenesis Factor2 is required for splicing of *ycf3*. J Integr Plant Biol 62:1741–1761

Watkins KP, Rojas M, Friso G, van Wijk KJ, Meurer J, Barkan A (2011) APO1 promotes the splicing of chloroplast group II introns and harbors a plant-specific zinc-dependent RNA binding domain. Plant Cell 23:1082–1092

Wissink EM, Vihervaara A, Tippens ND, Lis JT (2019) Nascent RNA analyses: tracking transcription and its regulation. Nat Rev Genet 20:705–723

Yan J, Zhang Q, Guan Z, Wang Q, Li L, Ruan F, Lin R, Zou T, Yin P (2017) MORF9 increases the RNA-binding activity of PLS- type pentatricopeptide repeat protein in plastid RNA editing. Nat Plants 3:17037

Yang EJ, Yoo CY, Liu J, Wang H, Cao J, Li FW, Pryer KM, Sun TP, Weigel D, Zhou P, Chen M (2019) NCP activates chloroplast transcription by controlling phytochrome-dependent dual nuclear and plastidial switches. Nat Commun 10:2630

Yang Z, Liu M, Ding S, Zhang Y, Yang H, Wen X, Chi W, Lu C, Lu Q (2021) Plastid deficient 1 is essential for the accumulation of plastid-encoded RNA polymerase core subunit b and chloroplast development in *Arabidopsis*. Int J Mol Sci 22:13648

Yao WB, Meng BY, Tanaka M, Sugiura M (1989) An additional promoter within the protein-coding region of the psb D-psb C gene cluster in tobacco chloroplast DNA. Nucleic Acids Res 17(23):9583–9591

Yi DK, Lee HL, Sun BY, Chung MY, Kim KJ (2012) The complete chloroplast DNA sequence of Eleutherococcus senticosus (Araliaceae); comparative evolutionary analyses with other three asterids. Mol Cells 33:497–508

Yi L, Liu B, Nixon PJ, Yu J, Chen F (2022) Recent advances in understanding the structural and functional evolution of FtsH proteases. Front Plant Sci 13:837528

Yu HD, Yang XF, Chen ST, Wang YT, Li JK, Shen Q, Liu XL, Guo FQ (2012) Downregulation of chloroplast RPS1 negatively modulates nuclear heat-responsive expression of HsfA2 and its target genes in Arabidopsis. PLoS Genet 8(5):e1002669. https://doi.org/10.1371/journal.pgen.1002669

Zhang L, Chen J, Zhang L, Wei Y, Li Y, Xu X, Wu H, Yang ZN, Huang J, Hu F et al (2021) The pentatricopeptide repeat protein EMB1270 interacts with CFM2 to splice specific group II introns in Arabidopsis chloroplasts. J Integr Plant Biol 63:1952–1966

Zhang Y, Tian L, Lu C (2023) Chloroplast gene expression: recent advances and perspectives. Plant Commun 4(5):100611. https://doi.org/10.1016/j.xplc.2023.100611

Zhang Y, Zhang A, Li X, Lu C (2020) The role of chloroplast gene expression in plant responses to environmental stress. Int J Mol Sci 21:6082. https://doi.org/10.3390/ijms21176082

Zheng M, Liu X, Liang S, Fu S, Qi Y, Zhao J, Shao J, An L, Yu F (2016) Chloroplast translation initiation factors regulate leaf variegation and development. Plant Physiol 172:1117–1130

Zhu J, Liu M, Liu X, Dong Z (2018) RNA polymerase II activity revealed by GRO-seq and pNET-seq in *Arabidopsis*. Nat Plants 4:1112–1123

Zhuang K, Kong F, Zhang S, Meng C, Yang M, Liu Z, Wang Y, Ma N, Meng Q (2019) Whirly1 enhances tolerance to chilling stress in tomato via protection of photosystem II and regulation of starch degradation. New Phytol 221(4):1998–2012

Zoschke R, Bock R (2018) Chloroplast translation: structural and functional organization, operational control, and regulation. Plant Cell 30:45–770

Zoschke R, Liere K, Börner T (2007) From seedling to mature plant: Arabidopsis plastidial genome copy number, RNA accumulation and transcription are differentially regulated during leaf development. Plant J 50:710–722

Zoschke R, Watkins KP, Miranda RG, Barkan A (2016) The PPR-SMR protein PPR53 enhances the stability and translation of specific chloroplast RNAs in maize. Plant J 85:594–606

[illegible]

Chloroplast-Mediated Retrograde Signaling in Plants

3

Tulika, Variyata Agrawal, Neha Singh, Vijetna Singh, and Bhumi Nath Tripathi

Abstract

Retrograde signaling establishes a feedback loop where signals travel backward from the chloroplasts to the nucleus, eliciting responses. This intricate regulatory network comprises specific sensor molecules that move from the chloroplast to the nucleus via the cytosol, managing the essential expression of nuclear genes and conveying information about these energy-transducing organelles' functional and physiological states. In plants, plastid-mediated retrograde signaling operates through two pathways: the biogenic signaling pathway, which activates during early development, and the operational signaling pathway, which modulates the functional status of chloroplasts amid environmental stress. Many researchers believe that GUN1 (GENOMES UNCOUPLED), a chloroplast-localized protein, plays a crucial role in biogenic retrograde communication. However, relatively little is known about the signaling components and activities involved in the operational pathways. This review focuses on the involvement of signaling components, sensor molecules, and processes in intracellular signaling pathways that coordinate gene expression between the chloroplast and nucleus during various environmental stressors.

Keywords

Biogenic · Chloroplast · Operational · Retrograde · Signaling · Stress

Tulika · V. Agrawal · N. Singh · V. Singh · B. N. Tripathi (✉)
Department of Biotechnology, Indira Gandhi National Tribal University, Amarkantak, Madhya Pradesh, India
e-mail: bhuminath.tripathi@igntu.ac.in

V. K. Dalal, A. N. Misra (eds.), *Chloroplast Biogenesis and Plastid Interconversions*,
https://doi.org/10.1007/978-981-95-0772-6_3

3.1 Introduction

Various environmental stresses pose a severe risk to plants and lead to considerable loss of productivity. Plants have evolved intricate protection mechanisms to endure these challenges and improve their adaptability and stress tolerance capacity (Sharma et al. 2022). Tightly regulated signaling events within and between plant cells is one of the strategies plants use to thrive well in extreme conditions. A well-coordinated network and communication between various cell organelles, e.g., chloroplast, mitochondria, and nucleus of the plant cell, is essential for operating vital processes of the plant to withstand extreme conditions (Fig. 3.1). Nucleus of the cell controls the integrity of cells by providing crucial proteins for metabolic processes. Several vital proteins involved in chloroplast functions are nuclear-encoded, and anterograde regulation (communication from nucleus to the organelles) is critical for regulating organellar characteristics during development (Bräutigam et al. 2007). A complex regulatory network involving a series of interconnected pathways to manage the quintessential expression of nuclear genes encoding organelle proteins and other cellular proteins to orchestrate organelle, cell development and metabolism in response to altered mitochondria and chloroplast function referred to as retrograde signaling (Chi et al. 2013). Fluctuation in the developmental or metabolic state of chloroplast or mitochondria intensively affects the plant cell and leads to extensive changes in the nuclear transcriptome (Leister et al. 2014). Plant organelles generate retrograde signals to alter nuclear gene expression to coordinate their biogenesis, maintain cell integrity, or optimize their performance under adverse conditions. Understanding complex signaling pathways may allow us to modify them in the future to improve crop productivity or respond to combined abiotic and biotic stresses. However, the signaling pathways and the molecular mechanisms of retrograde signaling are not well understood. On the other hand, the findings of the previous studies suggest that no single and linear pathways exist proposing the exceedingly complex interaction of several signaling molecules to perform retrograde signaling (Nott et al. 2006). The present chapter aims to provide an update on various components and processes involved in retrograde signaling from chloroplast to the nucleus of a plant cell.

3.2 Components and Processes Involved in Chloroplast-Mediated Retrograde Signaling

Plastids emerged about 1.6 million years ago by engulfing free-living photosynthetic cyanobacteria into ancestral eukaryotes (Gould et al. 2008; Irisarri et al. 2020). Many cyanobacterial genes were lost or shifted to the nuclear genome of the host cell through an endosymbiotic event during evolution (Green 2011). However, current plastids still possess a small amount of protein-encoding genes (50–200) varying from species to species (Woodson and Chory 2008; Ponce-Toledo et al. 2019). Many protein complexes in plastids are nuclear-encoded and, therefore, require a high degree of synchronization in gene expression between chloroplast

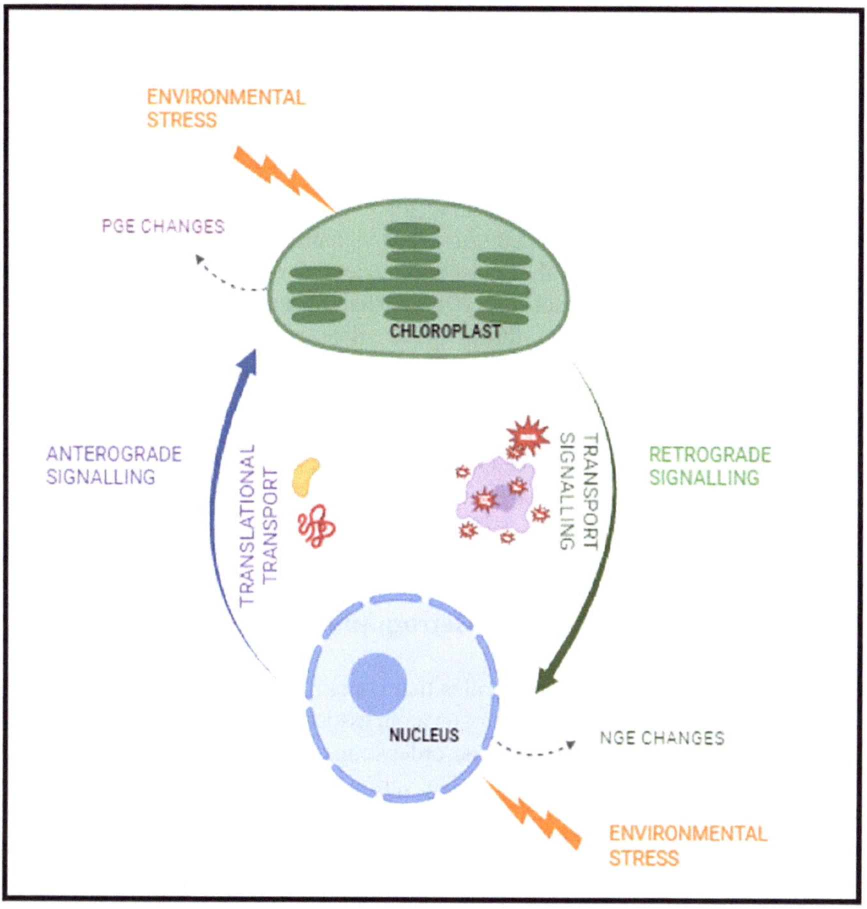

Fig. 3.1 A schematic representation of environmental challenges during plant growth and development. Plants respond to many environmental stressors, including extreme changes in ambient illumination, temperature, humidity, herbivory, drought, heavy metal toxicity, and soil salinity variations. All stimuli eventually contribute to the development of oxidative stress in plants. Plant cells covey information via two distinct signaling pathways to cope with oxidative stress: the anterograde and the retrograde signaling pathways. Plastid Gene Expression (PGE); Nuclear Gene Expression (NGE)

and nucleus (Bailey et al. 2021). Multiple signaling events occur between the nucleus and chloroplast to relay information between the genomes of these two organelles and to guide the proper formation and assembly of the molecules required for a functional and photosynthetically active chloroplast. Chloroplasts of higher plants are responsible for photosynthesis and lipid metabolism and the synthesis of several essential compounds, such as fatty acids, amino acids, phytohormones, nucleotides, vitamins, and pigments, secondary metabolites, and play critical roles in nitrogen and sulfur assimilation. Therefore, environment-mediated change in the overall status and subsequent readjustments in the chloroplasts require retrograde communication to encounter with the metabolic change in the chloroplast (Dobrogojski et al. 2020; Wu and Bock 2021).

Studies on two barley mutants (*Hordeum vulgare* L., albostrians and Saskatoon) with defective plastids that resulted in decline of cytoplasmically synthesized plastid proteins phosphoribulokinase and glyceraldehyde-3-phosphate dehydrogenase gave the first indication of retrograde signaling (Bradbeer et al. 1979). Plastid signals regulate the component responsible for encoding nuclear genes during this bilateral communication, so-called retrograde signaling. The major plastid signals known as the link between chloroplast and nucleus retrograde signaling pathways are carotenoids-derived (Ramel et al. 2012; Alsharafa et al. 2014), derivatives of the isoprenoid precursor methylerythritol cyclic diphosphate (MEcPP) (Xiao et al. 2012), β-cyclocitral and, tetrapyrrole derivatives such as heme, Mg-Proto-IX (Woodson et al. 2011; Terry and Smith 2013; Chi et al. 2015). Stress-induced changes in plastoquinone (PQ) pool redox state triggers the oxidation of β-carotene dissociating into β-cyclocitral (β-CC) and dihydroactinidiolide acting as retrograde signals. (Pfalz et al. 2012; Phua et al. 2021). These plastid signals are crucial in regulating photosynthesis-associated nuclear genes (PhANGs) (Table 3.1).

3.2.1 Tetrapyrrole-Mediated Retrograde Signaling

Tetrapyrrole are macrocyclic molecules that contain four pyrroles, aromatic rings with four carbon atoms, and one nitrogen atom associated with unsaturated methine groups in linear (e.g., bilins) or cyclic order (e.g., porphyrins) (Brzezowski et al. 2015; Battersby 2000). Porphyrins chelate various metal ions, such as Co_{2+}, $Fe_{2+,}$ or Fe_{3+}, and Mg2+ ions, possess a diverse molecular structure with specialized functions. Higher plants and algae possess many tetrapyrroles, including chlorophyll (Chl), heme, siroheme, phytochromobilin, and some of their precursors, which are synthesized in the chloroplast (Sylvestre-Gonon et al. 2020). The different chemical and physical properties of tetrapyrroles lead to essential functions like light signaling, detoxification of ROS, assimilation of nitrate and sulfate, respiration, photosynthesis light reaction, programmed cell death, electron transfer, and oxygen binding (Brzezowski et al. 2015; Tanaka and Tanaka 2007).

Genetic screening of *Arabidopsis* mutants revealed six mutants with defective bilateral communication between chloroplast and nuclei, termed GENOMES UNCOUPLED or GUN mutants (Wu and Bock 2021). Among the six GUN mutants, GUN-2, 3, 4, 5, and 6 mediate the tetrapyrrole biosynthetic pathway. However, a chloroplast localized pentatricopeptide repeat (PPR) protein GUN-1 interacts with several proteins of photosynthesis-associated nuclear genes, such as the gene for a light-harvesting chlorophyll-binding protein (LHCB) and rubisco small subunit (RBCS) that is involved in multiple retrograde signaling pathways like tetrapyrrole biosynthesis pathway, photosynthetic electron transport (PET)-machinery, plastid hormonal signaling pathway, and plastid gene expression (PGE)-mediated pathways (Shimizu and Masuda 2021; Kacprzak et al. 2019).

Table 3.1 List of molecules involved in retrograde signaling

			Sensor	Molecules		
Components	Processes	Stimulus	Chloroplast	Cytosol	Nucleus	References
Heme	Biogenic	Plastid biogenesis & oxidative stress	GUN-1	Unknown	GLK HY5 LHCB RBCS	Shimizu and Masuda (2021), Kacprzak et al. (2019), Li et al. (2020), and Takahashi et al. (2008)
PAP	Operational	Drought & high light	SAL1-PAP	PAP	APX2 ELIP2	Zhao et al. (2019)
β-Cyclocitral	Operational	High light	β-CC	MBS SCL14	NPR1	Shumbe et al. (2017), D'alessandro et al. (2018)
MEcPP	Operational	High light & Wounding	HDS MEcPP	Unknown	ICS-1 HPL	Seemann et al. (2005), Xiao et al. (2012), and Walley et al. (2015)
Singlet oxygen	Operational	Multiple stress stimulus	β-CC EX1 &EX2	Unknown	PRL1	Shao et al. (2013), Ramel et al. (2012), Dogra et al. (2017), and Baruah et al. (2009)
Hydrogen peroxide	Operational	Multiple stress stimulus	Thylakoid APX	Cytosolic APX MAPKs	Nuclear MAPKs	Maruta et al. (2016), Breeze and Mullineaux (2022), Kovtun et al. (2000), and Apel and Hirt (2004)

Abbreviations: *GUN-1* GENOMES UNCOUPLED, *GLK* transcription factor, *HY5* LONG HYPOCYTOL 5, *LHCB* light-harvesting complex of photosystem II, *RBCS* ribulose bisphosphate carboxylase, *SAL1-PAP* SAL1–3-phosphoadenosine 5-phosphate, *APX2* ascorbate peroxidase, *ELIP2* EARLY LIGHT INDUCIBLE PROTEIN1, *β-CC* β-cyclocitral, *MBS* methylene blue sensitive, *SCL14* SCARECROW-LIKE14, *NPR1* non-expressor of pathogenesis-related genes 1, *HDS* HMBPP synthase, *MEcPP* methylerythritol cyclic diphosphate, *ICS-1* isochorismate synthase 1, *HPL* hydroperoxide lyase, *EX1 &EX2* EXECUTER proteins, *PRL1* PLEIOTROPIC RESPONSE LOCUS 1 (PRL1), *MAPKs* mitogen-activated protein kinases

In *Arabidopsis*, a dominant gun-6 mutation leading to the overexpression of plastid ferrochelatase-1 represents an important signaling pathway that controls the photosynthesis associated with nuclear-encoded genes (PhANGs) (Page et al. 2020; Chow et al. 1998). The overexpression of ferrochelatase-1 leads to the production of heme, which controls the expression of the photosynthesis-associated nuclear-encoded genes, such as *GUN4*, *CA1*, *HEMA1*, *LHCB2.1,* and *CHLH*, in the presence of norflurazon (NF) a specific inhibitor of phytoene desaturase treatment as well as in the absence of NF treatment, shows that direct regulation of nuclear-encoded genes for chloroplast proteins during chloroplast biogenesis in the lack of a stress-responsive treatment and, therefore, act as a promotive signaling candidate during chloroplast-to-nucleus retrograde signaling pathway (Page et al. 2020). LHCB expression was suppressed in wild-type plants treated with norflurazon under continuous light due to plastid photobleaching. On the other hand, LHCB expression levels in the mutants were relatively high and less affected by plastid photobleaching.

In plants, the heme trafficking mechanism is still not very clear. Heme synthesized in the chloroplast was transferred to the nucleus via ER-chloroplast membrane contact site (MCSs) or cytosolic receptors, such as p22HBP/SOUL and *tau* glutathione transferases, and it may inactivate PIFs or activate transcription factors, GLK1 and HY5 (Sylvestre-Gonon et al. 2020; Li et al. 2020; Takahashi et al. 2008). GLK1, exhibits the "photosynthesis-type" transcript accumulation and helps activate photosynthesis-associated nuclear genes (PhANGs) expression and chloroplast biogenesis (Waters et al. 2009). GLK1 may work with another candidate, COL7, a transcription factor involved in light signaling that interacts with HY5 (Grübler et al. 2021). HY5 is an essential/leucine zipper (bZIP) transcription factor that participates in retrograde signaling during seedling de-etiolation (Bhatnagar et al. 2020). It regulates the tetrapyrrole-mediated plastid signal to control the expression of PhANG during the response to oxidative stress (Kindgren et al. 2012) (Fig. 3.2).

3.2.2 The Regulatory Role of Key Metabolites in the Control of Retrograde Signals

Metabolites are required for vital biological processes like glycolysis, the Krebs cycle, photosynthesis, and the related pathway. Plants develop a diverse repertory of metabolites under stress conditions to defend themselves from severe environmental conditions.

3.2.2.1 SAL1-PAP-Mediated Signaling

In plants, 3′-phosphoadenosine 5′-phosphate (PAP) accumulates during drought and high light (HL) stress (Zhao et al. 2019). The nucleotide phosphatase (SAL1), a bifunctional enzyme localized in both mitochondria and chloroplasts, regulates the PAP levels by dephosphorylating PAP to AMP. PAP accumulates due to redox inactivation of its catabolic phosphatase SAL1 during high light exposure and drought and travels from chloroplasts to the nucleus via a transporter (Estavillo et al. 2011; Chan et al. 2016a, b). In *Arabidopsis*, the SAL1-PAP (3′-phosphoadenosine 5′-phosphate) retrograde route interacts with ABA signaling to regulate stomatal closure and seed germination (Pornsiriwong et al. 2017). PAP plays broader roles in retrograde signaling, allowing chloroplasts to fine-tune physiological responses. A transgenic target to this mutant lowers the total PAP level and

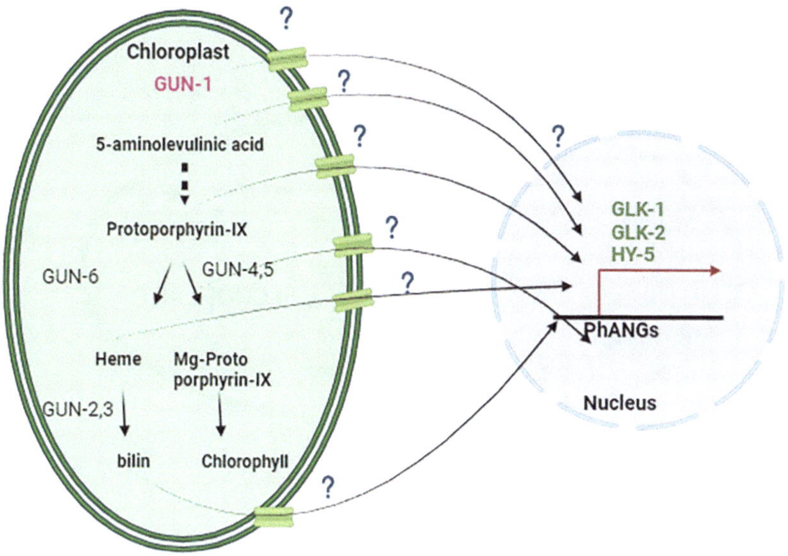

Fig. 3.2 Pictorial representation of tetrapyrrole-mediated retrograde signaling. Dashed arrow denotes multiple steps in between. Abbreviations: *GUN-1, 2, 3, 4, 5, 6* GENOMES UNCOUPLED, *GLK* transcription factor, *HY5* LONG HYPOCYTOL 5, *PhANGs* photosynthesis-associated nuclear genes

expression of the high light-inducible stress-sensitive APX2 gene, demonstrating that PAP must move within the cellular compartments. The SAL1-PAP retrograde pathway targets nuclear exoribonuclease (XRNs) (Estavillo et al. 2011) (Fig. 3.3). The ferritin genes AtFER1, AtFER3, and AtFER4 are upregulated in *Arabidopsis thaliana* mutant plants in the PAP-SAL1 pathway, according to a recent detailed transcriptome study, and, therefore, play a role in iron homeostasis (Balparda et al. 2020). PAP then favorably influences the synthesis of intracellular microRNAs, particularly miR398, enhancing plant heat tolerance (Fang et al. 2019).

3.2.2.2 β-Cyclocitral-Mediated Retrograde Signaling

β-cyclocitral is a volatile, short-chain lipid-soluble compound that enhances plant tolerance to high light-induced oxidative stress. Under high light stress, overproduced $^1O_2$ are quenched by their sensor molecule β-carotene (Ramel et al. 2012; Ramel et al. 2013). Upon oxidation, β-carotene dissociates into volatile compounds like β-cyclocitral (β-CC), β-ionone (P-I), and dihydroactinidiolide (Havaux 2020; Alsharafa et al. 2014) either derived from the enzymatic pathway by carotenoid cleavage dioxygenases (CCD) (Havaux 2020) or nonenzymatic pathway by the direct oxidation of β-carotene via reactive oxygen species, particularly singlet oxygen ($^1O_2$). This in turn induces several defense-responsive genes, such as glutathione-*S*-transferases (GST) and UDP glucosyltransferases. β-CC-dependent signaling involves several independent stress tolerance elements like methylene blue-sensitive (MBS), GST via the salicylic acid (SA) pathway, and the scarecrow-like 14 (SCL14)-dependent xenobiotic detoxifications pathway (D'alessandro et al. 2018).

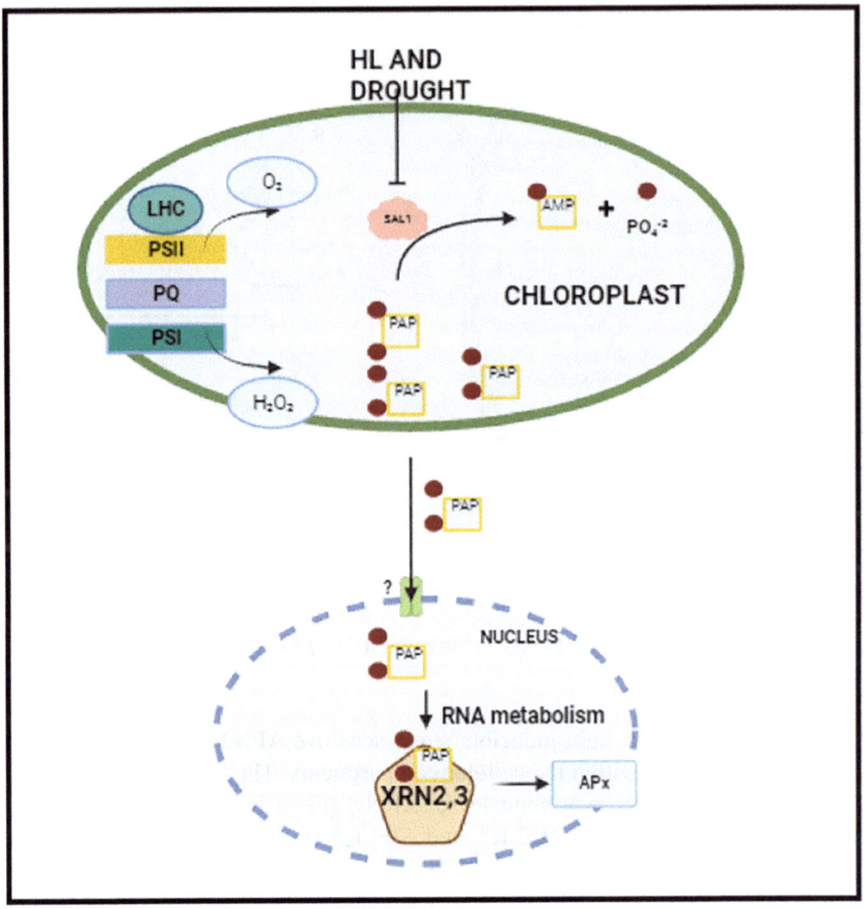

Fig. 3.3 SAL-PAP-XRN-mediated retrograde signaling pathway. Abbreviations: *LHC* light harvesting complex, *PS I,II* photosystem I and II, *PQ* plastoquinone, *XRNs* exoribonuclease

Under severe stress conditions, β-CC induces a small zinc finger protein, MBS-1 (METHYLENE BLUE SENSITIVITY 1), which plays a significant regulatory role in expressing the singlet oxygen-responsive genes (Shumbe et al. 2017). Under normal growth conditions, MBSs (METHYLENE BLUE SENSITIVITY 1) are present in the cytosol and nucleus. During high light stress conditions, β-CC elevates the level of MBS-1 protein and induces relocalization of the protein to the nucleus. In *Arabidopsis* mbs-1 mutant, the absence of MBS-1 led to the reduction of specific singlet oxygen marker genes and ROS-responsive transcription factors in response to β-cyclocitral, indicating that MBS1 is situated downstream of β-cyclocitral in the $^1O_2$signaling pathway, but the exact function of MBS1 is still needed to be identified. (Shumbe et al. 2017).

3.2.2.3 MEcPP-Induced Retrograde Signaling

The methylerythritol phosphate (MEP) pathway produced a signal metabolite methylerythritol cyclic diphosphate (MEcPP), a precursor of isoprenoids that elicits expression of selected stress-sensitive nuclear-encoded genes, for example, hydroperoxide-lyase and isochorismat-synthase-1 (Wang et al. 2017; Xiao et al. 2012; Xiao et al. 2013). MEcPP accumulates in response to high light or wounding stress and triggers salicylic acid (SA) and jasmonic acid (JA) biosynthesis (Lemos et al. 2016).

Hydroperoxide-lyase (HPL), a stress-inducible nuclear gene, encoded a plastid localized protein in the oxylipin pathway, which is involved in many developmental and defense paths, especially in herbivore resistance, disease resistance, and general stress responses (Fig. 3.4). A mutant ceh1 (constitutively expressed HPL1) was discovered during genetic screenings. This mutant contains a lesion in the enzyme 1-hydroxy-2-methyl-2-(E)-butenyl-4-diphosphate synthase (HDS), which includes a [4Fe–4S] cluster, which makes it highly sensitive to oxidative stress, leading to the

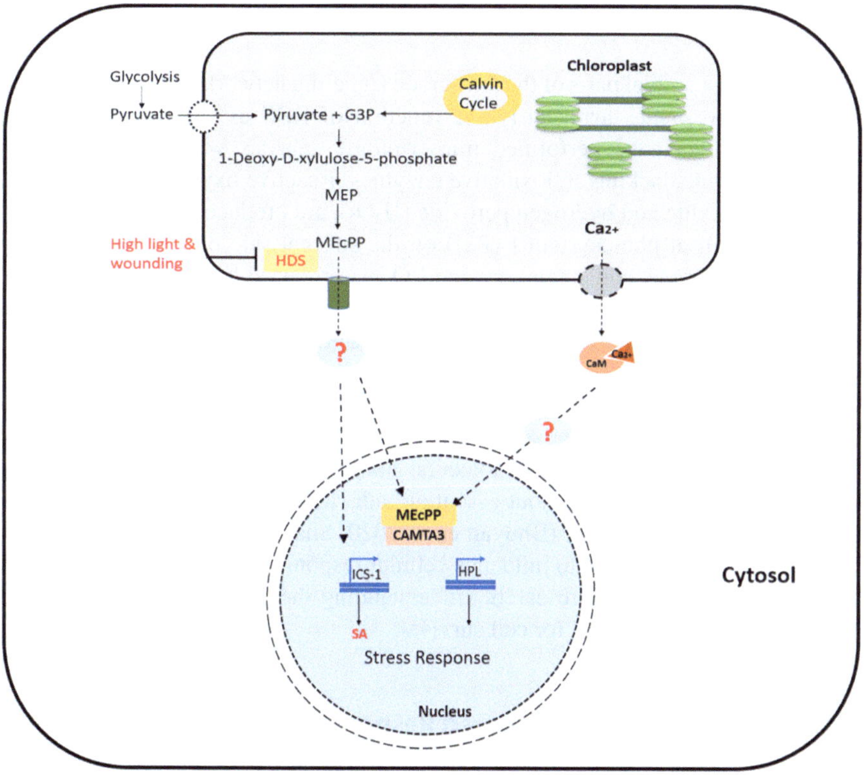

Fig. 3.4 Plastidial metabolite MEcPP-induced retrograde signaling. Abbreviations: *G3P* glyceraldehyde 3-phosphate, *MEP* methylerythritol phosphate, *MEcPP* methylerythritol cyclic diphosphate, *HDS* HMBPP synthase, *ICS-1* isochorismate synthase 1, *HPL* hydroperoxide lyase, *CaM* calmodulin, *CAMTA3* Ca2+/calmodulin-binding transcription factor

hyperaccumulation of MEcPP (Seemann et al. 2005; Xiao et al. 2012). However, genetic experiments have demonstrated that the upregulation of the HPL gene is not dependent on hormone abundance. A cross between *eds16–1*, SA-deficient mutant, and ceh-1, HPL-deficient mutant, proved that the upregulation of HPL is not due to hormonal abundance. In response to stress, MEcPP plays a role in potentiating unfolded protein response in the ER (Walley et al. 2015). In ER, aggregation of misfolded proteins occurs due to exposure to both abiotic and biotic stress (Liu and Howell 2016). MEcPP activates the UPR in the endoplasmic reticulum by activating the expression of UPR and stress genes (including *bZIP60* and possibly *HPL*) via the TF CAMTA3 in a calcium-dependent manner. MEcPP may also regulate the expression of the TF BBX19 and influence growth and development via an unknown mechanism (Walley et al. 2015; Benn et al. 2016). Nevertheless, there is evidence that MEcPP plays a role in chromatin remodeling (Grieshaber et al. 2004), though this has not yet been observed in plants.

3.3 Reactive Oxygen Species a Novel Candidate in Retrograde Signaling in Plant

One of the most crucial parts of the retrograde signaling network is ROS. To uncover ROS-responsive genes and clarify the function of ROS as signaling molecules, many researchers have performed transcriptome analysis with ROS-generating agents or mutants lacking antioxidative enzymes. Reactive oxygen species (ROS), such as superoxide and hydrogen peroxide (H_2O_2), are produced in chloroplasts by reduction of O_2 at photosystem I (PSI) on the stromal side of the non-appressed thylakoid membrane, and singlet oxygen ($^1O_2$) is produced by excessive excitation of photosystem II (PSII) in the grana core and non-appressed margins of the grana (Mittler et al. 2004; Foyer and Shigeoka 2011). Under ideal environmental circumstances, ROS are produced in low concentrations in the plant cell. On the other hand, extreme environmental conditions cause an overabundance of ROS (Maurya 2020). Due to the oxidation of biomolecules (lipids, carbohydrates, proteins, enzymes, DNA), ROS damages molecular and cellular components, inactivating enzymes participating throughout essential pathways and disrupting the signaling cascade, causing plant death (Bhuyan et al. 2020; Shah et al. 2019). The introduction of particular inhibitors to influence cellular responses and nuclear gene expression is the basis of future research. Understanding the ROS mechanism engaged during this signaling is vital for cell survival.

3.3.1 Singlet Oxygen-Triggered Responses

Singlet oxygen is a reactive oxygen species produced by excessive excitation of photosystem II in the chloroplast (Dogra et al. 2017), especially when plants are exposed to high light. $^1O_2$ has a toxic effect; reacts with several classes of biomolecules like protein, lipid, and DNA; and leads to ROS-induced cell death in leaves.

Apart from its toxic effects, $^1O_2$ can also trigger a signaling cascade, leading to programmed cell death or to acclimation. The lowest excited electronic state of molecular oxygen is singlet oxygen ($^1O_2$). It is harmful because it quickly oxidizes biological components (Dmitrieva et al. 2020). In plant cells, photosystem II reaction units produce $^1O_2$ primarily in the light in thylakoid membranes (Krieger-Liszkay and Shimakawa 2022). $^1O_2$ damages membranes, proteins, and DNA at high concentrations and inhibits protein synthesis in chloroplasts, resulting in photoinhibition of photosynthesis, and can cause cell death (Dmitrieva et al. 2020). Thylakoid membranes appear to be equipped with at least two putative $^1O_2$ sensors, grana core-localized β-carotene and grana margin-localized EX1 protein, which independently mediate distinct $^1O_2$ signaling pathways (Agrawal et al. 2023). Although the source of $^1O_2$ generation in the grana core is well understood, it is unclear how the grana margin generates $^1O_2$. β-carotene as a potential sensor of $^1O_2$ because of its volatile oxidized derivatives, such as β-cyclocitral (β-CC), dihydroactinidiolide (dhA), and β-ionone (β-I). Hence, these compounds easily diffuse from the chloroplast to the nucleus to mediate $^1O_2$ signaling (Ramel et al. 2012). A genetic screen using the unicellular green alga *C. reinhardtii* revealed methylene blue sensitivity 1 (MBS1), a small zinc finger protein located both in the cytosol and the nucleus, acts downstream of β-CC to the singlet oxygen-responsive genes (Shao et al. 2013). However, changes in nuclear expressions by MBS1 is not well known. In the *fluorescent* (*flu*) mutants, there is overaccumulation of Pchlide (responsible for $^1O_2$ generation) modulating nuclear gene expressions and programmed cell death via the nucleus-encoded chloroplast protein EXECUTER1 (EX1) (Fig. 3.5). In the *flu* mutants, loss of EX1 dramatically reduces $^1O_2$-induced stress responses and alterations in nuclear gene expression (Lee et al. 2007; Wagner et al. 2004). EX1 in the vicinity may detect $^1O_2$ via an unknown mechanism and activate $^1O_2$ signaling in an FtsH-dependent manner. EX1 proteolysis by FtsH appears to be required to trigger alterations in nuclear gene expression. The virtually exclusive location of EX1 proteins and Chl biosynthesis enzymes in the grana margin, as well as the oxidative suppression of the PSII repair machinery, would imply that unbound or free Chl released or generated during the PSII repair process may generate $^1O_2$ in the grana margin (Carmody et al. 2016; Dogra et al. 2017). However, the genuine signaling molecule(s) and downstream signaling components remain unknown.

3.3.2 Hydrogen Peroxide-Mediated Retrograde Signaling

The primary ROS in plants is hydrogen peroxide (H_2O_2). Previously, hydrogen peroxide was thought to be a damaging cellular product. However, current biochemical and genetic studies show that plants' unique signaling molecules modulate diverse responses to abiotic and biotic stimuli (Cerny et al. 2018). Mehler's reaction produces hydrogen peroxide in chloroplasts by reducing oxygen at photosystem I (Agrawal et al. 2023). H_2O_2 sensing within chloroplasts appears to be essential for its specialized signaling function because it is hypothesized to originate and transduce a retrograde signal in response to diverse stress situations during plastid gene

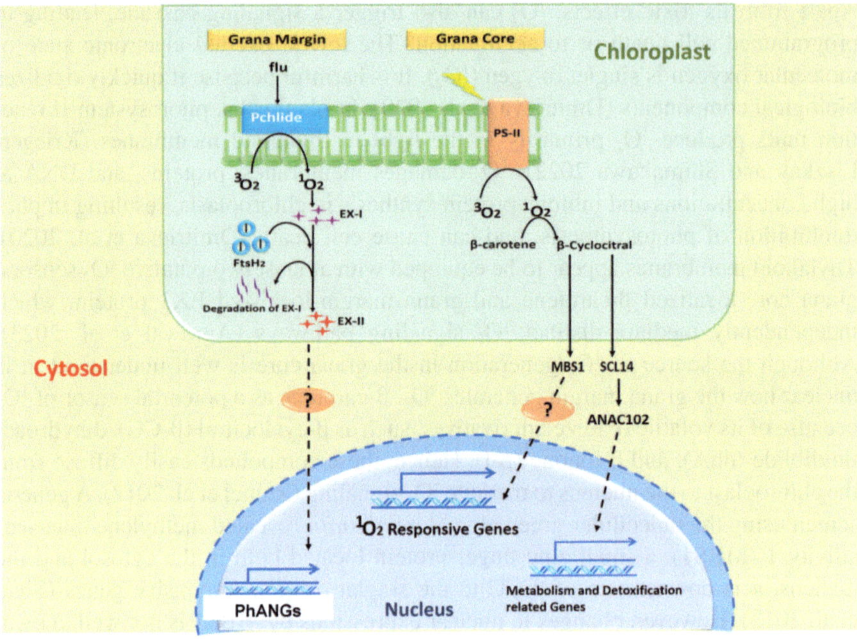

Fig. 3.5 Retrograde response by singlet oxygen in plants. Abbreviations: *flu* fluorescent mutant, *EX I and EX II* EXECUTER, *MBS1* methylene blue sensitive, *SCL14* SCARECROW-LIKE14, *ANAC102* RAO2/Arabidopsis NAC domain-containing protein

expression and the redox state of the photosynthetic electron transport chain to regulate nuclear gene expression (Carvalho and Silveira 2020). Understanding the characteristics of local and systemic redox/ H_2O_2 signaling in plants can aid in the identification of downstream signaling components. In chloroplasts, two important essential components, ascorbate peroxidase (APX) and thioredoxin peroxidase (TPXs), work together to remove H_2O_2 via NADPH and photosynthetic electron transfer via ferredoxin as the ultimate reductant (Maruta et al. 2016). In chloroplast, thylakoid membrane-bound ascorbate peroxidase (tAPX) plays a major role in the regulation of hydrogen peroxide. A recent microarray analysis revealed that tAPX silencing using estrogen-inducible RNAi affects the expression of many genes involved in the response to chilling and pathogens.

Additionally, tAPX silencing upregulates the salicylic acid responses. High light (HL) up- or downregulated tAPX silencing-responsive genes and tAPX silencing had a negative effect on the expression of ROS-responsive genes, implying both synergistic and antagonistic roles for chloroplasticH_2O_2in HL response. These findings shed new light on the role of H_2O_2-triggered retrograde signaling from chloroplasts in plant stress response (Maruta et al. 2012).

According to theories, H_2O_2 diffuses through the chloroplast envelope and stimulates a cytosolic mitogen-activated protein kinase (MAPK) cascade that phosphorylates H_2O_2 and gene activation (Breeze and Mullineaux 2022; Kovtun et al. 2000;

Apel and Hirt 2004). In *Arabidopsis*, H_2O_2 enables the MAPKK kinases ANP1 or MEKK1, to activate other downstream MAPKs (Kovtun et al. 2000). Future studies will undoubtedly reveal novel signaling roles for H_2O_2 and interplay with other signals.

3.4 Stromule-Mediated Retrograde Signaling

This refers to a process where stromules (stroma-filled tubules) facilitate communication between plastids and the nucleus (Brunkard et al. 2015). This signaling pathway allows plastids to convey information about their status, developmental changes, or environmental responses to the nucleus, influencing gene expression and plant responses. Recent researches have emphasized stromules' role in plant immunity and inter-organellar communication (Hanson and Conklin 2020). Investigation of stromule dynamics in *Arabidopsis thaliana* revealed their correlation with nuclear movement, suggesting a passive, mechanical mechanism highlighting the importance of spatial organization and organelle dynamics in plant cell signaling (Erickson et al. 2017) (Fig. 3.6). Photobleaching and photoconversion studies demonstrate protein exchange between plastids via stromules. Studies have also shown the interaction of stromules with plasma membrane, mitochondria, and peroxisomes as well under various conditions (Jung et al. 2024). However, detailed molecular mechanism behind the stromules formation remains unknown.

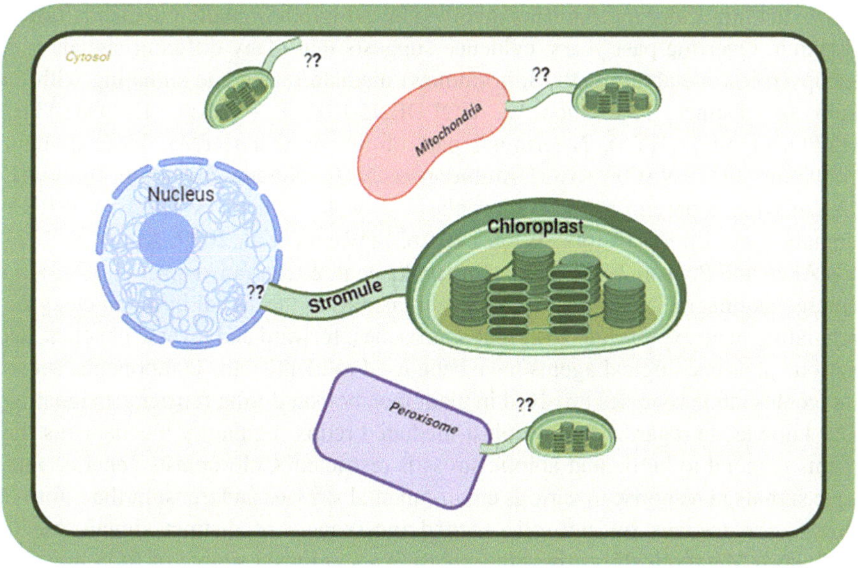

Fig. 3.6 Stromule-mediated retrograde signaling in plant cells. Stromules facilitate cross talk between chloroplasts, mitochondria, and peroxisomes to regulate plant cell responses to environmental stimuli

3.5 Cytosolic Signaling During Retrograde Pathway

The cytosolic translation product of the nuclear genome is critical for plant function and fitness. Recently, multiple components of a translation-dependent retrograde signaling pathway revealed that MPK6 and the AKIN10 subunit of SnRK1 regulate rapid stress acclimation responses after low light to high light transfer. The translatome reorganization initiated from low to high light involves a phosphorylation-dependent pathway that inhibits global translation and preferential transcript loading to the polysomes. This may serve as a very efficient local control mechanism and possibly outcompetes transcriptional control in terms of the time requirement needed to initiate acclimation by just provision of proteins required to the plastids (Agrawal et al. 2023). Another essential component is cytosolic HSP90–70 chaperons. HSP90 chaperone complex is a central regulator of Phang expression that determines the GUN phenotype. PAP most likely regulates gene expression by altering RNA metabolism mediated by 5–3exoribonucleases (XRNs). There are three XRNs in *Arabidopsis*; two encode the nuclear-localized XRN2 and XRN3, and the third one, XRN4, encodes for a cytosolic XRN (homologous to yeast Xrn2p) that targets the resulting three cleavage products of miRNA targets (Souret et al. 2004; Chan et al. 2016a, b); still very little evidence of the mechanisms of cytosolic control retrograde signaling is available.

3.6 Conclusion

The exact molecular mechanisms involved in retrograde signaling demands further research. Over the past years, evidence suggests that many different signals (i.e., tetrapyrroles, metabolites, ROS, hormones) mediate retrograde signaling with the help of distinct TFs, such as GOLDEN2-LIKE1 (GLKs), ELONGATED HYPOCOTYL5 (HY5), N-terminal part of the protein (PTM), NAC domain-containing TF (NAC), RAO2/Arabidopsis NAC domain-containing protein17 (ANAC017), a protein located chloroplasts, as well as nuclei (WHIRLY), WRKY-domain transcription factors (WRKY40&63) (Waters et al. 2009; Xin et al. 2021; VanAken and Pogson 2017; Świda-Barteczka et al. 2018; VanAken et al. 2013) that emerge as integration points, eventually transmitting the signal in the nucleus and regulating gene expression. Different approaches, forward and reverse genetics, and various pharmacological agents have been used to identify the components, sensor molecules, and processes involved in plant metabolism during retrograde signaling. Our knowledge regarding chloroplast-mediated redox regulatory mechanisms that plants respond to biotic and abiotic stress is restricted. Chloroplasts generate multiple signals in response to various environmental stresses and transmit the information to the nucleus by cytosolic second messengers or distinct signal cascade pathways. Metabolically active chloroplast is a significant source of ROS that triggers stress responses. Identifying the novel interconnecting components of retrograde signaling questions will help us better understand the communication between chloroplast to the nucleus and enable us to manipulate retrograde signaling through genetic approaches for crop improvement in the future.

References

Agrawal V, Singh V, Tripathi BN (2023) Components and processes involved in retrograde signalling from chloroplast to nucleus. Physiol Plant 175(4):e13987

Alsharafa K, Vogel MO, Oelze ML, Moore M, Stingl N, König K et al (2014) Kinetics of retrograde signalling initiation in the high light response of *Arabidopsis thaliana*. Philos Trans R Soc B Biol Sci 369(1640):20130424

Apel K, Hirt H (2004) Reactive oxygen species: metabolism, oxidative stress, and signal transduction. Annu Rev Plant Biol 55:373–399

Bailey M, Ivanauskaite A, Grimmer J, Akintewe O, Payne AC, Osborne R et al (2021) The *Arabidopsis* NOT4A E3 ligase promotes PGR3 expression and regulates chloroplast translation. Nat Commun 12(1):1–15

Balparda M, Armas AM, Estavillo GM, Roschzttardtz H, Pagani MA, Gomez-Casati DF (2020) The PAP/SAL1 retrograde signalling pathway is involved in iron homeostasis. Plant Mol Biol 102(3):323–337

Baruah A, Šimková K, Hincha DK, Apel K, Laloi C (2009) Modulation of 1O2-mediated retrograde signalling by the PLEIOTROPIC RESPONSE LOCUS 1 (PRL1) protein, a central integrator of stress and energy signalling. Plant J 60(1):22–32

Battersby A (2000) Tetrapyrroles: the pigments of life. Nat Prod Rep 17:507–526

Benn G, Bjornson M, Ke H, De Souza A, Balmond EI, Shaw JT, Dehesh K (2016) Plastidial metabolite MEcPP induces a transcriptionally centered stress-response hub via the transcription factor CAMTA3. Proc Natl Acad Sci USA 113:8855–8860

Bhatnagar A, Singh S, Khurana JP, Burman N (2020) HY5-COP1: the central module of light signalling pathway. J Plant Biochem Biotechnol:1–21

Bhuyan MB, Hasanuzzaman M, Parvin K, Mohsin SM, Al Mahmud J, Nahar K, Fujita M (2020) Nitric oxide and hydrogen sulfide: two intimate collaborators regulating plant defense against abiotic stress. Plant Growth Regul 90:409–424

Bradbeer JW, Atkinson YE, Börner T, Hagemann R (1979) Cytoplasmic synthesis of plastid polypeptides may be controlled by plastid-synthesised RNA. Nature 279(5716):816–817

Bräutigam K, Dietzel L, Pfannschmidt T (2007) Plastid-nucleus communication: anterograde and retrograde signalling in the development and function of plastids. In: Cell and molecular biology of plastids. Springer, pp 409–455

Breeze E, Mullineaux PM (2022) The passage of H2O2 from chloroplasts to their associated nucleus during retrograde signalling: reflections on the role of the nuclear envelope. Plants 11(4):552

Brunkard JO, Runkel AM, Zambryski PC (2015) Chloroplasts extend stromules independently and in response to internal redox signals. Proc Natl Acad Sci 112(32):10044–10049

Brzezowski P, Richter AS, Grimm B (2015) Regulation and function of tetrapyrrole biosynthesis in plants and algae. Biochim Biophys Acta 1847(9):968–985

Carmody M, Crisp PA, d'Alessandro S, Ganguly D, Gordon M, Havaux M et al (2016) Uncoupling high light responses from singlet oxygen retrograde signalling and spatial-temporal systemic acquired acclimation. Plant Physiol 171(3):1734–1749

Carvalho FE, Silveira JA (2020) H2O2-retrograde signalling as a pivotal mechanism to understand priming and cross stress tolerance in plants. In: Priming-mediated stress and cross-stress tolerance in crop plants. Academic, pp 57–78

Černý M, Habánová H, Berka M, Luklová M, Brzobohatý B (2018) Hydrogen peroxide: its role in plant biology and crosstalk with signalling networks. Int J Mol Sci 19(9):2812

Chan KX, Mabbitt PD, Phua SY, Mueller JW, Nisar N, Gigolashvili T et al (2016a) Sensing and signalling of oxidative stress in chloroplasts by inactivation of the SAL1 phosphoadenosine phosphatase. Proc Natl Acad Sci 113(31):E4567–E4576

Chan KX, Phua SY, Crisp P, McQuinn R, Pogson BJ (2016b) Learning the languages of the chloroplast: retrograde signalling and beyond. Annu Rev Plant Biol 67:25–53

Chi W, Sun X, Zhang L (2013) Intracellular signalling from plastid to nucleus. Annu Rev Plant Biol 64:559–582

Chi W, Feng P, Ma J, Zhang L (2015) Metabolites and chloroplast retrograde signalling. Curr Opin Plant Biol 25:32–38

Chow KS, Singh DP, Walker AR, Smith AG (1998) Two different genes encode ferrochelatase in *Arabidopsis*: mapping, expression and subcellular targeting of the precursor proteins. Plant J 15(4):531–541

D'alessandro S, Ksas B, Havaux M (2018) Decoding β-cyclocitral-mediated retrograde signalling reveals the role of a detoxification response in plant tolerance to photooxidative stress. Plant Cell 30(10):2495–2511

Dmitrieva VA, Tyutereva EV, Voitsekhovskaja OV (2020) Singlet oxygen in plants: generation, detection, and signalling roles. Int J Mol Sci 21(9):3237

Dobrogojski J, Adamiec M, Luciński R (2020) The chloroplast genome: a review. Acta Physiol Plant 42:1–13

Dogra V, Duan J, Lee KP, Lv S, Liu R, Kim C (2017) FtsH2-dependent proteolysis of EXECUTER1 is essential in mediating singlet oxygen-triggered retrograde signalling in *Arabidopsis thaliana*. Front Plant Sci 8:1145

Erickson JL, Kantek M, Schattat MH (2017) Plastid-nucleus distance alters the behavior of stromules. Front Plant Sci 8:1135

Estavillo GM, Crisp PA, Pornsiriwong W, Wirtz M, Collinge D, Carrie C et al (2011) Evidence for a SAL1-PAP chloroplast retrograde pathway that functions in drought and high light signalling in *Arabidopsis*. Plant Cell 23(11):3992–4012

Fang X, Zhao G, Zhang S, Li Y, Gu H, Li Y et al (2019) Chloroplast-to-nucleus signalling regulates microRNA biogenesis in *Arabidopsis*. Dev Cell 48(3):371–382

Foyer CH, Shigeoka S (2011) Understanding oxidative stress and antioxidant functions to enhance photosynthesis. Plant Physiol 155:93–100

Gould SB, Waller RF, McFadden GI (2008) Plastid evolution. Annu Rev Plant Biol 59:491–517

Green BR (2011) Chloroplast genomes of photosynthetic eukaryotes. Plant J 66(1):34–44

Grieshaber NA, Fischer ER, Mead DJ, Dooley CA, Hackstadt T (2004) Chlamydial histone–DNA interactions are disrupted by a metabolite in the methylerythritol phosphate pathway of isoprenoid biosynthesis. Proc Natl Acad Sci USA 101:7451–7456

Grübler B, Cozzi C, Pfannschmidt T (2021) A core module of nuclear genes regulated by biogenic retrograde signals from plastids. Plants 10(2):296

Hanson MR, Conklin PL (2020) Stromules, functional extensions of plastids within the plant cell. Curr Opin Plant Biol 58:25–32

Havaux M (2020) β-Cyclocitral and derivatives: emerging molecular signals serving multiple biological functions. Plant Physiol Biochem 155:35

Irisarri I, Strassert JF, Burki F (2020) Phylogenomic insights into the origin of primary plastids. BioRxiv

Jung S, Woo J, Park E (2024) Talk to your neighbors in an emergency: Stromule-mediated chloroplast-nucleus communication in plant immunity. Curr Opin Plant Biol 79:102529

Kacprzak SM, Mochizuki N, Naranjo B, Xu D, Leister D, Kleine T, Okamoto H, Terry MJ (2019) Plastid-to-nucleus retrograde signalling during chloroplast biogenesis does not require ABI4. Plant Physiol 179:18–23

Kindgren P, Noren L, Lopez JDDB, Shaikhali J, Strand A (2012) Interplay between HEAT SHOCK PROTEIN 90 and HY5 controls PhANG expression in response to the GUN5 plastid signal. Mol Plant 5:901–913

Kovtun Y, Chiu WL, Tena G, Sheen J (2000) Functional analysis of oxidative stress-activated mitogen-activated protein kinase cascade in plants. Proc Natl Acad Sci 97(6):2940–2945

Krieger-Liszkay A, Shimakawa G (2022) Regulation of the generation of reactive oxygen species during photosynthetic electron transport. Biochem Soc Trans 50(2):1025–1034

Lee KP, Kim C, Landgraf F, Apel K (2007) EXECUTER1-and EXECUTER2-dependent transfer of stress-related signals from the plastid to the nucleus of Arabidopsis thaliana. Proc Natl Acad Sci 104(24):10270–10275

Leister D, Romani I, Mittermayr L, Paieri F, Fenino E, Kleine T (2014) Identification of target genes and transcription factors implicated in translation-dependent retrograde signalling in *Arabidopsis*. Mol Plant 7(7):1228–1247

Lemos M, Xiao Y, Bjornson M, Wang JZ, Hicks D, Souza AD et al (2016) The plastidial retrograde signal methyl erythritol cyclopyrophosphate is a regulator of salicylic acid and jasmonic acid crosstalk. J Exp Bot 67(5):1557–1566

Li T, Xiao Z, Li H, Liu C, Shen W, Gao C (2020) A combinatorial reporter set to visualize the membrane contact sites between endoplasmic reticulum and other organelles in plant cell. Front Plant Sci 11:1280

Liu JX, Howell SH (2016) Managing the protein folding demands in the endoplasmic reticulum of plants. New Phytol 211:418–428

Maruta T, Noshi M, Tanouchi A, Tamoi M, Yabuta Y, Yoshimura K et al (2012) H2O2-triggered retrograde signalling from chloroplasts to nucleus plays specific role in response to stress. J Biol Chem 287(15):11717–11729

Maruta T, Sawa Y, Shigeoka S, Ishikawa T (2016) Diversity and evolution of ascorbate peroxidase functions in chloroplasts: more than just a classical antioxidant enzyme? Plant Cell Physiol 57(7):1377–1386

Maurya AK (2020) Oxidative stress in crop plants. In: Hasanuzzaman M (ed) Agronomic crops: stress responses and tolerance. Springer, Singapore, pp 349–380

Mittler R, Vanderauwera S, Gollery M, Van Breusegem F (2004) Reactive oxygen gene network of plants. Trends Plant Sci 9:490–498

Nott A, Jung HS, Koussevitzky S, Chory J (2006) Plastid-to-nucleus retrograde signalling. Annu Rev Plant Biol 57:739–759

Page MT, Garcia-Becerra T, Smith AG, Terry MJ (2020) Overexpression of chloroplast-targeted ferrochelatase 1 results in a genomes uncoupled chloroplast-to-nucleus retrograde signalling phenotype. Philos Trans R Soc B 375(1801):20190401

Pfalz J, Liebers M, Hirth M, Grübler B, Holtzegel U, Schröter Y et al (2012) Environmental control of plant nuclear gene expression by chloroplast redox signals. Front Plant Sci 3:257

Phua SY, De Smet B, Remacle C, Chan KX, Van Breusegem F (2021) Reactive oxygen species and organellar signalling. J Exp Bot 72(16):5807–5824

Ponce-Toledo RI, López-García P, Moreira D (2019) Horizontal and endosymbiotic gene transfer in early plastid evolution. New Phytol 224(2):618–624

Pornsiriwong W, Estavillo GM, Chan KX, Tee EE, Ganguly D, Crisp PA et al (2017) A chloroplast retrograde signal, 3′-phosphoadenosine 5′-phosphate, acts as a secondary messenger in abscisic acid signalling in stomatal closure and germination. elife 6:e23361

Ramel F, Birtic S, Ginies C, Soubigou-Taconnat L, Triantaphylidès C, Havaux M (2012) Carotenoid oxidation products are stress signals that mediate gene responses to singlet oxygen in plants. Proc Natl Acad Sci 109(14):5535–5540

Ramel F, Mialoundama AS, Havaux M (2013) Nonenzymic carotenoid oxidation and photooxidative stress signalling in plants. J Exp Bot 64(3):799–805

Seemann M, Wegner P, Schünemann V, Bui BT, Wolff M, Marquet A, Trautwein AX, Rohmer M (2005) Isoprenoid biosynthesis in chloroplasts via the methylerythritol phosphate pathway: the (E)-4-hydroxy-3-methylbut-2-enyl diphosphate synthase (GcpE) from *Arabidopsis thaliana* is a [4Fe-4S] protein. J Biol Inorg Chem 10:131–137

Shah K, Chaturvedi V, Gupta S (2019) Climate change and abiotic stress-induced oxidative burst in rice. In: Advances in rice research for abiotic stress tolerance. Woodhead Publishing, pp 505–535

Shao N, Duan GY, Bock R (2013) A mediator of singlet oxygen responses in Chlamydomonas reinhardtii and Arabidopsis identified by a luciferase-based genetic screen in algal cells. Plant Cell 25(10):4209–4226

Sharma M, Kumar P, Verma V, Sharma R, Bhargava B, Irfan M (2022) Understanding plant stress memory response for abiotic stress resilience: molecular insights and prospects. Plant Physiol Biochem 179:10

Shimizu T, Masuda T (2021) The role of tetrapyrrole-and GUN1-dependent signalling on chloroplast biogenesis. Plants 10(2):196

Shumbe L, d'Alessandro S, Shao N, Chevalier A, Ksas B, Bock R, Havaux M (2017) METHYLENE BLUE SENSITIVITY 1 (MBS1) is required for acclimation of Arabidopsis to singlet oxygen and acts downstream of β-cyclocitral. Plant Cell Environ 40(2):216–226

Souret FF, Kastenmayer JP, Green PJ (2004) AtXRN4 degrades mRNA in Arabidopsis and its substrates include selected miRNA targets. Mol Cell 15:173–183

Świda-Barteczka A, Krieger-Liszkay A, Bilger W, Voigt U, Hensel G, Szweykowska-Kulinska Z, Krupinska K (2018) The plastid-nucleus located DNA/ RNA binding protein WHIRLY1 regulates microRNA levels during stress in barley (*Hordeum vulgare* L.). RNA Biol 15:886–891

Sylvestre-Gonon E, Schwartz M, Girardet JM, Hecker A, Rouhier N (2020) Is there a role for tau glutathione transferases in tetrapyrrole metabolism and retrograde signalling in plants? Philos Trans R Soc B 375(1801):20190404

Takahashi S, Ogawa T, Inoue K, Masuda T (2008) Characterization of cytosolic tetrapyrrole-binding proteins in *Arabidopsis thaliana*. Photochem Photobiol Sci 7(10):1216–1224

Tanaka R, Tanaka A (2007) Tetrapyrrole biosynthesis in higher plants. Annu Rev Plant Biol 58:321–346

Terry MJ, Smith AG (2013) A model for tetrapyrrole synthesis as the primary mechanism for plastid-to-nucleus signalling during chloroplast biogenesis. Front Plant Sci 4:14

VanAken O, Pogson BJ (2017) Convergence of mitochondrial and chloroplastic ANAC017/PAP-dependent retrograde signalling pathways and suppression of programmed cell death. Cell Death Differ 24:955–960

VanAken O, Zhang B, Law S, Narsai R, Whelan J (2013) AtWRKY40 and AtWRKY63 modulate the expression of stress-responsive nuclear genes encoding mitochondrial and chloroplast proteins. Plant Physiol 162:254–271

Wagner D, Przybyla D, op den Camp R, Kim C, Landgraf F, Lee KP et al (2004) The genetic basis of singlet oxygen induced stress responses of *Arabidopsis thaliana*. Science 306(5699):1183–1185

Walley J, Xiao Y, Wang JZ, Baidoo EE, Keasling JD, Shen Z et al (2015) Plastid-produced interorgannellar stress signal MEcPP potentiates induction of the unfolded protein response in endoplasmic reticulum. Proc Natl Acad Sci 112(19):6212–6217

Wang JZ, Li B, Xiao Y, Ni Y, Ke H, Yang P et al (2017) Initiation of ER body formation and indole glucosinolate metabolism by the plastidial retrograde signalling metabolite, MEcPP. Mol Plant 10(11):1400–1416

Waters MT, Wang P, Korkaric M, Capper RG, Saunders NJ, Langdale JA (2009) GLK transcription factors coordinate expression of the photosynthetic apparatus in *Arabidopsis*. Plant Cell 21:1109–1128

Woodson JD, Chory J (2008) Coordination of gene expression between organellar and nuclear genomes. Nat Rev Genet 9(5):383–395

Woodson JD, Perez-Ruiz JM, Chory J (2011) Heme synthesis by plastid ferrochelatase I regulates nuclear gene expression in plants. Curr Biol 21(10):897–903

Wu GZ, Bock R (2021) GUN control in retrograde signalling: how GENOMES UNCOUPLED proteins adjust nuclear gene expression to plastid biogenesis. Plant Cell 33(3):457–474

Xiao Y, Savchenko T, Baidoo EE, Chehab WE, Hayden DM, Tolstikov V et al (2012) Retrograde signalling by the plastidial metabolite MEcPP regulates expression of nuclear stress-response genes. Cell 149(7):1525–1535

Xiao Y, Wang J, Dehesh K (2013) Review of stress specific organelles-to-nucleus metabolic signal molecules in plants. Plant Sci 212:102–107

Xin K, Pan T, Gao S, Yan S (2021) A transcription factor regulates gene expression in chloroplasts. Int J Mol Sci 22(13):6769

Zhao C, Wang Y, Chan KX, Marchant DB, Franks PJ, Randall D et al (2019) Evolution of chloroplast retrograde signalling facilitates green plant adaptation to land. Proc Natl Acad Sci 116(11):5015–5020

Chloroplast Protein Import: Signals, Machinery, Processing, and Regulation

4

Siddhartha Dutta

Abstract

Chloroplasts originated from an endosymbiotic event where a free-living cyanobacterium was engulfed by a eukaryotic host, leading to the transfer of many genes to the host nucleus. As a result, although modern chloroplasts contain 3000–4500 proteins, only about 120 are encoded by the chloroplast genome. The vast majority are encoded by nuclear genes and imported into the chloroplast posttranslationally. This import process involves several steps, beginning with guiding the preproteins in the cytosol to the chloroplast envelope, where they bind to specific receptors. The preproteins are then transported across the chloroplast's outer and inner envelopes by the TOC-TIC (translocon at the outer/inner envelope membrane of chloroplasts) machinery, a hetero-oligomeric complex. This chapter aims to provide an understanding of chloroplast biogenesis and protein import, covering the journey of preproteins from the cytosol to their final destination within the chloroplast, the components of the import apparatus, and the regulation of this process. Additionally, it explores current perspectives on the existence of multiple TOC-TIC complexes and their specific roles in dynamically regulating the chloroplast proteome.

Keywords

Chloroplast biogenesis · Protein targeting · Translocon · Preprotein · TOC · TIC · Thylakoid membrane

---

S. Dutta (✉)
Department of Biotechnology, School of Health Science and Translational Research, Sister Nivedita University, Kolkata, West Bengal, India
e-mail: siddhartha.d@snuniv.ac.in

V. K. Dalal, A. N. Misra (eds.), *Chloroplast Biogenesis and Plastid Interconversions*,
https://doi.org/10.1007/978-981-95-0772-6_4

4.1 Introduction

Plants, being sessile organisms, have evolved highly specialized adaptive mechanisms to survive and thrive under diverse environmental conditions. One of the primary adaptations that enable their survival is their ability to perform photosynthesis, categorizing them as photoautotrophs. This fundamental process is carried out by chloroplasts, which are indispensable organelles unique to plant cells and certain algae. Chloroplasts are responsible for converting light energy into chemical energy, thereby facilitating the fixation of carbon dioxide into organic compounds. Besides photosynthesis, chloroplasts play vital roles in the biosynthesis of amino acids, nucleotides, lipids, and hormones, as well as in nitrogen and sulfur assimilation, making them central to plant metabolism and growth. The evolution of autotrophic organisms dates back approximately 3.7 billion years. However, the chloroplasts we observe today are believed to have originated around 1.5 billion years ago through a significant endosymbiotic event. In this evolutionary milestone, a non-photosynthetic eukaryotic cell engulfed a photosynthetic cyanobacterium. This event led to a symbiotic relationship, with extensive lateral gene transfer of genetic material from the engulfed cyanobacterium to the host cell's nucleus (Blankenship 2010; Ahmad et al. 2016). Consequently, the chloroplasts lost a considerable degree of their functional autonomy and became semiautonomous organelles, retaining only a small fraction of their ancestral genome.

In modern-day chloroplasts, the genome is highly reduced, encoding a limited number of proteins necessary for chloroplast function. Most of the approximately 3000 proteins required for chloroplast function are encoded by nuclear genes and must be posttranslationally imported into the chloroplast (Richardson and Schnell 2020). The chloroplast genome itself retains about 120 genes. This compartmentalized genetic arrangement presents two major challenges: first, the need for coordinated regulation of gene expression between the nuclear and chloroplast genomes in response to light and developmental signals, and second, the efficient import of nuclear-encoded chloroplast proteins to ensure the proper functioning of the chloroplast. Approximately 95% of chloroplast proteins are synthesized in the cytosol on free ribosomes and subsequently imported into the chloroplast. These proteins are synthesized as precursor proteins (preproteins) containing an N-terminal targeting sequence known as the transit peptide. The transit peptide directs the preproteins to the chloroplast surface, where they are recognized by receptor components of the hetero-oligomeric protein import apparatus located on the chloroplast envelope. This apparatus includes the TOC (translocon at the outer membrane of chloroplasts) and TIC (translocon at the inner membrane of chloroplasts) complexes, which facilitate the translocation of proteins across the outer and inner chloroplast membranes, respectively. The TOC/TIC complexes have been extensively studied in peas and Arabidopsis, revealing detailed insights into their composition and function. The transit peptides of preproteins interact with receptors on the TOC complex, which facilitates their passage through the outer membrane channel component. Subsequently, the preproteins are transferred through the TIC complex across the inner membrane into the chloroplast stroma. This process is mediated by an array of additional proteins, including molecular chaperones, which ensure the proper targeting, translocation, and folding of the preproteins within the chloroplast (Chotewutmontri et al. 2017; Sjuts et al. 2017; Bölter 2018; Day and Theg 2018).

Chloroplasts are not only pivotal in photosynthesis and metabolic functions but also play active roles in the developmental processes of plants. They function as environmental sensors, helping plants adapt to changing environmental conditions through inter-organelle communication (Chan et al. 2016; Zhao et al. 2019). The dynamic regulation of protein abundance within the chloroplast necessitates continuous replenishment of degraded or damaged proteins. This highlights the significance of the protein import mechanism in chloroplast biogenesis and overall chloroplast function. Recent research has uncovered additional layers of complexity in chloroplast biogenesis and protein import dynamics. For instance, studies have identified specific chaperone proteins that assist in the proper folding and assembly of imported proteins, ensuring their functionality within the chloroplast. Moreover, the import machinery itself is subject to regulation by various factors, including redox status, ATP availability, and developmental signals, which modulate the efficiency of protein translocation in response to cellular needs. Furthermore, advancements in proteomics and bioinformatics have enabled the identification of novel components of the chloroplast protein import machinery, expanding our understanding of its intricate regulation. These findings have significant implications for biotechnological applications, such as the engineering of plants with enhanced photosynthetic efficiency and stress tolerance. This chapter provides an in-depth overview of the mechanisms involved in protein import into chloroplasts, detailing the core components of the TOC-TIC complexes, elucidating the mechanism of protein import, and discussing the dynamic regulation of this process. Understanding these mechanisms is crucial for comprehending how chloroplasts maintain their functions and adapt to both developmental cues and environmental changes, ultimately contributing to the overall fitness and productivity of plants.

4.2 Chloroplast Biogenesis and the General Protein Import Pathway: Brief Overview

Chloroplast development and function are intricately linked to the coordinated interaction between the nuclear and organelle genetic systems. Chloroplast proteins exhibit significant variation, with estimates ranging from about 2100 in Arabidopsis to 4800 in rice, though these figures remain uncertain. The chloroplast genome comprises approximately 120 genes, with roughly 80 coding for proteins (Richly and Leister 2004). Over time, most of the original chloroplast genome has been relocated to the nucleus of plant cells, making the majority of plastid proteins nuclear-encoded. These proteins are synthesized in the cytoplasm and then transported into the chloroplast in an energy-dependent process (Richardson et al. 2017). Chloroplasts, essential for photosynthesis, are encased by a double membrane known as the chloroplast envelope. Within the chloroplast, thylakoids form an extensive network of photosynthetic membranes. These membranes can be arranged as individual stromal thylakoids or organized into stacks known as grana. All thylakoids, whether part of the grana or existing individually, enclose an internal space called the lumen. The thylakoid membranes contain the primary protein complexes necessary for photosynthesis: photosystems I and II, the cytochrome b_6/f complex, and ATP synthase. The grana structure is critical for separating the two photosystems, with PSII located in granal membranes and PSI in stroma-exposed thylakoids. Additionally, chlorophyll, the principal pigment for photosynthesis, is bound within pigment-protein complexes embedded in the thylakoid membranes. Other pigments, such as carotenoids, are also integrated into these complexes, playing essential roles in light absorption and photoprotection (Jensen and Leister 2014).

Light stimulates the transformation of proplastids, which are undifferentiated plastids, into chloroplasts. Etioplasts, which are chloroplasts that have not yet been exposed to light, do not contain active pigment and can be classified as a type of leucoplast. Light signals trigger the expression of photosynthetic genes in both the chloroplast and the nucleus, with photoreceptors facilitating these signals to initiate specific physiological responses (Fankhauser and Chory 1997; Deng and Quail 1999). Key photosynthetic proteins encoded by the nuclear genome include the RbcS, PsbP, PsbO, PsbD, PsbQ, petC, PsaE, petF, and most proteins involved in chlorophyll biosynthesis. Etioplasts contain prolamellar bodies, which are membrane aggregates of semicrystalline tubules that hold chlorophyll precursors. When exposed to light, these etioplasts convert into chloroplasts, containing thylakoids and grana. The assembly of the photosynthetic machinery involves the coordinated expression of genes from both plastid and nuclear genomes, regulated by environmental factors, developmental processes, and plant hormones. This tightly regulated process is crucial for plants to maintain their photoautotrophic lifestyle and adapt to environmental changes (Richardson et al. 2017).

Nuclear-encoded chloroplast proteins are synthesized on free cytosolic ribosomes. Each of these proteins contains a transit peptide at its N-terminus, typically 30–150 amino acids long, which targets the preprotein to the chloroplast. The transit peptide is cleaved upon import into the chloroplast, allowing the protein to mature and become functional (Chotewutmontri et al. 2017; Park et al. 2018). The transit peptide is associated with various cytosolic factors, including 14-3-3, Hsp90, and Hsp70, which keep the preprotein in an import-competent state (Lee et al. 2013). These preproteins remain unfolded to facilitate their import into the chloroplast. The initial recognition and transport through the outer membrane are mediated by the TOC (translocon at the outer chloroplast membrane) complex, composed of Toc159, Toc34, and Toc75 (Richardson and Schnell 2020). Toc159 and Toc34, through their GTPase activity, act as primary receptors that recognize the incoming preprotein and facilitate its binding to the chloroplast outer envelope (Fig. 4.1). Arabidopsis has multiple isoforms in both the Toc34 and Toc159 families. The Toc159 family includes AtToc159, AtToc132, AtToc120, and AtToc90, while the Toc34 family consists of AtToc33 and AtToc34. Toc159 and Toc33 have been demonstrated to play a role in the import of photosynthetic preproteins, while Toc132 and Toc34 are shown to be specifically involved in targeting non-photosynthetic preproteins into the chloroplast (Jarvis et al. 1998; Bauer et al. 2000; Kubis et al.

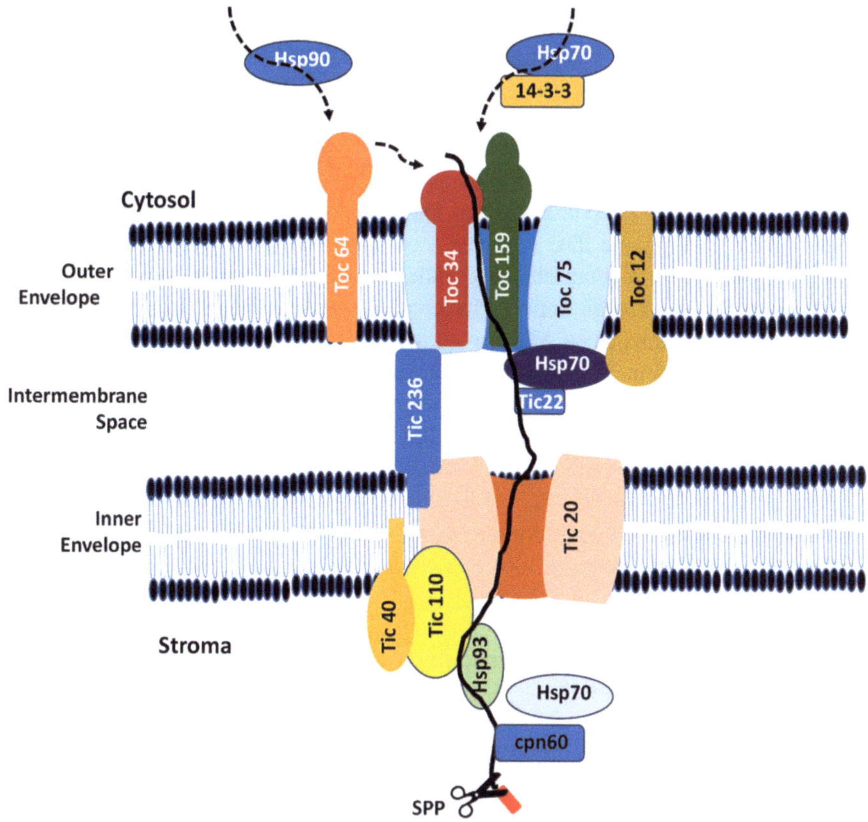

Fig. 4.1 Schematic representation of the key components involved in chloroplast protein import, illustrating the general working model of the TOC/TIC complex in this process. The newly synthesized preprotein is targeted to the TOC complex through two pathways: either by interaction with Hsp70/14–3-3 or Hsp93 chaperone-containing complexes, forming the import guidance complex, or via Toc64. The preprotein's transit peptide binds to the Toc34 and Toc159 receptors, facilitating its transfer through the outer envelope via the Toc75 channel and through the inner envelope via Tic20. Tic236 supports the formation of the TOC-TIC supercomplex, while Tic22, assisted by Toc12 and Hsp70, guides the preprotein through the intermembrane space to the TIC complex. Tic20 tightly associates with Tic110, which acts as a molecular scaffold alongside Tic40 and Hsp93, forming an ATP-dependent import motor that drives the preprotein into the stroma. Once imported, the N-terminal transit peptide is cleaved by the stromal processing peptidase (SPP), aided by chaperones like Hsp70 and Cpn60, allowing the preprotein to reach its mature form

2004; Constan et al. 2004). Toc75 functions as the channel protein that interacts with the transit peptide, working in coordination with the receptor components to transport the preprotein through the outer envelope. To traverse the inner envelope, the preprotein is guided by the TIC (translocon at the inner chloroplast membrane) complex (Kim et al. 2023) (Fig. 4.1). The passage through the intermembrane space is facilitated by Tic22. The TOC and TIC complexes associate to form a supercomplex, ensuring efficient transport of the preprotein from the cytosol to the stroma.

This coordination is mediated by Tic236. Tic20 and Tic110 are proposed to form the channel for preprotein transport through the inner envelope. Additionally, Tic110, along with Tic40, forms a scaffold for the binding of stromal chaperones. However, the composition of the TIC core complex is a subject of ongoing debate. A separate study in Arabidopsis has identified a distinct 1-MD translocon complex as the TIC complex, comprising components such as Tic20, Tic56, Tic100, Tic214, and Tic12 (Kikuchi et al. 2013; Zhao et al. 2022). This suggests the existence of two schools of thought regarding the TIC complex's composition. Resolving these differing views is essential for a comprehensive understanding of chloroplast protein import mechanisms. The stromal import chaperone complex, including cpHsp70, and ClpC/Hsp93, functions as an ATP-driven motor to assist the movement of the preprotein through the inner envelope into the stroma. Once in the stroma, the preprotein undergoes processing by stromal enzymes, which cleave the transit peptide, allowing the protein to fold into its mature form and interact with other stromal proteins to perform its functions. Some mature proteins may further engage in suborganellar targeting pathways, such as Tat-dependent, Sec-dependent, SRP-dependent, or spontaneous pathways, to reach the thylakoids and inner envelope membranes (Gutensohn et al. 2006).

4.3 Chloroplast Protein Import Machinery: Mechanisms and Processes

4.3.1 Targeting Signal and the Cytosolic Events

The "signal hypothesis" was first proposed to explain how proteins are targeted to the endoplasmic reticulum (Blobel and Sabatini 1971). Similar targeting signals are present in both prokaryotes and eukaryotes, such as the amino-terminal targeting signals, which help direct cytosolically synthesized proteins, to their specific subcellular destinations, including mitochondria and chloroplasts. This hypothesis laid the groundwork for understanding protein targeting mechanisms across different organelles. The initial discovery of the "envelope carrier hypothesis" emerged from studies on pea and *Chlamydomonas reinhardtii* chloroplasts using in vitro protein import assays of RbcS. This hypothesis suggests that all proteins synthesized on free cytosolic ribosomes destined for chloroplasts possess a common "site" recognized by receptors on the outer envelope of the chloroplast, facilitating their post-translational import (Chua and Schmidt 1978; Highfield and Ellis 1978). The targeting signal was identified as an N-terminal amino acid sequence of RbcS, which is now well-established as sufficient for interacting with the TOC (translocon at the outer chloroplast membrane) machinery (Bruce 2001).

Efficient import of chloroplast preproteins into the chloroplast, rather than mistargeting to other subcellular localizations, requires the transit peptides to have specific physicochemical properties derived from their amino acid composition and interactions with cytosolic components, including chaperons and 14-3-3 proteins (May and Soll 2000; Smith et al. 2004) (Fig. 4.1). The average length of these peptides ranges from 20 to over 100 amino acid residues. Numerous mutation and deletion studies have demonstrated that transit peptides contain certain functional domains or motifs. Despite their diversity, transit peptides share several conserved properties and specific motifs, predominantly including hydroxylated amino acids. One key property of transit peptides is their hydrophobicity. Certain hydrophobic sequence motifs within transit peptides are essential for their interaction with stromal Hsp70, which is crucial for the translocation process (Chotewutmontri and Bruce 2015). Additionally, transit peptides are rich in basic residues such as arginine (Arg) and lysine (Lys) and have a scarcity of acidic residues, resulting in an overall positive charge. Any alteration in the conserved basic residues of the transit peptide can impair the targeting and import of preproteins (Razzak et al. 2017). Generally, transit peptides are unstructured but may fold into amphipathic helices upon contact with a membrane. This structural flexibility is partly due to the presence of proline residues, which act as helix breakers and maintain the unstructured nature of transit peptides (Lee et al. 2018). Studies have also shown that transit peptides interact strongly with artificial membranes composed of chloroplast lipids, MGDG (Pinnaduwage and Bruce 1996). For example, a mutant of Arabidopsis deficient in a chloroplast-specific galactolipid exhibits defects in chloroplast protein import (Chen and li 1998). However, the precise mechanism of interaction between the transit peptide and the chloroplast envelope lipid remains to be established. Interestingly, the transit peptide contains distinct motifs for various functions, including binding to cytosolic AtHsc70-4, which mediates the proteasomal degradation of unimported preproteins to prevent cellular damage (Lee et al. 2016), as well as specific motifs for import regulation depending on the plant's age (Teng et al. 2012).

The posttranslational import of preproteins synthesized on free cytosolic ribosomes requires specific guiding factors to ensure proper targeting to the chloroplast, thereby avoiding mistargeting to other organelles. Similar to the targeting processes observed in peroxisomal and mitochondrial proteins, chloroplast preproteins engage cytosolic factors early in their journey for accurate localization. Chloroplast preproteins must remain in an unfolded state to be recognized and imported through the TOC-TIC (translocon at the outer/inner chloroplast membrane) complex. Chaperones play a critical role in maintaining this unfolded state, facilitating the import process. The Hsp70 family has been identified as one of the major chaperones involved in binding and stimulating the import of preproteins into chloroplasts (May and Soll 2000). Research indicates that both the transit peptide and the mature portion of the preprotein interact with Hsp70, enhancing the efficiency of the import process. Furthermore, in its Hsp70-bound form, the transit peptide also binds to the 14-3-3 protein (Fig. 4.1). The 14-3-3 protein preferentially binds to phosphorylated threonines or serines within the transit peptide, which are associated with Hsp70. This phosphorylation is mediated by specific STK kinases, including STY7, STY18, and STY46 (Lamberti et al. 2011). The interaction between the transit peptide, Hsp70, and 14-3-3 protein forms a "guidance complex," which is suggested to render the preprotein into an "import-competent" form, thereby increasing import efficiency. The heterooligomeric complex of preprotein with Hsp70 and 14-3-3 demonstrates translocation rates that are three to four times higher than those of the free precursor (May and Soll 2000). Since mitochondrial presequences are unable to form such "guidance complexes," it has been suggested that this mechanism may help distinguish the chloroplast transit peptides from structurally similar presequences, thereby ensuring the fidelity of protein trafficking within plant cells. This guidance complex hypothesis posits that the preprotein is directed toward docking with TOC receptor components on the chloroplast envelope.

In addition to Hsp70, another chaperone, Hsp90, has been shown to bind to the transit peptide and facilitate its guidance to the TOC complex (Qbadou et al. 2006). Intriguingly, the Hsp90-bound form of the transit peptide is suggested to guide the preprotein to an alternative docking site, the Toc64 receptor, on the outer envelope membrane (Sohrt and Soll 2000) (Fig. 4.1). However, mutant analyses in higher plants, such as Arabidopsis, and moss, such as *Physcomitrella patens*, have demonstrated that the absence of Toc64 does not significantly decrease chloroplast protein import efficiency (Aronsson et al. 2007). This suggests that the Hsp90/Toc64-mediated preprotein import pathway remains to be conclusively proven. Overall, the interactions of cytosolic factors, such as Hsp70, 14-3-3 proteins, and Hsp90, with chloroplast transit peptides are crucial for the efficient and accurate targeting of preproteins to the chloroplast. These interactions highlight the complex regulatory mechanisms involved in chloroplast protein import and the importance of chaperones in maintaining preprotein import competency.

4.3.2 The Stages of Import Process

The event following the arrival of a preprotein at the chloroplast surface is its recognition and binding to the TOC apparatus followed by its passage through the TOC-TIC complex into the stroma (Fig. 4.1). This import process is an active process and involves varying energy requirements at different stages, broadly categorized into three main steps: energy-independent binding, low energy depended intermediate stage, and high energy-dependent import (Theg et al. 1989; Kouranov and Schnell 1997; Young et al. 1999). Initially, the process is energy-independent, involving the reversible binding of the preprotein to the TOC receptor components. The TOC complex, which includes receptor proteins such as Toc159 and Toc34, specifically recognizes and binds the transit peptide of the preprotein (Gutensohn et al. 2000; Smith et al. 2004). This interaction is crucial as it positions the preprotein for subsequent stages of the import process without requiring energy input in the form of ATP or GTP. This initial binding phase ensures that the preprotein is correctly aligned for further translocation into the chloroplast. Following this, the preprotein enters what is known as the "intermediate stage," characterized by the requirement for a low concentration of ATP, less than 0.1 mM. During this stage, the preprotein undergoes deep insertion into the TOC complex and reaches the intermembrane space by passing through the channel protein Toc75. This step is also associated with the hydrolysis of GTP, making it an irreversible process. The involvement of GTP indicates its critical role in facilitating the irreversible insertion and ensuring the preprotein's commitment to translocation across the chloroplast envelope. This intermediate phase is a pivotal point in the import process as it transitions the preprotein from a loosely bound state to one that is fully engaged with the translocation machinery.

The final stage of the import process involves the translocation of the preprotein through the TIC complex into the chloroplast stroma. This phase is highly energy-dependent, requiring a high concentration of ATP, around 1 mM, which is provided by the chloroplast itself (Theg et al. 1989). The energy from ATP hydrolysis drives the movement of the preprotein through the TIC complex, ensuring its successful import into the chloroplast interior. Once inside the stroma, the transit peptide is recognized and cleaved by the stromal processing peptidase (SPP), thereby releasing the mature protein ready for its functional roles within the chloroplast. An interesting aspect of this import mechanism is the interaction between the TOC and TIC complexes. While these complexes are spatially discrete under ideal conditions, they interact to form a TOC-TIC supercomplex during the import of preproteins (Chen and Li 2017; Liu et al. 2023). This interaction occurs at specialized regions known as contact sites, where the outer and inner chloroplast membranes are closely apposed. These contact sites facilitate the efficient transfer of preproteins from the TOC complex to the TIC complex, ensuring a seamless translocation process. The formation of the TOC-TIC supercomplex underscores the coordinated nature of protein import into chloroplasts, highlighting the intricate interplay between various components of the translocation machinery. This elaborate import mechanism underscores the complexity and efficiency of intracellular protein targeting and translocation processes essential for chloroplast function and overall cellular homeostasis.

4.3.3 The Translocation Machinery

Various experimental approaches, including biochemical analysis of pea chloroplasts and binding assays of precursor proteins to the chloroplast envelope, have significantly contributed to our understanding of the TOC-TIC apparatus. Protein cross-linking studies have been particularly instrumental in identifying the key components of this complex system (Friedman and Keegstra 1989; Kessler et al. 1994; Schnell et al. 1994; Soll and Schleiff 2004; Bédard and Jarvis 2005) These studies, which involve creating covalent bonds between interacting proteins, have helped to map out the interactions between various TOC and TIC components, providing a clearer picture of how preproteins are recognized, translocated, and imported into chloroplasts. In addition to these biochemical techniques, mutant analysis in *Arabidopsis thaliana* has been invaluable in identifying new players involved in chloroplast protein import and elucidating their roles. Arabidopsis, a model organism in plant biology, has been extensively used to generate mutants with defects in specific genes encoding components of the TOC and TIC complexes (Jarvis et al. 1998; Bauer et al. 2000; Chou et al. 2003, Kikuchi et al. 2013; Richardson and Schnell 2020). By studying these mutants, researchers have been able to pinpoint the functions of individual proteins within the import machinery and understand the broader mechanistic framework governing chloroplast protein import.

The integration of these approaches has led to the discovery of several key proteins that play crucial roles in the Toc-Tic system. For example, the identification of Toc159 and Toc34 as GTPase receptors has shed light on the energy-dependent steps of the import process, while the discovery of Toc75 as the translocation channel has elucidated the pathway through which preproteins enter the chloroplast. Similarly, Tic110, Tic40, and Tic20 have been recognized as central components of the Tic complex, working in concert with stromal chaperones to facilitate the final stages of protein import. Moreover, these studies have revealed the dynamic and regulated nature of the TOC-TIC apparatus. It is now understood that the import machinery is not static but responds to various cellular signals and environmental cues (Richardson et al. 2017). For instance, the redox state of the chloroplast can influence the activity of certain TIC components, thereby modulating the efficiency of protein import under different physiological conditions (Stengel et al. 2008). Additionally, developmental signals can regulate the expression and assembly of TOC and TIC complexes, ensuring that chloroplast biogenesis is tightly coordinated with the developmental stage of the plant (Richardson et al. 2017; Wan et al. 2023). Advancements in proteomics and high-resolution imaging techniques have further extended our knowledge of the chloroplast import machinery. Proteomics has enabled the comprehensive identification of all the proteins associated with the TOC and TIC complexes, including transient and low-abundance components that were previously difficult to detect. High-resolution imaging, such as cryo-electron microscopy, has provided detailed structural insights into the TOC-TIC supercomplex, revealing how the outer and inner membrane translocation machineries interact at the molecular level (Schleiff et al. 2003; Jin et al. 2022).

4.3.3.1 Translocation across the TOC Complex

The translocon at the outer envelope of chloroplasts (TOC) complex serves as the initial contact point for precursor proteins with the chloroplast apparatus. Initial characterizations of TOC components were conducted using pea chloroplasts, identifying five components: Toc75, Toc159, Toc34, Toc64, and Toc12, named according to their molecular weights (Ling and Jarvis 2015). Among these, Toc159, Toc34, and Toc75 constitute the TOC core complex, with a molecular weight ranging from 500 to 1000 kDa (Schleiff et al. 2003) (Fig. 4.1). Proteomic analyses have predicted that this core complex has an apparent molecular mass of approximately 500 kDa, with a stoichiometry of 1:4:4–5 for Toc159, Toc75, and Toc34, respectively. Two-dimensional structural analysis using electron microscopy has revealed that the TOC complex forms roughly circular particles with a diameter of approximately 130 Å (Schleiff et al. 2003). These particles consist of a solid ring structure with a less dense interior. A three-dimensional map, derived from random conical tilt reconstruction of electron micrographs, suggests that a "finger"-like central region separates four curved translocation channels within a single complex. The estimated height of the complex (10–12 nm) is greater than the thickness of the outer chloroplast envelope, indicating its substantial size and complex structure. Alternate estimations suggest a TOC complex size of 800–1000 kDa, potentially due to the presence of six Toc34, six Toc75, and two Toc159 molecules (Chen and Li 2007). Another hypothesis proposes that the composition of the TOC complex undergoes dynamic changes. Such variability may reflect the functional plasticity of the TOC complex in response to different physiological conditions or developmental stages. In Arabidopsis, Toc34 has two paralogs, atToc33 and atToc34. Similarly, there are four paralogs for Toc75 (atToc75-I, atToc75-III, atToc75-IV, and atToc75-V/OEP80) and four paralogs for Toc159 (atToc159, atToc132, atToc120, and atToc90). These paralogs exhibit specialized or redundant functions, contributing to the versatility and robustness of the chloroplast protein import machinery (Jarvis 2008; Andrès et al. 2010; Ling and Jarvis 2015; Richardson and Schnell 2020).

The early and intermediate stages of preprotein transport are driven by GTP hydrolysis, facilitated by the receptor components Toc159 and Toc34. GTP is crucial for forming early-import intermediates but is not required for the translocation step of protein import into chloroplasts (Young et al. 1999). GTP's role is distinct from the ATP required for early-import intermediate formation. It has been shown that GTP does not affect translocation steps following the formation of the early-import intermediate. However, using GTP analogs such as GTP-γS (slowly hydrolyzable) and GMP-PNP (non-hydrolyzable) inhibits the formation of early-import intermediates (Young et al. 1999). Both Toc159 and Toc34 share domains responsible for GTPase activity. Toc159 is a tripartite protein consisting of a C-terminal membrane domain (M-domain) for anchoring in the outer membrane, a central GTP-binding domain (G-domain), and an N-terminal highly acidic domain (A-domain) that faces the cytosol. Interestingly, the A-domain is part of the intrinsically disordered protein (IDP) family, lacking a fixed three-dimensional structure, and extends into the cytosol, regulating the interaction between the preprotein and the Toc159 receptor (Richardson et al. 2009; Dutta et al. 2014). Toc34 has a simpler structure with a GTPase domain (G-domain) and a C-terminal hydrophobic membrane-spanning domain anchoring it to the outer membrane. Both Toc159 and Toc34 can homodimerize and heterodimerize to facilitate preprotein import (Yeh et al. 2007). The G-domain is critical for this dimerization process. Mutagenesis that disrupts dimerization impairs the initiation of preprotein translocation across the outer membrane. Additionally, preproteins stimulate the GTPase activity of the receptors, suggesting that preproteins may act as GTPase-activating factors (GAPs) during the early import process (Jelic et al. 2002). This elaborate interaction network within the TOC complex underscores the complexity and precision of chloroplast protein import mechanisms, reflecting an intricate evolutionary adaptation to the chloroplast's functional demands.

Toc159 was first identified in pea chloroplasts as an 86 kDa protein (named Toc86) (Bolter et al. 1998; Chen and Li 2007). It was later recognized as the Toc159 protein lacking the acidic A-domain. Genome and proteomic analyses in Arabidopsis identified the full-length Toc159 protein. The presence of the A-domain was shown to be essential for optimal protein import into chloroplasts, suggesting a crucial role of the cytosolic A-domain in the import process (Agne et al. 2010). Due to its acidic nature, it was proposed to interact with the transit peptide through electrostatic interactions. The A-domain is highly variable in amino acid sequence between species and among the Toc159 family members in Arabidopsis. Protein-protein interaction studies of the A-domain in the two primary Arabidopsis Toc159 family members, atToc159 and atToc132, suggest that it contributes to preprotein selectivity for the TOC complex. The A-domain of Toc159 is essential for providing functional diversity to the TOC complex under varying environmental and developmental conditions (Agne et al. 2010). Researchh in Arabidopsis also identified Toc159 in soluble form, leading to the hypothesis that it may act in transporting the preprotein from the cytosol to the TOC receptor (Hiltbrunner et al. 2001; Smith et al. 2002). However, this hypothesis was challenged, and it was believed that this soluble form might result from low-density fragments that occur during chloroplast fractionation.

The Toc159 in Arabidopsis is represented by four homologs: atToc159, atToc132, atToc120, and atToc90. The member atToc159 is the most abundant in green tissues. All four members share similar M- and G-domains, but they have highly dissimilar A-domains, which is completely lacking in atToc90. Genetic studies suggest that all members have distinct and/or overlapping roles in the protein transport process (Bauer et al. 2000; Ivanova et al. 2004; Kubis et al. 2004; Infanger et al. 2011; Inoue et al. 2010). Mutants of atToc159, *plastid protein import 2* (*ppi2*), exhibit an albino phenotype and do not grow on soil, requiring supplemented media for growth. Biochemical analysis revealed that the chloroplasts of *ppi2* mutants lack photosynthetic proteins, suggesting that atToc159 has a specific role in targeting photosynthetic preproteins (Class 1 import substrates). This also suggests a division of labor, with other members potentially targeting non-photosynthetic preproteins, including housekeeping proteins. The *ppi2* mutants accumulate photosynthetic preproteins in the cytosol, where they may be degraded, activating retrograde signaling and leading to the downregulation of expression of nuclear-encoded photosynthetic genes. Mutants of the other Toc159 family members do not show severe phenotypes and can grow on soil. Interestingly, the double mutant *toc120–1/toc132–1* shows defects in plant growth phenotype, indicating their essential roles and functional redundancy. Moreover, partial recovery of the *ppi2* mutant was achieved through atToc90 complementation, suggesting an overlapping role between atToc159 and atToc90.

In vitro and in vivo studies of the Toc159 receptor family, including mutations, protein-protein interactions, and domain swapping, suggest that the A-domain contributes to the selective recognition of photosynthetic (Class 1) and non-photosynthetic (Class 2) preproteins during early import stages (Bauer et al. 2000; Inoue et al. 2010; Dutta et al. 2014). Complementation of *ppi2* mutants with various atToc159 domains has shown that the presence of the M-domain alone was sufficient to rescue the lethal phenotype by restoring chloroplast preprotein import activity. However, chlorophyll content measurements demonstrated that this complementation could only partially rescue the phenotype, as chlorophyll content was only 50% of that of the wild type. This suggests the indispensable role of the full-length atToc159 in achieving complete and efficient chloroplast targeting (Lee et al. 2003). Moreover, the M-domain is required for the membrane insertion of atToc159. It has been shown to be highly resistant to protease treatment and extraction with alkaline and salt solutions, suggesting that this domain is deeply inserted into the outer envelope membrane. Interestingly, Toc159GM, consisting of the G- and M-domains, was also able to rescue the albino phenotype of *ppi2* mutants, indicating that the Toc86 form of Toc159 observed in earlier studies is likely an active protein (Lee et al. 2003; Agne et al. 2009).

Toc34 was the first member of the TOC family identified, primarily through co-isolation with preproteins during the initial stages of import. Like Toc159, Toc34 is considered a receptor in the TOC complex, crucial for the early recognition of preproteins (Hirsch et al. 1994; Kessler et al. 1994; Perry and Keegstra 1994; Jarvis et al. 1998; Constan et al. 2004). Toc34 features a short C-terminal membrane-anchoring region and an outer cytosolic GTPase domain, sharing 30% identity with the G-domain of Toc159. In Arabidopsis, two Toc34-type proteins are present: atToc33 and atToc34. Although they have similar sequences, they diverge at their C-terminal ends. Typically, atToc33 is the more prevalent isoform with higher expression levels. Extensive in vivo analyses using Arabidopsis mutants have confirmed Toc34's vital role in plastid import. The atToc33 loss-of-function mutant, *ppi1*, shows notable defects in chloroplast protein import, including reduced chlorophyll levels, compromised chloroplast ultrastructure, and decreased protein import efficiency. On the other hand, the *ppi3* mutant, which lacks atToc34, appears normal in shoot tissues but exhibits defective root development. This indicates that atToc33 mainly facilitates the import of photosynthetic preproteins, while atToc34 primarily handles non-photosynthetic preproteins. Despite this, their functions do overlap, as suggested by their high sequence similarity (around 80%) and genetic studies. For instance, the overexpression of atToc34 can compensate for the *ppi1* phenotype, and double knockout mutants of these two genes result in embryonic lethality (Constan et al. 2004). The crystal structure of pea Toc34 revealed that its G-domain contains residues that interact with specific residues of the G-domain of another monomer, facilitating homodimerization (Sun et al. 2002). It was initially predicted that one monomer functions as the GTPase activating protein (GAP) for the other, enhancing its GTPase activity. However, this observation was later challenged by a study where mutation of the residues responsible for dimerization (atToc33-R130A) reduced the ability to form homodimers but did not affect GTPase activity (Lee et al. 2009). Subsequent studies presented a different perspective, showing that homodimerization of Toc34 actually limits GDP-GTP exchange. Disruption of dimerization, which promotes nucleotide exchange, was also shown to be triggered by preprotein binding (Jelic et al. 2002, 2003). This supports the hypothesis that preproteins activate GTP hydrolysis on TOC receptors. Intriguingly, both in vitro and in vivo experiments suggest that Toc34 can heterodimerize with Toc159 through their G-domain interactions. Genetic studies have established the existence of two groups of receptors: one associating atToc159 with atToc33, primarily responsible for importing photosynthetic preproteins (class I), and another comprising atToc132/atToc120 with atToc34, which handles the recognition and import of non-photosynthetic preproteins (Class II) (Ivanova et al. 2004; Smith et al. 2004) (Fig. 4.2).

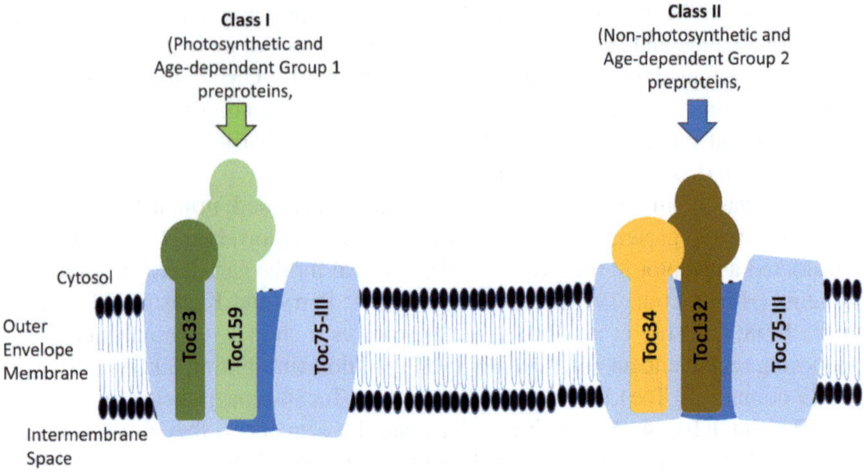

Fig. 4.2 Model depicting the operation of substrate-specific protein import pathways, illustrating the distinct routes taken by Class I (photosynthetic) and Class II (non-photosynthetic) preproteins during chloroplast import. The atToc159 preferentially associates with atToc33 to form a TOC complex that specifically targets Class 1 substrates, which are primarily abundant photosynthetic preproteins. Meanwhile, atToc132 and/or atToc120 tend to pair with atToc34 to create a different TOC complex that specializes in importing Class 2 substrates, mostly consisting of low-abundance housekeeping preproteins

The study of the two receptor groups in the TOC complex has led to two schools of thought on how preproteins interact with the TOC complex during the early stages of import (Kessler and Schnell 2004; Bedard and Jarvis 2005). One hypothesis, the "targeting hypothesis," considers Toc159 as the primary receptor of the TOC complex. According to this model, Toc159 initially interacts with the preprotein and then enters the TOC complex through its heterodimerization with Toc34. The GTPase activity of Toc159 is thought to be crucial for its interaction with the TOC complex, as mutations in the GTP-binding motif of Toc159 failed to complement the *ppi2* mutants and led to the accumulation of Toc159 in a soluble form. Once the preprotein is delivered to the TOC complex, Toc159 is free to dissociate from the outer envelope to continue its targeting role. This hypothesis is supported by observations that Toc159 and its homologs exist in both membrane-bound and cytosolic forms in roughly equal abundance. The alternate model, known as the "motor hypothesis," posits that Toc34 is the primary receptor of the TOC complex. In this model, Toc34 is permanently associated with the core TOC complex and does not rely on its GTPase activity for membrane interaction. Instead, the GTPase activity facilitates the interaction of preproteins with the TOC complex. Upon GTP hydrolysis, the preprotein is transferred to the centrally located Toc159, which acts as a "motor" driving the preprotein through the TOC complex via its multiple GTPase activities. Proponents of this model argue that the observed soluble form of Toc159 is an artifact, suggesting that Toc34 and Toc159 are firmly bound to the membrane as part of the TOC core complex. The "motor hypothesis" also gains support from previously predicted stoichiometric data of the TOC complex, proposing the existence of four channels, each containing one Toc75 and one Toc34 unit, with a single centrally located Toc159 molecule. This Toc159 molecule may rotate around its axis, delivering preproteins received from Toc34 receptors to the Toc75 channels like a motor through its GTPase activity.

A third member, Toc75, was identified as part of the TOC complex through interaction studies with the transit peptides of preproteins at intermediate stages of import (Perry and Keegstra 1994; Ma et al. 1996). Genome analysis suggests that it belongs to the Omp85 (Outer Membrane Protein 85) superfamily, which is found in mitochondria and gram-negative bacteria. Similar to Omp85, Toc75 contains an N-terminal domain with three POTRA (polypeptide transport-associated) repeats and a C-terminal β-barrel domain. The β-barrel domain, composed of 16–18 β-strands, forms the channel across the outer envelope, while the POTRA domain facilitates assembly with the TOC complex and interaction with incoming preproteins (Sánchez-Pulido et al. 2003; Ertel et al. 2005). Sequence analysis of the full-length Toc75 protein identified an N-terminal bipartite targeting sequence. The N-terminal part of the bipartite signal sequence targets Toc75 to the stroma, where it is cleaved by stromal processing peptidase. The C-terminal sequence of the bipartite signal sequence then directs Toc75 to its final location on the outer envelope, where it is cleaved by an envelope-bound type-I signal peptidase. Patch clamp analysis of planar lipid bilayers has shown that Toc75 forms a voltage-gated ion channel that interacts with preproteins. Reconstituted Toc75 exhibits a narrow restriction zone, corresponding to the selectivity filter, and a wider pore entrance, with diameters of approximately 14 Å and 26 Å, respectively (Hinnah et al. 1997; Hinnah et al. 2002).

In Arabidopsis, pea Toc75 has three orthologs: Toc75-III, Toc75-IV, and Toc75-V, with an additional member, Toc75-I, which is a pseudogene due to a transposon insertion (Jackson-Constan and Keegstra 2001). Among the three functional homologs, Toc75-III is the member identified as part of the TOC complex. Mutants of Toc75-III are embryo-lethal, indicating its indispensable role in chloroplast biogenesis. On the other hand, Toc75-IV mutants do not show any defects in protein import under normal conditions. However, the ultrastructure of chloroplasts isolated from dark-grown mutant plants is altered, suggesting a role in etioplast development. The third member, Toc75-V, now named OEP80, is involved in protein targeting to the chloroplast envelope independently of the TOC complex (Inoue and Potter 2004).

Another protein, Toc64, has been identified as part of the TOC complex (Sohrt and Soll 2000). Analysis of Toc64's sequence revealed that it comprises an N-terminal transmembrane region, a central region homologous to amidases, and a C-terminal domain with three tetratricopeptide repeat (TPR) motifs. Toc64 is proposed to facilitate an alternative pathway for preprotein delivery to the TOC complex. Specifically, it is suggested to act as a receptor for preproteins associated with Hsp90 rather than Hsp70 (Qbadou et al. 2006) (Fig. 4.1). In Arabidopsis, there are three paralogs of Toc64: Toc64-I, Toc64-II, and Toc64-III, with Toc64-III being the only member targeted to the chloroplast envelope. Mutant analyses in moss and Arabidopsis have shown that Toc64 is not essential for chloroplast import (Hofmann and Theg 2005; Aronsson et al. 2007). Additionally, triple mutants of all Toc64 homologs in Arabidopsis exhibit no defects in protein import, confirming its dispensable role in protein targeting to the chloroplast. Consequently, Toc64 has been alternatively named OEP64. However, a separate study using a double mutant of toc64-III and *ppi1* suggested that in Arabidopsis, Toc33 and Toc64-III are involved in the protein translocation process and cooperate in the insertion of Toc75-III, the pore-forming subunit of the TOC complex in *A. thaliana* (Sommer et al. 2013).

Toc12, a J-domain containing protein, has been identified as being associated with the TOC complex (Becker et al. 2004). It is located in the intermembrane space of the outer envelope, where it interacts with Toc64. Toc12 has been shown to interact with Hsp70 in the intermembrane space, acting as a cochaperone. Toc12, like other DnaJ family proteins, contains a conserved N-terminal domain called the J domain, which is composed of approximately 70 amino acids. This domain is structured into four helices, with the second helix featuring a charged surface that includes key basic residues essential for interacting with the ATPase domain of Hsp70 chaperones. In Toc12, the J domain is located at its C-terminal region, where it plays a critical role in stimulating the ATPase activity of Hsp70 chaperones. This interaction is particularly important in the intermembrane space of chloroplasts, where Toc12's J domain facilitates the proper functioning of Hsp70, ensuring efficient protein translocation across membranes. It is suggested to play a role in binding incoming preproteins and preventing them from slipping back into the cytosol at the early intermediate stage. Genome analysis has determined that Toc12 is a truncated clone of one of the pea DnaJ-J8 proteins, which are soluble proteins found in the chloroplast stroma. Arabidopsis mutants with T-DNA insertions in the DnaJ-J8 gene show no defects in chloroplast protein import (Becker et al. 2004). An early hypothesis proposed that Toc12 might be involved in forming contact sites, thereby bridging the TOC and TIC complexes.

4.3.3.2 Translocation Across the TIC Complex

The identification of the actual and putative members of the TIC complex has been a subject of ongoing debate, with considerable disagreement in the literature. Studies involving co-immunoprecipitation and transit-peptide interactions with the TIC complex have led to the identification of two major classes of TIC complexes, which remain contentious. One proposed TIC complex comprises Tic20, Tic110, Tic40, Tic22 and Tic21, in association with accessory members, Tic62, Tic55, and Tic32. In contrast, later studies identified Tic20 as the only common component, with a new TIC complex primarily consisting of Tic100, Tic56, and Tic214 (Fig. 4.3).

After translocation through the TOC complex, the successful delivery of preproteins into the stroma requires their guidance to the TIC complex via the intermembrane space and the formation of the TOC-TIC supercomplex. There is a consensus on the role of Tic22 and Tic236 in these processes. Tic22 has been identified as the first TIC component that the preprotein interacts with (Rudolf et al. 2013). It is proposed that Tic22 guides the preprotein through the intermembrane space, facilitated by its interaction with the Hsp70 present in this space, and to some extent by Toc12

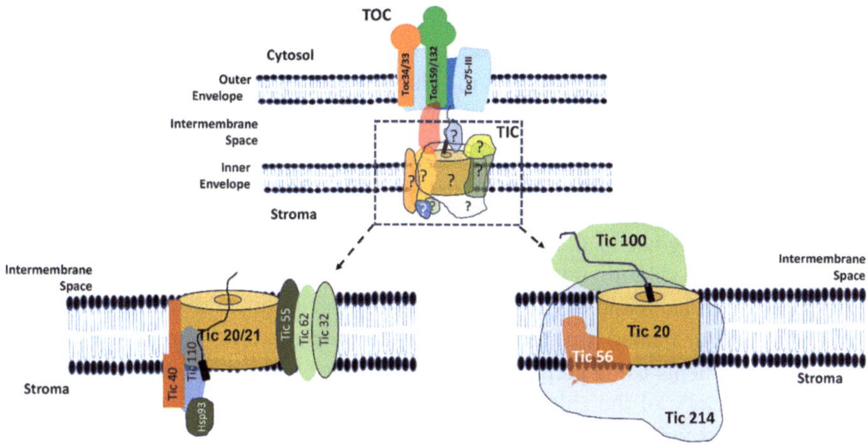

Fig. 4.3 Illustration of the two prevailing hypotheses on the composition of the TIC complex in chloroplast protein import. Both models highlight the involvement of distinct components within the TIC complex, with Tic20 being the shared membrane channel across both theories. The TIC complex consists of Tic20, Tic110, (and/or Tic21), Tic40, along with hsp93, Tic62, Tic55, and Tic32. Tic110 is thought to work together with Tic40 and Hsp93 in forming a stromal import motor complex. Tic62, Tic55, and Tic32, potentially in coordination with ferredoxin NAD(P) reductase (FNR) or calmodulin (CaM), may regulate protein import in response to redox signals, with their recruitment to import sites occurring only under specific conditions or for certain preproteins. The 1 MDa TIC complex is composed of Tic20, Tic56, Tic100, and Tic214, with Tic20 serving as the channel. The conserved N-terminal membrane-embedded region of Tic214 may also contribute to the channel structure. Tic56 and Tic100 are peripheral components located on the intermembrane space side, potentially involved in recognizing incoming preproteins and interacting with the intermembrane space domain of the TOC complex

(Fig. 4.1) (Becker et al. 2004; Flores-Pérez and Jarvis 2013). Tic22 is also suggested to play a role in the formation of the TOC-TIC supercomplex. In Arabidopsis, two closely related isoforms of Tic22, namely, *atTic22-III* and *atTic22-IV*, exist (Paila et al. 2016). While single mutants of both genes exhibited slight defects, the double mutant of both isoforms showed retarded growth, reduced photosynthetic performance, and a decreased import rate under high-light conditions. This suggests that both Tic22 isoforms function redundantly in chloroplast protein import. Moreover, these observations indicate that while Tic22 function is non-essential, it is required for normal chloroplast biogenesis, particularly during early plant development. Recently, Tic236 was identified as a key component of the TIC complex. Its primary 230-kDa domain extends into the intermembrane space, interacting with Toc75. Knockout mutants of Tic236 are embryo lethal, and in its absence, there is reduced association of Toc75 with Tic20 and a decrease in the TOC-TIC supercomplex population. This observation suggests that Tic236 may play an important role in the formation of the TOC-TIC supercomplex (Chen et al. 2018). The mutants show no effect on the preproteins' binding efficiency to the TOC complex but exhibit significant impairment in targeting the preprotein to the stroma, highlighting Tic236's critical role in the later phase of the import process.

Upon reaching the TIC complex, the primary function of the complex is to translocate the preprotein through the inner envelope into the chloroplast. To facilitate this function, the main component is the channel protein through which the preprotein must pass. Early studies on pea chloroplasts identified two major TIC components, Tic110 and Tic20, which act as channel proteins in the TIC complex. Tic110 was one of the first identified components of the TIC complex, with a proposed function as a channel protein (van den Wijngaard and Vredenberg 1999; Chen et al. 2002; Kovács-Bogdán et al. 2011). Proteomic analysis of intermediate translocon complexes, after solubilizing isolated pea chloroplasts with detergent, led to the identification of an abundant 100 kDa protein on the inner envelope. This 100 kDa protein was later named Tic110. Additionally, one research group, through cross-linking experiments using preproteins, identified Tic110 as an interacting partner at the inner envelope of the chloroplast. Interestingly, support for this observation is limited. Various independent studies were unable to detect Tic110 as the core components of the TIC translocon complex. Homozygous mutants lacking Tic110 are embryo-lethal, whereas heterozygous mutants show defects in growth and greening processes, suggesting the essential role of this component in chloroplast biogenesis. The role of Tic110 as a channel is primarily based on electrophysiological studies conducted using the protein reconstituted into artificial membranes. The structure of Tic110 is proposed to have a β-barrel structure with a large domain projected into the stroma. Tic110 is also suggested to serve as a molecular scaffold, recruiting molecular chaperones Hsp70, Hsp93/ClpC and a co-chaperon, Tic40 to the protein import site through its stromal domain (Kovacheva et al. 2005; Flores-Pérez et al. 2016). The chaperones and the co-chaperon play a role in ensuring the unidirectional movement of the preprotein at the later stage of the translocation process.

In contrast to the ambiguity surrounding the role of Tic110 in the TIC complex, Tic20 is well established as a TIC protein directly interacting with preproteins and forming a core component of the complex (Hirabayashi et al. 2011; Kasmati et al. 2011; Kovács-Bogdán et al. 2011). Tic20 is a hydrophobic membrane protein with four predicted α-helical transmembrane regions. The proposed role of Tic20 as a channel protein is primarily based on its predicted topology and its weak homology with channel components of the mitochondrial TIM machinery. Another study demonstrated that purified Tic20 alone is capable of forming a membrane channel, supporting its role as a channel protein in the TIC complex. Additionally, antisense downregulation of Tic20 expression in Arabidopsis caused defects in chloroplast biogenesis and impaired the translocation of preproteins through the inner membrane. In *Arabidopsis thaliana*, four homologues of the pea Tic20 protein have been identified: atTic20-I, atTic20-II, atTic20-IV, and atTic20-V. Homozygous knockout mutants of atTic20-I exhibited an albino phenotype, which was associated with abnormal chloroplast development and reduced levels of chloroplast proteins. In contrast, knockouts for atTic20-II, atTic20-IV, and atTic20-V did not display any phenotypic differences compared to the wild type. Further analysis of double homozygous mutants revealed that atTic20-I and atTic20-IV are the primary functional Tic20 isoforms in Arabidopsis, with partially overlapping roles. Genetic studies indicated that at least one of these two Tic20 proteins is essential for protein translocation into chloroplasts and suggested distinct substrate preferences between the isoforms. Interestingly, another protein, Tic21, with a similar topology to Tic20, was identified as a putative channel protein of the TIC complex (Teng et al. 2006). Arabidopsis mutants with a point mutation in the Tic21 gene showed defects in protein import into the chloroplast, suggesting its essential role in the translocation process, though the direct participation of Tic21 in this process remains an open question. An extensive cross-linking study identified another TIC component, Tic40, associated with the TIC complex at the stromal face (Chiu and Li 2008). Tic40 was also shown to interact with Tic110 through its TPR (tetratricopeptide repeat) domain and served as a scaffold for the binding of Hsp93 and Hsp70 through its Hip/Hop/Sti1 domain (Chou et al. 2003). The interaction between Tic110 and Tic40 is proposed to facilitate the release of preproteins from the channel, while the interaction of Tic40 with Hsp70/Hsp93 drives the import of preproteins into the stroma in an ATP-dependent manner. Arabidopsis null mutants of Tic40 exhibited a severe pale phenotype with slow growth but were not seedling lethal. These mutants also showed reduced transport of preproteins across the inner membrane (Chiu and Li 2008).

Three additional components—Tic62, Tic55, and Tic32—were identified as part of the TIC complex and are proposed to play a redox regulatory role. Tic62 is part of the TIC core complex along with Tic110 and Tic55 (Küchler et al. 2002; Hörmann et al. 2004; Stengel et al. 2009). The localization of Tic62 is regulated by the redox state of the chloroplast, shuttling between the inner envelope and the stroma depending on the NADP+/NADPH ratio (Küchler et al. 2002). The redox state also regulates its interaction with ferredoxin-NADP+ oxidoreductase (FNR), a key photosynthetic enzyme. Analysis of Tic62 knockout plants revealed that membrane attachment of FNR is severely disrupted in these mutants, and the overall amount of FNR in mutant chloroplasts is decreased. This redox-regulated interaction between Tic62 and FNR could be a mechanism for redox regulation of the transport of certain chloroplast preproteins. Tic55, another identified component along with Tic62, is a Rieske-type protein with a [2Fe-2S] cluster and an additional mononuclear Fe binding site. Its function is also linked to the redox regulation of chloroplast protein import. However, no detectable mutant phenotypes were found in T-DNA knockout Arabidopsis lines lacking Tic55, suggesting that Tic55 is not strictly required for protein import into chloroplasts (Boij et al. 2009). Tic32, the third component, tightly associates with Tic110, Tic40, Tic22, and Tic62. Arabidopsis null mutants of the atTic32 gene result in early seed abortion due to defects in embryo development. Using affinity chromatography, Tic32 was identified as the predominant calmodulin-binding protein of the inner envelope (Chigri et al. 2006). It was further established that Tic32 is a bona fide NADPH-dependent dehydrogenase. The dehydrogenase activity of Tic32 is affected by calcium binding, and NADPH was also shown to affect the interaction of Tic32 with Tic110 and Tic62. Thus, it is suggested that redox modulation and calcium regulation of chloroplast protein import may be mediated by Tic32 within the TIC complex.

Due to significant ambiguity regarding the acceptability of the actual and putative TIC components, a revised view of the TIC complex has emerged from a study employing a slightly modified approach involving affinity chromatography (Kikuchi et al. 2013). This approach led to the identification of a 1-megadalton (MDa) TIC complex at the inner envelope (Fig. 4.3). Blue native PAGE (BN-PAGE) separation of the isolated complex revealed that Tic20 forms the core of this TIC complex, while Tic21 is only peripherally associated with the 1 MDa TIC complex. Interestingly, no Tic110 or Tic40 proteins were observed in this 1 MDa complex. Additionally, three new components, Tic214, Tic100, and Tic56, were identified as part of the core 1 MDa TIC complex along with Tic20. Tic214 is encoded by the chloroplast gene, whereas the other two are nuclear-encoded. Tic214 is predicted to contain at least six transmembrane helices and three short MORN (membrane occupation and recognition nexus) motifs that may facilitate its interaction with the membrane. Tic56 does not contain any transmembrane regions and is found to be deeply embedded within the 1 MDa complex. Tic100 is peripherally associated, with a major part of the protein projected toward the intermembrane space. Electrophysiological analysis of the reconstituted 1 MDa complex on the planar lipid bilayer demonstrated that it interacts with preproteins and functions as a membrane channel. Stoichiometry analysis predicts the four members form a trimeric assembly with a 1:1:1:1 stoichiometry within the 1 MDa complex. The absence of either Tic56 or Tic100 caused significant reductions in the abundance of the 1 MDa TIC complex components, indicating their importance for the assembly of the complex. Furthermore, study of double mutants, *Tic20-I Tic100* and *Tic20-I Tic56*, showed similar phenotypes. Mutation analysis of Tic214 in tobacco and Chlamydomonas also suggests that this component of the TIC complex is essential for plant viability (Boudreau et al. 1997; Drescher et al. 2000). Thus, all four components of the 1 MDa TIC translocon are required for protein import into chloroplasts and are indispensable for plant viability. Based on these recent findings, a revised working model for the TIC complex has been proposed, suggesting the existence of a TIC core complex composed of Tic20, Tic214, Tic100, and Tic56 that interacts with the TOC complex to form the TOC-TIC supercomplex. According to this model, Tic110 and Tic40 may be involved at the very late stage of the import process.

4.3.3.3 Post-import Stromal Events

Two major stromal events following transport through the TIC complex involve the action of the "motor" that drives the movement of the TIC-associated preprotein into the chloroplast and the removal of the transit peptide to allow the preprotein to take its mature form. The initial step involving the unidirectional movement of the preprotein is believed to be driven by chaperones and cochaperones associated with Tic110, propelling the preprotein into the chloroplast (Inaba et al. 2003). These components include Hsp93, Hsp70, and Tic40 (Kovacheva et al. 2005, 2007) (Fig. 4.1). Genetic analyses involving mutations in Tic110, Hsp93, and Tic40 showed defects in chloroplast biogenesis and inefficiency in targeting both photosynthetic and non-photosynthetic preproteins (Kovacheva et al. 2005). Cross-linking experiments have demonstrated that Tic40 functions at the same import stage as Tic110 and Hsp93 (Chou et al. 2003). Moreover, analysis of double mutants of these components showed nonadditive interactions, suggesting that the three components cooperate during chloroplast import. The proposed model of this "motor" mechanism suggests that the interaction of the preprotein with Tic110 facilitates the interaction of Tic40 (through its TPR domain) with Tic110 (Chou et al. 2006). These interactions aid the delivery of the preprotein to Hsp93. Upon this event, the ATPase domain of Hsp93 is stimulated by Tic40, driving the preprotein into the chloroplast in an energy-dependent manner. Apart from these three components, stromal Hsp70 was also identified as an associated member through a co-immunoprecipitation study, showing that stromal Hsp70 simultaneously associates with the incoming preprotein along with Hsp93 and Tic40 (Su and Li 2010). Additional studies in moss demonstrated that stromal Hsp70 plays a critical role in the import process and acts as an energy-providing motor facilitating the irreversible movement of preproteins into the stroma through the TIC complex (Shi and Theg 2010).

Interestingly, genetic analyses using single and double mutants of Hsp70 and Hsp93 showed an additive effect in decreased import capacity, suggesting that these chaperones act partially in parallel as independent import players (Su and Li 2010). This led to a new understanding that Hsp70 functions as a motor protein, while Hsp93 associates with the Clp protease complex, possibly playing a role in degradation and quality control of preproteins rather than powering protein translocation. This new hypothesis is challenged by the observation that Hsp93 directly binds to both the N-terminal of transit peptide and the mature part of incoming preproteins, clearly indicating a role in early import stages (Huang et al. 2016). Based on these observations, it was hypothesized that both chaperones, stromal Hsp70 and Hsp93, might prefer different regions of the preprotein and thus provide different modes of translocation force.

Upon arrival into the stroma, the next event is the interaction of the preprotein with the stromal processing peptidase (SPP) through its transit peptide region (Richter and Lamppa 1999). This interaction occurs at a very early stage when the carboxyl terminal is still exposed on the chloroplast surface. The SPP recognizes and binds to approximately 10–15 residues at the carboxyl terminal of the transit peptide (Richter and Lamppa 2002). The SPP has a characterized zinc-binding motif that is critical for its peptidase activity. Binding of the SPP to the preprotein results in the cleavage of the N-terminal transit peptide, thereby releasing the mature protein into the stroma (Richter and Lamppa 1998). Transgenic approaches using antisense-mediated downregulation of SPP abundance result in a seedling-lethal and/or albino phenotype with strong defects in protein import into the chloroplast, suggesting that SPP plays an important role in protein import into the chloroplast (Wan et al. 1998; Zhong et al. 2003). To attain the functional state, the mature proteins are acted upon by stromal chaperones, including Hsp70 and Hsp60 (Cpn60), which are involved in modulating the folding of numerous chloroplast protein polypeptides (Jackson-Constan et al. 2001. Additionally, the imported preprotein may be targeted to various destinations within the chloroplast, such as the inner envelope, thylakoid, and thylakoid lumen. Targeting to the inner envelope may involve reinsertion into the inner membrane, whereas targeting to the thylakoid is mediated by the bipartite transit peptide and involves four sorting mechanisms: SRP-dependent transport, "spontaneous" transport, Sec-dependent pathway, and ΔpH/Tat-dependent pathways (Gutensohn et al. 2006).

4.4 Protein Transport Pathways Within Chloroplasts: Pathways in the Thylakoid Membrane System

The thylakoid membrane system is a major center for critical processes, including light capture, photosynthetic electron transport, and ATP synthesis. Consequently, it houses a large number of proteins involved in these functions, including four especially prominent complexes: photosystems I and II, the cytochrome b6/f complex, and ATP synthase. The targeting of these thylakoid proteins is well-studied and involves several independent pathways that are distinctly unique from each other and from the TIC-TOC pathways. Broadly, thylakoid targeting pathways can be categorized into two types: (i) targeting into the thylakoid membrane and (ii) translocation across the thylakoid membrane into the thylakoid lumen.

4.4.1 Targeting into the Thylakoid Membrane

Two pathways, the signal recognition particle (SRP)-dependent pathway and the "spontaneous' insertion pathway, have been identified as the mechanisms for protein targeting into the thylakoid membrane (Jarvis and Robinson 2004; Gutensohn et al. 2006). Among these, the SRP-dependent pathway is predominant, serving as the main route for membrane insertion. This pathway has been extensively studied using the light-harvesting chlorophyll a/b-binding protein (LHCP), which is a component of the light-harvesting complex associated with photosystem II. The LHCP preprotein possesses a typical stromal targeting transit peptide, which is cleaved by the stromal processing peptidase (SPP) upon entry into the stroma. Further targeting to the thylakoid membrane and its insertion is mediated by non-cleavable internal signals that are part of the mature protein. This targeting process is energy-dependent (powered by GTP hydrolysis) and relies on the stromal factors cpSRP54, cpSRP43, and cpFtsY, along with a thylakoid membrane protein, Alb3. In summary, LHCP binds to the heterodimeric cpSRP43/cpSRP54 complex to form a soluble transit complex, which then docks onto the SRP receptor cpFtsY and the Alb3 translocase at the membrane, leading to the release and integration of LHCP into the thylakoid membrane in a GTP-dependent manner (Ziehe et al. 2018).

An alternative method identified for the membrane insertion of thylakoid-targeted proteins is the "spontaneous" insertion (Schleiff and Klosgen 2001). This process is independent of stromal factors, receptors, translocases, and energy (Michl et al. 1994). This type of insertion is observed in single-spanning membrane proteins, which have a single transmembrane helix that spans the lipid bilayer. The single membrane-spanning segment is located near the NH_2-terminus of the mature polypeptide, which resides in the thylakoid lumen. Such targeting has been extensively studied in CFo-II and some subunits of photosystem II, including PsbW, PsbX, and PsbY (Kim et al. 1998; Thompson et al. 1999). In vitro protein targeting studies have shown that membrane insertion in this category of proteins relies on two hydrophobic domains flanking a hydrophilic NH2-terminal segment, which is translocated across the membrane. One hydrophobic domain is provided by the membrane anchor of the mature protein, while the other is provided by the transit peptide (Michl et al. 1999).

4.4.2 Targeting Across the Thylakoid Membrane

Two primary pathways, the Sec-dependent and Δ-pH/Tat-dependent pathways, facilitate the transport of preproteins across the thylakoid membrane into the lumen (Jarvis and Robinson 2004; Aldridge et al. 2009). Preproteins that utilize these transport mechanisms feature a characteristic bipartite transit peptide composed of two targeting signals: an N-terminal signal that directs the preprotein into the chloroplast, followed by a second signal that guides its translocation into the thylakoid lumen. The second translocation event begins after the chloroplast-targeting transit peptide is cleaved by the stromal processing peptidase (SPP). The remaining signal peptide is then cleaved by the thylakoid processing peptidase (TPP), a membrane-bound endopeptidase (Halpin et al. 1989).

The Sec-dependent pathway is energy-dependent and closely resembles the bacterial secretory pathway. In chloroplasts, it involves homologues of the bacterial Sec system components, including cpSecA, cpSecY, and cpSecE, and requires ATP. cpSecA is a stromal homolog of the bacterial SecA protein, while cpSecY and cpSecE are believed to form the translocation pore in the thylakoid membrane (Nakai et al. 1994; Roy and Barkan 1998; Schuenemann et al. 1999). Typical substrates that follow the Sec-dependent pathway include plastocyanin, the photosystem I subunit PsaF, and the 33 kDa protein of the oxygen-evolving complex (Gutensohn et al. 2006).

A unique transport mechanism was identified while studying the targeting of OE23, which can be directed to the thylakoid in a folded conformation, unlike other targeting pathways that typically require the protein to be unfolded. This targeting is independent of any stromal factors and nucleoside triphosphates, instead being driven by the proton gradient across the thylakoid membrane generated by photosynthetic electron flow (Mould and Robinson 1991; Cline et al. 1992. Hence, it is referred to as the Δ-pH-dependent pathway (Alder and Theg 2003). This Δ-pH/Tat-dependent pathway involves components Tha4, Hcf106, and cpTatC, which are closely related to their bacterial counterparts, TatA, TatB, and TatC, respectively (Settles et al. 1997; Walker et al. 1999; Motohashi et al. 2001). Preproteins transported by this pathway possess a thylakoid targeting signal featuring a pair of twin arginine residues, hence the name Tat (twin arginine translocation) pathway (Chaddock et al. 1995). The steps of the Tat-dependent pathway include: (1) Binding of the preprotein to the cpTatC-Hcf106 receptor complex, which triggers the assembly of Tha4 oligomers with the preprotein-receptor complex, forms a translocase.

(2) The preprotein is then transported through the lipid bilayer, and the thylakoid signal peptide is cleaved by the thylakoid processing peptidase (TPP), releasing the mature protein into the lumen. (3) After translocation, Tha4 dissociates from the receptor complex (Aldridge et al. 2009).

4.5 Regulation of Chloroplast Protein Import

4.5.1 Protein Degradation Pathways in Import Process Regulation

4.5.1.1 Ubiquitin-Proteasomal System (UPS) Regulation of Chloroplast Protein Import

The ubiquitin-proteasomal system (UPS) is a central protein regulatory mechanism in eukaryotes, playing a crucial role in protein degradation via a proteolytic pathway (Vierstra 2009). This system is composed of a series of enzymes that modify target proteins by attaching ubiquitin to them. The 26S proteasome is then responsible for breaking down these ubiquitin-tagged proteins. The UPS operates through a cascade involving three main types of enzymes: E1, E2, and E3. E1 enzymes, known as ubiquitin-activating enzymes, use energy from ATP hydrolysis to form a thioester bond between ubiquitin and themselves. This activated ubiquitin is subsequently transferred to E2 enzymes (ubiquitin-conjugating enzymes), forming another thioester bond. Finally, E2 enzymes, in cooperation with E3 enzymes (ubiquitin ligases), transfer the ubiquitin to the target substrate, marking it for degradation.

Early studies on the fate of preproteins under conditions where their targeting was defective led to the discovery of an E3 ligase known as the C-terminus of Hsc70-interacting protein (CHIP) E3 ligase (Shen et al. 2007). It was found that the interaction of Hsc70-4 (an isoform of Hsp70) with the transit peptide of preproteins recruits these preproteins to CHIP E3 ligase, resulting in their ubiquitination and subsequent degradation via the UPS. Further research into the role of the UPS in chloroplast import processes identified the role of DELLA proteins in regulating the levels of the Toc159 receptor (Shanmugabalaji et al. 2018). DELLA proteins were observed to interact with Toc159 before its incorporation into the TOC complex, initiating its degradation through the UPS. This posttranslational regulation of Toc159 reduces the abundance of Toc159-associated TOC complexes, thereby modulating the import of Class I preproteins during early germination stages.

Another significant discovery concerning the UPS's involvement in the regulation of the import process was the identification of the SP1 E3 ligase. SP1 was found to directly interact with components of the TOC complex, mediating their ubiquitination and degradation through the UPS (Ling et al. 2015). The posttranslational regulation of the TOC complex by SP1 suggests that SP1 selectively adjusts the abundance of these components in response to developmental cues and stress signals. The SP1 mediated TOC regulation was named chloroplast-associated protein degradation (CHLORAD). The selective degradation of Toc132/Toc120 and

Toc159 by CHLORAD determines the composition of different classes of TOC complexes (those containing either Toc130/120 or Toc159), thus facilitating the selective import of non-photosynthetic and photosynthetic preproteins.

4.5.1.2 Autophagy-Mediated Control of Import Processes

Autophagy is a crucial proteolytic pathway that controls protein levels in response to developmental cues and stress conditions (Marshall and Vierstra 2018). This process involves encapsulating the target protein in a double-membrane vesicle called an autophagosome, which then transports it to the vacuole or lysosome for degradation. The selection of these targets often relies on tagging them with specific signals, like ubiquitin, which are recognized by autophagy adapters that help deliver the target to the autophagic membrane. Recent studies on chlorophagy, or the autophagy of chloroplasts, have identified NBR1 as a selective autophagy adapter that mediates the degradation of the TOC complex in response to heat and UV stress (Wan et al. 2023). Confocal microscopy has shown that TOC proteins form puncta resembling autophagic bodies, suggesting that TOC components are selective targets for autophagy. NBR1, the autophagy adapter, has been found to interact with Toc33, Toc159, and Toc75. These findings indicate that selective autophagy plays a role in regulating the TOC complex during abiotic stress.

Moreover, polyubiquitination, depending on the specific linkage type, can direct protein degradation through either the UPS or autophagy. K48-linked polyubiquitin typically signals for proteasomal degradation, while K63-linked chains are associated with autophagic degradation. The K63 polyubiquitination of Toc159 under UVB stress suggests that ubiquitination triggers the autophagic degradation of TOC components. Thus, both the UPS and autophagy likely work together to regulate TOC abundance, thereby controlling the import of preproteins.

4.5.2 Phosphorylation-Mediated Control of Import Processes

Initial investigations into the number of import sites on a chloroplast employed strategies such as counting radiolabeled mature preproteins targeted inside the organelle and using immunogold-labeled antibodies to identify import components. These methods revealed discrepancies in the final count of import sites (Friedman and Keegstra 1989; Morin and Soll 1997). The difference between the actual number of import sites and the number of functional sites available for preprotein import led to the discovery of a phosphorylation regulatory mechanism within the TOC complex, which may control the import process (Aronsson et al. 2006; Oreb et al. 2007). It is known that the TOC receptor binds to the preprotein when it is in the GTP-bound form, and this interaction stimulates GTP hydrolysis. Studies showed that in pea plants, Toc34 can be phosphorylated by a kinase, and this phosphorylation significantly disrupts its ability to interact with GTP, thereby affecting its interaction with preproteins and regulating the import process. Further research revealed that in Arabidopsis, atToc33 can also be phosphorylated, while atToc34 cannot. The phosphorylation occurs on serine181, causing atToc33 to remain in an inactive,

GDP-loaded state. The signals triggering this phosphorylation are still not well understood. Therefore, phosphorylation is one of the posttranslational mechanisms that may regulate the TOC receptor's accessibility to preproteins, a process potentially influenced by developmental and environmental cues, thus regulating the overall import process.

4.5.3 Age-Dependent Regulation of Protein Import in Chloroplasts

Research on the import rates of various preproteins into the chloroplast has led to their classification into distinct groups based on their efficiency during different stages of chloroplast development (Teng et al. 2012). Group I preproteins are imported more efficiently during the early stages of chloroplast biogenesis, particularly in young chloroplasts found in young leaves. Group II consists of "housekeeping" preproteins that are imported efficiently throughout chloroplast development, independent of the chloroplast's age, and are essential for its ongoing maintenance. Group III preproteins are more efficiently imported during the later stages of chloroplast development. This age-dependent import is influenced by differences in the sequences of their transit peptides, suggesting that these organelle-targeting signals play a role in the cell's differential regulation networks based on age. Further analysis of this data led to the hypothesis that atToc159 serves as the receptor for Group I preproteins, while atToc132 is the receptor for Group II preproteins. Thus, the characteristics of the transit peptide are considered a key factor in regulating the import of specific subsets of preproteins through the TOC-TIC super-complex.

4.6 Unconventional Pathways for Protein Import

Although the TOC-TIC mediated pathway is the primary route for the import of most chloroplast-destined proteins, research has revealed a few significant alternative or noncanonical pathways that a small subset of preproteins may follow to reach the chloroplast. These pathways, though less common, provide important insights into the diversity and complexity of chloroplast protein import beyond the conventional TOC-TIC system. These alternative pathways include:

4.6.1 Endomembrane System Pathway

In this route, preproteins are delivered via the endoplasmic reticulum (ER) and Golgi before reaching the chloroplast. This mechanism has been observed in algae with complex plastids containing more than two envelope membranes (Nassoury and Morse 2005). The preproteins that follow this pathway have a bipartite targeting signal, consisting of an extreme N-terminal signal peptide fused to a chloroplast transit peptide. The signal peptide directs the preprotein to the ER, where it is processed and then passed along the secretory pathway through the Golgi (Villarejo et al. 2005; Radhamony and Theg 2006). Once the preprotein arrives at the plastid, the transit peptide facilitates its import into the chloroplast. A well-studied example of this pathway is Arabidopsis carbonic anhydrase 1 (CAH1), which was found to localize to the chloroplast stroma despite having a strong ER signal peptide. CAH1 could not be directly imported into isolated chloroplasts, suggesting the existence of a Golgi-mediated transport route for chloroplast targeting. Another protein, nucleotide pyrophosphatase/phosphodiesterase 1 (NPP1), was shown to follow a similar path (Nanjo et al. 2006).

4.6.2 Internal Targeting Signals

Some preproteins do not use the traditional transit peptide for chloroplast targeting. For example, the chloroplast envelope Quinone Oxidoreductase Homologue (ceQORH) and Tic32 lack canonical transit peptides. In ceQORH, studies have shown that its N-terminal sequence is not necessary for chloroplast targeting. Instead, an internal 40 amino acid sequence regulates its localization (Miras et al. 2002; Miras et al. 2007). This pathway also requires high energy in the form of ATP, independent of the TOC-TIC system, and relies on a proteinaceous import apparatus. Similarly, Tic32 is targeted to the chloroplast inner membrane without the assistance of key TOC machinery (Nada and Soll 2004).

4.6.3 Proteins Without Cleavable Transit Peptides

Another intriguing example of a targeting mechanism is seen in the localization of Toc34. Toc34 is anchored to the outer chloroplast envelope membrane via its C-terminal transmembrane domain (TMD), with its N-terminus, including the G-domain, exposed to the cytosol and a short C-terminal sequence (CTS) facing the intermembrane space. Proteins with this type of topology are referred to as tail-anchored (TA) proteins (Kim and Hwang 2013). They lack a cleavable targeting signal, and their targeting signals consist of the TMD and residues adjacent to it. In addition to the critical role of the TMD, the CTS and interactions with plastid-specific lipids are thought to contribute to the precise targeting of TA proteins to plastids (Hofmann and Theg 2005; Lee et al. 2011). The cytosolic factors, AKR2, is believed to recognize the TMD of newly synthesized Toc34 in the cytoplasm and assist in its delivery to the outer membrane (Dhanoa et al. 2010).

4.6.4 Substrate-Dependent Import

A unique mechanism was observed in the targeting of PORA, an isoform of NADPH oxidoreductase (POR) (Kim and Apel 2004). Unlike its counterpart PORB, which is imported through the standard TOC-TIC pathway, PORA uses a substrate-dependent import mechanism. PORA is imported only when it is bound to its substrate, protochlorophyllide. This pathway, termed "substrate-dependent PORA import," involves the outer plastid envelope protein OEP16, which plays a critical role in facilitating PORA's import into the plastid (Pollmann et al. 2007).

These noncanonical pathways reveal the diversity and complexity of chloroplast protein targeting, highlighting alternative mechanisms beyond the well-known TOC-TIC system.

4.7 Conclusion

In summary, the import of proteins into chloroplasts is a critical and well-coordinated process, essential for the proper functioning of chloroplasts and overall plant physiology. This chapter has explored the complex steps involved in this process, from the recognition of preproteins to their translocation through the TOC-TIC supercomplex. The identification of new components within the TOC and TIC complexes, as well as the emerging idea of alternative TOC and TIC complexes with specific or overlapping functions, has revealed that the import process is more intricate than previously thought, with significant regulatory aspects. Moreover, posttranslational modifications, such as phosphorylation and protein degradation pathways, have deepened our understanding of how the import process is finely tuned to align with the cellular needs. These insights highlight the vital role of chloroplast protein import in integrating environmental and developmental signals, enabling chloroplasts to adapt to changing conditions and contribute to the plant's overall response. As research continues, uncovering additional regulatory mechanisms and components will be key to further deciphering the complexities of this process. Additionally, it will be particularly interesting to explore how retrograde and/or anterograde signaling pathways coordinate the regulation of specific preprotein import, adding yet another layer of complexity to this already intricate process. Future discoveries in chloroplast protein import in crop plants are likely to unveil new aspects of plant biology, with promising applications for enhancing plant adaptability and productivity.

References

Agne B, Andrès C, Montandon C, Christ B, Ertan A, Jung F, Infanger S, Bischof S, Baginsky S, Kessler F (2010) The acidic A-domain of Arabidopsis TOC159 occurs as a hyperphosphorylated protein. Plant Physiol 153(3):1016–1030. https://doi.org/10.1104/pp.110.158048. Epub 2010 May 10. PMID: 20457805; PMCID: PMC2899928

Agne B, Infanger S, Wang F, Hofstetter V, Rahim G, Martin M, Lee DW, Hwang I, Schnell D, Kessler F (2009) A toc159 import receptor mutant, defective in hydrolysis of GTP, supports preprotein import into chloroplasts. J Biol Chem 284(13):8670–8679. https://doi.org/10.1074/jbc.M804235200. Epub 2009 Feb 2. PMID: 19188370; PMCID: PMC2659226

Ahmad N, Burgess SJ, Nielsen BL (2016) Editorial: advances in plastid biology and its applications. Front Plant Sci 7:1396. https://doi.org/10.3389/fpls.2016.01396. PMID: 27695471; PMCID: PMC5023683

Alder NN, Theg SM (2003) Energetics of protein transport across biological membranes. A study of the thylakoid DeltapH-dependent/cpTat pathway. Cell 112(2):231–242. https://doi.org/10.1016/s0092-8674(03)00032-1

Aldridge C, Cain P, Robinson C (2009) Protein transport in organelles: protein transport into and across the thylakoid membrane. FEBS J 276(5):1177–1186. https://doi.org/10.1111/j.1742-4658.2009.06875.x

Aronsson H, Boij P, Patel R, Wardle A, Töpel M, Jarvis P (2007) Toc64/OEP64 is not essential for the efficient import of proteins into chloroplasts in Arabidopsis thaliana. Plant J 52(1):53–68. https://doi.org/10.1111/j.1365-313X.2007.03207.x. Epub 2007 Jul 26

Aronsson H, Combe J, Patel R, Jarvis P (2006) In vivo assessment of the significance of phosphorylation of the Arabidopsis chloroplast protein import receptor, atToc33. FEBS Lett 580(2):649–655. https://doi.org/10.1016/j.febslet.2005.12.055. Epub 2005 Dec 28

Bauer J, Chen K, Hiltbunner A, Wehrli E, Eugster M, Schnell D, Kessler F (2000) The major protein import receptor of plastids is essential for chloroplast biogenesis. Nature 403(6766):203–207. https://doi.org/10.1038/35003214

Becker T, Hritz J, Vogel M, Caliebe A, Bukau B, Soll J, Schleiff E (2004) Toc12, a novel subunit of the intermembrane space preprotein translocon of chloroplasts. Mol Biol Cell 15(11):5130–5144. https://doi.org/10.1091/mbc.e04-05-0405. Epub 2004 Aug 18. PMID: 15317846; PMCID: PMC524789

Bédard J, Jarvis P (2005) Recognition and envelope translocation of chloroplast preproteins. J Exp Bot 56(419):2287–2320. https://doi.org/10.1093/jxb/eri243. Epub 2005 Aug 8

Blankenship RE (2010) Early evolution of photosynthesis. Plant Physiol 154(2):434–438. https://doi.org/10.1104/pp.110.161687. PMID: 20921158; PMCID: PMC2949000

Blobel G, Sabatini DD (1971) Ribosome–membrane interactions in eukaryotic cells. In: Manson LA (ed) Biomembranes. Plenum, New York, pp 193–195

Boij P, Patel R, Garcia C, Jarvis P, Aronsson H (2009) In vivo studies on the roles of Tic55-related proteins in chloroplast protein import in Arabidopsis thaliana. Mol Plant 2(6):1397–1409. https://doi.org/10.1093/mp/ssp079. Epub 2009 Oct 11

Bölter B, May T, Soll J (1998) A protein import receptor in pea chloroplasts, Toc86, is only a proteolytic fragment of a larger polypeptide. FEBS Lett 441(1):59–62. https://doi.org/10.1016/s0014-5793(98)01525-7

Bölter B (2018) En route into chloroplasts: preproteins' way home. Photosynth Res 138(3):263–275. https://doi.org/10.1007/s11120-018-0542-8. Epub 2018 Jun 26

Boudreau E, Turmel M, Goldschmidt-Clermont M, Rochaix JD, Sivan S, Michaels A, Leu S (1997) A large open reading frame (orf1995) in the chloroplast DNA of Chlamydomonas reinhardtii encodes an essential protein. Mol Gen Genet 253(5):649–653. https://doi.org/10.1007/s004380050368

Bruce BD (2001) The paradox of plastid transit peptides: conservation of function despite divergence in primary structure. Biochim Biophys Acta 1541(1–2):2–21. https://doi.org/10.1016/s0167-4889(01)00149-5

Chaddock AM, Mant A, Karnauchov I, Brink S, Herrmann RG, Klösgen RB, Robinson C (1995) A new type of signal peptide: central role of a twin-arginine motif in transfer signals for the delta pH-dependent thylakoidal protein translocase. EMBO J 14(12):2715–2722. https://doi.org/10.1002/j.1460-2075.1995.tb07272.x. PMID: 7796800; PMCID: PMC398390

Chan KX, Phua SY, Crisp P, McQuinn R, Pogson BJ (2016) Learning the languages of the chloroplast: retrograde signaling and beyond. Annu Rev Plant Biol 67:25–53. https://doi.org/10.1146/annurev-arplant-043015-111854. Epub 2015 Dec 21

Chen KY, Li HM (2007) Precursor binding to an 880-kDa Toc complex as an early step during active import of protein into chloroplasts. Plant J 49(1):149–158. https://doi.org/10.1111/j.1365-313X.2006.02944.x. Epub 2006 Nov 28. PMID: 17144891; PMCID: PMC1804235

Chen LJ, Li HM (1998) A mutant deficient in the plastid lipid DGD is defective in protein import into chloroplasts. Plant J 16(1):33–39. https://doi.org/10.1046/j.1365-313x.1998.00270.x

Chen LJ, Li HM (2017) Stable megadalton TOC-TIC supercomplexes as major mediators of protein import into chloroplasts. Plant J 92(2):178–188. https://doi.org/10.1111/tpj.13643. Epub 2017 Sep 15

Chen X, Smith MD, Fitzpatrick L, Schnell DJ (2002) In vivo analysis of the role of atTic20 in protein import into chloroplasts. Plant Cell 14(3):641–654. https://doi.org/10.1105/tpc.010336. PMID: 11910011; PMCID: PMC150586

Chen YL, Chen LJ, Chu CC, Huang PK, Wen JR, Li HM (2018) TIC236 links the outer and inner membrane translocons of the chloroplast. Nature 564(7734):125–129. https://doi.org/10.1038/s41586-018-0713-y. Epub 2018 Nov 21

Chigri F, Hörmann F, Stamp A, Stammers DK, Bölter B, Soll J, Vothknecht UC (2006) Calcium regulation of chloroplast protein translocation is mediated by calmodulin binding to Tic32. Proc Natl Acad Sci USA 103(43):16051–16056. https://doi.org/10.1073/pnas.0607150103. Epub 2006 Oct 11. PMID: 17035502; PMCID: PMC1635125

Chiu CC, Li HM (2008) Tic40 is important for reinsertion of proteins from the chloroplast stroma into the inner membrane. Plant J 56(5):793–801. https://doi.org/10.1111/j.1365-313X.2008.03638.x. Epub 2008 Jul 23. PMID: 18657235; PMCID: PMC2667645

Chotewutmontri P, Bruce BD (2015) Non-native, N-terminal Hsp70 molecular motor recognition elements in transit peptides support plastid protein translocation. J Biol Chem 290(12):7602–7621. https://doi.org/10.1074/jbc.M114.633586. Epub 2015 Feb 2. PMID: 25645915; PMCID: PMC4367265

Chotewutmontri P, Holbrook K, Bruce BD (2017) Plastid protein targeting: preprotein recognition and translocation. Int Rev Cell Mol Biol 330:227–294. https://doi.org/10.1016/bs.ircmb.2016.09.006. Epub 2017 Jan 23

Chou ML, Chu CC, Chen LJ, Akita M, Li HM (2006) Stimulation of transit-peptide release and ATP hydrolysis by a cochaperone during protein import into chloroplasts. J Cell Biol 175(6):893–900. https://doi.org/10.1083/jcb.200609172. Epub 2006 Dec 11. PMID: 17158958; PMCID: PMC2064699

Chou ML, Fitzpatrick LM, Tu SL, Budziszewski G, Potter-Lewis S, Akita M, Levin JZ, Keegstra K, Li HM (2003) Tic40, a membrane-anchored co-chaperone homolog in the chloroplast protein translocon. EMBO J 22(12):2970–2980. https://doi.org/10.1093/emboj/cdg281. PMID: 12805212; PMCID: PMC162133

Chua NH, Schmidt GW (1978) In vitro synthesis, transport, and assembly of ribulose 1,5-bisphosphate carboxylase subunits. Basic Life Sci 11:325–347. https://doi.org/10.1007/978-1-4684-8106-8_20

Cline K, Ettinger WF, Theg SM (1992) Protein-specific energy requirements for protein transport across or into thylakoid membranes. Two lumenal proteins are transported in the absence of ATP. J Biol Chem 267(4):2688–2696

Constan D, Patel R, Keegstra K, Jarvis P (2004) An outer envelope membrane component of the plastid protein import apparatus plays an essential role in Arabidopsis. Plant J 38(1):93–106. https://doi.org/10.1111/j.1365-313X.2004.02024.x

Charles, Andrès Birgit, Agne Felix, Kessler (2010) The TOC complex: Preprotein gateway to the chloroplast Biochimica et Biophysica Acta (BBA) - Molecular Cell Research 1803(6) 715-723 10.1016/j.bbamcr.2010.03.004

Day PM, Theg SM (2018) Evolution of protein transport to the chloroplast envelope membranes. Photosynth Res 138(3):315–326. https://doi.org/10.1007/s11120-018-0540-x. Epub 2018 Oct 5

Deng XW, Quail PH (1999) Signalling in light-controlled development. Semin Cell Dev Biol 10(2):121–129. https://doi.org/10.1006/scdb.1999.0287

Dhanoa PK, Richardson LG, Smith MD, Gidda SK, Henderson MP, Andrews DW, Mullen RT (2010) Distinct pathways mediate the sorting of tail-anchored proteins to the plastid outer envelope. PLoS One 5(4):e10098. https://doi.org/10.1371/journal.pone.0010098. PMID: 20418952; PMCID: PMC2854689

Drescher A, Ruf S, Calsa T Jr, Carrer H, Bock R (2000) The two largest chloroplast genome-encoded open reading frames of higher plants are essential genes. Plant J 22(2):97–104. https://doi.org/10.1046/j.1365-313x.2000.00722.x

Dutta S, Teresinski HJ, Smith MD (2014) A split-ubiquitin yeast two-hybrid screen to examine the substrate specificity of atToc159 and atToc132, two Arabidopsis chloroplast preprotein import receptors. PLoS One 9(4):e95026. https://doi.org/10.1371/journal.pone.0095026. PMID: 24736607; PMCID: PMC3988174

Ertel F, Mirus O, Bredemeier R, Moslavac S, Becker T, Schleiff E (2005) The evolutionarily related beta-barrel polypeptide transporters from Pisum sativum and Nostoc PCC7120 contain two distinct functional domains. J Biol Chem 280(31):28281–28289. https://doi.org/10.1074/jbc.M503035200. Epub 2005 Jun 10

Fankhauser C, Chory J (1997) Light control of plant development. Annu Rev Cell Dev Biol 13:203–229. https://doi.org/10.1146/annurev.cellbio.13.1.203

Flores-Pérez Ú, Bédard J, Tanabe N, Lymperopoulos P, Clarke AK, Jarvis P (2016) Functional analysis of the Hsp93/ClpC chaperone at the chloroplast envelope. Plant Physiol 170(1):147–162. https://doi.org/10.1104/pp.15.01538. Epub 2015 Nov 19. PMID: 26586836; PMCID: PMC4704595

Flores-Pérez Ú, Jarvis P (2013) Molecular chaperone involvement in chloroplast protein import. Biochim Biophys Acta 1833(2):332–340. https://doi.org/10.1016/j.bbamcr.2012.03.019. Epub 2012 Apr 12

Friedman AL, Keegstra K (1989) Chloroplast protein import: quantitative analysis of precursor binding. Plant Physiol 89(3):993–999. https://doi.org/10.1104/pp.89.3.993. PMID: 16666654; PMCID: PMC1055956

Gutensohn M, Fan E, Frielingsdorf S, Hanner P, Hou B, Hust B, Klösgen RB (2006) Toc, Tic, Tat et al.: structure and function of protein transport machineries in chloroplasts. J Plant Physiol 163(3):333–347. https://doi.org/10.1016/j.jplph.2005.11.009. Epub 2005 Dec 28

Gutensohn M, Schulz B, Nicolay P, Flügge UI (2000) Functional analysis of the two Arabidopsis homologues of Toc34, a component of the chloroplast protein import apparatus. Plant J 23(6):771–783. https://doi.org/10.1046/j.1365-313x.2000.00849.x

Halpin C, Elderfield PD, James HE, Zimmermann R, Dunbar B, Robinson C (1989) The reaction specificities of the thylakoidal processing peptidase and Escherichia coli leader peptidase are identical. EMBO J 8(12):3917–3921. https://doi.org/10.1002/j.1460-2075.1989.tb08572.x. PMID: 2684650; PMCID: PMC402083

Highfield PE, Ellis RJ (1978) Synthesis and transport of the small subunit of chloroplast ribulose bisphosphate carboxylase. Nature 271:420–424

Hiltbrunner A, Bauer J, Vidi PA, Infanger S, Weibel P, Hohwy M, Kessler F (2001) Targeting of an abundant cytosolic form of the protein import receptor at Toc159 to the outer chloroplast membrane. J Cell Biol 154(2):309–316. https://doi.org/10.1083/jcb.200104022. PMID: 11470820; PMCID: PMC2150772

Hinnah SC, Hill K, Wagner R, Schlicher T, Soll J (1997) Reconstitution of a chloroplast protein import channel. EMBO J 16(24):7351–7360. https://doi.org/10.1093/emboj/16.24.7351. PMID: 9405364; PMCID: PMC1170335

Hinnah SC, Wagner R, Sveshnikova N, Harrer R, Soll J (2002) The chloroplast protein import channel Toc75: pore properties and interaction with transit peptides. Biophys J 83(2):899–911. https://doi.org/10.1016/S0006-3495(02)75216-8. PMID: 12124272; PMCID: PMC1302194

Hirabayashi Y, Kikuchi S, Oishi M, Nakai M (2011) In vivo studies on the roles of two closely related Arabidopsis Tic20 proteins, AtTic20-I and AtTic20-IV. Plant Cell Physiol 52(3):469–478. https://doi.org/10.1093/pcp/pcr010. Epub 2011 Jan 25

Hirsch S, Muckel E, Heemeyer F, von Heijne G, Soll J (1994) A receptor component of the chloroplast protein translocation machinery. Science 266(5193):1989–1992. https://doi.org/10.1126/science.7801125

Hofmann NR, Theg SM (2005) Protein- and energy-mediated targeting of chloroplast outer envelope membrane proteins. Plant J 44(6):917–927. https://doi.org/10.1111/j.1365-313X.2005.02571.x

Hörmann F, Küchler M, Sveshnikov D, Oppermann U, Li Y, Soll J (2004) Tic32, an essential component in chloroplast biogenesis. J Biol Chem 279(33):34756–34762. https://doi.org/10.1074/jbc.M402817200. Epub 2004 Jun 5

Huang PK, Chan PT, Su PH, Chen LJ, Li HM (2016) Chloroplast Hsp93 directly binds to transit peptides at an early stage of the preprotein import process. Plant Physiol 170(2):857–866. https://doi.org/10.1104/pp.15.01830. Epub 2015 Dec 16. PMID: 26676256; PMCID: PMC4734592

Inaba T, Li M, Alvarez-Huerta M, Kessler F, Schnell DJ (2003) atTic110 functions as a scaffold for coordinating the stromal events of protein import into chloroplasts. J Biol Chem 278(40):38617–38627. https://doi.org/10.1074/jbc.M306367200. Epub 2003 Jul 21

Infanger S, Bischof S, Hiltbrunner A, Agne B, Baginsky S, Kessler F (2011) The chloroplast import receptor Toc90 partially restores the accumulation of Toc159 client proteins in the Arabidopsis thaliana ppi2 mutant. Mol Plant 4(2):252–263. https://doi.org/10.1093/mp/ssq071. Epub 2011 Jan 10

Inoue H, Rounds C, Schnell DJ (2010) The molecular basis for distinct pathways for protein import into Arabidopsis chloroplasts. Plant Cell 22(6):1947–1960. https://doi.org/10.1105/tpc.110.074328. Epub 2010 Jun 18. PMID: 20562235; PMCID: PMC2910967

Inoue K, Potter D (2004) The chloroplastic protein translocation channel Toc75 and its paralog OEP80 represent two distinct protein families and are targeted to the chloroplastic outer envelope by different mechanisms. Plant J 39(3):354–365. https://doi.org/10.1111/j.1365-313X.2004.02135.x

Ivanova Y, Smith MD, Chen K, Schnell DJ (2004) Members of the Toc159 import receptor family represent distinct pathways for protein targeting to plastids. Mol Biol Cell 15(7):3379–3392. https://doi.org/10.1091/mbc.e03-12-0923. Epub 2004 Apr 16. PMID: 15090618; PMCID: PMC452591

Jackson-Constan D, Akita M, Keegstra K (2001) Molecular chaperones involved in chloroplast protein import. Biochim Biophys Acta 1541(1–2):102–113. https://doi.org/10.1016/s0167-4889(01)00148-3

Jackson-Constan D, Keegstra K (2001) Arabidopsis genes encoding components of the chloroplastic protein import apparatus. Plant Physiol 125(4):1567–1576. https://doi.org/10.1104/pp.125.4.1567. PMID: 11299338; PMCID: PMC88814

Jarvis P, Chen LJ, Li H, Peto CA, Fankhauser C, Chory J (1998) An Arabidopsis mutant defective in the plastid general protein import apparatus. Science 282(5386):100–103. https://doi.org/10.1126/science.282.5386.100

Jarvis P, Robinson C (2004) Mechanisms of protein import and routing in chloroplasts. Curr Biol 14(24):R1064–R1077. https://doi.org/10.1016/j.cub.2004.11.049

Jelic M, Soll J, Schleiff E (2003) Two Toc34 homologues with different properties. Biochemistry 42(19):5906–5916. https://doi.org/10.1021/bi034001q

Jelic M, Sveshnikova N, Motzkus M, Hörth P, Soll J, Schleiff E (2002) The chloroplast import receptor Toc34 functions as preprotein-regulated GTPase. Biol Chem 383(12):1875–1883. https://doi.org/10.1515/BC.2002.211

Jensen PE, Leister D (2014) Chloroplast evolution, structure and functions. F1000Prime Rep 6:40. https://doi.org/10.12703/P6-40. PMID: 24991417; PMCID: PMC4075315

Jin Z, Wan L, Zhang Y, Li X, Cao Y, Liu H, Fan S, Cao D, Wang Z, Li X, Pan J, Dong MQ, Wu J, Yan Z (2022) Structure of a TOC-TIC supercomplex spanning two chloroplast envelope membranes. Cell 185(25):4788–4800.e13. https://doi.org/10.1016/j.cell.2022.10.030. Epub 2022 Nov 21

Kasmati AR, Töpel M, Patel R, Murtaza G, Jarvis P (2011) Molecular and genetic analyses of Tic20 homologues in Arabidopsis thaliana chloroplasts. Plant J 66(5):877–889. https://doi.org/10.1111/j.1365-313X.2011.04551.x. Epub 2011 Apr 4

Kessler F, Blobel G, Patel HA, Schnell DJ (1994) Identification of two GTP-binding proteins in the chloroplast protein import machinery. Science 266(5187):1035–1039. https://doi.org/10.1126/science.7973656

Kessler F, Schnell DJ (2004) Chloroplast protein import: solve the GTPase riddle for entry. Trends Cell Biol 14(7):334–338. https://doi.org/10.1016/j.tcb.2004.05.004

Kikuchi S, Bédard J, Hirano M, Hirabayashi Y, Oishi M, Imai M, Takase M, Ide T, Nakai M (2013) Uncovering the protein translocon at the chloroplast inner envelope membrane. Science 339(6119):571–574. https://doi.org/10.1126/science.1229262

Kim C, Apel K (2004) Substrate-dependent and organ-specific chloroplast protein import in planta. Plant Cell 16(1):88–98. https://doi.org/10.1105/tpc.015008. Epub 2003 Dec 19. PMID: 14688290; PMCID: PMC301397

Kim DH, Hwang I (2013) Direct targeting of proteins from the cytosol to organelles: the ER versus endosymbiotic organelles. Traffic 14(6):613–621. https://doi.org/10.1111/tra.12043. Epub 2013 Feb 6

Kim DB, Na C, Hwang I, Lee DW (2023) Understanding protein translocation across chloroplast membranes: Translocons and motor proteins. J Integr Plant Biol 65(2):408–416. https://doi.org/10.1111/jipb.13385. Epub 2022 Dec 31

Kim SJ, Robinson C, Mant A (1998) Sec/SRP-independent insertion of two thylakoid membrane proteins bearing cleavable signal peptides. FEBS Lett 424(1–2):105–108. https://doi.org/10.1016/s0014-5793(98)00148-3

Kouranov A, Schnell DJ (1997) Analysis of the interactions of preproteins with the import machinery over the course of protein import into chloroplasts. J Cell Biol 139(7):1677–1685. https://doi.org/10.1083/jcb.139.7.1677. PMID: 9412463; PMCID: PMC2132644

Kovacheva S, Bédard J, Patel R, Dudley P, Twell D, Ríos G, Koncz C, Jarvis P (2005) In vivo studies on the roles of Tic110, Tic40 and Hsp93 during chloroplast protein import. Plant J 41(3):412–428. https://doi.org/10.1111/j.1365-313X.2004.02307.x

Kovacheva S, Bédard J, Wardle A, Patel R, Jarvis P (2007) Further in vivo studies on the role of the molecular chaperone, Hsp93, in plastid protein import. Plant J 50(2):364–379. https://doi.org/10.1111/j.1365-313X.2007.03060.x. Epub 2007 Mar 21

Kovács-Bogdán E, Benz JP, Soll J, Bölter B (2011) Tic20 forms a channel independent of Tic110 in chloroplasts. BMC Plant Biol 11:133. https://doi.org/10.1186/1471-2229-11-133. PMID: 21961525; PMCID: PMC3203047

Kubis S, Patel R, Combe J, Bédard J, Kovacheva S, Lilley K, Biehl A, Leister D, Ríos G, Koncz C, Jarvis P (2004) Functional specialization amongst the Arabidopsis Toc159 family of chloroplast protein import receptors. Plant Cell 16(8):2059–2077. https://doi.org/10.1105/tpc.104.023309. Epub 2004 Jul 23. PMID: 15273297; PMCID: PMC519198

Küchler M, Decker S, Hörmann F, Soll J, Heins L (2002) Protein import into chloroplasts involves redox-regulated proteins. EMBO J 21(22):6136–6145. https://doi.org/10.1093/emboj/cdf621. PMID: 12426385; PMCID: PMC137210

Lamberti G, Gügel IL, Meurer J, Soll J, Schwenkert S (2011) The cytosolic kinases STY8, STY17, and STY46 are involved in chloroplast differentiation in Arabidopsis. Plant Physiol 157(1):70–85. https://doi.org/10.1104/pp.111.182774. Epub 2011 Jul 28. PMID: 21799034; PMCID: PMC3165899

Lee DW, Jung C, Hwang I (2013) Cytosolic events involved in chloroplast protein targeting. Biochim Biophys Acta 1833(2):245–252. https://doi.org/10.1016/j.bbamcr.2012.03.006. Epub 2012 Mar 16

Lee DW, Kim SJ, Oh YJ, Choi B, Lee J, Hwang I (2016) Arabidopsis BAG1 functions as a cofactor in Hsc70-mediated proteasomal degradation of Unimported plastid proteins. Mol Plant 9(10):1428–1431. https://doi.org/10.1016/j.molp.2016.06.005. Epub 2016 Jun 22

Lee DW, Yoo YJ, Razzak MA, Hwang I (2018) Prolines in transit peptides are crucial for efficient preprotein translocation into chloroplasts. Plant Physiol 176(1):663–677. https://doi.org/10.1104/pp.17.01553. Epub 2017 Nov 20. PMID: 29158328; PMCID: PMC5761803

Lee J, Wang F, Schnell DJ (2009) Toc receptor dimerization participates in the initiation of membrane translocation during protein import into chloroplasts. J Biol Chem 284(45):31130–31141. https://doi.org/10.1074/jbc.M109.053751. Epub 2009 Sep 10. PMID: 19744928; PMCID: PMC2781512

Lee J, Lee H, Kim J, Lee S, Kim DH, Kim S, Hwang I (2011) Both the hydrophobicity and a positively charged region flanking the C-terminal region of the transmembrane domain of signal-anchored proteins play critical roles in determining their targeting specificity to the endoplasmic reticulum or endosymbiotic organelles in Arabidopsis cells. Plant Cell 23(4):1588–1607. https://doi.org/10.1105/tpc.110.082230. Epub 2011 Apr 22. PMID: 21515817; PMCID: PMC3101543

Lee KH, Kim SJ, Lee YJ, Jin JB, Hwang I (2003) The M domain of atToc159 plays an essential role in the import of proteins into chloroplasts and chloroplast biogenesis. J Biol Chem 278(38):36794–36805. https://doi.org/10.1074/jbc.M304457200. Epub 2003 Jul 9

Ling Q, Jarvis P (2015) Functions of plastid protein import and the ubiquitin-proteasome system in plastid development. Biochim Biophys Acta 1847(9):939–948. https://doi.org/10.1016/j.bbabio.2015.02.017. Epub 2015 Mar 8

Ling Q, Jarvis P et al (2015) Curr Biol 25(19):2527–2534. https://doi.org/10.1016/j.cub.2015.08.015. Epub 2015 Sep 17. PMID: 26387714; PMCID: PMC4598742

Liu H, Li A, Rochaix JD, Liu Z (2023) Architecture of chloroplast TOC-TIC translocon supercomplex. Nature 615(7951):349–357. https://doi.org/10.1038/s41586-023-05744-y. Epub 2023 Jan 26

Ma Y, Kouranov A, LaSala SE, Schnell DJ (1996) Two components of the chloroplast protein import apparatus, IAP86 and IAP75, interact with the transit sequence during the recognition and translocation of precursor proteins at the outer envelope. J Cell Biol 134(2):315–327. https://doi.org/10.1083/jcb.134.2.315. PMID: 8707818; PMCID: PMC2120871

Marshall RS, Vierstra RD (2018) Autophagy: the master of bulk and selective recycling. Annu Rev Plant Biol 69:173–208. https://doi.org/10.1146/annurev-arplant-042817-040606. Epub 2018 Mar 14

May T, Soll J (2000) 14-3-3 proteins form a guidance complex with chloroplast precursor proteins in plants. Plant Cell 12(1):53–64. https://doi.org/10.1105/tpc.12.1.53. PMID: 10634907; PMCID: PMC140214

Michl D, Karnauchov I, Berghöfer J, Herrmann RG, Klösgen RB (1999) Phylogenetic transfer of organelle genes to the nucleus can lead to new mechanisms of protein integration into membranes. Plant J 17(1):31–40. https://doi.org/10.1046/j.1365-313x.1999.00348.x

Michl D, Robinson C, Shackleton JB, Herrmann RG, Klösgen RB (1994) Targeting of proteins to the thylakoids by bipartite presequences: CFoII is imported by a novel, third pathway. EMBO J 13(6):1310–1317. https://doi.org/10.1002/j.1460-2075.1994.tb06383.x. PMID: 8137815; PMCID: PMC394946

Miras S, Salvi D, Ferro M, Grunwald D, Garin J, Joyard J, Rolland N (2002) Non-canonical transit peptide for import into the chloroplast. J Biol Chem 277(49):47770–47778. https://doi.org/10.1074/jbc.M207477200. Epub 2002 Oct 3

Miras S, Salvi D, Piette L, Seigneurin-Berny D, Grunwald D, Reinbothe C, Joyard J, Reinbothe S, Rolland N (2007) Toc159- and Toc75-independent import of a transit sequence-less precursor into the inner envelope of chloroplasts. J Biol Chem 282(40):29482–29492. https://doi.org/10.1074/jbc.M611112200. Epub 2007 Jul 18

Morin XK, Soll J (1997) Immunogold labelling of cryosectioned pea chloroplasts and initial localization of the proteins associated with the protein import machinery. Planta 201:119–127. https://doi.org/10.1007/BF01007696

Motohashi R, Nagata N, Ito T, Takahashi S, Hobo T, Yoshida S, Shinozaki K (2001) An essential role of a TatC homologue of a Delta pH- dependent protein transporter in thylakoid membrane formation during chloroplast development in Arabidopsis thaliana. Proc Natl Acad Sci USA 98(18):10499–10504. https://doi.org/10.1073/pnas.181304598. PMID: 11526245; PMCID: PMC56989

Nada A, Soll J (2004) Inner envelope protein 32 is imported into chloroplasts by a novel pathway. J Cell Sci 117(Pt 17):3975–3982. https://doi.org/10.1242/jcs.01265

Nakai M, Goto A, Nohara T, Sugita D, Endo T (1994) Identification of the SecA protein homolog in pea chloroplasts and its possible involvement in thylakoidal protein transport. J Biol Chem 269(50):31338–31341

Nanjo Y, Oka H, Ikarashi N, Kaneko K, Kitajima A, Mitsui T, Muñoz FJ, Rodríguez-López M, Baroja-Fernández E, Pozueta-Romero J (2006) Rice plastidial N-glycosylated nucleotide pyrophosphatase/phosphodiesterase is transported from the ER-golgi to the chloroplast through the secretory pathway. Plant Cell 18(10):2582–2592. https://doi.org/10.1105/tpc.105.039891. Epub 2006 Oct 6. PMID: 17028208; PMCID: PMC1626603

Nassoury N, Morse D (2005) Protein targeting to the chloroplasts of photosynthetic eukaryotes: getting there is half the fun. Biochim Biophys Acta 1743(1–2):5–19. https://doi.org/10.1016/j.bbamcr.2004.09.017

Oreb M, Zoryan M, Vojta A, Maier UG, Eichacker LA, Schleiff E (2007) Phospho-mimicry mutant of atToc33 affects early development of Arabidopsis thaliana. FEBS Lett 581(30):5945–5951. https://doi.org/10.1016/j.febslet.2007.11.071. Epub 2007 Dec 3

Park MH, Zhong R, Lamppa G (2018) Chloroplast stromal processing peptidase activity is modulated by transit peptide determinants that include inhibitory roles for its N-terminal domain and initial Met. Biochem Biophys Res Commun 503(4):3149–3154. https://doi.org/10.1016/j.bbrc.2018.08.108. Epub 2018 Aug 24

Perry SE, Keegstra K (1994) Envelope membrane proteins that interact with chloroplastic precursor proteins. Plant Cell 6(1):93–105. https://doi.org/10.1105/tpc.6.1.93. PMID: 8130644; PMCID: PMC160419

Pinnaduwage P, Bruce BD (1996) In vitro interaction between a chloroplast transit peptide and chloroplast outer envelope lipids is sequence-specific and lipid class-dependent. J Biol Chem 271(51):32907–32915. https://doi.org/10.1074/jbc.271.51.32907

Pollmann S, Springer A, Buhr F, Lahroussi A, Samol I, Bonneville JM, Tichtinsky G, von Wettstein D, Reinbothe C, Reinbothe S (2007) A plant porphyria related to defects in plastid import of protochlorophyllide oxidoreductase A. Proc Natl Acad Sci USA 104(6):2019–2023. https://doi.org/10.1073/pnas.0610934104. Epub 2007 Jan 29. Erratum in: Proc Natl Acad Sci USA. 2010;107(12):5693. PMID: 17261815; PMCID: PMC1794315

Paul, Jarvis (2008) Targeting of nucleus-encoded proteins to chloroplasts in plants Summary New Phytologist 179(2) 257-285 10.1111/nph.2008.179.issue-2 10.1111/j.1469-8137.2008.02452.x

Qbadou S, Becker T, Mirus O, Tews I, Soll J, Schleiff E (2006) The molecular chaperone Hsp90 delivers precursor proteins to the chloroplast import receptor Toc64. EMBO J 25(9):1836–1847. https://doi.org/10.1038/sj.emboj.7601091. Epub 2006 Apr 13. PMID: 16619024; PMCID: PMC1456943

Radhamony RN, Theg SM (2006) Evidence for an ER to Golgi to chloroplast protein transport pathway. Trends Cell Biol 16(8):385–387. https://doi.org/10.1016/j.tcb.2006.06.003. Epub 2006 Jun 30

Razzak MA, Lee DW, Yoo YJ, Hwang I (2017) Evolution of rubisco complex small subunit transit peptides from algae to plants. Sci Rep 7(1):9279. https://doi.org/10.1038/s41598-017-09473-x. PMID: 28839179; PMCID: PMC5571161

Richardson LG, Jelokhani-Niaraki M, Smith MD (2009) The acidic domains of the Toc159 chloroplast preprotein receptor family are intrinsically disordered protein domains. BMC Biochem 10:35. https://doi.org/10.1186/1471-2091-10-35. PMID: 20042108; PMCID: PMC2805684

Richardson LGL, Schnell DJ (2020) Origins, function, and regulation of the TOC-TIC general protein import machinery of plastids. J Exp Bot 71(4):1226–1238. https://doi.org/10.1093/jxb/erz517. PMID: 31730153; PMCID: PMC7031061

Richardson LGL, Singhal R, Schnell DJ (2017) The integration of chloroplast protein targeting with plant developmental and stress responses. BMC Biol 15(1):118. https://doi.org/10.1186/s12915-017-0458-3. PMID: 29216893; PMCID: PMC5721484

Richly E, Leister D (2004) An improved prediction of chloroplast proteins reveals diversities and commonalities in the chloroplast proteomes of Arabidopsis and rice. Gene 329:11–16. https://doi.org/10.1016/j.gene.2004.01.008

Richter S, Lamppa GK (1998) A chloroplast processing enzyme functions as the general stromal processing peptidase. Proc Natl Acad Sci USA 95(13):7463–7468. https://doi.org/10.1073/pnas.95.13.7463. PMID: 9636172; PMCID: PMC22651

Richter S, Lamppa GK (2002) Determinants for removal and degradation of transit peptides of chloroplast precursor proteins. J Biol Chem 277(46):43888–43894. https://doi.org/10.1074/jbc.M206020200. Epub 2002 Sep 13

Richter S, Lamppa GK (1999) Stromal processing peptidase binds transit peptides and initiates their ATP-dependent turnover in chloroplasts. J Cell Biol 147(1):33–44. https://doi.org/10.1083/jcb.147.1.33. PMID: 10508853; PMCID: PMC2164977

Roy LM, Barkan A (1998) A SecY homologue is required for the elaboration of the chloroplast thylakoid membrane and for normal chloroplast gene expression. J Cell Biol 141(2):385–395. https://doi.org/10.1083/jcb.141.2.385. PMID: 9548717; PMCID: PMC2148445

Rudolf M, Machettira AB, Groß LE, Weber KL, Bolte K, Bionda T, Sommer MS, Maier UG, Weber AP, Schleiff E, Tripp J (2013) In vivo function of Tic22, a protein import component of the intermembrane space of chloroplasts. Mol Plant 6(3):817–829. https://doi.org/10.1093/mp/sss114. Epub 2012 Nov 30

R.M., Mould C., Robinson (1991) A proton gradient is required for the transport of two lumenal oxygen-evolving proteins across the thylakoid membrane Journal of Biological Chemistry 266(19) 12189-12193 10.1016/S0021-9258(18)98879-4

Sánchez-Pulido L, Devos D, Genevrois S, Vicente M, Valencia A (2003) POTRA: a conserved domain in the FtsQ family and a class of beta-barrel outer membrane proteins. Trends Biochem Sci 28(10):523–526. https://doi.org/10.1016/j.tibs.2003.08.003

Schleiff E, Klösgen RB (2001) Without a little help from 'my' friends: direct insertion of proteins into chloroplast membranes? Biochim Biophys Acta 1541(1–2):22–33. https://doi.org/10.1016/s0167-4889(01)00152-5

Schleiff E, Soll J, Küchler M, Kühlbrandt W, Harrer R (2003) Characterization of the translocon of the outer envelope of chloroplasts. J Cell Biol 160(4):541–551. https://doi.org/10.1083/jcb.200210060. PMID: 12591914; PMCID: PMC2173740

Schnell DJ, Kessler F, Blobel G (1994) Isolation of components of the chloroplast protein import machinery. Science 266(5187):1007–1012. https://doi.org/10.1126/science.7973649

Schuenemann D, Amin P, Hartmann E, Hoffman NE (1999) Chloroplast SecY is complexed to SecE and involved in the translocation of the 33-kDa but not the 23-kDa subunit of the oxygen-evolving complex. J Biol Chem 274(17):12177–12182. https://doi.org/10.1074/jbc.274.17.12177

Settles AM, Yonetani A, Baron A, Bush DR, Cline K, Martienssen R (1997) Sec-independent protein translocation by the maize Hcf106 protein. Science 278(5342):1467–1470. https://doi.org/10.1126/science.278.5342.1467

Shanmugabalaji V, Chahtane H, Accossato S, Rahire M, Gouzerh G, Lopez-Molina L, Kessler F (2018) Chloroplast biogenesis controlled by DELLA-TOC159 interaction in early plant development. Curr Biol 28(16):2616–2623.e5. https://doi.org/10.1016/j.cub.2018.06.006. Epub 2018 Aug 2

Shen G, Adam Z, Zhang H (2007) The E3 ligase AtCHIP ubiquitylates FtsH1, a component of the chloroplast FtsH protease, and affects protein degradation in chloroplasts. Plant J 52(2):309–321. https://doi.org/10.1111/j.1365-313X.2007.03239.x. Epub 2007 Aug 21

Shi LX, Theg SM (2010) A stromal heat shock protein 70 system functions in protein import into chloroplasts in the moss Physcomitrella patens. Plant Cell 22(1):205–220. https://doi.org/10.1105/tpc.109.071464. Epub 2010 Jan 8. PMID: 20061551; PMCID: PMC2828695

Sjuts I, Soll J, Bölter B (2017) Import of soluble proteins into chloroplasts and potential regulatory mechanisms. Front Plant Sci 8:168. https://doi.org/10.3389/fpls.2017.00168. PMID: 28228773; PMCID: PMC5296341

Smith MD, Hiltbrunner A, Kessler F, Schnell DJ (2002) The targeting of the atToc159 preprotein receptor to the chloroplast outer membrane is mediated by its GTPase domain and is regulated by GTP. J Cell Biol 159(5):833–843. https://doi.org/10.1083/jcb.200208017. Epub 2002 Dec 9. PMID: 12473690; PMCID: PMC2173378

Smith MD, Rounds CM, Wang F, Chen K, Afitlhile M, Schnell DJ (2004) atToc159 is a selective transit peptide receptor for the import of nucleus-encoded chloroplast proteins. J Cell Biol 165(3):323–334. https://doi.org/10.1083/jcb.200311074. PMID: 15138290; PMCID: PMC2172197

Sohrt K, Soll J (2000) Toc64, a new component of the protein translocon of chloroplasts. J Cell Biol 148(6):1213–1221. https://doi.org/10.1083/jcb.148.6.1213. PMID: 10725334; PMCID: PMC2174300

Soll J, Schleiff E (2004) Protein import into chloroplasts. Nat Rev Mol Cell Biol 5(3):198–208. https://doi.org/10.1038/nrm1333

Sommer M, Rudolf M, Tillmann B, Tripp J, Sommer MS, Schleiff E (2013) Toc33 and Toc64-III cooperate in precursor protein import into the chloroplasts of Arabidopsis thaliana. Plant Cell Environ 36(5):970–983. https://doi.org/10.1111/pce.12030. Epub 2013 Jan 7

Stengel A, Benz JP, Buchanan BB, Soll J, Bölter B (2009) Preprotein import into chloroplasts via the toc and tic complexes is regulated by redox signals in Pisum sativum. Mol Plant 2(6):1181–1197. https://doi.org/10.1093/mp/ssp043. Epub 2009 Jul 6

Stengel A, Benz P, Balsera M, Soll J, Bölter B (2008) TIC62 redox-regulated translocon composition and dynamics. J Biol Chem 283(11):6656–6667. https://doi.org/10.1074/jbc.M706719200. Epub 2008 Jan 7

Su PH, Li HM (2010) Stromal Hsp70 is important for protein translocation into pea and Arabidopsis chloroplasts. Plant Cell 22(5):1516–1531. https://doi.org/10.1105/tpc.109.071415. Epub 2010 May 18. PMID: 20484004; PMCID: PMC2899880

Sun YJ, Forouhar F, Li Hm HM, Tu SL, Yeh YH, Kao S, Shr HL, Chou CC, Chen C, Hsiao CD (2002) Crystal structure of pea Toc34, a novel GTPase of the chloroplast protein translocon. Nat Struct Biol 9(2):95–100. https://doi.org/10.1038/nsb744

Teng YS, Chan PT, Li HM (2012) Differential age-dependent import regulation by signal peptides. PLoS Biol 10(10):e1001416. https://doi.org/10.1371/journal.pbio.1001416. Epub 2012 Oct 30. PMID: 23118617; PMCID: PMC3484058

Teng YS, Su YS, Chen LJ, Lee YJ, Hwang I, Li HM (2006) Tic21 is an essential translocon component for protein translocation across the chloroplast inner envelope membrane. Plant Cell 18(9):2247–2257. https://doi.org/10.1105/tpc.106.044305. Epub 2006 Aug 4. PMID: 16891400; PMCID: PMC1560916

Theg SM, Bauerle C, Olsen LJ, Selman BR, Keegstra K (1989) Internal ATP is the only energy requirement for the translocation of precursor proteins across chloroplastic membranes. J Biol Chem 264(12):6730–6736

Thompson SJ, Robinson C, Mant A (1999) Dual signal peptides mediate the signal recognition particle/sec-independent insertion of a thylakoid membrane polyprotein. PsbY J Biol Chem 274(7):4059–4066. https://doi.org/10.1074/jbc.274.7.4059

van den Wijngaard PW, Vredenberg WJ (1999) The envelope anion channel involved in chloroplast protein import is associated with Tic110. J Biol Chem 274(36):25201–25204. https://doi.org/10.1074/jbc.274.36.25201

Vierstra RD (2009) The ubiquitin-26S proteasome system at the nexus of plant biology. Nat Rev Mol Cell Biol 10(6):385–397. https://doi.org/10.1038/nrm2688. Epub 2009 May 8

Villarejo A, Burén S, Larsson S, Déjardin A, Monné M, Rudhe C, Karlsson J, Jansson S, Lerouge P, Rolland N, von Heijne G, Grebe M, Bako L, Samuelsson G (2005) Evidence for a protein transported through the secretory pathway en route to the higher plant chloroplast. Nat Cell Biol 7(12):1224–1231. https://doi.org/10.1038/ncb1330. Epub 2005 Nov 13. Erratum in: Nat Cell Biol. 2006 Jan;8(1):100

Walker MB, Roy LM, Coleman E, Voelker R, Barkan A (1999) The maize tha4 gene functions in sec-independent protein transport in chloroplasts and is related to hcf106, tatA, and tatB. J Cell Biol 147(2):267–276. https://doi.org/10.1083/jcb.147.2.267. PMID: 10525534; PMCID: PMC2174219

Wan C, Zhang H, Cheng H, Sowden RG, Cai W, Jarvis RP, Ling Q (2023) Selective autophagy regulates chloroplast protein import and promotes plant stress tolerance. EMBO J 42(14):e112534. https://doi.org/10.15252/embj.2022112534. Epub 2023 May 30. PMID: 37248861; PMCID: PMC10350842

Wan J, Bringloe D, Lamppa GK (1998) Disruption of chloroplast biogenesis and plant development upon down-regulation of a chloroplast processing enzyme involved in the import pathway. Plant J 15:459

Yeh YH, Kesavulu MM, Li HM, Wu SZ, Sun YJ, Konozy EH, Hsiao CD (2007) Dimerization is important for the GTPase activity of chloroplast translocon components atToc33 and psToc159. J Biol Chem 282(18):13845–13853. https://doi.org/10.1074/jbc.M608385200. Epub 2007 Mar 1

Young ME, Keegstra K, Froehlich JE (1999) GTP promotes the formation of early-import intermediates but is not required during the translocation step of protein import into chloroplasts. Plant Physiol 121(1):237–244. https://doi.org/10.1104/pp.121.1.237. PMID: 10482679; PMCID: PMC59372

Yamuna D, Paila Lynn GL, Richardson Hitoshi, Inoue Elizabeth S, Parks James, McMahon Kentaro, Inoue Danny J, Schnell (2016) Multi-functional roles for the polypeptide transport associated domains of Toc75 in chloroplast protein import eLife 510.7554/eLife.12631

Zhao C, Wang Y, Chan KX, Marchant DB, Franks PJ, Randall D, Tee EE, Chen G, Ramesh S, Phua SY, Zhang B, Hills A, Dai F, Xue D, Gilliham M, Tyerman S, Nevo E, Wu F, Zhang G, Wong GK, Leebens-Mack JH, Melkonian M, Blatt MR, Soltis PS, Soltis DE, Pogson BJ, Chen ZH (2019) Evolution of chloroplast retrograde signaling facilitates green plant adaptation to land. Proc Natl Acad Sci USA 116(11):5015–5020. https://doi.org/10.1073/pnas.1812092116. Epub 2019 Feb 25. PMID: 30804180; PMCID: PMC6421419

Zhao X, Higa T, Nakai M (2022) Tic12, a 12-kDa essential component of the translocon at the inner envelope membrane of chloroplasts in Arabidopsis. Plant Cell 34(11):4569–4582. https://doi.org/10.1093/plcell/koac240. PMID: 35929083; PMCID: PMC9614449

Zhong R, Wan J, Jin R, Lamppa G (2003) A pea antisense gene for the chloroplast stromal processing peptidase yields seedling lethals in Arabidopsis: survivors show defective GFP import in vivo. Plant J 34(6):802–812. https://doi.org/10.1046/j.1365-313x.2003.01772.x

Ziehe D, Dünschede B, Schünemann D (2018) Molecular mechanism of SRP-dependent light-harvesting protein transport to the thylakoid membrane in plants. Photosynth Res 138(3):303–313. https://doi.org/10.1007/s11120-018-0544-6. Epub 2018 Jun 28. PMID: 29956039; PMCID: PMC6244792

Thylakoids Membrane Organization: Its Assembly and Disintegration

5

Joanna Grzyb and Beata Myśliwa-Kurdziel

Abstract

Thylakoid membranes are indispensable parts of chloroplasts and other cyanobacterial cells. They are a separate membrane system, manifesting as closed vesicles (sacks), separated from the chloroplast envelope or cyanobacterial plasma membrane. They host photosynthetic processes and provide a unique environment for their optimal performance. The first important advantage of thylakoid's presence is high photosynthetically active area, packed in relatively small space of chloroplasts. Thylakoids composition, concerning lipids, proteins, and pigments makes them a unique type of membrane, with high structural and functional conservation upon evolution. In this Chapter, we describe processes that lead to the formation of thylakoid architecture in both chloroplasts (especially of angiosperm) and cyanobacterial cells. We also discuss how the architecture is maintained and tuned upon varying environmental conditions, and how the thylakoid disassembly, which occurs during senescence or transformation to another type of plastid, is governed.

Keywords

Thylakoids · Biogenesis · Plasticity · Assembly · Disassembly · Aging · Maturation

J. Grzyb (✉)
Department of Biophysics, Faculty of Biotechnology, University of Wroclaw, Wroclaw, Poland
e-mail: joanna.grzyb@uwr.edu.pl

B. Myśliwa-Kurdziel
Department of Plant Physiology and Biochemistry, Faculty of Biochemistry, Biophysics and Biotechnology, Jagiellonian University, Krakow, Poland
e-mail: b.mysliwa-kurdziel@uj.edu.pl

V. K. Dalal, A. N. Misra (eds.), *Chloroplast Biogenesis and Plastid Interconversions*,
https://doi.org/10.1007/978-981-95-0772-6_5

5.1 Introduction

Very recently, we have been asked for a definition of a chloroplast. It came out, that despite of obvious sentence, that the chloroplast is the semi-autonomous organelle, hosting all photosynthetic processes, the definition should be extended to "containing thylakoid membranes." Simply, the thylakoids are the system of internal membranes of chloroplasts, divergent from the outer and inner chloroplast membrane envelope, built as closed vesicles. They are described usually as flattened sacks, with photosynthetic pigment-protein complexes embedded into membranes, containing photosynthetically active interior, lumen (Farci and Schröder 2023), and surrounded by stroma, which is also important for photosynthesis. The conversion of solar radiation energy into chemical energy stored in ATP and NADPH molecules takes place in thylakoid membranes, in the so-called light reactions. The products of these light reactions, ATP and NADPH, are subsequently consumed for carbon assimilation and other energy-requiring biochemical reactions in the so-called dark reactions.

Chloroplasts, as defined above, are present in the green tissue of higher plants, mosses, ferns, and algae. Cyanobacteria, which are also photosynthetically active, do not contain chloroplasts but do also have membrane systems known as thylakoids. These membranes are of homological structure and function to higher plants thylakoids, although in detail they differ. In other photosynthetic bacteria, the thylakoids are not distinguished, and the photosynthetic complexes are present in the plasma membrane. However, in certain conditions, the plasma membrane of purple photosynthetic bacteria may form invaginations, known as intracytoplasmic membranes (ICM). These membrane systems can be highly developed, and in transmission microscopy resemble thylakoids of cyanobacteria (Varga and Staehelin 1983; Adams and Hunter 2012). ICM might be densely packed inside a cell and as thylakoids organized in stacked (grana) and non-stacked (lamellae) regions (Varga and Staehelin 1983), with dedicated proteins involved in their organization (see, e.g., the role of mitochondrial crista-developing protein Mic60) (Muñoz-Gómez et al. 2023). Upon bacterial cell disruption, ICM can be separated and isolated as chromatophores. The ICM structure and composition might be altered in response to expression of e.g. light harvesting complexes (Satoh et al. 2023), what may resemble changes comparable to induced by varying environmental conditions (Howley et al. 2023). In green sulfur bacteria, however, there is no invagination of the plasma membrane, although photosynthetic antennae (chlorosomes) may resemble vesicles located in a connection to the plasma membrane (Matsubara and Tamiaki 2020). The photosystems with reaction centers, however, are still present in a plasma membrane, not in invaginations or vesicles (Orf and Blankenship 2013; Chen et al. 2020). Interestingly, membrane invagination of bacterial cells can serve not only photosynthesis, but also nitrogen assimiliation. The example here is the annamnosome membrane system of the bacteria *Planctomycetes*, where the ammonia monooxygenase enzyme is present (Van Niftrik et al. 2004).

It is widely accepted, from the very beginning of the study on the subject, that thylakoid membranes evolved for optimization of photosynthetic performance

(Wolfe and Hoober 1996); it is realized by a physical increase of photosynthetically active membrane area, what is true also for ICM. What is more, the presence of thylakoids allows the inhomogeneous location of photosystem I (PSI) and photosystem II (PSII), which is believed to be a mechanism for compensation of relatively slower trapping kinetics of PSII. The early studies of thylakoids arrangement recognized already the presence of flattened vesicles; however, their arrangement was discussed for a long time; early models assumed thylakoids arranged like a role of coins and emerged by invaginations of inner chloroplast envelope (Mechela et al. 2019). With great improvement in microscopic methods, especially modern electron tomography (Engel et al. 2015; Mai et al. 2019; Shimoni et al. 2005), the currently accepted model of grana stacks with helically arranged interconnecting stromal lamellae, was elaborated (Mustárdy and Garab 2003).

The 3D architecture of thylakoid membranes is an outcome of membrane composition, including lipids (Demé et al. 2014; Mazur et al. 2019), proteins, and photosynthetic pigments. What is more, thylakoid structure, once obtained, is not valid for all the conditions; it needs to be precisely controlled and adapted to varying abiotic and biotic conditions, especially light intensity and temperature. On the other hand, in seed cotyledons, a non-photosynthetic stage of plant development, there are no fully developed chloroplasts. This might be explained as a consequence of endosymbiothic origin of a chloroplast. As most of primary endosymbiont genes were transferred to the nucleus, their expression is precisely regulated in connection to whole cell/organism performance. Therefore, chloroplast nucleus-encoded proteins expression is light triggered. Presence of such mechanisms might be schematically explained using economy terms—keeping such a complicated organelle, as chloroplast, is costly, as it needs several proteins, lipids, and pigments. When there is no light, and chloroplast are cannot capture light energy, there is no reason to keep chloroplasts in their fully developed form. Light triggers selected pathways, leading then to the thylakoids assembly or rebuilt.

The lipid composition of the thylakoid membrane, in chloroplasts and cyanobacteria, is similar to each other and different form the composition of the chloroplast envelope or cyanobacterial plasma membrane (Wada and Murata 2007). The main components are galactolipids, mono galactosyl diacyl glycerol (MGDG) and digalactosy diacyl glycerol (DGDG), supplemented by phosphatidyl glycerol (PG) and sulfoquinovosyl diacyl glycerol (SQDG). The fatty acid chains of lipids are mostly 18- and 16-carbon in length, with the possibility of 1–3 unsaturated bonds (Zhu et al. 2008). The lipid composition of thylakoids is inhomogenous laterally, with a microdomain present, as well as vertically, with a different outer and inner leaflet of the bilayer (Dlouhý et al. 2021; Garab et al. 2017, 2022). When isolated, they can be reconstituted in a bilayer even without proteins presence. The most convenient experimental model of thylakoids is a suspension of liposomes, as this may include not only a membrane but also a representation of the thylakoid interior. Formation of a membrane from a lipid suspension is a spontaneous process, thermodynamically favored, with a driving force being mostly hydrophobic effect—the overall entropy increases due to the reorganization of water molecules. This is true for any type of amphipathic lipids, including those forming thylakoids (Garidel et al. 2015;

Marsh 2012). Charged PG and SQDG may contribute to additional electrostatic interactions (Demé et al. 2014). Thylakoid galactolipids may form stable monolayers, and therefore, their properties can be assessed by the Langmuir-Blodget technique. MGDG, DGDG, PG, and SQDG were shown to have different compression isotherms, which reflects their structure, namely, packing parameter and headgroup area (Perczyk and Broniatowski 2022). Due to its packing parameter above 1 (namely, small polar headgroup), MGDG alone does not form a stable bilayer and is not enough to create liposomes; instead, it forms reversed micelles, known also as reversed hexagonal structure or HII phase. The first hypothesis proposed that grana margins are enriched in MGDG to stabilize curvature (Murphy 1982; Seiwert et al. 2018); however, the actual testing showed of the fragmented thylakoids lipid composition showed that this is not the case (Dlouhý et al. 2021; Duchêne and Siegenthaler 2000). The role of MGDG is still under debate (Garab et al. 2016; Iwai et al. 2021; Lee 2000; Rast et al. 2015; Yu et al. 2020); it is assumed currently that it facilitates membrane curvature, thylakoid stacking, and general membrane fluidity, allowing protein complexes to rearrange laterally. Tietz et al. (2020) showed, for example, that MGDG controls light harvesting by LHCII, modulating lateral membrane pressure, sensed by peripheral LHCII pigments. The cited work is one of many research, done on model membranes with incorporated pigment-protein complexes, isolated form native thylakoid membranes. A lot of studies with such proteoliposomes model of thylakoids include light-harvesting complex II (LHCII) (Wilson et al. 2022); however, there are also proteoliposomes with reconstituted PSI (Niroomand et al. 2017) and PSII (Joly et al. 2011). The proteoliposome model was used to recreate grana stacks upon the action of magnesium ions (Rumak et al. 2010). The thylakoid membranes can be reconstituted in vivo in other systems. An example is the reconstitution of spinach thylakoids in a 2D membrane in a patterned scaffold of polymeric bilayer. There, membrane complexes preserved their lateral mobility. In such a reconstituted system, both PSI and PSII were found to be functional, with an electron flow between them and NADPH reduction activity (Yoneda et al. 2020).

The lipid composition of thylakoid membranes changes in response to environmental conditions, enabling a change in membrane fluidity (Yu et al. 2021). For example, the MGDG to DGDG ratio may decrease form 3:1 to 2:1 for Arabidopsis plants grown under low and high light, respectively. The fine-tuning of lipid composition was proposed to facilitate, for example, the reorganization of membranal supercomplexes, stabilizing chlorophyll dimer (Saini and Debnath 2023), as well as repair (e.g. D1 turnover). Adaptation to heat (Abdelrahman et al. 2020; Kulke et al. 2023) or chill/cold (Liu et al. 2018; Mazur et al. 2020) conditions are also reflected in lipid composition. Changes in membrane fluidity may impact the thylakoid membrane curvature and the resulting thylakoid network (Bykowski et al. 2021).

Although the assembly of lipids into a bilayer is mostly spontaneous, the incorporation of proteins into membranes—and, more importantly, their proper folding and precise incorporation into the membrane or transfer through a thylakoid membrane into a lumen (Zhu et al. 2022)—demands the assistance of whole assisting chaperone machinery (Klasek et al. 2020; Rochaix 2011). In vitro proteoliposome experiments show that isolated photosynthetic complexes may be spontaneously incorporated into a membrane; however, it is believed to depend on the preservation of some lipid molecules with the isolated complex. That way the thermodynamics favors lipids integration with a liposome, dragging proteins to the bilayer. However, this is not true in vivo. For de novo synthesis of multiple subunit complexes, such as PSII, there are several substeps of precomplexes assembly (Rokka et al., 2005). With D1 and D2 proteins coded by chloroplastic DNA, or for cyanobacterial photosynthetic proteins coded in bacterial nucleoid, the folding of PSII core may be supported by the presence of membrane right after translation on bacterial/chloroplastic ribosomes (Lu 2016), while subunits coded by nuclear DNA and synthetized on plasmatic ribosomes must be additionally translocated by the chloroplast envelopes by TIC-TOC import system (Gao et al. 2023; Liu et al. 2023; Thomson et al. 2020). TIC-TOC complex may distinguish between chloroplastic photosynthetic proteins and non-photosynthetic proteins, what is dependent on the translocon composition (with main subunit being TOC159 or TOC132 protein, deciding on the translocated protein types). The replacement of TOC159 by its isoform TOC132, guided by chloroplast-associated protein degradation (CHLORAD) complex (Li and Jarvis 2024), is correlated, e.g., with chloroplast transformation into chromoplast (Sun and Jarvis 2023). TIC-TOC translocation demand presence of a chloroplast targeting peptide in an N-terminal part of a preprotein. There is also minority of proteins that are translocated by non-conventional pathways. Those proteins are characterized by a different localization of a targeting peptide or even lack of recognized targeting sequence. Suborganelle targeting, which basically means protein translocation into the thylakoid membrane or into a lumen (therefore, passing through the thylakoid membrane), has its specific mechanisms that may also include specific targeting peptides. Other possibilities consist of twin-arginine translocation (TAT) system or Sec-dependent systems, with helper proteins related to the bacterial secretion system. The topic of protein translocation to the chloroplast and within chloroplast is detail described in Chap. 4 of this book.

As pointed out above, the specific assembly of thylakoids is a consequence of their structure and function, which are further outcomes of evolution and current environmental conditions. The basic divergence occurred between eukaryotes (carrying chloroplasts in their cells) and cyanobacteria. Therefore, first, we focus on eukaryotic, higher plants thylakoids and their assembly and disassembly, as they are the best studied, to come back to other thylakoid systems in the second part of the Chapter.

5.2 Membrane Structure of Higher Plant Thylakoid

Higher plant thylakoids, as isolated, are closed, spherical vesicles with a diameter of 200–300 nm, below the resolution limit of light microscopy (Raven 2021). While in the chloroplast, the vesicles are flattened and organized in a connected system of grana and stroma lamellae. Higher plant's thylakoid membranes, like all other thylakoids, host photosynthetic multicomponent transmembrane complexes, which constitute the energy conversion machinery, composed of light-harvesting antennae, two photosystems, PSI and PSII, cyt b_6f, and ATP synthase. Figure 5.1 presents the schematic organization of higher plants thylakoids.

The energy absorbed by photosynthetic pigments accumulated in the antenna complexes, mainly LHC complexes, is transferred to the PSI/PSII reaction centers, where it is converted into chemical energy. The electronically excited chlorophyll in the PSI/PSII reaction centers becomes one of the electron carriers in the photosynthetic electron transport chain. The transport of electrons in these redox reactions in the thylakoid membranes is coupled with H^+ transport across the membrane, leading to an increase in H+ concentration in the thylakoid lumen, which in turn powers ATP synthase. In the basic version of the linear electron transport chain, electrons are gained from water oxidation by protein components of the PSII complex that form an oxygen-evolving complex (OEC) and then transported through the cyt b_6f and PSI complexes to reduce $NADP^+$. Two mobile electron carriers, plastoquinone (PQ, a prenyl lipid in the lipid bilayer) and plastocyanin (PC, a protein in the thylakoid lumen), complete this chain between PSII and cyt b_6f and cyt b_6f and PSI, respectively (Fig. 5.1).

Electron and proton transport reactions and their associated protein complexes have remained essentially conserved in all organisms that perform oxygenic photosynthesis. However, the organization of thylakoids and the distribution of these complexes in membranes vary greatly among organisms. Despite these differences,

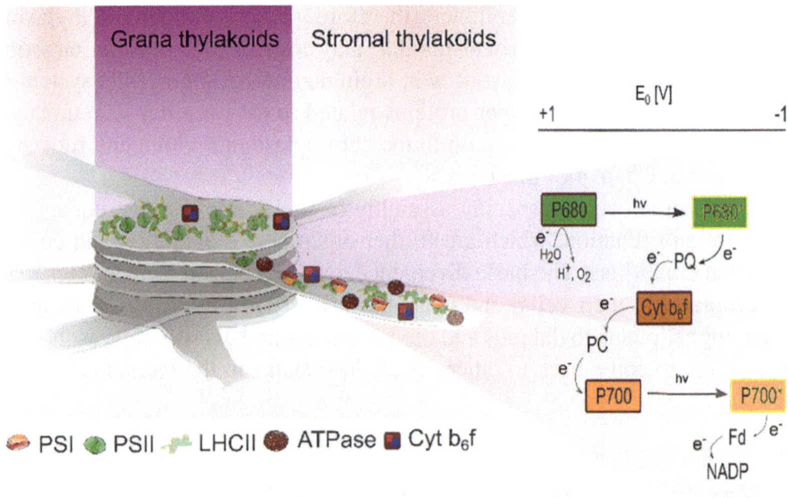

Fig. 5.1 Schematic representation of higher plants thylakoid architecture, with granal and stromal regions of thylakoids and heterogeneity of main photosynthetic complexes localization in relation to main reactions of linear photosynthetic electron transfer chain and electrochemical potential (E0). PSI photosystem I, PSII photosystem II, LHCII light-harvesting complex II, Cyt b6f cytochrome b6f, P680/P680* reaction center of PSII, P700/P700* reaction center of PSI, PQ plastoquinone, PC plastoquinone, Fd ferredoxin, hv light absorption event, e electron transfer event

the key to optimizing energy conversion in specific environmental conditions is the effective interaction between individual protein complexes involved in the absorption of radiation and the transport of protons and electrons.

As mentioned in the introduction, the thylakoid system in chloroplasts consists of appressed membrane disks, called grana, which are connected to each other by nonappressed membrane regions, known as stroma lamellae (Figs. 5.1 and 5.2).

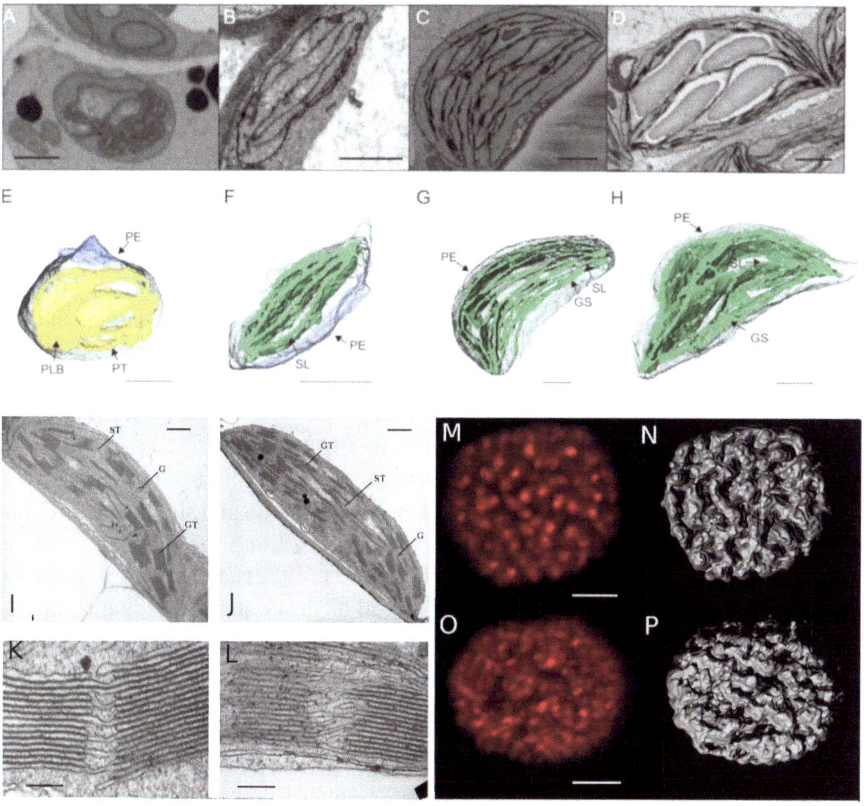

Fig. 5.2 Thylakoid development during greening in comparison to various arrangements of thylakoids in mature chloroplasts. (**a–d**) Scanning electron microscopy micrographs and partial 3D reconstruction of thylakoid membranes (green) and envelope (blue) of etioplasts and chloroplasts of Arabidopsis thaliana etiolated seedlings illuminated for 0 hr. (**a**, **e**), 4 h (**b**, **f**), 24 h (**c**, **g**), and 96 h (**d**, **h**). (**i-l**) Transmission electron microscopy micrographs of pea (**i**, **k**) and bean (**j**, **l**) mature chloroplasts, with zoom on the selected granal and integranal regions (**k**, **l**). (**m-p**) confocal scanning microscopy images of isolated, intact native chloroplasts of pea (**m**) and bean (**o**), with thylakoid membrane visualized by chlorophyll fluorescence, and 3D reconstituted model of the thylakoid network in the same chloroplasts (**n**, **p**). PLB prolamellar body, PT prothylakoid, PE plastid envelope, SG starch grain, GS grana stack, SL single lamella, ST stroma thylakoid, GT grana thylakoids, G plastoglobule. Scale bars: A–H, 1 μm, I, J, 500 nm, K, L, 100 nm, M, P –2 μm. (Adapted from (Pipitone et al. 2021) (**a–h**) (Creative Commons Attribution License) and (Rumak et al. 2010) (**i–p**) (Creative Commons CC BY license). The only changes made included a selection of original image fragments and combinations into the final one, with new subfigures labeling)

Additionally, the nonuniform distribution of photosynthetic protein complexes between grana and stroma thylakoids is characteristic features of plant thylakoid membranes (Andersson and Anderson 1980). PSII and LHCII complexes are primarily located in grana thylakoids, while most PSI complexes and all ATP synthases are found in stroma thylakoids, with cytochrome b_6f complexes distributed equally across both compartments. The dimer of the PSII core forms a supercomplex with its LHC antennae, namely, LHCII antennae (Müh and Zouni 2020; Nield et al. 2000a, b). Some of the LHCII proteins, i.e., lhcb1, lhcb2, and lhcb3, form trimers that are bound to the core with different strengths, while the lhcb4, lhcb5, and lhcb6 proteins occur as monomers (Caffarri et al. 2004, 2009; Dainese and Bassi 1991). The structures of PSII-LHCII supercomplexes have recently been revealed (Cao et al. 2018; Graça et al. 2021; Su et al. 2017; Wei et al. 2016). PSII-LHCII supercomplexes aggregate under low light conditions and form larger structures in the thylakoid membranes, described as a semicrystalline array (Kim et al. 2020; Kirchhoff et al. 2007). PSI, on the other hand, occurs as monomers surrounded by LHCI antennae on one side and has a docking site for phosphorylated LHCII trimers that are attached as a result of a light-induced state transition (Amunts et al. 2007; Pan et al. 2018). The state 1–state 2 transition is one of the mechanisms for increasing the excitation energy supply to PSI (Pan et al. 2018; Sheng et al. 2021).

5.3 Basis of Thylakoids Membrane Plasticity

Higher plant thylakoids are flexible and adapt to both subtle and significant environmental changes through various regulatory processes. Long-scale changes include modification in lipid composition, described already (Selstam 1998). It has long been thought that the lipid composition, as well as the content and organization of the transmembrane photosynthetic complexes, determines the shape and packing of the thylakoid grana. The discovery of the Curt1 family of proteins (Armbruster et al. 2013) added new factor to our understanding of thylakoid membrane shaping, as these proteins located at the edges of grana modulate their curvature and diameter and contribute to thylakoid plasticity (Pribil et al. 2018; Trotta et al. 2019). It should be remembered that all other photosynthetic complexes, and their interactions, also matter for thylakoid network formation. Table 5.1 summarizes the most important proteins, shown to influence thylakoid assembly. More detailed elaboration of those protein function can be also found in recent review by Ostermeier et al. (2024).

Thylakoid remodeling is found, for example, during a plant's adaptation to drought stress. For rice leaves under drought stress (Xu et al. 2022), thylakoid went from significantly stacked grana and well-developed stroma lamellae to much smaller, shrinked grana stacks ~~shrank~~ (although grana were still compact), and stroma lamellae cracking and curving. This happens together with a change of chloroplast shape change to a smaller oval. Meanwhile, both chlorophyll and carotenoids content were reduced, and a statistically significant change in lipids composition was found; there was a decrease of 18:1 and 18:3 lipids, while 18:0, 16:0, and 18:2 lipids content increased. Overall, the unsaturation index decreased in

Table 5.1 Proteins involved in maintaining the structure and plasticity of thylakoid membranes

	Protein	Occurrence	Specific function
Direct structural function	CURT1A	*A. thaliana*	Present in margins of PLB, required for pre-granal thylakoid assembly from PLB tubules during de-etiolation. Induce curvature of grana (Armbruster et al. 2013; Liang et al. 2022; Pribil et al. 2018)
	CURT1B		Interacts with PSII-LHCII in grana margins, regulated by phosphorylation, optimizes thylakoids under light intensity variation (Messant et al. 2021; Trotta et al. 2019)
	CURT1C		Required for proper PLB crystal-like structure formation in the dark (Liang et al. 2021), required for grana stacking in mature chloroplasts (Armbruster et al. 2013)
	CURT1D		Necessary for grana formation (Armbruster et al. 2013)
	CURT homologs	*Anabena, Nostoc*	Thylakoids formation and reorganization during cell division (Santamaria-Gomez et al. 2018)
		Synechocystis	Influence grana curvature (Heinz et al. 2016)
	AncM	*Synechocystis*	Influence grana curvature, antagonist of CURT proteins (Ostermeier et al. 2022)
	Rab protein, CPRabA5e	*A. thaliana*	Interacts with CURT1A and LHCB1.5, influence thylakoid curvature (Lindquist and Aronsson 2014)
	FZL, dynamin-like Fzo-like membrane fusion protein		Participate in the formation of interconnection of grana and lamellae during thylakoid development (Liang et al. 2018) Required for proper thylakoid morphology, including grana stacking and the proportion of grana and stroma thylakoids (Gao et al. 2006)
	CP29		Phosphorylation of CP29 by STN7 was one of the crucial features necessary for proper thylakoid membrane assembly (Poudyal et al. 2020)
	RIQ1 and RIQ2		Affect grana stacking (Yokoyama et al. 2016)
	VIPP1		Triggers membrane fusion in an Mg2 + -dependent manner (Ohnishi et al. 2018)
	CPSAR1		Essential for formation of proper thylakoids via involvement of lipid transport to chloroplasts (Garcia et al. 2010)

(continued)

Table 5.1 (continued)

	Protein	Occurrence	Specific function
Modification and indirect structural function	Albino3	*A. thaliana.*	Necessary for LHC protein incorporation into thylakoids (Spence et al. 2004)
	Albino3/oxal	*Synechocystis*	Thylakoid biogenesis (Klostermann et al. 2002)
	STN7 TAP38 - LHCII kinase and phosphatase	*A. thaliana*	Arrangement of thylakoids, grana thickness, grana stacking, and interaction between adjacent thylakoid membranes in grana (Armbruster et al. 2013; Cutolo et al. 2019; Dukic et al. 2023; Vainonen et al. 2005; Wood et al. 2019; Zhang et al. 2022)
	STN8—PBCB kinase and phosphatase of PSII core and internal PSII antennae		
	Magnesium transporters KEA3, VCCN1, MGT10		
	WATER-SOAKED SPOT1	*Rice*	Influence thylakoid assembly and disassembly via control of redox-state (Xu et al. 2022)
	CPSFL1	*Chlamydomonas reinhardti*	Binds carotenoids and promotes membrane adhesion under light (García-Cerdán et al. 2020)
	VIPP2		Supports membrane adaptation to stress conditions (Theis et al. 2020)

response to drought. Also, an increased accumulation of plastoglobules (PG) and starch granules was noted (Xu et al. 2022).

Mentioned PGs are specialized structures, connected directly to the thylakoid margins. PGs are spherical, filled with hydrophobic interiors, containing lipids and hydrophobic proteins, and enveloped by a single leaflet (monolayer) of amphipathic lipids with stabilizing proteins. PGs are several to dozen nanometers in diameter. After their discovery, PGs were considered inert trash for currently not-used lipophilic materials. Deeper studies suggest that their role is active; for example, their composition may vary between chill-resistant and chill-sensitive plants (Wójtowicz et al. 2021), and they are a place for the synthesis of lipids.

Fast reactions of thylakoid reorganization, necessary in response to unpredictable changes in the environment, particularly fluctuations in sunlight intensity, are challenging for land plants and require precise regulation of light absorption processes and the electron transport path, based on reversible changes. One of the key regulatory mechanisms is protein phosphorylation which in plant thylakoid membranes regulates energy conversion and repair processes of damaged photosynthetic machinery by restructuring the thylakoid ultrastructure and rearranging protein complexes. Second, the fast adaptive mechanism is the fastest pH-dependent qE component of non-photochemical quenching (Holt et al. 2004; Ruban 2016; Ruban and Wilson 2021). Reversible phosphorylation regulates dynamic macroscopic changes in the thylakoid network, involving their folding and changes in the size of grana and stroma thylakoids, as well as the degree of their packing, (Johnson and Wientjes 2020; Kirchhoff 2014). Low light induces grana stacking and high light induces partial grana destacking, resulting in a vertical separation of grana membrane pairs at the stromal side, lateral reduction of the grana diameter, and thylakoid swelling (Herbstová et al. 2012; Khatoon et al. 2009; Kim et al. 2020; Kirchhoff et al. 2011). A similar swelling phenomenon is observed upon heat, drought, and heavy metal stress, explaining partially the general mechanism governing thylakoid rearrangement.

Grana stacking means simply shortening the distance between two membranes of adjacent grana thylakoids. The distance might be as small as 3.5 nm. The process of grana stacking is favored by the formation of semi-crystalline protein arrays consisting of PSII–LHCII supercomplexes located in grana thylakoids (Daum et al. 2010). Interactions between stromal protein moieties of LHCIIs localized in adjacent stacked membranes are important for grana assembly (Mullet 1983). PSII-LHCII complexes form a semicrystalline array in grana thylakoids, stabilizing and shaping grana packing through electrostatic interactions among LHC proteins embedded in the adjacent grana layers (Dekker and Boekema 2005; Kouřil et al. 2012). At low light, when photon flow becomes a limiting factor for photosynthesis, these complexes are densely packed, which contributes to efficient energy transfer toward the PSII reaction center. In contrast, under high light, the array of PSII-LHCII complexes is reduced and less tightly packed (Kouřil et al. 2013). The high-light-induced disassembly of PSII-LHCII complexes facilitates efficient energy dissipation, which constitutes the initial step of the state 1–state 2 transition, one of the mechanisms for dissipating excess excitation energy, which is triggered by phosphorylation of LHCII by STN7 kinase and reversed by dephosphorylation by TAP38 phosphatase (Wood et al. 2019). The height of granum (i.e., number of stacked thylakoids) is not constant and may be an additional adaptation of the thylakoid membrane system for improvement of photosynthetic performance.

Under high light, the photoinhibition of the PSII complex occurs (Li et al. 2018), and this PSII damage is associated with the impairment of the D1 protein (Jansen et al. 1999), one of the PSII core proteins. The PSII repair cycle (Järvi et al. 2015; Nath et al. 2013; Nixon et al. 2005; J. Su et al. 2024; Theis and Schroda 2016), aimed at exchanging the degraded D1 protein with a newly synthesized copy, is a multistep process that enables efficient recovery from photooxidative PSII damage. The replacement of the D1 protein buried in the middle of the LHCII–PSII megacomplex is challenging due to the fact that the LHCII-PSII complexes are located in appressed grana thylakoid regions. The repair process begins with the phosphorylation of PSII core subunits and LHCII antennae proteins by the STN8 and STN7 kinases (Bonardi et al. 2005; Fristedt and Vener 2011; Pesaresi et al. 2011; Rochaix et al. 2012; Vainonen et al. 2005), which triggers the monomerization and detachment of LHCII from the PSII core. Subsequently, the phosphorylated PSII core migrates from stacked thylakoids to unstacked regions, where the phosphorylated proteins are dephosphorylated by PBCP phosphatase, and the impaired D1 subunit is degraded by FtsH, an ATP-dependent zinc metalloprotease (Kato et al. 2009, 2018; Nixon et al. 2005; Sakamoto et al. 2003), and by Deg proteases (Itzhaki et al. 1998; Kapri-Pardes et al. 2007; Luciński et al. 2011). A newly synthesized D1 copy, properly processed by CtpA, is inserted in the PSII core. Finally, the PSII monomers reassemble and migrate back to stacked grana for dimerization and the assembly of the supercomplex. What should be noted, PSII assembly is precisely regulated not only during D1 turnover process. The dedicated chaperone, RESISTANCE TO PHYTOPHTHORA1 (RPH1) protein, takes care of cyt b559 subunit of PSII li-ping (Che et al. 2024). RPH1 is proposed to facilitate heme insertion, and the lack of this process reduces the properly folded PSII content to about 20%. Assembly of PSI is also assisted by chaperones (Rolo et al. 2024a, b), such as CO-EXPRESSED WITH PSI ASSEMBLY1 (CEPA1) protein and its interaction factor, PHOTOSYSTEM I ASSEMBLY3 (PSA3). CEPA1 specific role might be anchoring the soluble components of PSI and/or insertion of core PSI subunits (Rolo et al. 2024a).

The mentioned earlier PSII repair process takes advantage of the subcompartmentalization into stacked grana and unstacked thylakoid regions and reversible protein phosphorylation and is closely related to light-induced structural changes in the thylakoid membrane architecture (Kirchhoff 2014). Thylakoid unstacking favors the diffusion of protein complexes and is crucial for D1 protein protection and repair (Khatoon et al. 2009). The plasticity of thylakoid membranes is ensured by CURT1 proteins that accumulate at the edges of the grana and modulate their shape, i.e., their diameter, the number of layers, and the curvature of the membranes (Armbruster et al. 2013; Pribil et al. 2018). The change in the diameter of the grana, in turn, affects the distance between the grana thylakoids and the stromal lamellae, facilitating access for proteases that degrade damaged D1 protein and enabling effective PSII repair (Kirchhoff 2013; Kirchhoff et al. 2011).

5.4 De-etiolation and Thylakoids Development

As mentioned, the production of chlorophyll and chloroplasts is costly, and when there is no light, it is more economical for plants not to produce them. However, when light is present, it is reasonable to utilize it as quickly as possible. This is the paradox of germinating seedlings and young plant shots. Moreover, improperly assembled photosynthetic complexes pose a risk of photoreactions, initiated by the excitation of unbound chlorophyll molecules. To avoid this scenario, plants strictly regulate the chlorophyll synthesis pathway and the subsequent assembly of thylakoid membranes. The regulation includes retrograde signaling (a messenger passage from chloroplast to nucleus), leading to expression of photosynthesis associated genes, leading to production of proteins necessary for thylakoids assembly mostly under light availability (Loudya et al. 2024).

In angiosperms, which are the main point of interest in greening studies, chlorophyll biosynthesis is entirely dependent on light availability. The penultimate reaction in this biosynthesis, the reduction of protochlorophyllide, is catalyzed by the photoenzyme light-dependent NADPH:protochlorophyllide oxidoreductase (EC 1.3.1.33; LPOR) and does not occur in the absence of light (Gabruk et al. 2015; Gabruk and Mysliwa-Kurdziel 2020; Scrutton et al. 2012; Vedalankar and Tripathy 2024). It is a checkpoint in chlorophyll biosynthesis that requires light to complete the process. The reaction catalyzed by LPOR is also a key regulatory point in morphogenesis, directing development toward skotomorphogenesis or photomorphogenesis in darkness and light, respectively (Heyes et al. 2021). Each of these patterns leads to the development of specific plastids: etioplasts in darkness and chloroplasts in light (Armarego-Marriott et al. 2020; Cackett et al. 2022; Hernández-Verdeja et al. 2020; Liebers et al. 2022).

5.4.1 The Assembly of Thylakoids During Angiosperm Seedling De-etiolation

Etiolated seedlings of dark-grown angiosperms, subjected to de-etiolation under controlled light conditions, have gained recognition as a good model for studying thylakoid biogenesis (Armarego-Marriott et al. 2020; Solymosi and Aronsson 2013; Sundqvist and Dahlin 1997). A characteristic feature of etioplasts, formed in etiolated seedlings from proplastids, is the presence of a distinctive lipid-protein structure referred to as the prolamellar body (PLB) (Fig. 5.2), surrounded by prothylakoids (Mechela et al. 2019; Solymosi and Aronsson 2013; Solymosi and Mysliwa-Kurdziel 2021; Solymosi and Schoefs 2010). The PLB structure is highly ordered and varies depending on the plant species and etiolation conditions. It typically exhibits cubic, hexagonal, or open symmetry, with varying sizes, densities and packing arrangements (Bykowski et al. 2020; Gunning 2001; Gunning 1965; Kowalewska et al. 2019; Liang et al. 2022; Murakami et al. 1985; Osumi et al. 1984; Selstam et al. 2007; Williams et al. 1998). CURT1 proteins are required for the assembly of PLBs (Sandoval-Ibáñez et al. 2021). PLBs already have some thylakoid protein accumulated (Blomqvist et al. 2008; Plöscher et al. 2011; von Zychlinski et al. 2005). However, its main protein component is LPOR (Blomqvist et al. 2008), which forms ternary complexes with its substrates Pchlide and NADPH Schoefs et al. 2000 and arranges them in a helical array surrounding PLB tubules (Floris and Kühlbrandt 2021). These complexes are ready to be activated by light to convert Pchlide to Chlide, enabling rapid chlorophyll synthesis under light (*see for review of the Chl biosynthesis and its regulation* (Shimizu and Masuda 2021; Tanaka et al. 2011; Wang et al. 2022; Wang and Grimm 2021; Wittmann et al. 2021)).

Concerning the lipid composition, PLBs and prothylakoids consist mainly of galactolipids, similar to the main content of thylakoid membranes, although the relative lipid proportions are different than those found in mature thylakoids (*reviewed in* Solymosi and Mysliwa-Kurdziel 2021; Sundqvist and Dahlin 1997). Galactolipids are crucial for the proper organization of etioplast inner membranes and protochlorophyllide accumulation (Fujii et al. 2014, 2017, 2018, 2019a) with a significant role in anionic lipids also (Yoshihara et al. 2024). Etioplasts already contain carotenoids, and their content in relation to chlorophyll remains high in the early steps of thylakoid assembly and young chloroplast formation (Bykowski et al. 2020; Jedynak et al. 2022; Sun et al. 2018). The high content of carotenoids together with their de novo synthesis (Banas et al. 2025) suggests that the etioplast carotenoids pool is not fully incorporated into newly created thylakoid membranes.

The etioplast-to-chloroplast transformation (Fig. 5.2) in etiolated angiosperm seedlings is started immediately after the onset of light, and the Pchlide-to-Chlide photoreduction, catalyzed by LPOR, is the starting reaction (Gabruk et al. 2015; Heyes et al. 2021; Vedalankar and Tripathy 2024). The components of PLBs and prothylakoids serve as building blocks for developing thylakoids. This multiphasic process involves the consumption of existing components within these structures, along with de novo synthesis of others. This process has been intensively investigated for almost a century using numerous mono- and dicotyledonous plant species, and the progress has been summarized in numerous valuable papers (Adam et al. 2011; Adamson et al. 2006; Armarego-Marriott et al. 2020; Solymosi and Aronsson 2013; Sundqvist and Dahlin 1997; Vothknecht and Westhoff 2001) and the works cited therein. Our recent results (Banaś et al. 2024) suggest that the formation of mature thylakoids might be fine-tuned by the color of light, implicating the involvement of cryptochrome/phytochrome, although the main trigger is not color dependent. This is in agreement with several research, showing that cryptochrome and phytochrome mutants are developing functional chloroplasts (Neff and Chory 1998).

Structural changes include, in the first stage, the loosening of the PLB structure to form a lamellar system of membranes, followed by their folding and overlapping, leading to the formation of grana. The loosening of the structure, accompanied by vesicle formation at the PLB surface, has been shown with 2D TEM microscopy as early as 30 min after Pchlide photoreduction, as summarized in (Solymosi and Aronsson 2013). Studies on isolated PLBs suggest spontaneous disassembly, related to POR activity only (Grzyb et al. 2013). An ET analysis of de-etiolating PLBs in runner bean cotyledons revealed that PLB tubules directly convert into planar thylakoid elements, and the helical arrangement of granum inter-disk connections appears at early stages, while no vesicle formation was observed on the PLB surface (Kowalewska et al. 2016). Recently, the formation of grana stacks directly from PLBs, governed by the presence of CURT1A, has been confirmed (Liang et al. 2022).

Recent studies reveal the sequence of events during the etioplast-to-chloroplast transformation in *A. thaliana* cotyledons almost hour by hour (Liang et al. 2022; Pipitone et al. 2021). However, the results of these studies cannot be directly compared because de-etiolation was performed under different light intensities, and the seedlings were etiolated for either 3 or 7 days before the experiment. It has been shown that the length of the etiolation period (seedling age) affects the content of pigments and LPOR transcripts (Jedynak et al. 2022), which in turn can influence the speed of greening.

Pipitone et al. (2021) distinguished two phases of chloroplast biogenesis during *A. thaliana* seedling de-etiolation. First, the transformation of PLBs toward photosynthetically active thylakoids was observed within a 24-h illumination period. The process included structural changes that started with the disassembly of the PLB structure, smoothly transitioning into the gradual formation of thylakoid lamellae alongside the synthesis of galactolipids, photosynthesis-related proteins (PSII, PSI, and Cyt b_6f), and photosynthetic pigments (chlorophylls and carotenoids). It took only 14 h to develop a fully functional photosynthetic apparatus, with PSII assembly preceding that of PSI. In the second phase, an increase in chloroplast numbers per cell was observed until the end of the experiment, i.e., till 96 h, in parallel with cell expansion. During this process, massive synthesis of galactolipids and photosynthesis-related proteins and pigments, as well as more layers per grana (up to 10 lamellae), and an increase in lamella thickness were observed. The structural transformations during the de-etiolation of *A. thaliana* cotyledons studied by Liang et al. (2022) were similar. After 12 h of greening, the chloroplasts contained typical thylakoid networks consisting of initial grana (6–8 disks) and lamella system. They also contained starch particles, indicating photosynthetic activity. Curt1 proteins are also crucial for etioplast-to-chloroplast transformation, as CURT1C contributes to the formation of PLB cubic crystal structure, while CURT1A is needed for the assembly of "pre-granal thylakoid assembly" at the initial phase of deetiolation (Liang et al. 2022).

The accumulation of galactolipids during de-etiolation takes advantage of two biochemical pathways for their de novo synthesis, namely, the prokaryotic-type and the eukaryotic-type pathways (Hölzl and Dörmann 2019). During the first phase of deetiolation, MGDG and DGDG derived from the eukaryotic-type pathway were accumulated, and later on, the process continued using lipids synthesized via a prokaryotic-type pathway (Pipitone et al. 2021). It was noticed that the MGDG/DGDG ratio first decreased and then increased to the constant value, reaching the minimum at 8 h of greening. This decrease and subsequent increase were related to the transition from PLB (cubic lipid phase) to thylakoid membrane (lamellar structure). The significance of galactolipids during the etioplast-to-chloroplast transformation has been documented (Fujii et al. 2019a, b; Kobayashi and Wada 2016).

The continuous, multiphase process of converting etioplasts into chloroplasts has recently been observed during the de-etiolation of alfalfa (Ni et al. 2024). Differences in the accumulation and activity evolution of PSI and PSII (Pipitone et al. 2021) have also been observed in etiolated monocotyledons seedlings, which exhibit a linear gradient of plastid development along the leaf blade (Pogson et al. 2015; Solymosi and Aronsson 2013). PSII protein abundance was higher in the early stages of thylakoid formation before the grana organization. The exact molecular mechanisms behind this differential accumulation are unknown, but delaying PSI formation until grana formation and PSII relocalization might prevent energy spill-over between the two photosystems.

5.4.2 The Assembly of Thylakoids in Seedlings Germinating Under Light

The study of thylakoid formation during de-etiolation of etiolated seedlings with developed cotyledons is somehow artificial. In nature, there is rarely such a stage and during seed germination; even deep in soli, there is already some light. Liang et al. (2018) examined, therefore, the process of thylakoid formation in *A. thaliana* cotyledons during seed germination under white light (120 μmol photons m^{-2} s^{-1}) and described this process as a linear developmental sequence with five stages of thylakoid assembly. First, within 24 h after seed imbibition, tubulo-vesicular prothylakoids were formed. These then enlarged and flatted, converting into sheetlike pregranal thylakoids observed 36 h after seed imbibition. Subsequently, the proliferation of pregranal stacks with wide tubular connections, followed by the structural differentiation of pregranal stacks and expanded stroma thylakoids, and finally, the conversion of pregranal stacks into mature grana stacks was observed at 60, 84, and 120 h after seed imbibition, respectively. Concerning the assembly of photosynthetic complexes, PSII complex subunits appeared at the stage of sheetlike pregranal thylakoids, whereas the ATP synthase, cytochrome b_6f, and LHCII proteins were detected at the pregrana thylakoid stage. PSI proteins and the curvature-inducing CURT1A protein appear much later, in coincidence with the structural differentiation of prograna stacks and expanded stroma thylakoids.

Development of the planar pregranal thylakoids and the pro-grana membrane stacks coincide with the appearance of thylakoid-bound polysomes (Liang et al. 2018). A dynamin-like Fzo-like (FZL) membrane fusion protein has been shown to be involved in the formation of interconnection between progranal stacks, facilitating the fusion of thylakoid membrane compartments (Liang et al. 2018).

5.4.3 The Assembly of Thylakoids in Non-seedling Systems

In angiosperms, chloroplasts in true leaves develop in the light from proplastids and the proplastid-to-chloroplast transition occur at the shoot apex within the apical meristem (Adam et al. 2011). It has been shown that this region contains heterogeneous plastids at different stages of development, depending on the meristem cell layer (Charuvi et al. 2012). In some layers, the thylakoid network is quite well developed, especially near the leaf buds. During shoot growth, plastids of individual meristem layers undergo distinct differentiation processes, leading either to the development or loss of thylakoid membranes. A new experimental system was recently developed to study the de-etiolation of leaves in a dicotyledonous plant grown in soil (Armarego-Marriott et al. 2019). In this model, mature tobacco (*N. tabacco*) plants were grown under long-day conditions, optimized so that the stored photosynthetic starch would fuel subsequent plant growth in darkness. Then, the dark period was significantly extended, yet without inducing senescence, and etiolated apical leaves containing plastids with structured, paracrystalline PLBs appeared. The leaves were then subjected to de-etiolation under controlled conditions for almost 5 days. Two phases of leaf greening were distinguished in this system: the first de-etiolation phase, in which the plastid PLB is disassembled, and the second building phase, which mostly involved the accumulation of thylakoid membranes with associated photosynthetic complexes showing the photosynthetic activity. The development of the thylakoid network proceeded similarly to that described above in de-etiolated cotyledons.

Among the systems used so far to study the development of thylakoids, the following should also be mentioned: partly shaded leaf primordia in buds (Solymosi et al. 2006, 2012), sunflower cotyledons partly covered by the pericarp (Solymosi et al. 2007), soil-shaded lowermost hypocotyl segments of bean (Kakuszi et al. 2017), and leaf layers of cabbage heads (Solymosi et al. 2004), where different stages of PLBs-to-thylakoids transformation and etio-chloroplast formation were observed due to varying light access (Solymosi and Schoefs 2010).

5.4.4 Chloroplast Growth and New Thylakoids Formation in Light-Exposed Leaves

Although the first chloroplasts of a seedling are those transformed from proplastids present in seed tissue, the chloroplasts change both in size and the total number. During the first 24 h of greening, chloroplasts increase their size, and later they divide. The chloroplast division is independent of the cell division, and one may assume that new chloroplasts still appear in mesophyll cells and new thylakoid membranes are formed. Although precise studies of this process are missing due to the relatively low amount of new chloroplasts, we may still translate results obtained for de novo thylakoid assembly and thylakoid repair.

5.5 Thylakoid Decomposition in Higher Plants

There is the point in plants' life cycle that thylakoids are being disassembled. The first such process is the senescence of leaves and other photosynthetically active parts of plants, which manifests as yellowing, resulting from the decomposition of chlorophyll and uncovering of carotenoid colors. The process is accompanied by the transformation of chloroplasts into gerontoplasts. Those organelles are characterized by a reduced number of thylakoid membranes and degenerated outer chloroplast membrane; the disassembly of thylakoids is then one of the senescence features. Severe damage of thylakoids, as well as the dysfunction of adaptive mechanisms, as state transition, leads to a retrograde signal (putatively, H_2O_2 in the first place), triggering plant cell senescence. (Zentgraf et al. 2022) proposed that formally not an H_2O_2 concentration but the ratio between cytoplasmic and chloroplastic content of hydrogen peroxide is the actual senescence trigger. In line with that are the results showing that WATER-SOAKED SPOT1, a chloroplastic transmembranal kinase involved in the regulation of ROS homeostasis, also impairs senescence (Xu et al. 2022). Chloroplast decomposition might be the event starting programmed cell death (Woodson 2022; Zapata et al. 2005) and aims at the redistribution of nutrients.

Jaijic et al. (Jajić et al. 2015) investigated levels of specific ROS, namely, H_2O_2, O_2^{-*}, and $^1O_2$ in thylakoids isolated from aging barley leaves. The authors found a gradual increase in H_2O_2 and O_2^{-*} concentrations until the 29th day after sowing, which also correlated with an increase in the maximum chlorophyll content. $^1O_2$ concentration was stable during this growth phase. Later, $^1O_2$ concentration slightly increased, while O_2^{-*} content decreased, whereas H_2O_2 level showed some oscillation. The most important for the evaluation of the thylakoid state appeared the gradual accumulation of lipid peroxidation products (Jajić et al. 2015).

In naturally senescing *Arabidopsis thaliana* plants grown in long-day conditions (Poudyal et al. 2020), the tendency to increase grana thickness was still present even in leaves older than 18 days, bearing already phenotypic features of aging. The grana thickness increased from about 91 nm on the 18th day to 141 nm on the 30th day after emergence. Grana was thicker in some mutants (stn7, cp29); however, the tendency to increase its high was conserved during the aging process. Interestingly, after 4 constitutive days, grana were not detectable. This transition is accompanied by a significant decrease in the chlorophyll a/b ratio (enhanced in mutants with faster senescence phenotypes), as well as the reduction of Fv/Fm ratio—taken together, it shows the decomposition of photosynthetic light-harvesting complexes. The decomposition of PSI and PSII was not drastic in WT Arabidopsis, although increased in stn7 and cp29 mutants (Poudyal et al. 2020).

Senescence of grape leaves chloroplasts manifested by increase of chloroplast volume, corresponding to high accumulation of plastoglobules. Thylakoid lamellae were degraded and less grana were observed. The decrease in the chlorophyll a/b ratio indicated the gradual decomposition of membrane photosynthetic complexes (Wang et al. 2016).

The appearance of the high number of plastoglobules is an indication of thylakoid membrane remodeling. In general, these subchloroplastic structures contain spare lipids, including as well as carotenoids and prenylolipids (Arzac et al. 2022). The decomposition of the thylakoid membrane is strongly correlated with the downregulation of genes, coding photosynthetic complexes as well as thylakoid membrane lipids synthases, as shown for flag barley leaves (Cohen et al. 2022).

Senescence of chloroplasts may be induced by darkening of leaves, even plants attached (Paluch-Lubawa et al. 2021). For dark-induced senescence of barley, a gradual destacking and reduction of thylakoid membrane number was observed. Interestingly, such conditions increased expression of plastid transglutaminases, that was mostly localized on thylakoid membranes. After 5 days of aging, the transglutaminases were also found in the stroma (Sobieszczuk-Nowicka et al. 2015). In a detailed analysis, polyamines were found to be attached to multiple proteins, including ATPase and PSI/PSII subunits. Another significant feature of gerontoplats is the increasing number and size of plastoglubeles (Delfosse et al. 2016; Sobieszczuk-Nowicka et al. 2015). Stroma of gerontoplast may be secluded in senescence-associated vesicles (Delfosse et al. 2016).

Transition to gerontoplasts is a natural process, found also during the development of seeds' inner integument. In the study on Jatropha seeds (Shah et al. 2016), the primary tissue development stage (mature ovulae) showed plastids with an internal membrane system, starch granules, and plastoglobules. On fifth day after pollination, there are already no plastoglobules. During the following days (up to the 5th), internal membrane disassembles, which is accompanied by the degradation of matrix. Some plastids were accompanied by stroma-derived vesicles. The observed degradation can be correlated with high content of peptidases, found in proteome analysis (Shah et al. 2016).

Chloroplast transformation may also lead to a chromoplast. This organelle is characterized by a lack of photosynthetic activity, degradation of thylakoids (disassembly of grana, random localization of membrane vesicles), and high accumulation of carotenoids. Chromoplasts cannot divide. The chloroplast-to-chromoplast transformation can be a natural process of fruit ripening (e.g., in tomato fruits at the beginning containing fully active chloroplasts) and can be triggered by virus-mediated expression of bacterial phytoene synthase (Llorente et al. 2020). This enzyme catalyzes the conversion of geranylgeranyl diphosphate to phytoene, a substrate for following carotenoids. At the same time, the reduced pool of geranylgeranyl diphosphate means a decreased substrate pool for chlorophyll synthesis.

A chromoplast may contain also huge grains of starch, in such case being called a chromoamyloplast. Those organelles are found, e.g., in apples (Schaeffer et al. 2017). The transformation is accompanied by de novo synthesis of carotenoids, fatty acid chains, and aromatic compounds. Structurally, plastid intermediates still contain some vesiculate lamellae. In mature chromoplast, the reduced number of membranous structures is localized in the central parts of the organelle, accompanied by a high number of plastoglobuli. In apples of green peel after ripening (e.g., Granny Smith) the thylakoids are still present; however, they do not form grana anymore (Schaeffer et al. 2017).

Chloroplast to chromoplast transformation involves mentioned earlier chloroplast-associated degradation pathway (CHLORAD); it was shown that overexpression of plastid ubiquitin E3 ligase SP1 enhances tomato ripening, while lack of this protein delays the process (Ling et al. 2021).

5.6 Variation of Thylakoid Membranes

In general, cyanobacterial thylakoids host the same photosynthetic electron transfer chain, as higher plants. Localization of several proteins within thylakoids was shown to be evolutionarily conserved, from *Synechocystis* to *Arabidopsis* (Baers et al. 2019) The main difference is the type of main photosynthetic antennae that in higher plants are integral membrane pigment-protein complex of LHCI and LHCII, while in cyanobacteria, there are phycobilisome (PBS) complexes. PBS are water-soluble, do not contain chlorophyll or carotenoids but phycocyanobilins, and are peripheral membrane proteins able to diffuse between PSI and PSII.

Formally, thylakoid membranes cannot be connected physically with a cell membrane. In most photosynthetic organisms, this condition is fulfilled, as thylakoid vesicles are inside the cell. However, there are primitive cyanobacteria, belonging to *Gloebacterales*, that are "thylakoid-less." In this case, photosynthetic pigments and proteins are embedded into the plasma membrane, although already in separated regions. "Paddle-" or "bundle"-shaped phycobilisomes (in comparison to a predominant hemidiscoidal) are located directly on the plasma membrane (Jiang et al. 2023). Early works with *Gloebacter* membrane fractionation distinguish green domains and orange domains. The first type seemed to represent bioenergetically active regions (both in photosynthesis and respiration), while the second contained proteins needed for the development of the green region (Rexroth et al. 2011). Separation within the plasma membrane, resulting in such specialized domains, is considered as the first step in evolutionary development of thylakoids membranes (Nickelsen and Zerges 2013).

In all other cyanobacteria groups, the cells contain proper thylakoid vesicles in their interior. These thylakoids, however, are greatly varied by shape and organization. Summarizing, there can be parietal, radial, fascicular, parallel, and irregular arrangement (Fig. 5.3). Mareš et al. (2019) analyzed cyanobacterial morphology in connection to their phylogenetic relationship, concluding that the use of thylakoid architecture for taxonomy is strongly limited. There seems to be no one particular protein, responsible for all architecture types; CurT homologs are missing in several cyanobacterial clades (Mareš et al. 2019), although it is present in *Synechocystis*, and its deletion resulted in thylakoid organization disruption (Magnuson and Cardona 2016).

Thylakoids were identified in fossils dated back to 0.152–0.157 Ga, being linear or concerted, and of a thickness of 35–70 nm. Another study showed microfossil *Navifusa majensis* from the *1.75 Ga McDermott Formation* contained thylakoids, already prone to organized in stacks as thick as 97–200 nm, although the thickest might result from compression artifact occurring during fossilization (Demoulin et al. 2024).

Chloroplasts of diatoms are believed to be the results of double endosymbiosis (i.e., endosymbiosis of the red alga, the already had, and endosymbiotic chloroplasts). The thylakoids are organized in stacks, and also there are parallel connections between adjacent layers (Flori et al. 2017). There are also domains within diatoms thylakoid membranes, responsible for spatial separation of PSI and

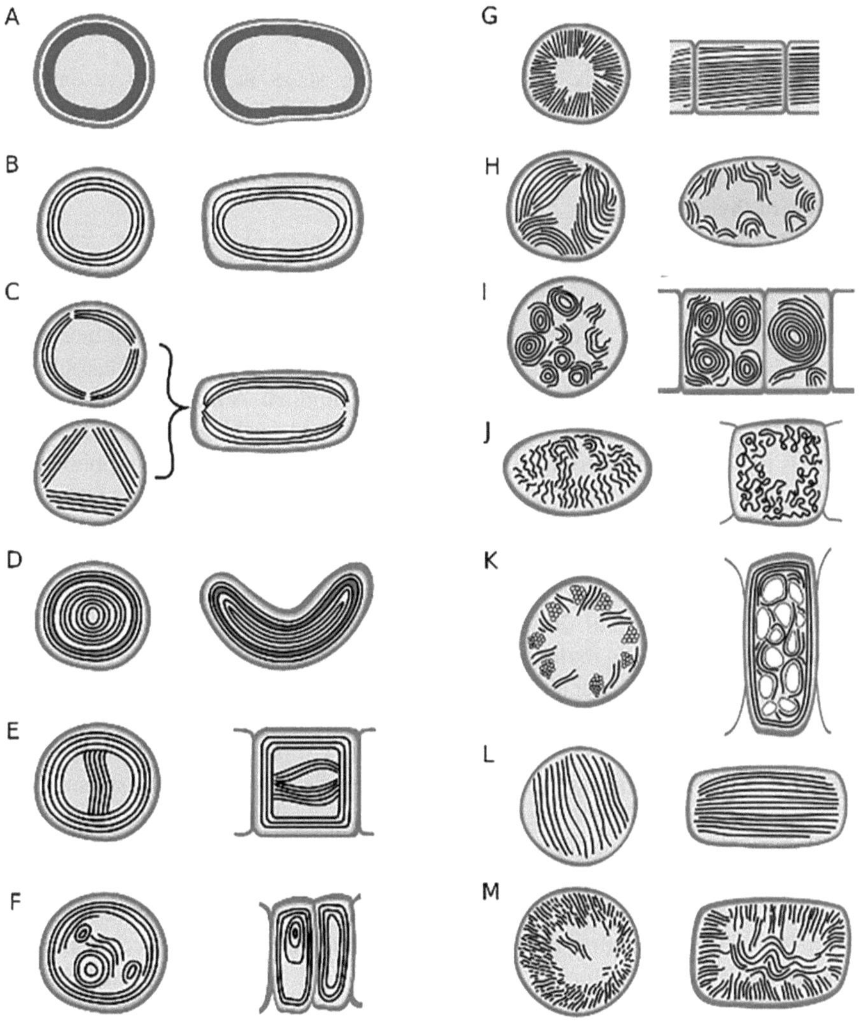

Fig. 5.3 Various types of thylakoid arrangement within cyanobacterial cells: cross-section and longitudinal section. (**a**) thylakoid less (layer of pigments represented as broad grey line), (**b**) parietal, (**c**) arietal composed of peripheral fascicles, (**d**) with a subperipheral dense layer, (**e**) parietal thylakoids with a central fascicle, (**f**) with spherical formations, (**g**) radial, (**h**) fascicular, (**i**) fascicular with dominant spherical formations, (**j**) irregular in heterocysts, (**k**) irregular, displacemed by cellular inclusions, (**l**) parallel, (**m**) cyanothece-like arrangement. (Adapted from (Mareš et al. 2019) under Creative Commons Attribution License (CC BY) under http://creativecommons.org/licenses/by/4.0/)

PSII. The functional architecture of thylakoids is believed to be different than those of higher plants. Thylakoids are less structured and observed parallel connections facilitate diffusion of plastoquinone and cyt c6, promoting faster reaching of redox equilibrium (Flori et al. 2017).

5.7 Thylakoids Development in Cyanobacteria

In cyanobacteria, there is no formal greening stage or structure resembling PLB. However, there is always a necessity to build new thylakoid membranes during cell growth and division, as well as after recovery from the darkening stage. In cyanobacteria cells, there is a specific subcellular compartment, responsible for biogenesis of thylakoids.

Huokko et al. (2021) investigated thylakoid biogenesis in unicellular cyanobacteria, *Synechococus* PCC 7922, applying modulated light intensity. Such conditions allowed first to reduce thylakoid content in the cell in high light and later to detect of clear thylakoid formation under lower light intensity. New membranes were formed independently of the plasma membrane and already existing thylakoids and were concentrated asymmetrically on one longitudinal cell side. Later, newly formed vesicles expanded, filling available space and pushing existing thylakoid towards the cell center. Formed thylakoids are gradually populated with photosynthetic complexes, PSI, PSII, cyt b_6f, PBS, and ATPase. PSI assembly was faster than the assembly of PSII. Interestingly, complexes distribution seemed to be inhomogeneous, with PSI trimers clustering at one pole of the cell (Fig. 5.4). Newly formed thylakoids are discontinued (perforated), which resembles typical *Synechococus* morphology (Huokko et al. 2021).

Another way to study thylakoid biogenesis is to adapt cells to darkness and then expose it to light (Barthel et al. 2013). Dark (chemotropically) grown *Synechocystis* cells contain a reduced number of thylakoid vesicles, though a relatively high number of PSI, PSII, and phycobilisome components are present in the cytoplasm. A formation of typical thylakoid morphology demanded 24 h of light exposure; the first 8–6 h were somehow a lag phase, without observable photosynthetic activity. Newly synthetized membranes were first cloudy and not structured properly; the desired organization started to be visible after 12 h of light exposure. Oxygen evolution activity of PSII was absent in chemotropically grown cells, existing but low for the 6–8 h of lag phase after light exposure, with a drastic increase occurring later. The same was apparent for chlorophyll content. Full restoration of liner electron transfer (ET) demanded a longer time (10–12 h), which indicates that the formal structure of the thylakoid membrane is necessary for proper connection between ET components. Interestingly, the thylakoid architecture remains unchanged in PSII-lacking mutants (Barthel et al. 2013).

Assembly of active thylakoids membranes demands specialized protein factors. The Albino3 protein of *Synechocystiss* PCC6803 was proposed to generally maintain thylakoids but not plasma membrane, organization (Spence et al. 2003). This protein is a homolog of higher plants Alb3 protein, responsible for membrane insertion of internal antennae of photosystems.

Synthesis of proteins, building photosynthetic pigment-protein complexes, may be already correlated with their incorporation into thylakoid membranes, as in higher plants thylakoids. For mRNA coding core subunits of PSI (psaA) and PSII (psbA), the fluorescence in situ hybridization (FISH) method showed localization on the thylakoid membrane (Mahbub et al. 2020). This anchoring was first realized

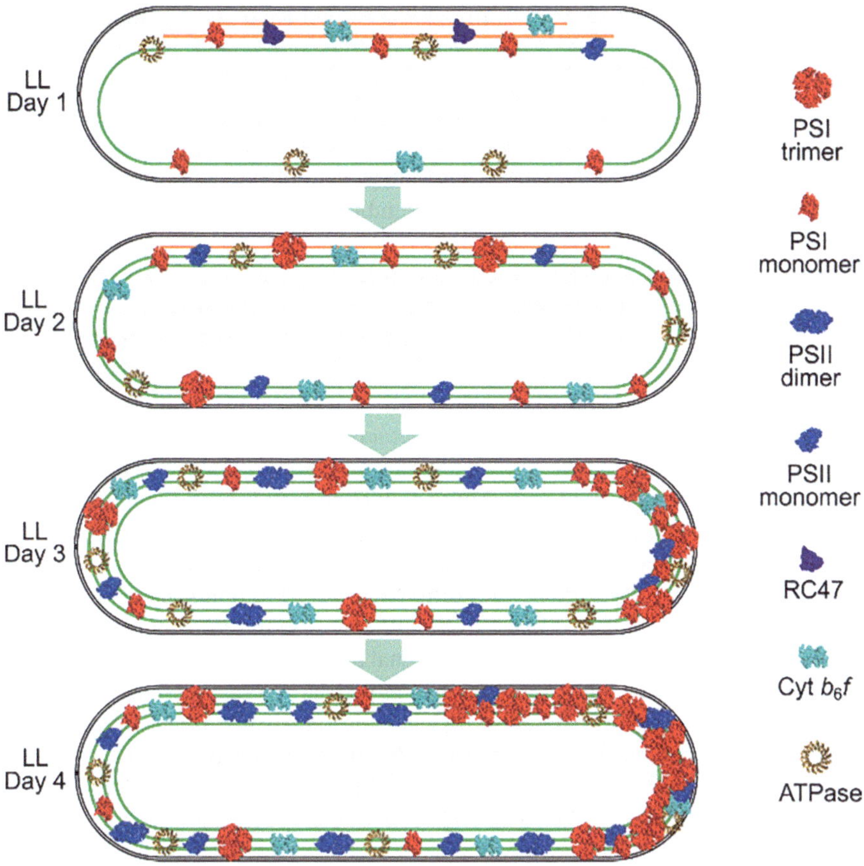

Fig. 5.4 Thylakoids assembly in *Synechococus* cells, during adaptation to low light (LL, from day 1 to day 4) upon previous growth in high light conditions. New thylakoid membranes are drawn as red. Photosynthetic complex's inhomogeneous distribution is represented in the scheme. (Reproduced from (Huokko et al. 2021) under Creative Commons Attribution 4.0 International License (http://creativecommons.org/licenses/by/4.0/))

by RNA-binding protein (RBP), independently of ribosomes, that showed high affinity to thylakoid membrane; later, ribosomes recognized mRNA and proceeded with translation, producing the N-terminal part of the protein, containing the leader sequence. The latter is then recognized by a translocon complex in a specialized region of the insertion zone. In *Synechococuss* cells, both psaA and psbA were concentrated in clear foci, scattered among all thylakoid membranes; under high light stress, damaging PSII, psbA production increased, and its distribution was more even. Synechocystis cells have a more localized thylakoid biogenesis zone (Mahbub et al. 2020).

Assembly of cyanobacterial thylakoid membranes needs also pigments: chlorophylls and carotenoids. In the absence of tetrapyrole synthesis pathways, which was realized by ΔbtpA mutation, the thylakoids were absent (Skotnicová et al. 2024).

Also, the photosystems were not fully produced, although some components (D1 protein of PSII, coded by psbA) were present. Some other thylakoid proteins, such as ATP synthase, NDH1 complex, and cyt b6f, were found in a relatively high amount. The BtpA protein was identified as the oligomerizing homo-oligomers, stabilizing glutamyl-tRNA reductase, that catalyzes the first step leading to chlorophyll ring (Skotnicová et al. 2024). The BptA is, therefore, a component of a broad network precisely regulating thylakoid assembly, even without physical involvement in that process.

The organization of thylakoid membranes is altered in heterocysts (Magnuson and Cardona 2016). These are specialized cells, dedicated to nitrogen assimilation. For optimal nitrogenase activity, anoxygenic conditions are required; therefore, the OEC is missing.

However, thylakoid membranes are still necessary because of PSI involvement in the powering of nitrogen assimilation by supplementation of reduced ferredoxin. In Nostoc PCC 7120, vegetative cells contain layered thylakoids, while heterocyst is characterized by less ordered ones. Those membranes contain large amounts of PSI and remnants of PSII. Phycobilisomes are being degraded to a low level by a specific protease; however, in mature heterocysts, there is still energy transfer from phycobilisomes to PSI. There are differences in phycobilisome composition between vegetative cells and heterocysts; the last is believed to be smaller, with shorter rods. In heterocysts, there are additional membrane domains, known as honeycomb, and they are supposed to consume molecular oxygen.

It might be interesting to note that thylakoid decomposition is also a natural process during bioaugmentation of soil, although this is not related already to thylakoid performance in a live cell. This process was studied for fungal lipase (lecitase) as a catalyst (Perczyk and Broniatowski 2022), indeed showing that thylakoid galactolipids are the target of the enzyme.

5.8 Concluding Remarks

Thylakoid assembly and disassembly are indispensable parts of photosynthetic organisms' growth and senescence. Thylakoid role is to maintain optimal conditions for photosynthesis in given environmental conditions, which demands optimization of the light-harvesting area and PSI-PSII reaction balance. This cannot be achieved without short- and long-scale thylakoid remodeling. As described above, the process of thylakoid biogenesis is complex and based on simple thermodynamics of lipids organization in the bilayer, combined with precisely controlled insertion of membrane proteins, pigment, and cofactors and their assembly in functional multisubunit systems. Fine-tuning includes precise control of chlorophyll and carotenoid synthesis. The whole process seems to be evolutionarily conserved, although with understandable differences between higher plants and ancestors of their chloroplasts, cyanobacteria. The complexity and precision of this construction make us again to be astonished by nature.

References

Abdelrahman M, Ishii T, El-Sayed M, Tran L-SP (2020) Heat sensing and lipid reprograming as a signaling switch for heat stress responses in wheat. Plant Cell Physiol 61(8):1399–1407

Adam Z, Charuvi D, Tsabari O, Knopf RR, Reich Z (2011) Biogenesis of thylakoid networks in angiosperms: knowns and unknowns. Plant Mol Biol 76(3–5):221–234. https://doi.org/10.1007/s11103-010-9693-5

Adams PG, Hunter CN (2012) Adaptation of intracytoplasmic membranes to altered light intensity in Rhodobacter sphaeroides. Biochim Biophys Acta Bioenerg 1817(9):1616–1627. https://doi.org/10.1016/J.BBABIO.2012.05.013

Adamson HY, Hiller RG, Walmsley J (2006) New Trends in Photobiology (Invited Review) Protochlorophyllide reduction and greening in angiosperms: an evolutionary perspective. J Photochem Photobiol 41:201

Amunts A, Drory O, Nelson N (2007) The structure of a plant photosystem I supercomplex at 3.4 Å resolution. Nature 447(7140). https://doi.org/10.1038/nature05687

Andersson B, Anderson JM (1980) Lateral heterogeneity in the distribution of chlorophyll-protein complexes of the thylakoid membranes of spinach chloroplasts. Biochim Biophys Acta Bioenerg 593(2):427–440. https://doi.org/10.1016/0005-2728(80)90078-X

Armarego-Marriott T, Kowalewska Ł, Burgos A, Fischer A, Thiele W, Erban A, Strand D, Kahlau S, Hertle A, Kopka J, Walther D, Reich Z, Schöttler MA, Bock R (2019) Highly resolved systems biology to dissect the Etioplast-to-chloroplast transition in tobacco leaves. Plant Physiol 180(1):654–681. https://doi.org/10.1104/PP.18.01432

Armarego-Marriott T, Sandoval-Ibañez O, Kowalewska Ł (2020) Beyond the darkness: recent lessons from etiolation and de-etiolation studies. J Exp Bot 71(4):1215–1225. https://doi.org/10.1093/jxb/erz496

Armbruster U, Labs M, Pribil M, Viola S, Xu W, Scharfenberg M, Hertle AP, Rojahn U, Jensen PE, Rappaport F, Joliot P, Dörmann P, Wanner G, Leister D (2013) Arabidopsis CURVATURE THYLAKOID1 proteins modify thylakoid architecture by inducing membrane curvature. Plant Cell 25:2661–2678. https://doi.org/10.1105/tpc.113.113118

Arzac MI, Fernández-Marín B, García-Plazaola JI (2022) More than just lipid balls: quantitative analysis of plastoglobule attributes and their stress-related responses. Planta 255(3):62

Baers LL, Breckels LM, Mills LA, Gatto L, Deery MJ, Stevens TJ, Howe CJ, Lilley KS, Lea-Smith DJ (2019) Proteome mapping of a cyanobacterium reveals distinct compartment organization and cell-dispersed metabolism. Plant Physiol 181(4):1721–1738

Banaś AK, Leja K, Zgłobicki P, Jedynak P, Kowalska E, Strzałka W, Grzyb J, Myśliwa-Kurdziel B (2024) De-etiolation is almost colour blind: the study of photosynthesis awakening under blue and red light. Plant Cell Physiol 65:pcae119

Banaś AK, Leja K, Zgłobicki P, Jedynak P, Kowalska E, Strzałka W, Grzyb J, Myśliwa-Kurdziel B (2024) De-etiolation is Almost Color Blind: The Study of Photosynthesis Awakening under Blue and Red Light Abstract Plant And Cell Physiology, 65(12), 1993–2017. https://doi.org/10.1093/pcp/pcae119

Barthel S, Bernát G, Seidel T, Rupprecht E, Kahmann U, Schneider D (2013) Thylakoid membrane maturation and PSII activation are linked in greening Synechocystis sp. PCC 6803 cells. Plant Physiol 163(2):1037–1046

Blomqvist LA, Ryberg M, Sundqvist C (2008) Proteomic analysis of highly purified prolamellar bodies reveals their significance in chloroplast development. Photosynth Res 96(1):37–50. https://doi.org/10.1007/s11120-007-9281-y

Bonardi V, Pesaresi P, Becker T, Schleiff E, Wagner R, Pfannschmidt T, Jahns P, Leister D (2005) Photosystem II core phosphorylation and photosynthetic acclimation require two different protein kinases. Nature 437(7062). https://doi.org/10.1038/nature04016

Bykowski M, Mazur R, Buszewicz D, Szach J, Mostowska A, Kowalewska Ł (2020) Spatial Nanomorphology of the Prolamellar body in etiolated Arabidopsis thaliana plants with disturbed pigment and Polyprenol composition. Front Cell Dev Biol 8:586628. https://doi.org/10.3389/fcell.2020.586628

Bykowski M, Mazur R, Wójtowicz J, Suski S, Garstka M, Mostowska A, Kowalewska Ł (2021) Too rigid to fold: carotenoid-dependent decrease in thylakoid fluidity hampers the formation of chloroplast grana. Plant Physiol 185(1):210–227

Cackett L, Luginbuehl LH, Schreier TB, Lopez-Juez E, Hibberd JM (2022) Chloroplast development in green plant tissues: the interplay between light, hormone, and transcriptional regulation. New Phytologist 233(5). https://doi.org/10.1111/nph.17839

Caffarri S, Croce R, Cattivelli L, Bassi R (2004) A look within LHCII: differential analysis of the Lhcb1-3 complexes building the major trimeric antenna complex of higher-plant photosynthesis. Biochemistry 43(29). https://doi.org/10.1021/bi036265i

Caffarri S, Kouřil R, Kereïche S, Boekema EJ, Croce R (2009) Functional architecture of higher plant photosystem II supercomplexes. EMBO J 28(19). https://doi.org/10.1038/emboj.2009.232

Cao P, Su X, Pan X, Liu Z, Chang W, Li M (2018) Structure, assembly and energy transfer of plant photosystem II supercomplex. Biochim Biophys Acta Bioenerg 1859(9). https://doi.org/10.1016/j.bbabio.2018.03.007

Charuvi D, Kiss V, Nevo R, Shimoni E, Adam Z, Reich Z (2012) Gain and loss of photosynthetic membranes during plastid differentiation in the shoot apex of arabidopsis. Plant Cell 24(3):1143–1157. https://doi.org/10.1105/tpc.111.094458

Che L-P, Ruan J, Xin Q, Zhang L, Gao F, Cai L, Zhang J, Chen S, Zhang H, Rochaix J-D, Peng L (2024) RESISTANCE TO PHYTOPHTHORA1 promotes cytochrome b559 formation during early photosystem II biogenesis in Arabidopsis. Plant Cell 36(10):4143–4167. https://doi.org/10.1093/PLCELL/KOAE196

Chen JH, Wu H, Xu C, Liu XC, Huang Z, Chang S, Wang W, Han G, Kuang T, Shen JR, Zhang X (2020) Architecture of the photosynthetic complex from a green sulfur bacterium. Science 370(6519). https://doi.org/10.1126/SCIENCE.ABB6350/SUPPL_FILE/ABB6350_S6.MP4

Cohen M, Hertweck K, Itkin M, Malitsky S, Dassa B, Fischer AM, Fluhr R (2022) Enhanced proteostasis, lipid remodeling, and nitrogen remobilization define barley flag leaf senescence. J Exp Bot 73(19):6816–6837

Cutolo E, Parvin N, Ruge H, Pirayesh N, Roustan V, Weckwerth W, Teige M, Grieco M, Larosa V, Vothknecht UC (2019) The high light response in Arabidopsis requires the calcium sensor protein CAS, a target of STN7-and STN8-mediated phosphorylation. Front Plant Sci 10:974

Dainese P, Bassi R (1991) Subunit stoichiometry of the chloroplast photosystem II antenna system and aggregation state of the component chlorophyll a/b binding proteins. J Biol Chem 266(13). https://doi.org/10.1016/s0021-9258(18)92952-2

Daum B, Nicastro D, Austin J, Richard McIntosh J, Kühlbrandt W (2010) Arrangement of photosystem II and ATP synthase in chloroplast membranes of spinach and pea. Plant Cell 22(4):1299–1312. https://doi.org/10.1105/TPC.109.071431

Dekker JP, Boekema EJ (2005) Supramolecular organization of thylakoid membrane proteins in green plants. Biochim Biophys Acta Bioenerg 1706(1–2):12–39. https://doi.org/10.1016/J.BBABIO.2004.09.009

Delfosse K, Wozny MR, Jaipargas E-A, Barton KA, Anderson C, Mathur J (2016) Fluorescent protein aided insights on plastids and their extensions: a critical appraisal. Front Plant Sci 6:1253

Demé B, Cataye C, Block MA, Maréchal E, Jouhet J (2014) Contribution of galactoglycerolipids to the 3-dimensional architecture of thylakoids. FASEB J 28(8):3373–3383

Demoulin CF, Lara YJ, Lambion A, Javaux EJ (2024) Oldest thylakoids in fossil cells directly evidence oxygenic photosynthesis. Nature 625(7995):529–534. https://doi.org/10.1038/s41586-023-06896-7

Dlouhý O, Javornik U, Zsiros O, Šket P, Karlický V, Špunda V, Plavec J, Garab G (2021) Lipid polymorphism of the subchloroplast—granum and stroma thylakoid membrane—particles. I. 31P-NMR spectroscopy. Cells 10(9):2354

Duchêne S, Siegenthaler P-A (2000) Do glycerolipids display lateral heterogeneity in the thylakoid membrane? Lipids 35(7):739–744

Dukic E, van Maldegem KA, Shaikh KM, Fukuda K, Töpel M, Solymosi K, Hellsten J, Hansen TH, Husted S, Higgins J, Sano S, Ishijima S, Spetea C (2023) Chloroplast magnesium transporters play essential but differential roles in maintaining magnesium homeostasis. Front Plant Sci 14:1221436. https://doi.org/10.3389/FPLS.2023.1221436/BIBTEX

Engel BD, Schaffer M, Kuhn Cuellar L, Villa E, Plitzko JM, Baumeister W (2015) Native architecture of the Chlamydomonas chloroplast revealed by in situ cryo-electron tomography. elife 4:e04889

Farci D, Schröder WP (2023) Thylakoid Lumen; from "proton bag" to photosynthetic functionally important compartment. Front Plant Physiol 1:1310167

Flori S, Jouneau P-H, Bailleul B, Gallet B, Estrozi LF, Moriscot C, Bastien O, Eicke S, Schober A, Bártulos CR (2017) Plastid thylakoid architecture optimizes photosynthesis in diatoms. Nat Commun 8(1):15885

Floris D, Kühlbrandt W (2021) Molecular landscape of etioplast inner membranes in higher plants. Nat Plants 7(4). https://doi.org/10.1038/s41477-021-00896-z

Fristedt R, Vener AV (2011) High light induced disassembly of photosystem II supercomplexes in Arabidopsis requires STN7-Dependent phosphorylation of CP29. PLoS One 6(9). https://doi.org/10.1371/journal.pone.0024565

Fujii S, Kobayashi K, Nakamura Y, Wada H (2014) Inducible knockdown of Monogalactosyldiacylglycerol Synthase1 reveals roles of Galactolipids in organelle differentiation in Arabidopsis cotyledons. Plant Physiol 166(3):1436–1449. https://doi.org/10.1104/pp.114.250050

Fujii S, Kobayashi K, Nagata N, Masuda T, Wada H (2017) Monogalactosyldiacylglycerol facilitates synthesis of photoactive protochlorophyllide in etioplasts. Plant Physiol 174(4):2183–2198. https://doi.org/10.1104/pp.17.00304

Fujii S, Kobayashi K, Nagata N, Masuda T, Wada H (2018) Digalactosyldiacylglycerol is essential for organization of the membrane structure in etioplasts. Plant Physiol 177:1487–1497. https://doi.org/10.1104/pp.18.00227

Fujii S, Nagata N, Masuda T, Wada H, Kobayashi K (2019a) Galactolipids are essential for internal membrane transformation during etioplast-to-chloroplast differentiation. Plant Cell Physiol 60(6):1224–1238. https://doi.org/10.1093/pcp/pcz041

Fujii S, Wada H, Kobayashi K (2019b) Role of galactolipids in plastid differentiation before and after light exposure. Plants 8(10):357. https://doi.org/10.3390/plants8100357

Gabruk M, Mysliwa-Kurdziel B (2020) The origin, evolution and diversification of multiple isoforms of light-dependent protochlorophyllide oxidoreductase (LPOR): focus on angiosperms. Biochem J 477:2221–2236. https://doi.org/10.1042/BCJ20200323

Gabruk M, Stecka A, Strzałka W, Kruk J, Strzałka K, Mysliwa-Kurdziel B (2015) Photoactive protochlorophyllide-enzyme complexes reconstituted with PORA, PORB and PORC proteins of A. thaliana: fluorescence and catalytic properties. PLoS One 10(2):e0116990. https://doi.org/10.1371/journal.pone.0116990

Gao H, Sage TL, Osteryoung KW (2006) FZL, an FZO-like protein in plants, is a determinant of thylakoid and chloroplast morphology. Proc Natl Acad Sci 103(17):6759–6764. https://doi.org/10.1073/pnas.0507287103

Gao LL, Hong ZH, Wang Y, Wu GZ (2023) Chloroplast proteostasis: a story of birth, life, and death. Plant Commun 4(1). https://doi.org/10.1016/j.xplc.2022.100424

Garab G, Ughy B, Goss R (2016) Role of MGDG and non-bilayer lipid phases in the structure and dynamics of chloroplast thylakoid membranes. In: Lipids in plant and algae development, pp 127–157

Garab G, Ughy B, de Waard P, Akhtar P, Javornik U, Kotakis C, Šket P, Karlický V, Materová Z, Špunda V (2017) Lipid polymorphism in chloroplast thylakoid membranes–as revealed by 31P-NMR and time-resolved merocyanine fluorescence spectroscopy. Sci Rep 7(1):13343

Garab G, Yaguzhinsky LS, Dlouhý O, Nesterov SV, Špunda V, Gasanoff ES (2022) Structural and functional roles of non-bilayer lipid phases of chloroplast thylakoid membranes and mitochondrial inner membranes. Progress Lipid Res 86. https://doi.org/10.1016/j.plipres.2022.101163

Garcia C, Khan NZ, Nannmark U, Aronsson H (2010) The chloroplast protein CPSAR1, dually localized in the stroma and the inner envelope membrane, is involved in thylakoid biogenesis. Plant J 63(1):73–85. https://doi.org/10.1111/J.1365-313X.2010.04225.X

García-Cerdán JG, Schmid EM, Takeuchi T, McRae I, McDonald KL, Yordduangjun N, Hassan AM, Grob P, Xu CS, Hess HF, Fletcher DA, Nogales E, Niyogi KK (2020) Chloroplast Sec14-like 1 (CPSFL1) is essential for normal chloroplast development and affects carotenoid accumulation in Chlamydomonas. Proc Natl Acad Sci USA 117(22):12452–12463

Garidel P, Kaconis Y, Heinbockel L, Wulf M, Gerber S, Munk A, Vill V, Brandenburg K (2015) Self-organisation, thermotropic and lyotropic properties of glycolipids related to their biological implications. Open Biochem J 9:49

Graça AT, Hall M, Persson K, Schröder WP (2021) High-resolution model of Arabidopsis Photosystem II reveals the structural consequences of digitonin-extraction. Sci Rep 11(1). https://doi.org/10.1038/s41598-021-94914-x

Grzyb JM, Solymosi K, Strzałka K, Mysliwa-Kurdziel B (2013) Visualization and characterization of prolamellar bodies with atomic force microscopy. J Plant Physiol 170(14):1217–1227. https://doi.org/10.1016/j.jplph.2013.04.017

Gunning BES (1965) The greening process in plastids – 1. The structure of the prolamellar body. Protoplasma 60(1):111–130. https://doi.org/10.1007/BF01248133

Gunning B (2001) Membrane geometry of "open" prolamellar bodies. Protoplasma 215(1–4):4–15. https://doi.org/10.1007/BF01280299

Heinz S, Rast A, Shao L, Gutu A, Gügel IL, Heyno E, Labs M, Rengstl B, Viola S, Nowaczyk MM (2016) Thylakoid membrane architecture in Synechocystis depends on CurT, a homolog of the granal CURVATURE THYLAKOID1 proteins. Plant Cell 28(9):2238–2260

Herbstová M, Tietz S, Kinzel C, Turkina MV, Kirchhoff H (2012) Architectural switch in plant photosynthetic membranes induced by light stress. Proc Natl Acad Sci USA 109(49). https://doi.org/10.1073/pnas.1214265109

Hernández-Verdeja T, Vuorijoki L, Strand Å (2020) Emerging from the darkness: interplay between light and plastid signaling during chloroplast biogenesis. Physiol Plant 169:397–406. https://doi.org/10.1111/ppl.13100

Heyes DJ, Zhang S, Taylor A, Johannissen LO, Hardman SJO, Hay S, Scrutton NS (2021) Photocatalysis as the 'master switch' of photomorphogenesis in early plant development. Nat Plants 7(3):268–276. https://doi.org/10.1038/s41477-021-00866-5

Holt NE, Fleming GR, Niyogi KK (2004) Toward an understanding of the mechanism of non-photochemical quenching in green plants. Biochemistry 43(26):8281–8289. https://doi.org/10.1021/bi0494020

Hölzl G, Dörmann P (2019) Chloroplast lipids and their biosynthesis. Ann Rev Plant Biol 70. https://doi.org/10.1146/annurev-arplant-050718-100202

Howley E, Mangus A, Williams D, Torres CI (2023) Intracytoplasmic membranes develop in Geobacter sulfurreducens under thermodynamically limiting conditions. Npj Biofilms Microbiomes 9(1):1–8. https://doi.org/10.1038/s41522-023-00384-6

Huokko T, Ni T, Dykes GF, Simpson DM, Brownridge P, Conradi FD, Beynon RJ, Nixon PJ, Mullineaux CW, Zhang P (2021) Probing the biogenesis pathway and dynamics of thylakoid membranes. Nat Commun 12(1):3475

Itzhaki H, Naveh L, Lindahl M, Cook M, Adam Z (1998) Identification and characterization of DegP, a serine protease associated with the luminal side of the thylakoid membrane. J Biol Chem 273(12). https://doi.org/10.1074/jbc.273.12.7094

Iwai M, Yamada-Oshima Y, Asami K, Kanamori T, Yuasa H, Shimojima M, Ohta H (2021) Recycling of the major thylakoid lipid MGDG and its role in lipid homeostasis in Chlamydomonas reinhardtii. Plant Physiol 187(3):1341–1356

Jajić I, Sarna T, Szewczyk G, Strzałka K (2015) Changes in production of reactive oxygen species in illuminated thylakoids isolated during development and senescence of barley. J Plant Physiol 184:49–56. https://doi.org/10.1016/j.jplph.2015.06.009

Jansen MAK, Mattoo AK, Edelman M (1999) D1-D2 protein degradation in the chloroplast: complex light saturation kinetics. Eur J Biochem 260(2). https://doi.org/10.1046/j.1432-1327.1999.00196.x

Järvi S, Suorsa M, Aro EM (2015) Photosystem II repair in plant chloroplasts – regulation, assisting proteins and shared components with photosystem II biogenesis. Biochim Biophys Acta Bioenerg 1847(9). https://doi.org/10.1016/j.bbabio.2015.01.006

Jedynak P, Trzebuniak KF, Chowaniec M, Zgłobicki P, Banaś AK, Mysliwa-Kurdziel B (2022) Dynamics of etiolation monitored by seedling morphology, carotenoid composition, antioxidant level, and Photoactivity of Protochlorophyllide in Arabidopsis thaliana. Front Plant Sci 12. https://doi.org/10.3389/fpls.2021.772727

Jiang H-W, Wu H-Y, Wang C-H, Yang C-H, Ko J-T, Ho H-C, Tsai M-D, Bryant DA, Li F-W, Ho M-C (2023) A structure of the relict phycobilisome from a thylakoid-free cyanobacterium. Nat Commun 14(1):8009

Johnson MP, Wientjes E (2020) The relevance of dynamic thylakoid organisation to photosynthetic regulation. Biochim Biophys Acta Bioenerg 1861(4). https://doi.org/10.1016/J.BBABIO.2019.06.011

Joly D, Govindachary S, Fragata M (2011) Photosystem II reconstitution into proteoliposomes and methodologies for structure–function characterization. Methods Mol Biol . 684:217–245

Kakuszi A, Solymosi K, Böddi B (2017) Transformation of plastids in soil-shaded lowermost hypocotyl segments of bean (Phaseolus vulgaris) during a 60-day cultivation period Physiologia Plantarum, 159(4), 483–491. https://doi.org/10.1111/ppl.2017.159. issue-4 https://doi.org/10.1111/ppl.12519

Kapri-Pardes E, Naveh L, Adam Z (2007) The thylakoid lumen protease Deg1 is involved in the repair of photosystem II from photoinhibition in Arabidopsis. Plant Cell 19(3). https://doi.org/10.1105/tpc.106.046573

Kato Y, Miura E, Ido K, Ifuku K, Sakamoto W (2009) The variegated mutants lacking chloroplastic FtsHs are defective in D1 degradation and accumulate reactive oxygen species. Plant Physiol 151(4). https://doi.org/10.1104/pp.109.146589

Kato Y, Hyodo K, Sakamoto W (2018) The photosystem II repair cycle requires FTSH turnover through the ENGA GtPase. Plant Physiol 178(2). https://doi.org/10.1104/pp.18.00652

Khatoon M, Inagawa K, Pospišil P, Yamashita A, Yoshioka M, Lundin B, Horie J, Morita N, Jajoo A, Yamamoto Y, Yamamoto Y (2009) Quality control of photosystem II: thylakoid unstacking necessary to avoid further damage to the D1 protein and to facilitate D1 degradation under light stress in spinach thylakoids. J Biol Chem 284(37). https://doi.org/10.1074/jbc.M109.007740

Kim E, Watanabe A, Duffy CDP, Ruban AV, Minagawa J, Sciences C, Mary Q, Road M (2020) Multimeric and monomeric photosystem II supercomplexes represent structural adaptations to low- and high-light conditions. J Biol Chem. https://doi.org/10.1074/jbc.RA120.014198

Kirchhoff H (2013) Structural constraints for protein repair in plant photosynthetic membranes. Plant Signaling Behav 8(4). https://doi.org/10.4161/psb.23634

Kirchhoff H (2014) Structural changes of the thylakoid membrane network induced by high light stress in plant chloroplasts. Philos Trans R Soc B: Biol Sci 369(1640):3–9. https://doi.org/10.1098/rstb.2013.0225

Kirchhoff H, Haase W, Wegner S, Danielsson R, Ackermann R, Albertsson PA (2007) Low-light-induced formation of semicrystalline photosystem II arrays in higher plant chloroplasts. Biochemistry 46(39). https://doi.org/10.1021/bi700748y

Kirchhoff H, Hall C, Wood M, Herbstová M, Tsabari O, Nevo R, Charuvi D, Shimoni E, Reich Z (2011) Dynamic control of protein diffusion within the granal thylakoid lumen. Proc Natl Acad Sci USA 108(50). https://doi.org/10.1073/pnas.1104141109

Klasek L, Inoue K, Theg SM (2020) Chloroplast chaperonin-mediated targeting of a thylakoid membrane protein. Plant Cell 32(12):3884–3901

Klostermann E, DROSTE gen. HELLING, I, Carde J-P, Schünemann D (2002) The thylakoid membrane protein ALB3 associates with the cpSecY-translocase in Arabidopsis thaliana. Biochem J 368(3):777–781

Kobayashi K, Wada H (2016) Role of lipids in chloroplast biogenesis. In: Nakamura Y, Li-Beisson Y (eds) Lipids in plant and algae development. Subcellular biochemistry, vol 86. Springer International Publishing, pp 103–125. https://doi.org/10.1007/978-3-319-25979-6_5

Kouřil R, Dekker JP, Boekema EJ (2012) Supramolecular organization of photosystem II in green plants. Biochim Biophys Acta Bioenerg 1817(1):2–12. https://doi.org/10.1016/J.BBABIO.2011.05.024

Kouřil R, Wientjes E, Bultema JB, Croce R, Boekema EJ (2013) High-light vs. low-light: effect of light acclimation on photosystem II composition and organization in Arabidopsis thaliana. Biochim Biophys Acta Bioenerg 1827(3):411–419. https://doi.org/10.1016/J.BBABIO.2012.12.003

Kowalewska Ł, Mazur R, Suski S, Garstka M, Mostowska A (2016) Three-dimensional visualization of the tubular-lamellar transformation of the internal plastid membrane network during runner bean chloroplast biogenesis. Plant Cell 28(4):875–891

Kowalewska Ł, Bykowski M, Mostowska A (2019) Spatial organization of thylakoid network in higher plants. Bot Lett 166(3):326–343. https://doi.org/10.1080/23818107.2019.1619195

Kulke M, Weraduwage SM, Sharkey TD, Vermaas JV (2023) Nanoscale simulation of the thylakoid membrane response to extreme temperatures. Plant Cell Environ 46(8):2419–2431

Kakuszi et al. 2017 - Kakuszi, A., Solymosi, K., & Böddi, B. (2017). Transformation of plastids in soil-shaded lowermost hypocotyl segments of bean (Phaseolus vulgaris) during a 60-day cultivation period. Physiologia Plantarum, 159(4), 483-491. doi: 10.1111/ppl.12519.

Lee AG (2000) Membrane lipids: it's only a phase. Curr Biol 10(10):R377–R380

Li N, Jarvis RP (2024) Recruitment of Cdc48 to chloroplasts by a UBX-domain protein in chloroplast-associated protein degradation. Nat Plants 10(9):1400–1417. https://doi.org/10.1038/s41477-024-01769-x

Li L, Aro EM, Millar AH (2018) Mechanisms of photodamage and protein turnover in photoinhibition. Trends Plant Sci 23(8). https://doi.org/10.1016/j.tplants.2018.05.004

Liang Z, Zhu N, Mai KK, Liu ZY, Tzeng D, Osteryoung KW, Zhong S, Staehelin LA, Kang BH (2018) Thylakoid-bound Polysomes and a dynamin-related protein, FZL, mediate critical stages of the linear chloroplast biogenesis program in greening Arabidopsis cotyledons. Plant Cell 30(7):1476–1495. https://doi.org/10.1105/TPC.17.00972

Liang Z, Yeung WT, Ki Mai KK, Ma J, Liu Z, Chong Y-LF, Kang B-H (2021) CURT1A and CURT1C mediate distinct stages of plastid conversion in Arabidopsis. BioRxiv, 2012–2021

Liang Z, Yeung W-T, Ma J, Mai KKK, Liu Z, Chong Y-LF, Cai X, Kang B-H (2022) Electron tomography of prolamellar bodies and their transformation into grana thylakoids in cryofixed Arabidopsis cotyledons. Plant Cell 34(10):3830–3843. https://doi.org/10.1093/plcell/koac205

Liebers M, Cozzi C, Uecker F, Chambon L, Blanvillain R, Pfannschmidt T (2022) Biogenic signals from plastids and their role in chloroplast development. In. J Exp Bot 73, Issue 21. https://doi.org/10.1093/jxb/erac344

Lindquist E, Aronsson H (2014) Proteins affecting thylakoid morphology–the key to understanding vesicle transport in chloroplasts? Plant Signal Behav 9(12):e977205

Ling Q, Sadali NM, Soufi Z, Zhou Y, Huang B, Zeng Y, Rodriguez-Concepcion M, Jarvis RP (2021) The chloroplast-associated protein degradation pathway controls chromoplast development and fruit ripening in tomato. Nat Plants 7(5):655–666

Liu X, Zhou Y, Xiao J, Bao F (2018) Effects of chilling on the structure, function and development of chloroplasts. Front Plant Sci 9:1715

Liu H, Li A, Rochaix JD, Liu Z (2023) Architecture of chloroplast TOC–TIC translocon supercomplex. Nature 615(7951):349–357. https://doi.org/10.1038/s41586-023-05744-y

Llorente B, Torres-Montilla S, Morelli L, Florez-Sarasa I, Tomás Matus J, Ezquerro M, D'andrea L, Houhou F, Picó B, Cebolla J, Troncoso A, Fernie AR, Daròs J-A, Rodriguez-Concepcion M (2020) Synthetic conversion of leaf chloroplasts into carotenoid-rich plastids reveals mechanistic basis of natural chromoplast development. Proc Natl Acad Sci USA. https://doi.org/10.1073/pnas.2004405117/-/DCSupplemental

Loudya N, Barkan A, López-Juez E (2024) Plastid retrograde signaling: a developmental perspective. Plant Cell 36(10):3903–3913. https://doi.org/10.1093/PLCELL/KOAE094

Lu Y (2016) Identification and roles of photosystem II assembly, stability, and repair factors in Arabidopsis. Front Plant Sci 7:168

Luciński R, Misztal L, Samardakiewicz S, Jackowski G (2011) The thylakoid protease Deg2 is involved in stress-related degradation of the photosystem II light-harvesting protein Lhcb6 in Arabidopsis thaliana. New Phytol 192(1). https://doi.org/10.1111/j.1469-8137.2011.03782.x

Magnuson A, Cardona T (2016) Thylakoid membrane function in heterocysts. Biochim Biophys Acta Bioenerg 1857(3):309–319. https://doi.org/10.1016/j.bbabio.2015.10.016

Mahbub M, Hemm L, Yang Y, Kaur R, Carmen H, Engl C, Huokko T, Riediger M, Watanabe S, Liu LN, Wilde A, Hess WR, Mullineaux CW (2020) mRNA localization, reaction Centre biogenesis and thylakoid membrane targeting in cyanobacteria. Nat Plants 6(9):1179–1191. https://doi.org/10.1038/s41477-020-00764-2

Mai KKK, Yeung W-T, Han S-Y, Cai X, Hwang I, Kang B-H (2019) Electron tomography analysis of thylakoid assembly and fission in chloroplasts of a single-cell C4 plant, Bienertia sinuspersici. Sci Rep 9(1):19640

Mareš J, Strunecký O, Bučinská L, Wiedermannová J (2019) Evolutionary patterns of thylakoid architecture in cyanobacteria. Front Microbiol 10:277

Marsh D (2012) Thermodynamics of phospholipid self-assembly. Biophys J 102(5):1079–1087

Matsubara S, Tamiaki H (2020) Supramolecular chlorophyll aggregates inspired from specific light-harvesting antenna "chlorosome": static nanostructure, dynamic construction process, and versatile application. J Photochem Photobiol C: Photochem Rev 45:100385. https://doi.org/10.1016/J.JPHOTOCHEMREV.2020.100385

Mazur R, Mostowska A, Szach J, Gieczewska K, Wójtowicz J, Bednarska K, Garstka M, Kowalewska Ł (2019) Galactolipid deficiency disturbs spatial arrangement of the thylakoid network in Arabidopsis thaliana plants. J Exp Bot 70(18):4689–4704

Mazur R, Gieczewska K, Kowalewska Ł, Kuta A, Proboszcz M, Gruszecki WI, Mostowska A, Garstka M (2020) Specific composition of lipid phases allows retaining an optimal thylakoid membrane fluidity in plant response to low-temperature treatment. Front Plant Sci 11:723

Mechela A, Schwenkert S, Soll J (2019) A brief history of thylakoid biogenesis. Open Biol 9. https://doi.org/10.1098/rsob.180237

Messant M, Krieger-Liszkay A, Shimakawa G (2021) Dynamic changes in protein-membrane association for regulating photosynthetic electron transport. Cells 10(5):1216

Müh F, Zouni A (2020) Structural basis of light-harvesting in the photosystem II core complex. Protein Sci pro.3841. https://doi.org/10.1002/pro.3841

Mullet JE (1983) The amino acid sequence of the polypeptide segment which regulates membrane adhesion (grana stacking) in chloroplasts. J Biol Chem 258(16). https://doi.org/10.1016/s0021-9258(17)44588-1

Muñoz-Gómez SA, Cadena LR, Gardiner AT, Leger MM, Sheikh S, Connell LB, Bilý T, Kopejtka K, Beatty JT, Koblížek M, Roger AJ, Slamovits CH, Lukeš J, Hashimi H (2023) Intracytoplasmic-membrane development in alphaproteobacteria involves the homolog of the mitochondrial crista-developing protein Mic60. Curr Biol 33(6):1099–1111.e6. https://doi.org/10.1016/J.CUB.2023.02.059/ASSET/0602AA22-F5BA-4759-BFEA-BDECA162E592/MAIN.ASSETS/GR6_LRG.JPG

Murakami S, Yamada N, Nagano M, Osumi M (1985) Three-dimensional structure of the prolamellar body in squash etioplasts. Protoplasma 128(2–3). https://doi.org/10.1007/BF01276336

Murphy DJ (1982) The importance of non-planar bilayer regions in photosynthetic membranes and their stabilisation by galactolipids. FEBS Lett 150(1):19–26

Mustárdy L, Garab G (2003) Granum revisited. A three-dimensional model–where things fall into place. Trends Plant Sci 8(3):117–122

Nath K, Jajoo A, Poudyal RS, Timilsina R, Park YS, Aro EM, Nam HG, Lee CH (2013) Towards a critical understanding of the photosystem II repair mechanism and its regulation during stress conditions. FEBS Lett 587(21). https://doi.org/10.1016/j.febslet.2013.09.015

Neff MM, Chory J (1998) Genetic interactions between phytochrome A, phytochrome B, and cryptochrome 1 during Arabidopsis development. Plant Physiol 118(1):27–35

Ni H, Song R, Liu B, Hu H, Liu J, Wang Q, Wang R, Mao P, Jia S (2024) Temporal dynamics of chloroplast biogenesis revealed initiation of photosynthesis-related gene expression and protein complexes during alfalfa seed germination. Plant Physiol Biochem 213. https://doi.org/10.1016/J.PLAPHY.2024.108868

Nickelsen J, Zerges W (2013) Thylakoid biogenesis has joined the new era of bacterial cell biology. Frontiers Plant Sci 4:458. Frontiers Media SA

Nield J, Funk C, Barber J (2000a) Supermolecular structure of photosystem II and location of the PsbS protein. Philos Trans R Soc Lond Ser B Biol Sci 355(1402):1337–1344. https://doi.org/10.1098/rstb.2000.0695

Nield J, Orlova EV, Morris EP, Gowen B, Van Heel M, Barber J (2000b) 3D map of the plant photosystem II supercomplex obtained by cryoelectron microscopy and single particle analysis. Nat Struct Biol 7(1). https://doi.org/10.1038/71242

Niroomand H, Mukherjee D, Khomami B (2017) Tuning the photoexcitation response of cyanobacterial Photosystem I via reconstitution into Proteoliposomes. Sci Rep 7(1):2492

Nixon PJ, Barker M, Boehm M, De Vries R, Komenda J (2005) FtsH-mediated repair of the photosystem II complex in response to light stress. J Exp Bot 56(411). https://doi.org/10.1093/jxb/eri021

Ohnishi N, Zhang L, Sakamoto W (2018) VIPP1 involved in chloroplast membrane integrity has GTPase activity in vitro. Plant Physiol 177(1):328–338

Orf GS, Blankenship RE (2013) Chlorosome antenna complexes from green photosynthetic bacteria. Photosynth Res 116(2–3):315–331. https://doi.org/10.1007/S11120-013-9869-3/FIGURES/9

Ostermeier M, Heinz S, Hamm J, Zabret J, Rast A, Klingl A, Nowaczyk MM, Nickelsen J (2022) Thylakoid attachment to the plasma membrane in Synechocystis sp. PCC 6803 requires the AncM protein. Plant Cell 34(1):655–678. https://doi.org/10.1093/PLCELL/KOAB253

Ostermeier M, Garibay-Hernández A, Holzer VJC, Schroda M, Nickelsen J (2024) Structure, biogenesis, and evolution of thylakoid membranes. Plant Cell 36(10):4014–4035. https://doi.org/10.1093/PLCELL/KOAE102

Osumi M, Yamada N, Nagano M, Murakami S, Baba N, Oho E, Kanaya K (1984) Three-dimensional observation of the prolamellar bodies in etioplasts of squash Cucurbita moschata. Scan Electron Microsc 1:111

Paluch-Lubawa E, Stolarska E, Sobieszczuk-Nowicka E (2021) Dark-induced barley leaf senescence—a crop system for studying senescence and autophagy mechanisms. Front Plant Sci 12:635619. https://doi.org/10.3389/FPLS.2021.635619/BIBTEX

Pan X, Ma J, Su X, Cao P, Chang W, Liu Z, Zhang X, Li M (2018) Structure of the maize photosystem I supercomplex with light-harvesting complexes I and II. Science 360(6393). https://doi.org/10.1126/science.aat1156

Perczyk P, Broniatowski M (2022) Membrane composition and successful bioaugmentation. Studies of the interactions of model thylakoid and plasma cyanobacterial and bacterial membranes with fungal membrane-lytic enzyme Lecitase ultra. Biochim Biophys Acta Biomembr 1864(6):183888

Pesaresi P, Pribil M, Wunder T, Leister D (2011) Dynamics of reversible protein phosphorylation in thylakoids of flowering plants: the roles of STN7, STN8 and TAP38. Biochim Biophys Acta Bioenerg 1807(8). https://doi.org/10.1016/j.bbabio.2010.08.002

Pipitone R, Eicke S, Pfister B, Glauser G, Falconet D, Uwizeye C, Pralon T, Zeeman SC, Kessler F, Demarsy E (2021) A multifaceted analysis reveals two distinct phases of chloroplast biogenesis during de-etiolation in arabidopsis. elife 10:1–32. https://doi.org/10.7554/ELIFE.62709

Plöscher M, Reisinger V, Eichacker LA (2011) Proteomic comparison of etioplast and chloroplast protein complexes. J Proteome 74(8):1256–1265. https://doi.org/10.1016/j.jprot.2011.03.020

Pogson BJ, Ganguly D, Albrecht-Borth V (2015) Insights into chloroplast biogenesis and development. Biochim Biophys Acta 1847(9):1017–1024. https://doi.org/10.1016/j.bbabio.2015.02.003

Poudyal RS, Rodionova MV, Kim H, Lee S, Do E, Allakhverdiev SI, Nam HG, Hwang D, Kim Y (2020) Combinatory actions of CP29 phosphorylation by STN7 and stability regulate leaf age-dependent disassembly of photosynthetic complexes. Sci Rep 10(1):10267

Pribil M, Sandoval-Ibáñez O, Xu W, Sharma A, Labs M, Liu Q, Galgenmüller C, Schneider T, Wessels M, Matsubara S, Jansson S, Wanner G, Leister D (2018) Fine-tuning of photosynthesis requires CURVATURE THYLAKOID1-mediated THYLAKOID plasticity 1. Plant Physiol 176:2351–2364. https://doi.org/10.1104/pp.17.00863

Rast A, Heinz S, Nickelsen J (2015) Biogenesis of thylakoid membranes. Biochim Biophys Acta Bioenerg 1847(9):821–830

Raven JA (2021) Determinants, and implications, of the shape and size of thylakoids and cristae. J Plant Physiol 257:153342

Rexroth S, Mullineaux CW, Ellinger D, Sendtko E, Rögner M, Koenig F (2011) The plasma membrane of the cyanobacterium Gloeobacter violaceus contains segregated bioenergetic domains. Plant Cell 23(6):2379–2390

Rochaix J-D (2011) Assembly of the photosynthetic apparatus. Plant Physiol 155(4):1493–1500

Rochaix JD, Lemeille S, Shapiguzov A, Samol I, Fucile G, Willig A, Goldschmidt-Clermont M (2012) Protein kinases and phosphatases involved in the acclimation of the photosynthetic apparatus to a changing light environment. Philos Trans R Soc B: Biol Sci 367(1608). https://doi.org/10.1098/rstb.2012.0064

Rokka A, Suorsa M, Saleem A, Battchikova N, Aro E-M (2005) Synthesis and assembly of thylakoid protein complexes: multiple assembly steps of photosystem II. Biochem J 388(1):159–168

Rolo D, Sandoval-Ibáñez O, Thiele W, Schöttler MA, Gerlach I, Zoschke R, Schwartzmann J, Meyer EH, Bock R (2024a) CO-EXPRESSED WITH PSI ASSEMBLY1 (CEPA1) is a photosystem I ASSEMBLY factor in Arabidopsis. Plant Cell 36(10):4179–4211. https://doi.org/10.1093/PLCELL/KOAE042

Rolo D, Schöttler MA, Sandoval-Ibáñez O, Bock R (2024b) Structure, function, and assembly of PSI in thylakoid membranes of vascular plants. Plant Cell 36(10):4080–4108. https://doi.org/10.1093/PLCELL/KOAE169

Ruban AV (2016) Nonphotochemical chlorophyll fluorescence quenching: mechanism and effectiveness in protecting plants from photodamage. Plant Physiol 170:1903–1916. https://doi.org/10.1104/pp.15.01935

Ruban AV, Wilson S (2021) The mechanism of non-photochemical quenching in plants: localization and driving forces. Plant Cell Physiol 62(7). https://doi.org/10.1093/pcp/pcaa155

Rumak I, Gieczewska K, Kierdaszuk B, Gruszecki WI, Mostowska A, Mazur R, Garstka M (2010) 3-D modelling of chloroplast structure under (Mg2+) magnesium ion treatment. Relationship between thylakoid membrane arrangement and stacking. Biochim Biophys Acta Bioenerg 1797(10):1736–1748

Saini R, Debnath A (2023) Thylakoid composition facilitates chlorophyll a dimerization through stronger Interlipid interactions. J Phys Chem B 127(42):9082–9094

Sakamoto W, Zaltsman A, Adam Z, Takahashi Y (2003) Coordinated regulation and complex formation of yellow Variegated1 and yellow Variegated2, Chloroplastic FtsH metalloproteases involved in the repair cycle of photosystem II in Arabidopsis thylakoid membranes. Plant Cell 15(12). https://doi.org/10.1105/tpc.017319

Sandoval-Ibáñez O, Sharma A, Bykowski M, Borràs-Gas G, Behrendorff JBYH, Mellor S, Qvortrup K, Verdonk JC, Bock R, Kowalewska Ł, Pribil M (2021) Curvature thylakoid 1 proteins modulate prolamellar body morphology and promote organized thylakoid biogenesis in Arabidopsis thaliana. Proc Natl Acad Sci USA 118(42). https://doi.org/10.1073/PNAS.2113934118

Santamaria-Gomez J, Mariscal V, Luque I (2018) Mechanisms for protein redistribution in thylakoids of Anabaena during cell differentiation. Plant Cell Physiol 59(9):1860–1873

Satoh I, Gotou K, Nagatsuma S, Nagashima KVP, Kobayashi M, Yu LJ, Madigan MT, Kimura Y, Wang-Otomo ZY (2023) Selective expression of light-harvesting complexes alters phospholipid composition in the intracytoplasmic membrane and core complex of purple phototrophic bacteria. Biochim Biophys Acta Bioenerg 1864(4):149001. https://doi.org/10.1016/J.BBABIO.2023.149001

Schaeffer SM, Christian R, Castro-Velasquez N, Hyden B, Lynch-Holm V, Dhingra A (2017) Comparative ultrastructure of fruit plastids in three genetically diverse genotypes of apple (Malus × domestica Borkh.) during development. Plant Cell Rep 36(10):1627–1640. https://doi.org/10.1007/S00299-017-2179-Z/FIGURES/9

Scrutton NS, Louise Groot M, Heyes DJ (2012) Excited state dynamics and catalytic mechanism of the light-driven enzyme protochlorophyllide oxidoreductase. Phys Chem Chem Phys 14(25):8818. https://doi.org/10.1039/c2cp23789j

Seiwert D, Witt H, Ritz S, Janshoff A, Paulsen H (2018) The nonbilayer lipid MGDG and the major light-harvesting complex (LHCII) promote membrane stacking in supported lipid bilayers. Biochemistry 57(15):2278–2288

Selstam E (1998) Development of thylakoid membranes with respect to lipids. In: Lipids in photosynthesis: structure, function and genetics. Springer, pp 209–224

Selstam E, Schelin J, Williams WP, Brain APR (2007) Structural organisation of prolamellar bodies (PLB) isolated from Zea mays. Parallel TEM, SAXS and absorption spectra measurements on samples subjected to freeze-thaw, reduced pH and high-salt perturbation. Biochim Biophys Acta Biomembr 1768(9):2235–2245. https://doi.org/10.1016/j.bbamem.2007.05.005

Shah M, Soares EL, Lima MLB, Pinheiro CB, Soares AA, Domont GB, Nogueira FCS, Campos FAP (2016) Deep proteome analysis of gerontoplasts from the inner integument of developing seeds of Jatropha curcas. J Proteome 143:346–352

Sheng X, Liu Z, Kim E, Minagawa J (2021) Plant and algal PSII–LHCII Supercomplexes: structure, evolution and energy transfer. Plant Cell Physiol 62(7):1108–1120. https://doi.org/10.1093/PCP/PCAB072

Shimizu T, Masuda T (2021) The role of tetrapyrrole-and gun1-dependent signaling on chloroplast biogenesis. Plants 10(2):1–19. https://doi.org/10.3390/PLANTS10020196

Shimoni E, Rav-Hon O, Ohad I, Brumfeld V, Reich Z (2005) Three-dimensional organization of higher-plant chloroplast thylakoid membranes revealed by electron tomography. Plant Cell 17(9):2580–2586

Skotnicová P, Srivastava A, Aggarwal D, Talbot J, Karlínová I, Moos M, Mareš J, Bučinská L, Koník P, Šimek P (2024) A thylakoid biogenesis BtpA protein is required for the initial step of tetrapyrrole biosynthesis in cyanobacteria. New Phytol 241(3):1236–1249

Sobieszczuk-Nowicka E, Zmienko A, Samelak-Czajka A, Łuczak M, Pietrowska-Borek M, Iorio R, Del Duca S, Figlerowicz M, Legocka J (2015) Dark-induced senescence of barley leaves involves activation of plastid transglutaminases. Amino Acids 47(4):825–838. https://doi.org/10.1007/s00726-014-1912-y

Solymosi K, Aronsson H (2013) Etioplasts and their significance in chloroplast biogenesis. In: Biswal B, Krupińska K, Biswal AK (eds) Plastid development in leaves during growth and senescence: advances in photosynthesis and respiration, vol 36. Springer, Dordrecht, pp 39–71. https://doi.org/10.1007/978-94-007-5724-0

Solymosi K, Mysliwa-Kurdziel B (2021) The role of membranes and lipid-protein interactions in the Mg-branch of Tetrapyrrole biosynthesis. Front Plant Sci 12:663309. https://doi.org/10.3389/fpls.2021.663309

Solymosi K, Schoefs B (2010) Etioplast and etio-chloroplast formation under natural conditions: the dark side of chlorophyll biosynthesis in angiosperms. Photosynth Res 105(2):143–166. https://doi.org/10.1007/s11120-010-9568-2

Solymosi K, Vitányi B, Hideg E, Böddi B (2007) Etiolation symptoms in sunflower (Helianthus annuus) cotyledons partially covered by the pericarp of the achene. Annals of Botany, 99(5), 857–867. https://doi.org/10.1093/aob/mcm034

Solymosi K, Lenti K, Myśliwa-Kurdziel B, Fidy J, Strzałka K, Böddi B (2004) Hg2+ reacts with different components of the NADPH: protochlorophyllide oxidoreductase macrodomains. Plant Biology, 6, 358–368. https://doi.org/10.1055/s-2004-817893

Solymosi K, Morandi D, Bóka K, Böddi B, Schoefs B (2012) High biological variability of plastids photosynthetic pigments and pigment forms of leaf primordia in buds Planta, 235(5), 1035–1049. https://doi.org/10.1007/s00425-011-1559-9

Solymosi K, Myśliwa-Kurdziel B, Bóka K, Strzałka K, Böddi B (2006) Disintegration of the prolamellar body structure at high concentrations of Hg2+. Plant Biology, 8(05), 627–635. https://doi.org/10.1055/s-2006-924110

Spence E, Bailey S, Nenninger A, Geir SM, Robinson C (2004) A homolog of albino3/oxai is essential for thylakoid biogenesis in the cyanobacterium synechocystis sp. PCC6803. J Biol Chem, 279(53), 55792–55800. https://doi.org/10.1074/jbc.M411041200
Su X, Ma J, Wei X, Cao P, Zhu D, Chang W, Liu Z, Zhang X, Li M (2017) Structure and assembly mechanism of plant C2S2M2-type PSII-LHCII supercomplex. Science 357(6353):815–820. https://doi.org/10.1126/science.aan0327
Su J, Jiao Q, Jia T, Hu X (2024) The photosystem-II repair cycle: updates and open questions. Planta 259(1). https://doi.org/10.1007/s00425-023-04295-w
Sun Y, Jarvis RP (2023) Chloroplast Proteostasis: import, sorting, ubiquitination, and proteolysis. Ann Rev Plant Biol 74:259–283. https://doi.org/10.1146/ANNUREV-ARPLANT-070122-032532/CITE/REFWORKS
Sun T, Yuan H, Cao H, Yazdani M, Tadmor Y, Li L (2018) Carotenoid Metabolism in Plants: The Role of Plastids Molecular Plant, 11(1), 58–74. https://doi.org/10.1016/j.molp.2017.09.010
Sundqvist C, Dahlin C (1997) With chlorophyll pigments from prolamellar bodies to light-harvesting complexes. Physiol Plant 100(4):748–759. https://doi.org/10.1034/j.1399-3054.1997.1000402.x
Tanaka R, Kobayashi K, Masuda T (2011) Tetrapyrrole metabolism in Arabidopsis thaliana. Arabidopsis Book 9(9):e0145. https://doi.org/10.1199/tab.0145
Theis J, Schroda M (2016) Revisiting the photosystem II repair cycle. Plant Signal Behav 11(9). https://doi.org/10.1080/15592324.2016.1218587
Theis J, Niemeyer J, Schmollinger S, Ries F, Rütgers M, Gupta TK, Sommer F, Muranaka LS, Venn B, Schulz-Raffelt M, Willmund F, Engel BD, Schroda M (2020) VIPP2 interacts with VIPP1 and HSP22E/F at chloroplast membranes and modulates a retrograde signal for HSP22E/F gene expression. Plant Cell Environ 43(5):1212–1229. https://doi.org/10.1111/PCE.13732
Thomson SM, Pulido P, Jarvis RP (2020) Protein import into chloroplasts and its regulation by the ubiquitin-proteasome system. Biochem Soc Trans 48(1):71–82
Tietz S, Leuenberger M, Höhner R, Olson AH, Fleming GR, Kirchhoff H (2020) A proteoliposome-based system reveals how lipids control photosynthetic light harvesting. J Biol Chem 295(7):1857–1866
Trotta A, Bajwa AA, Mancini I, Paakkarinen V, Pribil M, Aro E-M (2019) The role of phosphorylation dynamics of CURVATURE THYLAKOID 1B in plant thylakoid membranes 1[OPEN]. Plant Physiol 181:1615–1631. https://doi.org/10.1104/pp.19.00942
Vainonen JP, Hansson M, Vener AV (2005) STN8 protein kinase in Arabidopsis thaliana is specific in phosphorylation of photosystem II core proteins. J Biol Chem 280(39). https://doi.org/10.1074/jbc.M505729200
Van Niftrik LA, Fuerst JA, Sinninghe Damsté JS, Kuenen JG, Jetten MSM, Strous M (2004) The anammoxosome: an intracytoplasmic compartment in anammox bacteria. FEMS Microbiol Lett 233(1):7–13. https://doi.org/10.1016/J.FEMSLE.2004.01.044
Varga AR, Staehelin L (1983) Spatial differentiation in photosynthetic and non-photosynthetic membranes of Rhodopseudomonas palustris. J Bacteriol 154(3):1414–1430
Vedalankar P, Tripathy BC (2024) Light dependent protochlorophyllide oxidoreductase: a succinct look. Physiol Mol Biol Plants 30(5):719–731. https://doi.org/10.1007/s12298-024-01454-5
von Zychlinski A, Kleffmann T, Krishnamurthy N, Sjölander K, Baginsky S, Gruissem W (2005) Proteome analysis of the rice etioplast: metabolic and regulatory networks and novel protein functions. Mol Cell Proteomics 4(8):1072–1084. https://doi.org/10.1074/mcp.M500018-MCP200
Vothknecht UC, Westhoff P (2001) Biogenesis and origin of thylakoid membranes. Biochim Biophys Acta, Mol Cell Res 1541(1–2):91–101. https://doi.org/10.1016/S0167-4889(01)00153-7
Wada H, Murata N (2007) The essential role of phosphatidylglycerol in photosynthesis. Photosynth Res 92:205–215
Wang P, Grimm B (2021) Connecting chlorophyll metabolism with accumulation of the photosynthetic apparatus. Trends Plant Sci 26(5):484–495. https://doi.org/10.1016/J.TPLANTS.2020.12.005

Wang S, Wang X, Shi X, Wang B, Zheng X, Wang H, Liu F (2016) Red and blue lights significantly affect photosynthetic properties and ultrastructure of mesophyll cells in senescing grape leaves. Hortic Plant J 2(2):82–90

Wang P, Ji S, Grimm B (2022) Post-translational regulation of metabolic checkpoints in plant tetrapyrrole biosynthesis. J Exp Bot 73(14):4624–4636. https://doi.org/10.1093/JXB/ERAC203

Wei X, Su X, Cao P, Liu X, Chang W, Li M, Zhang X, Liu Z (2016) Structure of spinach photosystem II-LHCII supercomplex at 3.2 Å resolution. Nature 534(7605):69–74. https://doi.org/10.1038/nature18020

Williams WP, Selstam E, Brain T (1998) X-ray diffraction studies of the structural organisation of prolamellar bodies isolated from Zea mays. FEBS Lett 422(2):252–254. https://doi.org/10.1016/S0014-5793(98)00019-2

Wilson S, Li D-H, Ruban AV (2022) The structural and spectral features of light-harvesting complex II proteoliposomes mimic those of native thylakoid membranes. J Phys Chem Lett 13(24):5683–5691

Wittmann D, Sinha N, Grimm B (2021) Thioredoxin-dependent control balances the metabolic activities of tetrapyrrole biosynthesis. Biol Chem 402(3):379–397. https://doi.org/10.1515/HSZ-2020-0308/ASSET/GRAPHIC/J_HSZ-2020-0308_FIG_002.JPG

Wójtowicz J, Grzyb J, Szach J, Mazur R, Gieczewska KB (2021) Bean and Pea plastoglobules change in response to chilling stress. Int J Mol Sci 22(21):11895

Wolfe GR, Hoober JK (1996) Evolution of thylakoid structure. In: Ort DR, Yocum CF, Heichel IF (eds) Oxygenic photosynthesis: the light reactions, Advances in photosynthesis and respiration, vol 4. Springer, Dordrecht. https://doi.org/10.1007/0-306-48127-8_3

Wood WHJ, Barnett SFH, Flannery S, Hunter CN, Johnson MP (2019) Dynamic thylakoid stacking is regulated by LHCII phosphorylation but not its interaction with PSI. Plant Physiol 180(4). https://doi.org/10.1104/pp.19.00503

Woodson JD (2022) Control of chloroplast degradation and cell death in response to stress. Trends Biochem Sci 47(10):851–864

Xu J, Ji Z, Wang C, Xu F, Wang F, Zheng Y, Tang Y, Wei Z, Zhao T, Zhao K (2022) WATER-SOAKED SPOT1 controls chloroplast development and leaf senescence via regulating reactive oxygen species homeostasis in rice. Front Plant Sci 13:918673

Yokoyama R, Yamamoto H, Kondo M, Takeda S, Ifuku K, Fukao Y, Kamei Y, Nishimura M, Shikanai T (2016) Grana-localized proteins, RIQ1 and RIQ2, affect the organization of light-harvesting complex II and grana stacking in Arabidopsis. Plant Cell 28(9):2261–2275

Yoshihara A, Kobayashi K, Nagata N, Fujii S, Wada H, Kobayashi K (2024) Anionic lipids facilitate membrane development and protochlorophyllide biosynthesis in etioplasts. Plant Physiol 194(3). https://doi.org/10.1093/plphys/kiad604

Yu C, Lin Y, Li H (2020) Increased ratio of galactolipid MGDG: DGDG induces jasmonic acid overproduction and changes chloroplast shape. New Phytol 228(4):1327–1335

Yu L, Fan J, Zhou C, Xu C (2021) Chloroplast lipid biosynthesis is fine-tuned to thylakoid membrane remodeling during light acclimation. Plant Physiol 185(1):94–107

Yoneda T, Tanimoto Y, Takagi D, Morigaki K (2020) Photosynthetic model membranes of natural plant thylakoid embedded in a patterned polymeric lipid bilayer langmuir, 36(21), 5863–5871. https://doi.org/10.1021/acs.langmuir.0c00613

Zapata JM, Guera A, Esteban-Carrasco A, Martin M, Sabater B (2005) Chloroplasts regulate leaf senescence: delayed senescence in transgenic ndhF-defective tobacco. Cell Death & Differentiation, 12(10), 1277–1284. https://doi.org/10.1038/sj.cdd.4401657

Zentgraf U, Andrade-Galan AG, Bieker S (2022) Specificity of H_2O_2 signaling in leaf senescence: is the ratio of H_2O_2 contents in different cellular compartments sensed in Arabidopsis plants? Cell Mol Biol Lett 27(1):4

Zhang B, Zhang C, Tang R, Zheng X, Zhao F, Fu A, Lan W, Luan S (2022) Two magnesium transporters in the chloroplast inner envelope essential for thylakoid biogenesis in Arabidopsis. New Phytol 236(2):464–478. https://doi.org/10.1111/NPH.18349

Zhu S, Zhao H, Liang J, Ji B, Jiao D (2008) Relationships between phosphatidylglycerol molecular species of thylakoid membrane lipids and sensitivities to chilling-induced photoinhibition in rice. J Integr Plant Biol 50(2):194–202

Zhu D, Xiong H, Wu J, Zheng C, Lu D, Zhang L, Xu X (2022) Protein targeting into the thylakoid membrane through different pathways. Front Physiol 12:802057

Plastids in Variegated Leaves

6

Marija Vidović

Abstract

The chloroplast is a unique organelle in photosynthetic eukaryotes, essential for fulfilling the complex and diverse functions required in green plant tissues. To support these functions, chloroplasts must be fully developed and operational. Their development is regulated by an intricate network of external and intrinsic factors.

Despite the necessity of functional chloroplasts, many plant species exhibit variegated phenotypes characterized by leaves containing pale green, yellowish, or white segments, which often lack functional chloroplasts. The physiological and ecological advantages of this phenomenon remain unclear, as leaf variegation is typically perceived as detrimental to optimal photosynthetic efficiency. However, under specific adverse environmental conditions, the adaptive benefits of variegation may outweigh the drawbacks associated with the absence of fully developed chloroplasts.

To gain a deeper understanding of the molecular origins and physiological implications of variegation, it is essential to investigate the ultrastructure of plastids in non-photosynthetically active leaf cells under physiological and various stress conditions. Therefore, this chapter focuses on the histological differences between the mesophyll tissue of uniformly green leaves and the yellow or white sectors of variegated leaves. Numerous ultrastructural studies suggest that the plastids in the non-green leaf sectors represent a distinct type of plastid, referred to as "albinoplasts."

The primary objective of this chapter is to highlight the key ultrastructural differences between normal plastids and the albinoplasts observed in the non-

M. Vidović (✉)
Group for Plant Molecular Biology, Institute of Molecular Genetics and Genetic Engineering, University of Belgrade, Belgrade, Serbia
e-mail: marija.vidovic@imgge.bg.ac.rs

V. K. Dalal, A. N. Misra (eds.), *Chloroplast Biogenesis and Plastid Interconversions*,
https://doi.org/10.1007/978-981-95-0772-6_6

green leaf sectors. This analysis considers a broad spectrum of molecular factors contributing to variegated phenotypes in both dicotyledonous and monocotyledonous plant species.

Keywords

Leaf Variegation · Variegated · Stress · Albino · Photosynthesis · Chloroplast biogenesis · Chlorophyll

6.1 Introduction

The chloroplast is a unique cellular compartment for photosynthetic eukaryotes. Together with other members of the plastid family, they fulfil an impressive array of tasks in addition to photosynthesis, including the synthesis of fatty acids, certain amino acids, photosynthetic pigments, phytochormones and lipids—all of which are essential for plant cell growth and function.

Chloroplasts are semi-autonomous organelles that have their own DNA (plastome; 120–160 kb), RNA and protein translation machinery. Most of the approximately 3000 proteins of the chloroplasts' proteome are nuclear-encoded and are imported into the chloroplast post-translationally (Nakai 2018). For instance, the barley (*Hordeum vulgare*) plastome encodes only 78 of 3000 chloroplast proteins (Saski et al. 2007).

Chloroplasts constantly adapt their energy conversion processes and metabolic functions to meet the demands of the plant metabolism under fluctuating environmental conditions. Furthermore, chloroplasts have developed various mechanisms to minimise and cope with toxic photochemical side products (e.g. reactive oxygen species, ROS), which are unavoidable generated through conversion and transfer of light energy in an oxygen-rich environment.

Chloroplasts must be fully developed and functional to perform their essential functions in green plant organs. Their maturation is regulated by a complex network of external and internal factors. However, a variety of plant species have evolved variegated phonotypes, containing leaves with pale green, yellowish, or white segments that, in general, typically lack functional chloroplasts.

It is still unclear what physiological and ecological benefits this phenomenon brings, as leaf variegation is often thought to be unfavourable for optimal photosynthetic efficiency. However, under certain challenging environmental conditions, the adaptive advantages of variegation for the overall leaf physiology may prevail over the potential disadvantage of having less chlorophyll. To answer these questions, a better understanding of the molecular causes and physiological consequences of variegation on the metabolism of non-photosynthetically active leaf cells is required. Therefore, the histological distinctions between the mesophyll tissue of plain morphs and the yellow/white leaf sectors of variegated leaves are the main topic of this chapter. The aim of this chapter is to highlight the main ultrastructural differences between normal and abnormal plastids of the non-green variegated leaf sectors, taking into account a wide variety of molecular causes of variegated phenotypes in dicots and monocots.

6.2 Chloroplasts-Physiological Roles and Ultrastructure

Chloroplasts are unique plant organelles. They are essential for the conversion of light energy, carbon dioxide and water into various chemical components, required for the synthesis of carbohydrates and other biomolecules (Gutiérrez-Nava et al. 2004). Along with other members of the plastid family, chloroplasts play a key role in the synthesis of fatty acids, pigments, phytochormones, particular amino acids and the storage of glucose in the form of starch (Rolland et al. 2018).

In plant cells, the chloroplasts act as a central hub for signalling transduction. They are the most important organelles involved in the perception of environmental stimuli and conducting the signals to other cell compartments, which can then lead to adaptations in plant growth and development (Chan et al. 2016). These organelles are sensitive to environmental fluctuations. For example, a sudden increase in sunlight leads to a corresponding increase in energy uptake by the photosynthetic apparatus, but this additional energy may not be used by the plant's metabolism. Such energy excess can trigger the formation of toxic by-products (i.e. ROS) by the primary photochemical reactions in thylakoid membranes (Foyer 2018). One of the fastest protective mechanisms against excess light is the development of the proton gradient across the thylakoid membrane (Krause 1973; Briantais et al. 1980) and violaxanthin deepoxidation to zeaxanthin (Demmig-Adams 1990), which leads to dissipation of the excess excitation energy in photosystem II (PSII) by heat, resulting in non-photochemical quenching (NPQ) of chlorophyll *a* fluorescence (Krause and Weis 1991). In addition to the various mechanisms that regulate light capture, photorespiration rate and ROS scavenging (Allen et al. 1981; Foyer and Hanke 2022; Mekala et al. 2015), in response to the intensity and direction of blue light, chloroplasts can relocate within mesophyll cells (Kasahara et al. 2002; Wada 2013).

The development of chloroplasts in plants is regulated by the synchronised expression of nuclear and chloroplast genomes and by a series of coordinated biological processes (Stern et al. 2004). Chloroplasts develop from proplastids, undeveloped predecessors located in shoot meristematic tissue and root tip. Proplastids are non-photosynthetic, small, roundish organelles that contain only a few internal membranes usually in the form of vesicles or small sac-like structures. Proplastids are also progenitors of amyloplasts (starch storage), chromoplasts (pigment-containing plastids), leucoplasts (storage), elaioplasts (lipid storage) and finally etioplasts (developing plastids) (Rolland et al. 2018). Juvenile chloroplasts are characterised by lenticular flattened appearances with parallel layered grana, but are smaller than mature chloroplasts, which contain disc-shaped structures that coalesce into grana stacks.

Mature chloroplasts from the mesophyll tissue of green leaves are large organelles (5–10 μm long). They contain unique structures such as thylakoids, starch, dense stroma, and plastoglobuli. Moreover, the structure of mature chloroplasts is defined by three membrane systems: the outer and inner envelope membranes and the thylakoid membrane system (Staehelin 2003). The chloroplasts of higher plants are circumscribed by a double membrane called the chloroplast envelope. The stroma of the chloroplast is bounded by the inner envelope membrane and the thylakoid membrane and contains DNA, RNA, and 70S ribosomes and harbours all so-called light-independent metabolic reactions. In electron micrographs, the stroma often appears unstructured and packed with scattered particles, which mainly represent the CO_2-fixing enzyme Rubisco (Kirchhoff 2019). The thylakoid membrane system, located in the stroma, is the site for the light-dependent reactions: it functions like a biological solar cell by converting sunlight into electron transport and a proton gradient across the thylakoid membrane. Thylakoids may appear either as single stacks or as grana stacks. The two other aqueous compartments in chloroplasts are the intermembrane space between the two envelope membranes and the thylakoid lumen, which is enclosed by the thylakoid membrane (Staehelin 2003).

Depending on the amount of glucose produced by the chloroplast and required in the cell, the chloroplasts may or may not contain starch granules (Thalmann and Santelia 2017). They appear as electron translucent round or ellipsoidal grains inside the chloroplasts and are not surrounded by a membrane. They often shrink during chemical fixation and are therefore surrounded by an electron translucent region that appears brighter than the starch grain itself (Zechmann 2019). Plastoglobuli are electron opaque round objects inside the plastids and are physically connected to the thylakoid membrane (Kirchhoff 2019). The current model assumes that they are bounded by a lipid monolayer which is a continuation of the outer thylakoid monolayer (Austin and Staehelin 2011; van Wijk and Kessler 2017). Therefore, plastoglobuli can be considered as thylakoid-associated lipid microcompartments filled with lipophilic metabolites similar to those in thylakoids, as well as enzymes that synthesise lipids in thylakoids (van Wijk and Kessler 2017; Zechmann 2019). Plastoglobuli are involved in the synthesis, repair, and disposal of metabolites that are important during plastid development, ageing, and adaptation to stress (Rottet et al. 2015).

Mesophyll tissue is a primary site of photosynthesis, as the mesophyll cells possess the majority of chloroplasts and it is widely accepted that, with the exception of guard cells, the epidermal layer in most higher plants does not contain chloroplasts (Zeiger et al. 1981; Zemel and Gepstein 1985). In most, but not all species, the guard cells also contain photosynthetically active smaller chloroplasts with less granum, with a reduced thylakoid network and lower chlorophyll content compared to mesophyllT (Lawson et al. 2002). However, recent observations in higher plants, has shown that epidermal pavement cells can also contain chloroplasts, contrary to previous assumptions. These chloroplasts, termed pavement cell chloroplasts, are smaller than those in mesophyll cells and have a higher stroma to grana ratio (Barton et al. 2018).

Chloroplast biogenesis depends on a network of environmental, temporal, and cellular factors, including the expression of both plastid and nuclear genes (Pogson and Albrecht 2011). When green tissue develops in the presence of light, chloroplasts are normally formed from undifferentiated proplastids that lack internal membranes. The presence of developed thylakoid grana is a marker for developed chloroplasts. The organisation of ribosomes into polysomes and their further attachment to the developing thylakoid membranes is an essential step in membrane biogenesis, as this could facilitate membrane flattening, which is required for grana formation (Liang et al. 2018).

Many mutants defective in chloroplast development have been identified and classified into four classes in terms of phenotype severity in relation to plant growth and development: embryo lethal, seedling lethal or albino, seedling lethal but conditional survival, and autotrophy with pale green, virescent, or variegated leaves (Kim et al. 2009). The white or yellow sections of these variegated leaves generally contain plastids different than normal chloroplasts.

6.3 Foliar Variegation

Leaf variegation is a common phenotype observed in many angiosperms. Variegated leaves typically contain green and white or yellowish sectors, and provide a model to study the expression of genes important for chloroplast development. A total of 1710 angiosperm species belonging to 356 genera and 78 families possess variegated leaves. Orchidaceae (252 species), Bromeliaceae (245 species), Begoniaceae (221 species), Araceae (127 species), and Asparagaceae (80 species) are the families with the most species having variegated leaves (Zhang et al. 2020).

The macroscopic patterns of variegated leaves are numerous: herringbone-shaped, dotted, V-shaped, spotted, striped, reticulate, and pinnate. Depending on the colour pattern distribution, site of colour origin, and organisation across the leaf section, Zhang et al. (2020) defined five causes of leaf variegation. Variegation can be observed due to: (i) lower chlorophyll content; (ii) presence of intercellular spaces between epidermal and palisade cells and among palisade cells; (iii) thicker epidermal layer; (iv) accumulation of non-photosynthetic pigments, such as anthocyanins; and (v) uneven distribution of coloured multicellular outgrowths on the adaxial epidermis of leaves (Zhang et al. 2020).

Several mechanisms, such as somatic chimerism, mutations in plastidial and mitochondrial genomes, plastome mutators, virus infections, plastid-nucleus incompatibility, transposable element activity, or epigenetic phenomena, are involved in the generation of cells with different genotypes in a single plant (Yu et al. 2007). In all these cases, the cells in the green leaf sectors have a wild-type genotype, while the cells in the white leaf sectors have an albino genotype. Additionally, variegation can also occur in mutants with identical genotypes in the green and the chlorotic tissue sectors.

The original site of mutation causing a variegated phenotype in the shoot apical meristem is responsible for the distribution of cells with normal and aberrant chloroplasts, which is reflected as the final leaf colouration pattern. Clonally distinct cell layers of the shoot apical meristem contribute to specific tissues within the newly formed lateral organs. In the leaves of most angiosperms, the outer meristem layer (L1) forms the epidermis, the second meristem layer (L2) forms the subepidermal palisade mesophyll and the abaxial spongy mesophyll tissue, and the inner corpus (L3) forms the deep mesophyll and vascular tissue (Fig. 6.1; overview in Tilney-Bassett 1986). If the mutation leading to abnormal chloroplasts occurs in the L2 layer, where the cells divide anticlinally, it spreads throughout the meristem layer, forming a uniform, genetically distinct cell layer (Fig. 6.1).

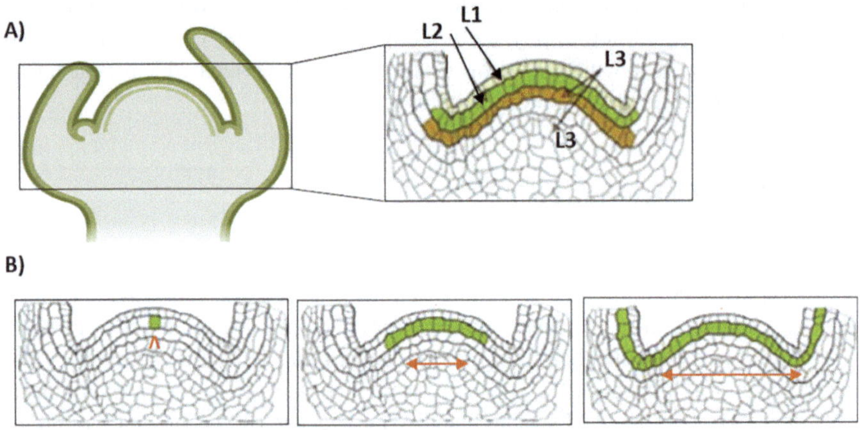

Fig. 6.1 Generation of periclinal chimaeras. (**a**) Shoot apical meristem of dicots and position of the L1, L2 and L3 layers. (**b**) The development of a periclinal chimaera begins with a specific mutation in a single cell in the L2 layer, from which the mutation spreads to the entire L2 layer by anticlinal cell divisions

Fig. 6.2 Periclinal GWG variegated chimaera. (**a**) Variegated *Plectranthus coleoides* plant. (**b**) Variegated *Pelargonium zonale* cv. Frank Headley plant. (**c**) A schematic presentation of leaf sectors affected by a mutation causing non-functional, albino, plastids in the L2 meristem layer leading to a non-chlorophyllous phenotype in the mesophyll tissue of the leaf margin

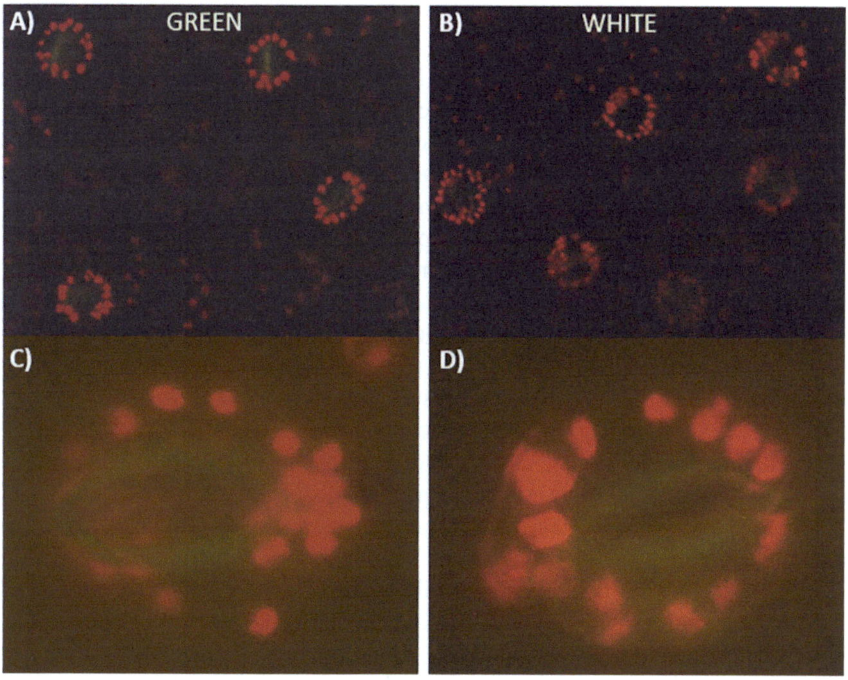

Fig. 6.3 Chlorophyll autofluorescence in the chloroplasts of stomata cells in the adaxial epidermal strips of green (**a**, **c**) and white leaf sectors (**b**, **d**) of *P. zonale* cv. Frank Headley plants. (**a**, **b**) (×20 magnification); (**c**, **d**) (×100 magnification). Courtesy of Dr. Ljiljana Prokić (protocol described in Veljović-Jovanović et al. 2016)

In variegated Swedish ivy (*Plectranthus coleoides*, Fig. 6.2a) and pelargonium (*Pelargonium zonale* cv. Frank Headley, Fig. 6.2b), for example, the mutation that causes the variegated phenotype occurs in the L2 layer, which is responsible for the development of the meristem at the margin of the leaf blade (Fig. 6.2c), resulting in a typical, stable GWG periclinal chimaera.

The variegated *P. zonale* cv. Frank Headley is a stable GWG periclinal chimaera, with a green sector in the centre of the leaf blade and a white leaf margin. This is further evidenced by the presence of unaffected chloroplasts in the stomata cells of *P. zonale* cv. Frank Headley, as they originate from the wild-type L1 layer (Fig. 6.3). Moreover, the epidermis of both leaf sectors of variegated *P. zonale* showed a similar number and size of stomata and a similar chlorophyll content (Veljović-Jovanović et al. 2016).

Although variegation can occur for various reasons, it ultimately leads to the loss or abnormality of chloroplasts, which is reflected in the frequently occurring non-green leaf colouration in different plant species. Many authors have studied variegated plants because they can provide information on the inheritance of the plastids and on the relationships between the structure of the apical meristem and the structure of the leaf blade layers (Tilney-Bassett 1963; Kirk and Tilney-Basset 1967; Walles 1971).

As an illustration, several studies have been conducted using the green-white leaf variegation of *P. zonale* for genetic research related to the inheritance of plastids and mitochondria (Tilney-Bassett 1976; Tilney-Bassett and Almouslem 1989; Tilney-Bassett et al. 1992; Ming-Yin 2005; Grieger 2007). In addition, Toshoji et al. (2012) investigated morphological differences and mitochondrial DNA levels in green and white leaf sectors of twelve variegated species, including *P. zonale*. Only in variegated *P. zonale* the plastid DNA content in the white leaf sections was twice as high as in the green leaf sections. Nevertheless, almost all species analysed, together with *P. zonale*, showed increased mitochondrial DNA contents in the white sectors, which were not correlated with the lower respiration rates compared to the green leaf sectors. Toshoji et al. (2012) suggested that chloroplast dysfunction in white leaf sectors could induce higher mitochondrial DNA levels, probably due to impaired ROS production, and recommended this phenotype for studying the effects of developmental and/or physiological status of plastids on mitochondria.

Moreover, variegated plants are an excellent model system to study cross-communication between the nucleus and the other DNA-containing compartments such as plastids and mitochondria. Several mutations affecting mitochondrial development can impair chloroplast development, function and gene expression, leading to variegated phenotypes (Quesada et al. 2011). In the *chm* mutant of Arabidopsis, for example, variegation is caused by the absence of the mitochondrial AtMSH1 protein, which is thought to be required for the mitochondrial genome maintenance and prevention of impaired recombination and/or replication of abnormal genomes (Abdelnoor et al. 2003). In addition, loss of function of another mitochondria-targeted protein, RUGOSA2 (RUG2), leads to variegation in *RUG2* Arabidopsis (Quesada et al. 2011).

Variegated leaves are useful resources for studying the advantages and the molecular mechanism of variegation, due to their unique property of having both normal and defective plastids in different sectors of the same tissue (Sakamoto 2003). Many variegations are caused by mutations in nuclear genes that produce defective plastids in some, but not all, cells of the plant (Rodermel 2002). These variegations are mendelian inherited. On the other hand, permanently defective plastids can result from the generation of mutations in the plastid genome by defective nuclear gene products, as in some "chloroplast mutator" lines (Kirk and Tilney-Basset 1967). For example, impairment of plastid RNA polymerases can lead to altered chloroplasts and tobacco plants with pale or variegated leaves (De Santis-Maciossek et al. 1999).

Moreover, mutations leading to variegation are of great interest for research to understand the mechanism of development and maintenance of plant plastids. Chlorophyll-deficient mutants have largely served as valuable genetic tools for the identification of nuclear genes involved in chloroplast biogenesis and regulation (Barkan 1998). In the Arabidopsis stock centres of Ohio State and Nottingham, there are about 200 mutants responsible for variegation (Rodermel 2002). Besides Arabidopsis, the molecular mechanisms responsible for the variegation phenotype have been studied in several other species, but no single and definite cause of variegation has been found (Yu et al. 2007). Despite progress in identifying the underlying genes in different variegated plant species, there is little understanding of the molecular mechanisms controlling this phenomenon (Sakamoto 2003). Several genes involved in variegation pattern formation have been cloned in monocot and dicot species (Wu et al. 1999; Takechi et al. 2000; Prikryl et al. 2008; Hayashi-Tsugane et al. 2014; Wang et al. 2016; Zheng et al. 2016; Guan et al. 2017; Zagari et al. 2017). Examples of extensively studied variegated mutants exhibiting normally developed chloroplasts in the green leaf sectors and aberrant plastids in the non-green sectors are listed in Table 6.1.

Variegated Arabidopsis *im* genotype carries the nuclear *immutans* mutation (Aluru et al. 2001). In this mutant, variegation is caused by the loss-of-function of a nuclear-encoded plastid terminal oxidase (PTOX), a chloroplastic homologous to the alternative oxidase of inner mitochondrial membranes (Aluru et al. 2007). In the early stages of early chloroplast biogenesis, the loss of PTOX leads to an over-reduction of the plastoquinone pool and impairs the activity of phytoene desaturase, an enzyme involved in the biosynthesis of carotenoids (Rosso et al. 2009). This leads to phytoene accumulation, which blocks carotenoid synthesis. Under high light intensity and the lack of protective carotenoids, the newly synthesised chlorophylls in the white leaf sectors of the *immutans* are photooxidised. Finally, the absence of IM leads to an irregular organisation of the lamellar structures of thylakoids (Aluru et al. 2009). The authors showed that in both tissue types of *immutans* leaves there is a strong induction of genes encoding a number of enzymes involved in the antioxidant protection, as well as in phenylpropanoid and flavonoid biosynthesis, when compared with wild-type leaves (Aluru et al. 2009).

The other example of an extensively studied variegated Arabidopsis mutant is *yellow variegated2* (*var2*), which lacks protein Filamentous temperature-sensitive H2 (FtsH2), a thylakoid-bound ATP-dependent zinc metalloprotease that is homologous to *E. coli* FtsH. AtFtsH2 is involved in the degradation of the photooxidised D1 subunit of PSII (Kato et al. 2009; Wang et al. 2022) and functions as a scaffolding protein during thylakoid development (Kato and Sakamoto 2018). Yu et al. (2004) suggested that two pairs of FtsH proteins—AtFtsH1 and 5 and AtFtsH2 and 8—form oligomeric complexes in the thylakoid membrane and that a threshold level of complexes is needed for physiological chloroplast function and green sector formation. When complex levels fall below the threshold, this leads to increased PSII photoinhibition and oxidative stress, which in turn results in chloroplast degradation. AtFtsH5 (genome encoded) is involved in the regulation of D1 turnover regulation in *variegated1* (*var1*) (Sakamoto et al. 2002). Both *var1* and *var2* mutants are susceptible to photodamage under intense light, leading to a yellow variegated phenotype (Sakamoto et al. 2003).

Deletion of the chloroplast-localised *Thylakoid formation1* (*Thf1*) gene in Arabidopsis leads to deficient thylakoid formation and variegated leaves (Wang et al. 2004). The Thf1 protein is chloroplast-localised protein. It has been suggested that Thf1 plays a role in membrane vesicle transport and their fusion into thylakoid membranes and its impairment hinders chloroplast development.

Table 6.1 Studied variegated mutants showing aberrant plastid morphology

Species	Variegated mutant	Affected gene/protein	Affected process	References
Arabidopsis	*immutans (im)*	IM—Similar to alternative oxidase of mitochondrial inner membranes	Carotenoid biosynthesis	Wetzel et al. (1994), Rodermel (2002), Aluru et al. (2001, 2007, 2009), Rosso et al. (2009)
Arabidopsis	*yellow variegated1 (var1)*	VAR1 (AtFtsH5)—member of the plastid FtsH metalloprotease gene family	PSII repair cycle	Martínez-Zapater (1993), Sakamoto et al. (2002)
Arabidopsis	*yellow variegated2 (var2)*	VAR2—Chloroplast homolog of *E. coli* FtsH, AN ATP dependent zinc metalloprotease	D1 turnover (PSII)	Martínez-Zapater (1993), Chen et al. (2000), Sakamoto et al. (2003), Kato et al. (2009), Kato and Sakamoto (2018) and Wang et al. (2022)
Arabidopsis	*yellow variegated3 (var3)*	VAR3—89.5 kDa Zn-finger protein required for chloroplast and palisade cell development	Regulation of carotenoid biosynthesis	Næsted et al. (2004)
Arabidopsis	*thylakoid formation1 (Thf1)*	Thf1— controls the fusion of chloroplast inner envelope vesicles into mature thylakoid stacks	Thylakoid membrane development leaf development	Wang et al. (2004)
Arabidopsis	*chloroplast mutator (chm)*	CHM— AtMSH1 mitochondrial-targeted protein	Control of mitochondrial genome "Substoichiometric shifting"	Abdelnoor et al. (2003)
Arabidopsis	*cloroplastos alterados1 (cla1)*	*CLA1* encodes 1-deoxy-D-xylulose 5-phosphate (DXP) synthase	Isoprenoid biosynthesis	Mandel et al. (1996)
Arabidopsis	*snowy cotyledon 2 (sco2)*	SCO2—DNAJ-like protein involved in thylakoid membrane biogenesis	Accumulation of PSII–LHCII complexes	Zagari et al. (2017)

(continued)

Table 6.1 (continued)

Species	Variegated mutant	Affected gene/protein	Affected process	References
Arabidopsis	*snowy cotyledon3 (sco3-1)*	At3g19570 — peroxisomal microtubule-associated protein with unknown function	Chloroplast biogenesis	Albrecht et al. (2010)
Arabidopsis	*RUGOSA2 (RUG2)*	RUG2—stabilizes subunits of mitochondrial respiratory chain	Over-reduction of the stroma of chloroplasts	Quesada et al. (2011)
Ficus microcarpa	cv. Milky Stripe Fig	mTERF—mitochondrial transcription termination factor	Chloroplast development	Shih et al. (2019)
Maize	*iojap*	Iojap—protein component of the 50S subunit of the plastid ribosome	Regulation of plastid protein translation	Han and Martienssen (1995)
Barley	*albostrians*	HvCMF7, (CCT Motif Family 7)— a transcriptional regulator for nuclear genes encoding chloroplast proteins	Expression of nuclear genes essential for chloroplast biogenesis	Hess et al. (1993), Zhelyazkova et al. (2012) and Li et al. (2019)
Tomato	*variegated-leaf (vg)* cv. Alisa Craig	*VG encodes* thylakoid protein for chloroplast development	Expression of chlorophyll synthesis and photosynthesis genes	Song et al. (2023)
Clivia miniata	var. *variegata*,	CG hypermethylation of *CPSAR1* and *ycf2 (orthologue to AtFtsH2)* genes; ycf2—plastid protein translocon; CPSAR1—chloroplastic protein required for the organization of mature thylakoid stacks	Thylakoid biogenesis	Wang et al. (2016)

FtsH2 Filamentous temperature-sensitive H2, *LHCII* light harvesting complex II, *PSII* Photosystem II

Snowy cotyledon 2 (SCO2) is a DNAJ-like protein involved in the biogenesis of thylakoid membranes. It interacts with the light-harvesting chlorophyll-binding protein (LHCB1), which enables adequate accumulation of PSII–LHCII complexes. Disruption of the *SCO2* gene in *A. thaliana* and *Lotus japonicus* disturbing the balance required for the assembly of photosynthetic mega complexes causing variegation (Zagari et al. 2017).

Mutations in the genes whose products are related to mitochondria or peroxisomes can also lead to variegated phenotypes. Recently, it has been suggested that suppression of the mitochondrial transcription termination factor (mTERF) alters chloroplast development in albino leaf sectors of the variegated *Ficus microcarpa* (cv. Milky Stripe Fig) (Shih et al. 2019). In an additional variegated Arabidopsis mutant, *snowy cotyledon3* (*sco3-1*), the absence of a peroxisomal microtubule-associated protein, SCO3, causes the albino phenotype. SCO3 is thought to alter actin fine structure and alter protein import into chloroplasts through the stromules, leading to impaired chloroplast development (Albrecht et al. 2010). Another example of the mutation of a gene coding for a mitochondrial protein as a cause of variegation was demonstrated in the aforementioned RUG2 mutant of Arabidopsis. Quesada et al. (2011) suggest that RUG2-1 may stabilise respiratory chain subunits and/or act as a protein and lipid scaffold and control the organisation and integrity of the inner membrane. In this mutant, the mitochondrial respiratory chain possesses a decreased potential to oxidise chloroplast reductants, leading to over-reduction of chloroplast stroma and PSII photooxidation.

Another way in which permanently defective plastids can arise is the loss of plastid 70S ribosomes, which are required for the translation of the 70S ribosomal proteins encoded by the plastid genome (Hess et al. 1993; Han and Martienssen 1995). This is the case for the *iojap* variegated mutant of maize and the *albostrians* mutant of barley. Both genotypes, *iojap* and *albostrians*, belong to the group of mutants with identical genotypes in green and non-green tissues, whereby the phenotype of the mutant is only expressed in a subset of cells (white leaf sectors). The *iojap* gene encodes a component of the 50S subunit of the plastid ribosome, and its loss of function impairs the regulation of plastid protein translation (Han and Martienssen 1995).

The green-white variegated barley *albostrians* mutant served as a model for investigation of the transcriptional machinery of the chloroplast genome (Hess et al. 1993; Zhelyazkova et al. 2012). *Albostrians* was also used as a model to investigate retrograde signalling and regulatory interactions between plastids, mitochondria and the nucleus (Bradbeer et al. 1979; Hess et al. 1994; Hedtke et al. 1999). This barley mutant lacks 70S ribosomes in the plastids of the albinotic leaf sectors (Hess et al. 1993). Leaf variegation and impaired chloroplast development in the *albostrians* leaves are caused by a truncated CCT domain gene (Li et al. 2019). The CCT domain is named after fairly well-studied plant transcription factors: CO, CO-LIKE, and TOC1 (Strayer et al. 2000). *Hordeum vulgare CCT Motif Family 7* (*HvCMF7*) gene encodes a transcriptional regulator for nuclear genes encoding ribosomal chloroplast proteins and for the chloroplast protein translocon (Sun et al. 2009). The exact molecular function of HvCMF7 is still elusive. One of the two closest homologs of HvCMF7 in Arabidopsis is CHLOROPLAST IMPORT APPARATUS2 (AtCIA2), a nuclear-localised transcription factor involved in the expression of both protein translocon and ribosomal protein genes by binding to their respective promoters (Sun et al. 2001).

In addition to mutations in the plastome or genome related to chloroplast development, variegation can also be caused by epigenetic changes and transposons (Sakamoto et al. 2003; Tsai et al. 2017). The example is a variegated mutant of *Clivia miniata* var. *variegata*, which has attractive green-yellow striped leaves. A sectorial decline in chlorophyll content is associated with CG hypermethylation in the *CPSAR1* and *ycf2* genes, which could lead to gene silencing, causing leaf green-yellow stripes (Wang et al. 2016). In the second mechanism, transposon insertion disrupts a nuclear gene required for normal chloroplast biogenesis (white sectors), while element excision restores wild-type gene expression (green sectors) (Rodermel 2002). In Arabidopsis, variegations of this type were detected in transposon-tagging studies with heterologous elements, such as the autonomous En-1 element from maize (Wisman et al. 1998).

In addition to being used to study chloroplast development and retrograde signalling, natural variegation of *P. zonale* and *P. coleoides* leaves proved to be an excellent model for comparing photosynthetically active and photosynthetically non-active tissues in terms of H_2O_2 production within the same leaf sharing the same microenvironmental conditions (Vidović et al. 2015a, b, 2016a, b; Milić et al. 2023a). In green leaves, ROS are mainly produced by photorespiration and the photosynthetic electron transport chain, even under physiological growth conditions (Foyer and Shigeoka 2011). Since white leaf sectors lack functional chloroplasts and peroxisomes (see below), the main sources of ROS in this tissue are the electron transport chain in the mitochondria and apoplast (Møller 2001; Sierla et al. 2013). White leaf sectors of *P. zonale* show constitutively higher activities of enzymes involved in the ascorbate–glutathione cycle and Cu/Zn superoxide dismutase compared to green leaf sectors (Vidović et al. 2016a). In addition, higher cytosolic ascorbate and total cellular glutathione levels were detected in the white leaf sectors than in the green leaf sectors. On the other hand, a higher total ascorbate content and higher activities of catalase and thylakoid ascorbate peroxidase were observed in the green leaf sectors of *P. zonale*).

Additionally, variegated pelargonium served as an impeccable model system to study *source-sink* interactions in terms of carbon and nitrogen allocation conditions related to sugars, polyphenols, and amino acids within the same leaf, providing the same microenvironment for both sectors (Abadie et al. 2015; Milić et al. 2023b). A higher content of soluble sugars was demonstrated in the autotrophic leaf tissue, while a higher abundance of nitrogenous compounds, especially free amino acids, was detected in heterotrophic leaf tissue.

Finally, variegated leaf phenotypes frequently contribute to the higher economic value of ornamental plants (Tsai et al. 2017; Xie et al. 2021). Pelargoniums are popular ornamental garden plants with considerable economic importance in

Europe. The United Kingdom, Germany, France, and Belgium are the leading countries that import pelargonium (Janson 2000).

6.4 Plastids in Variegated Leaves

In variegated leaves, impaired chloroplast development is considered a hallmark of the non-green phenotype. While the cells in the green sectors contain normally developed chloroplasts, the cells in the non-chlorophyllous sectors lack pigments and appear to be held at various stages of chloroplast biogenesis (Yu et al. 2007). Adequate differentiation of chloroplasts is a complex process, coordinated with peroxisomes, mitochondria, and the cytoskeleton, and involves regulation and cooperation between nuclear and chloroplast genes (involving retrograde signalling) (Cackett et al. 2022). Cytoskeletal components are important for the development of stromules, mobile and transitory chloroplast extensions involved in the interorganellar transport of proteins and metabolites (Reski 2020).

Ultrastructural analysis of the mesophyll cells of variegated *P. zonale* cv. Frank Headley by transmission electron microscopy (TEM) within the green leaf sectors showed that the chloroplasts contained a well-developed thylakoid system with recognisable grana, well-defined large plastoglobuli and dominant starch grains (Fig. 6.4). In contrast, the plastids of the white leaf sectors lacked thylakoid membranes and starch granules, while several membrane vesicles of different sizes and smaller, coupled plastoglobuli were observed in the inner soluble region (Fig. 6.4). In addition, no peroxisomes were detected in the cells of the white leaf sections (Vidović et al. 2015b).

As expected, a clearly different subplastid organisation was observed in green and white leaf tissue of variegated *P. coleoides* plants (Vidović et al. 2015a). In the green leaf sectors, TEM analysis showed that the sectioned chloroplasts contained a well-developed thylakoid system with well-defined grana stacks, distinct plastoglobuli and starch granules (Fig. 6.4). In contrast, the plastids in the white leaf tissue lacked thylakoid membranes and starch grains (Fig. 6.4). At the same time, the plastids in non-photosynthetically active cells accumulated numerous membrane vesicles of different sizes and plastoglobuli in the inner soluble region (Vidović et al. 2015a).

The mesophyll cells of the green leaves of another herbaceous perennial ornamental plant, *Dianella tasmanica* (Asphodelaceae), contain fully developed chloroplasts with properly organised grana and stroma lamellae (Zhou et al. 2024). In addition, the green mesophyll tissue of the intermediate and green sectors of variegated *D. tasmanica* leaves harbours chloroplasts with a similar ultrastructure to those in the leaves of green individuals. In contrast, the mesophyll cells in the albino sector of variegated *D. tasmanica* morphs exhibit only weakly developed plastids that lack both grana and stroma lamellae. In the guard cells of the lower epidermis of *D. tasmanica*, chloroplasts with loosely arranged thylakoids and large starch

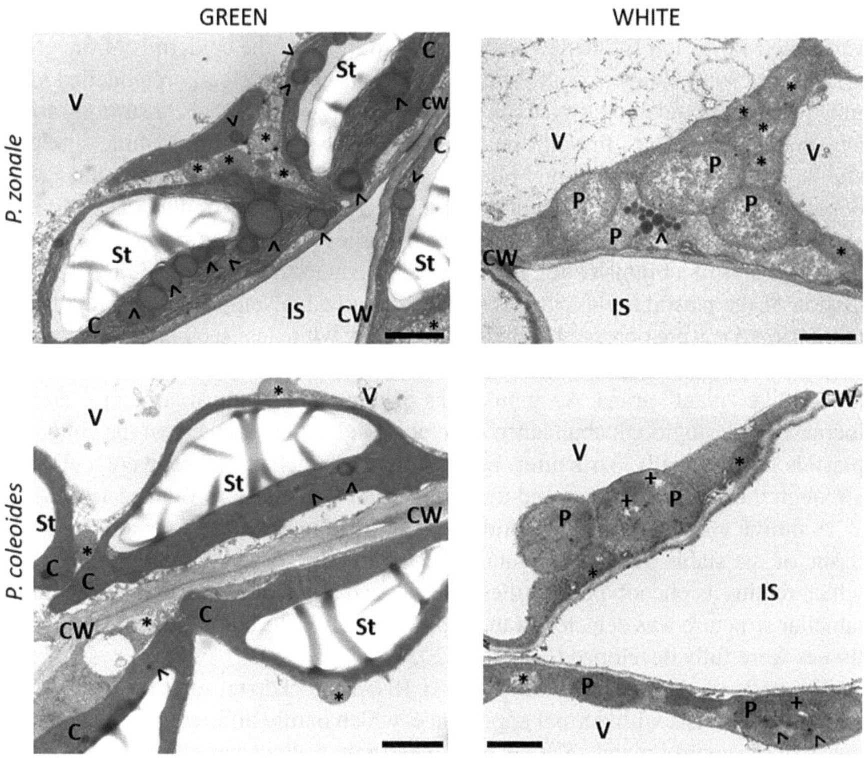

Fig. 6.4 Transmission electron micrographs of the ultrastructure of photosynthetically active (left) and non-photosynthetically active leaf cells (right) of variegated *P. zonale* (top) and *P. coleoides* (bottom). Microscopic images of green leaf sectors show chloroplasts (C) with large starch granules (St) and numerous plastoglobuli (arrowheads). Plastids from white leaf sectors contain numerous plastoglobuli (arrowheads) and vesicles (plus), but no starch grains. CW cell walls, IS intercellular spaces, * mitochondria (asterisks), P plastid, V vacuoles. Bars = 1 µm. The experimental conditions are described in Vidović et al. 2015a, b). With the kind permission of Dr. Bernd Zechmann

granules detected in green morphs exhibited a similar ultrastructure to those from the epidermis of albino and green sectors of variegated morphs (Zhou et al. 2024), similar to that reported for the epidermis of variegated *P. zonale* (Vidović et al. 2016a, b).

The epidermal cells of green leaf sectors of the variegated *Weigelia florida* DC cultivar contain small chloroplasts with small grana and slightly observed osmiophilic globules, similar to those present in senescent plastids (Grandi et al. 1981). The osmiophilic globules appeared in the TEM as electron-dense, dark coloured structures mostly consisting of degraded thylakoids in senescence (Wilson et al. 2002). Several vesicles connected to the inner membrane were observed inside these plastids. The mesophyll cells of the green leaf sectors contain normal chloroplasts with thylakoids, grana and starch granules. Similar to *P. zonale* and *D.*

tasmanica, the plastids of the epidermal cells of the white-yellowish areas of the variegated *W. florida* DC leaves were similar to those of the epidermis of the green area. All mesophyll layers of the white leaf sectors contain clearly remodelled plastids with a reduced lamellar system consisting of single thylakoids or strongly swollen, bundled lamellae. In addition, several slightly visible osmiophilic globules similar to those of the epidermal plastids and several vesicles of various sizes were observed in the mesophyll cells of the albinotic leaf sectors (Grandi et al. 1981).

The plastid ultrastructure has even been studied in the very rare natural variegated mutants of a Finnish birch population (Valanne and Valanne 1972). The vesiculation of the plastid thylakoids, as well as disturbed development and orientation of the lamellae were observed in the plastids of the white leaf sectors. These plastids contained several large vesicles or numerous smaller vesicles and structures that looked like "giant" grana. Accumulations of tubular and helical structures and an increase in plastoglobuli abundance were observed at the periphery of the abnormal plastids. Sporadically, structures resembling prolamellar bodies were detected, although the plants were exposed to optimal light (Valanne and Valanne 1972).

A similar aberration of the plastids was observed in the cells of the albinotic leaf tissue of the stable variegated mutant *Camellia sinensis* cv. "Yanlingyinbiancha," which retains its phenotype regardless of light and temperature variations. No intact lamellar structure was detected in these plastids, while the chloroplasts of the green tissues were fully developed (Gao et al. 2021).

The cells of the green sectors of FtsH RNAi knockdown tobacco leaves contained chloroplasts with normal appearance, which hardly differed from the chloroplasts of the control plants (Kato et al. 2012). In the yellow leaf sectors, the plastids contained various abnormal membrane structures and no thylakoid membranes. Instead, they contained electron-dense structures that most likely correspond to plastoglobuli, as well as fragmented membranes and small vesicle structures (Kato et al. 2012).

Recently, a new spontaneous *variegated-leaf* (*vg*) variegated mutant of tomato cv. Alisa Craig was characterised (Song et al. 2023). The defective expression of *vg* gene encodes a thylakoid protein for chloroplast development, impaired chlorophyll synthesis and reduced photosynthesis. This mutant contains pale yellow immature leaves while the white colouration occurs only on the newly mature leaves. The plastids of the mesophyll cells of the pale yellow leaves of the *vg* mutant lacked the typical thylakoid membranes and grana stacks, whereas no or only a few starch granules were detected (Song et al. 2023). More recently, Dechkrong et al. (2024) described a phenotype of a new variegated tomato M2 population of EMS mutagenised seeds. In contrast to the green leaf sectors, where mature chloroplasts in the mesophyll tissue contained large starch granules with abundant thylakoid membranes and dense grana stacking, mesophyll cells of the white leaf sectors contained abnormal plastids. These plastids were described as fat structures attached to the cell wall, completely lacking thylakoid membranes and grana structure, but with small and large vacuoles inside (Dechkrong et al. 2024).

Regarding the most analysed variegated Arabidopsis mutants, the mesophyll cells of the non-green leaf segments show a similar ultrastructure regardless of the molecular cause of the mutation. In the *Thf1* Arabidopsis mutant, some chloroplasts in the cells of the green leaf sectors have normally developed thylakoids in terms of their quantity and structure (Wang et al. 2004). However, many plastids in the same tissue showed a smaller amount of thylakoid stacks, and the distances between the grana stacks are larger than in the wild-type chloroplasts. Although a few chloroplasts with relatively normal appearance were found, most plastids in the cells of the yellow/white leaf sectors lacked thylakoid membranes, grana or starch granules. However, the structure of the envelope membrane, the size and the general shape of these plastids appear to be normal. These plastids accumulated numerous membrane vesicles of different sizes in the inner soluble region, but the vesicles do not fuse to form thylakoids (Wang et al. 2004).

The white leaf sectors of Arabidopsis *im* plants grown under normal light conditions exhibited normal leaf thickness at the mesophyll level, although the palisade cells did not expand normally (Aluru et al. 2001). These cells contain abnormal vacuolated plastids without organised lamellar structures and rare, normal-appearing chloroplasts (Wetzel et al. 1994). Similarly, the green leaf sectors in *var2* Arabidopsis mutants developed chloroplasts that were nearly identical to those of the wild type plants (Sakamoto et al. 2009). In contrast, developed thylakoid membranes and grana stacks were not found in the plastids of white leaf sections, while spherical vacuolated membrane structures consisting of a single membrane and varying in size were detected. These plastids contained structures that look like prolamellar bodies (Sakamoto et al. 2009).

As for the ultrastructure of the divergent plastids of the non-green leaf areas of variegated monocots, similar changes were observed. Due to the different leaf structure between monocots and dicots, variegation in monocots is also referred to as striping. In variegated temple bamboo plants (*Sinobambusa tootsik* f. *luteolo albostriata*), TEM analysis revealed a physiologically normal cell structure in the mesophyll tissue of the green leaves (Chen et al. 2018). The elliptical chloroplasts retained their structural integrity with developed thylakoid membranes. However, the mesophyll cells of the completely yellow and white leaves contained plastids that more closely resembled etioplasts with unconsolidated thylakoids and fragmented or absent chloroplast membranes without a clearly recognisable, well-organised thylakoid membrane (Chen et al. 2018).

The development of the plastids in the albino sectors of the *albostrians* barley mutant was also impaired (Li et al. 2019). In contrast to the well-developed crescent-shaped chloroplasts of the wild type barley, smaller and irregularly shaped plastids containing only a few vesicle-like structures were observed in the white sectors of the *albostrians* mutant (Li et al. 2019). Similar to *albostrians*, the mesophyll cells in the white leaf strips of another allelic variegated barley mutant, called *grandpa 1* (*gpa1*), contain only a few undeveloped plastids without developed thylakoid membranes, lamellae and stacked grana (Yang et al. 2021). On the other hand, numerous typical chloroplasts with folded thylakoids arranged in grana stacks and connected by stroma lamellae were detected in the mesophyll cells of the parental, wildtype barley line Bowman (Yang et al. 2021).

Finally, although various patterns and mutations have been found among a wide range of variegated species, the white leaf sectors show common features that represent arrested thylakoid differentiation. Cells in green sectors exhibit a wild-type genotype with functional chloroplasts, while chlorophyll-deficient sectors contain aberrant plastids without developed thylakoids, grana stack, or starch granules. Instead, these plastids contain small coupled plastoglobuli and vacuolar/vesicular membrane structures, in some cases followed by prolamellar body-like structures, making them more similar to proplastids. Accordingly, the differentiation of plastids into chloroplasts in the non-chlorophyllous leaf segments appears to be blocked at the early stage of thylakoid development (Sakamoto et al. 2009). Overall, plastids within chlorophyll-deficient sectors of variegated leaves exhibit consistent structural features, defining them as a distinct plastid type —here designated as albinoplasts.

6.5 Conclusion

Variegated plants represent a unique model system to study not only processes related to chloroplast inheritance and development per se, but they are also a perfect model to study *source-sink* interactions and processes related to photosynthesis and associated ROS regulation within cells belonging to the same leaf that share the same microenvironmental conditions (in contrast to the usual shoot-root studies). Although leaf variegation reduces plant growth rate, biomass production and, in some species, photosynthetic capacity, it is naturally occurring in almost 2000 angiosperms. Therefore, the question of what maintains the variegation phenotype remains open. Briefly, based on our extensive biochemical investigations (Vidović et al. 2015a, b, 2016a, b; Veljović Jovanović et al. 2023; Milić et al. 2023a, b) and several ecophysiological studies (Abadie et al. 2015; Sullivan and Koski 2024), the efficient antioxidant system, redox homeostasis, higher content of nitrogenous compounds, lower ROS accumulation in white leaf sectors compared to the green ones and dynamic metabolite reallocation between these two tissue types demonstrate the benefits of variegation for overall plant metabolism. These adaptive values may outweigh the potential disadvantages of variegation under challenging environmental conditions.

Investigating the chloroplast ultrastructure under different environmental conditions can provide reliable markers for different metabolic scenarios in response to stress. Moreover, the lack of starch, the appearance of large plastoglobuli, and the decrease of thylakoids can be considered as reliable stress indicators in plants and predict possible functional consequences on photosynthesis. Indeed, it is interesting to study the changes of specific components of the impaired plastids in non-green leaf sectors of variegated leaves under various conditions. Numerous ultrastructural studies (some of them are presented in this chapter) indicate that the plastids in these leaf sectors can be regarded as a specific, distinct type of plastids —albinoplast. So far, there is little insight into the response of albinoplasts to different unfavourable environmental conditions and their dynamic interactions with the chloroplasts in the neighbouring photosynthetically active cells.

Cryo-electron microscopy and cryo-electron tomography represent the cutting edge of EM technology, which has created exciting new opportunities for 3D visualisation of the ultrastructure of cells and chloroplasts at molecular resolution. It will be feasible to study structure–function relationships from the level of single molecule to the entire thylakoid system and beyond. This already heralds an exciting decade for the study of albinoplast components that respond to external stimuli and their connections with neighbouring cells and the wild-type chloroplast.

Acknowledgement This research was funded by the Ministry of Science, Technological Development and Innovation of the Republic of Serbia (Contract No: 451-03-47/2025-01/200042). The Author would like to acknowledge Dr. Ljiljana Prokic, Faculty of Agriculture, University of Belgrade, Serbia and Dr. Bernd Zechmann, Centre for Microscopy and Imaging, Baylor University, Waco, TX 76798-7046, USA, for obtaining the fluorescence images and TEM images of the variegated *P. zonale* and *P. coleoides* plants.

References

Abadie C, Lamothe M, Mauve C, Gilard F, Tcherkez G (2015) Leaf green-white variegation is advantageous under N deprivation in Pelargonium hortorum. Funct Plant Biol 42:543–551

Abdelnoor RV, Yule R, Elo A, Christensen AC, Meyer-Gauen G, Mackenzie SA (2003) Substoichiometric shifting in the plant mitochondrial genome is influenced by a gene homologous to muts. Proc Natl Acad Sci USA 100:5968–5973

Ahchige MW (2022) Canalization of plant metabolism and yield. PhD thesis, Mathematisch+naturwissen-schaftlichen Fakultät Institut für Biologie und Biochemie der Universität Potsdam, Potsdam, Germany

Albrecht V, Ingenfeld A, Apel K (2006) Characterization of the snowy cotyledon 1 mutant of Arabidopsis thaliana: the impact of chloroplast elongation factor G on chloroplast development and plant vitality. Plant Mol Biol 60:507–518

Albrecht V, Simková K, Carrie C, Delannoy E, Giraud E, Whelan J, Small ID, Apel K, Badger MR, Pogson BJ (2010) The cytoskeleton and the peroxisomal-targeted snowy cotyledon3 protein are required for chloroplast development in Arabidopsis. Plant Cell 22:3423–3438

Allen JF, Bennett J, Steinback KE, Arntzen CJ (1981) Chloroplast protein phosphorylation couples plastoquinone redox state to distribution of excitation energy between photosystems. Nature 291:25–29

Aluru M, Bae H, Wu D, Rodermel S (2001) The Arabidopsis immutans mutation affects plastid differentiation and the morphogenesis of white and green sectors in variegated plants. Plant Physiol 127:67–77

Aluru MR, Stessman D, Spalding MH, Rodermel SR (2007) Alterations in photosynthesis in Arabidopsis lacking IMMUTANS, a chloroplast terminal oxidase. Photosynth Res 91:11–23

Aluru MR, Zola J, Foudree A, Rodermel SR (2009) Chloroplast photooxidation-induced transcriptome reprogramming in Arabidopsis immutans white leaf sectors. Plant Physiol 150:904–923

Austin JR, Staehelin LA (2011) Three-dimensional architecture of grana and stroma thylakoids of higher plants as determined by electron tomography. Plant Phys 155:1601–1611

Barkan A (1998) Approaches to investigating nuclear genes that function in chloroplast biogenesis in land plants. Methods Enzymol 297:38–57

Barton KA, Wozny MR, Mathur N, Jaipargas EA, Mathur J (2018) Chloroplast behaviour and interactions with other organelles in Arabidopsis thaliana pavement cells. J Cell Science 131(2):jcs202275

Beck M, Baumeister W (2016) Cryo-electron tomography: can it reveal the molecular sociology of cells in atomic detail? Trends Cell Biol 26:825–837

Bradbeer JW, Atkinson YE, Börner T, Hagemann R (1979) Cytoplasmic synthesis of plastid polypeptides may be controlled by plastid-synthesised RNA. Nature 279:816–817

Briantais JM, Vernotte C, Picaud M, Krause GH (1980) Chlorophyll fluorescence as a probe for the determination of the photoinduced proton gradient in isolated chloroplasts. Biochim Biophys Acta Bioenerg 591:198–202

Cackett L, Luginbuehl LH, Schreier TB, Lopez-Juez E, Hibberd JM (2022) Chloroplast development in green plant tissues: the interplay between light, hormone, and transcriptional regulation. New Phytol 233:2000–2016

Chan KX, Phua SY, Crisp P, Mc Quinn R, Pogson BJ (2016) Learning the languages of the chloroplast: retrograde signalling and beyond. Annu Rev Plant Biol 67:25–53

Chen M, Choi Y, Voytas DF, Rodermel SR (2000) Mutations in the Arabidopsis VAR2 locus cause leaf variegation due to the loss of a chloroplast FtsH protease. Plant J 22:303–313

Chen L, Lai J, He T, Rong J, Tarin MWK, Zheng Y (2018) Differences in photosynthesis of variegated temple bamboo leaves with various levels of variegation are related to chlorophyll biosynthesis and chloroplast development. J Am Soc Hortic Sci 143:144–153

De Santis-Maciossek G, Kofer W, Bock A, Schoch S, Maier RM, Wanner G et al (1999) Targeted disruption of the plastid RNA polymerase genes rpoA, B and C1: molecular biology, biochemistry and ultrastructure. Plant J 18:477–489

Dechkrong P, Srima S, Sukkhaeng S, Utkhao W, Thanomchat P, de Jong H, Tongyoo P (2024) Mutation mapping of a variegated EMS tomato reveals an FtsH-like protein precursor potentially causing patches of four phenotype classes in the leaves with distinctive internal morphology. BMC Plant Biol 24:265

Demmig-Adams B (1990) Carotenoids and photoprotection in plants: a role for the xanthophyll zeaxanthin. Biochim Biophys Acta Bioenerg 1020:1–24

Foyer CH (2018) Reactive oxygen species, oxidative signalling and the regulation of photosynthesis. Environ Exp Bot 154:134–142

Foyer CH, Hanke G (2022) ROS production and signalling in chloroplasts: cornerstones and evolving concepts. Plant J 111:642–661

Foyer CH, Shigeoka S (2011) Understanding oxidative stress and antioxidant functions to enhance photosynthesis. Plant Physiol 155:93–100

Gao X, Zhang C, Lu C, Wang M, Xie N, Chen J, Li Y, Chen J, Shen C (2021) Disruption of photomorphogenesis leads to abnormal chloroplast development and leaf variegation in Camellia sinensis. Front Plant Sci 12:720800

Grandi GT, Severi A, Manenti G (1981) Plastidial ultrastructure in variegated plants of Weigelia Florida DC. Caryologia 34:289–296

Grieger P (2007) Untersuchungen zur Züchtung variegater Pelargonium x zonale Hybriden auf tetraploider Stufe. PhD thesis, Agricultural-Horticultural Faculty, Humboldt University of Berlin, Berlin, Germany

Guan X, Li Z, Zhang Z, Wei X, Xie J, Chen J, Chen Q (2017) Overexpression of an EaZIP gene devoid of transit peptide sequence induced leaf variegation in tobacco. PLoS One 12:e0175995

Gutiérrez-Nava MDLL, Gillmor CS, Jiménez LF, Guevara-García A, León P (2004) CHLOROPLAST BIOGENESIS genes act cell and non-cell autonomously in early chloroplast development. Plant Physiol 135:471–482

Han C-D, Martienssen RA (1995) The iojap protein (IJ) is associated with 50S chloroplast ribosomal subunits. Maize Coop Newsletter 691:32

Hayashi-Tsugane M, Takahara H, Ahmed N, Himi E, Takagi K, Iida S et al (2014) A mutable albino allele in rice reveals that formation of thylakoid membranes requires the SNOW-WHITE LEAF1 gene. Plant Cell Physiol 55:3–15

Hedtke B, Wagner I, Börner T, Hess WR (1999) Interorganellar crosstalk in higher plants: impaired chloroplast development affects mitochondrial gene and transcript levels. Plant J 19:635–643

Hess WR, Prombona A, Fieder B, Subramanian AR, Börner T (1993) Chloroplast rps15 and the rpoB/C1/C2 gene cluster are strongly transcribed in ribosome-deficient plastids: evidence for a functioning non-chloroplast-encoded RNA polymerase. EMBO J 12:563–571

Hess WR, Müller A, Nagy F, Börner T (1994) Ribosome deficient plastids affect transcription of light-induced nuclear genes: genetic evidence for a plastid-derived signal. Mol Gen Genet 242:305–312

Janson C (2000) Ornamental plants market study the Western European market from a Nicaraguan perspective. Chemonics International Inc, Washington, DC

Kasahara M, Kagawa T, Oikawa K, Suetsugu N, Miyao M, Wada M (2002) Chloroplast avoidance movement reduces photodamage in plants. Nature 420:829–832

Kato Y, Sakamoto W (2018) FtsH protease in the thylakoid membrane: physiological functions and the regulation of protease activity. Front Plant Sci 20:9855

Kato Y, Miura E, Ido K, Ifuku K, Sakamoto W (2009) The variegated mutants lacking chloroplastic FtsHs are defective in D1 degradation and accumulate reactive oxygen species. Plant Physiol 151:1790–1801

Kim J, Rudella A, Ramirez Rodriguez V, Zybailov B, Olinares PDB, van Wijk KJ (2009) Subunits of the plastid ClpPR protease complex have differential contributions to embryogenesis, plastid biogenesis, and plant development in Arabidopsis. Plant Cell 21:1669–1692

Kim SH, Kwon CT, Song G, Koh HJ, An G, Paek NC (2018) The rice zebra3 (z3) mutation disrupts citrate distribution and produces transverse dark-green/green variegation in mature leaves. Rice 11:1–15

Kirchhoff H (2019) Chloroplast ultrastructure in plants. New Phytol 223:565–574

Kirk JTO, Tilney-Basset RAE (1967) The plastids: their chemistry, structure, growth, and inheritance, WH Freeman & Co, London, San Francisco

Krause GH (1973) The high-energy state of the thylakoid system as indicated by chlorophyll fluorescence and chloroplast shrinkage. Biochim Biophys Acta Bioenerg 292:715–728

Krause AG, Weis E (1991) Chlorophyll fluorescence and photosynthesis: the basics. Ann Rev Plant Biol 42:313–349

Lawson T, Oxborough K, Morison JI, Baker NR (2002) Responses of photosynthetic electron transport in stomatal guard cells and mesophyll cells in intact leaves to light, CO_2, and humidity. Plant Physiol 128:52–62

Li M, Hensel G, Mascher M, Melzer M, Budhagatapalli N, Rutten T et al (2019) Leaf variegation and impaired chloroplast development caused by a truncated CCT domain gene in albostrians barley. Plant Cell 31:1430–1445

Liang Z, Zhu N, Mai KK, Liu Z, Tzeng D, Osteryoung KW et al (2018) Thylakoid-bound polysomes and a dynamin-related protein, FZL, mediate critical stages of the linear chloroplast biogenesis program in greening Arabidopsis cotyledons. Plant Cell 30:1476–1495

Mandel MA, Feldmann KA, Herrera-Estrella L, Rocha-Sosa M, León P (1996) CLA1, a novel gene required for chloroplast development, is highly conserved in evolution. Plant J 9:649–658

Martínez-Zapater JM (1993) Genetic analysis of variegated mutants in Arabidopsis. J Hered 84:138–140

Mekala NR, Suorsa M, Rantala M, Aro EM, Tikkanen M (2015) Plants actively avoid state transitions upon changes in light intensity: role of light-harvesting complex II protein dephosphorylation in high light. Plant Physiol 168:721–734

Milić D, Živanović B, Samardžić J, Nikolić N, Cukier C, Limami AM, Vidović M (2023a) Carbon and nitrogen allocation between the sink and source leaf tissue in response to the excess excitation energy conditions. Int J Mol Sci 24:2269

Milić D, Pantelić A, Banović-Đeri B, Samardžić J, Vidović M (2023b) Contrasting metabolisms in green and white leaf sectors of variegated Pelargonium zonale – an integrative transcriptomic and metabolomic study. Int J Mol Sci 24(6):5288

Ming-Yin L (2005) Anatomical, cytological and histological investigation of somatic variability of the muted clone of Pelargonium zonale 'Kleiner Liebling'. PhD thesis, Humboldt University of Berlin, Berlin, Germany

Møller IM (2001) Plant mitochondria and oxidative stress: electron transport, NADPH turnover, and metabolism of reactive oxygen species. Annu Rev Plant Physiol 52:561–591

Næsted H, Holm A, Jenkins T, Nielsen HB, Harris CA, Beale MH, Andersen M, Mant A, Scheller H, Camara B, Mattsson O, Mundy J (2004) Arabidopsis VARIEGATED 3 encodes a chloroplast-targeted, zinc-finger protein required for chloroplast and palisade cell development. J Cell Sci 117:4807–4818

Nakai M (2018) New perspectives on chloroplast protein import. Plant Cell Physiol 59:1111–1119

Pogson BJ, Albrecht V (2011) Genetic dissection of chloroplast biogenesis and development: an overview. Plant Physiol 155:1545–1551

Prikryl J, Watkins KP, Friso G, van Wijk KJ, Barkan A (2008) A member of the Whirly family is a multifunctional RNA- and DNA-binding protein that is essential for chloroplast biogenesis. Nucleic Acids Res 36:5152–5165

Quesada V, Sarmiento-Mañús R, González-Bayón R, Hricová A, Pérez-Marcos R, Graciá-Martínez E, Medina-Ruiz L, Leyva-Díaz E, Ponce MR, Micol JL (2011) Arabidopsis RUGOSA2 encodes an mTERF family member required for mitochondrion, chloroplast and leaf development. Plant J 68(4):738–753

Reski R (2020) Challenges to our current view on chloroplasts. Biol Chem 390:731–738

Rodermel S (2002) Arabidopsis variegation mutants. The Arabidopsis Book/American Society of Plant Biologists, p 1

Rolland N, Bouchnak I, Moyet L, Salvi D, Kuntz M (2018) The main functions of plastids. In: Plastids: methods and protocols, pp 73–85

Rosso D, Bode R, Li W, Krol M, Saccon D, Wang S et al (2009) Photosynthetic redox imbalance governs leaf sectoring in the Arabidopsis thaliana variegation mutants immutans, spotty, var1, and var2. Plant Cell 21:3473–3492

Rottet S, Besagni C, Kessler F (2015) The role of plastoglobules in thylakoid lipid remodeling during plant development. Biochim Biophys Acta Bioenerg 1847:889–899

Sakamoto W (2003) Leaf-variegated mutations and their responsible genes in Arabidopsis thaliana. Genes Genet Sys 78:1–9

Sakamoto W, Tamura T, Hanba-Tomita Y, Sodmergen MM (2002) The VAR1 locus of Arabidopsis encodes a chloroplastic FtsH and is responsible for leaf variegation in the mutant alleles. Gen Cells 7:769–780

Sakamoto W, Uno Y, Zhang Q, Miura E, Kato Y, Sodmergen. (2009) Arrested differentiation of proplastids into chloroplasts in variegated leaves characterized by plastid ultrastructure and nucleoid morphology. Plant Cell Physiol 50:2069–2083

Saski C, Lee SB, Fjellheim S, Guda C, Jansen RK, Luo H et al (2007) Complete chloroplast genome sequences of Hordeum vulgare, Sorghum bicolor and Agrostis stolonifera, and comparative analyses with other grass genomes. Theor Appl Genetics 115:571–590

Shih TH, Lin SH, Huang MY, Huang WD, Yang CM (2019) Transcriptome profile of the variegated Ficus microcarpa c.v. Milky Stripe Fig Leaf. Int J Mol Sci 20:1338

Sierla M, Rahikainen M, Salojärvi J, Kangasjärvi J, Kangasjärvi S (2013) Apoplastic and chloroplastic redox signalling networks in plant stress responses. Antioxid Redox Signal 18:2220–2239

Staehelin LA (2003) Chloroplast structure: from chlorophyll granules to supra-molecular architecture of thylakoid membranes. Photosynth Res 76:185–196

Stern DB, Hanson MR, Barkan A (2004) Genetics and genomics of chloroplast biogenesis: maize as a model system. Trends Plant Sci 9:293–301

Strayer C, Oyama T, Schultz TF, Raman R, Somers DE, Más P, Panda S, Kreps JA, Kay SA (2000) Cloning of the Arabidopsis clock gene TOC1, an autoregulatory response regulator homolog. Science 289:768–771

Sullivan CN, Koski MH (2024) An elevational cline in leaf variegation: testing anti-herbivory and abiotic heterogeneity hypotheses in maintaining a polymorphism. Am J Bot 111:e16411

Sun CW, Chen LJ, Lin LC, Li HM (2001) Leaf-specific upregulation of chloroplast translocon genes by a CCT motif-containing protein, CIA 2. Plant Cell 13:2053–2061

Sun CW, Huang YC, Chang HY (2009) CIA2 coordinately up-regulates protein import and synthesis in leaf chloroplasts. Plant Physiol 150:879–888

Takechi K, Sodmergen N, Murata M, Motoyoshi F, Sakamoto W (2000) The YELLOW VARIEGATED (VAR2) locus encodes a homologue of FtsH, an ATP-dependent protease in Arabidopsis. Plant Cell Physiol 41:1334–1346

Taylor WC, Barkan A, Martienssen RA (1987) Use of nuclear mutants in the analysis of chloroplast development. Dev Genet 8:305–320

Thalmann M, Santelia D (2017) Starch as a determinant of plant fitness under abiotic stress. New Phytol 214:943–951

Tilney-Bassett RAE (1963) The structure of periclinal chimeras. Heredity 18:265–285

Tilney-Bassett RAE (1976) The control of plastid inheritance in pelargonium IV. Heredity 37:95–107

Tilney-Bassett RAE (1986) Plant chimeras. Cambridge University Press, Cambridge

Tilney-Bassett RAE, Almouslem AB (1989) Variation in plastid inheritance between pelargonium cultivars and their hybrids. Heredity 63:145–153

Tilney-Bassett RAE, Almouslem AB, Amoatey HM (1992) Complementary genes control biparental plastid inheritance in Pelargonium. Theor Appl Genet 85:317–324

Toshoji H, Katsumata T, Takusagawa M, Yusa Y, Sakai A (2012) Effects of chloroplast dysfunction on mitochondria: white sectors in variegated leaves have higher mitochondrial DNA levels and lower dark respiration rates than green sectors. Protoplasma 249:805–817

Tsai CC, Wu YJ, Sheue CR, Liao PC, Chen YH, Li SJ, Liu JW, Chan HT, Liu WL, Ko YZ et al (2017) Molecular basis underlying leaf variegation of a moth orchid mutant (Phalaenopsis aphrodite subsp. formosana). Front Plant Sci 8:1333

Valanne N, Valanne T (1972) Structure of plastids of a variegated Betula pubescens mutant. Can J Bot 50:1835–1839

van Wijk KJ, Kessler F (2017) Plastoglobuli: plastid microcompartments with integrated functions in metabolism, plastid developmental transitions, and environmental adaptation. Ann Rev Plant Biol 68:253–289

Veljović Jovanović S, Kasalica B, Miletić K, Vidović M, Šušić N, Jeremić D, Belča I (2023) Red-light transmittance changes in variegated Pelargonium zonale—diurnal variation in chloroplast movement and photosystem ii efficiency. Int J Mol Sci 24:14265

Veljović-Jovanović S, Vidović M, Morina F, Prokić L, Todorović DM (2016) Comparison of photoacoustic signals in photosynthetic and nonphotosynthetic leaf tissues of variegated Pelargonium zonale. Int J Thermophys 37:91–101

Vidović M, Morina F, Milić S, Zechmann B, Albert A, Winkler JB, Veljović Jovanović S (2015a) UV-B component of sunlight stimulates photosynthesis and flavonoid accumulation in variegated Plectranthus coleoides leaves depending on background light. Plant Cell Environ 38:968–979

Vidović M, Morina F, Milić S, Albert A, Zechmann B, Tosti T, Winkler JB, Veljović-Jovanović S (2015b) Carbon allocation from source to sink leaf tissue in relation to flavonoid biosynthesis in variegated Pelargonium zonale under UV-B radiation and high PAR intensity. Plant Physiol Biochem 93:44–55

Vidović M, Morina F, Milić S, Vuleta A, Zechmann B, Prokić L, Veljović-Jovanović S (2016a) Characterisation of antioxidants in photosynthetic and non-photosynthetic leaf tissues of variegated Pelargonium zonale plants. Plant Biol 18:669–680

Vidović M, Morina F, Prokić L, Milić-Komić S, Živanović B, Veljović JS (2016b) Antioxidative response in variegated Pelargonium zonale leaves and generation of extracellular H_2O_2 in (peri) vascular tissue induced by sunlight and paraquat. J Plant Physiol 206:25–39

Wada M (2013) Chloroplast movement. Plant Sci 210:177–182

Walles B (1971) Chromoplast development in a carotenoid mutant of maize. Protoplasma 73:159–175

Wang Q, Sullivan RW, Kight A, Henry RL, Huang J, Jones AM, Korth KL (2004) Deletion of the chloroplast-localized thylakoid formation1 gene product in Arabidopsis leads to deficient thylakoid formation and variegated leaves. Plant Physiol 136:3594–3604

Wang QM, Wang L, Zhou Y, Cui J, Wang Y, Zhao C (2016) Leaf patterning of Clivia miniata var. variegata is associated with differential DNA methylation. Plant Cell Rep 35:167–184

Wang X, Li Q, Zhang Y, Pan M, Wang R, Sun Y, An L, Liu X, Yu F, Qi Y (2022) VAR2/AtFtsH2 and EVR2/BCM1/CBD1 synergistically regulate the accumulation of PSII reaction centre D1 protein during de-etiolation in Arabidopsis. Plant Cell Environ 45:2395–2409

Wetzel CM, Jiang C-Z, Meehan LJ, Voytas DF, Rodermel SR (1994) Nuclear-organelle interactions: the immutans variegation mutant of Arabidopsis is plastid autonomous and impaired in carotenoid biosynthesis. Plant J 61:161–175

Wilson KA, McManus MT, Gordon ME, Jordan TW (2002) The proteomics of senescence in leaves of white clover, Trifolium repens (L.). Proteomics 2:1114–1122

Wisman E, Hartmann U, Sagasser M, Baumann E, Palme K, Hahlbrock K, Saedler H, Weisshaar B (1998) Knock-out mutants from an En-1 mutagenized Arabidopsis thaliana population generate phenylpropanoid biosynthesis phenotypes. Proc Natl Acad Sci USA 951:12432–12437

Wu D, Wright DA, Wetzel C, Voytas DF, Rodermel S (1999) The IMMUTANS variegation locus of Arabidopsis defines a mitochondrial alternative oxidase homolog that functions during early chloroplast biogenesis. Plant Cell 11:43–55

Xie N, Zhang C, Zhou P, Gao X, Wang M, Tian S, Lu C, Wang C, Shen C (2021) Transcriptomic analyses reveal variegation induced metabolic changes leading to high L-theanine levels in albino sectors of variegated tea (Camellia sinensis). Plant Physiol Biochem 169:29–39

Yang S, Overlander M, Fiedler J (2021) Genetic analysis of the barley variegation mutant, grandpa1.A. BMC Plant Biol 21:134

Yu F, Park S, Rodermel SR (2004) The Arabidopsis FtsH metalloprotease gene family: interchangeability of subunits in chloroplast oligomeric complexes. Plant J 37:864–876

Yu FEI, Fu A, Aluru M, Park S, Xu Y, Liu H, Rodermel S (2007) Variegation mutants and mechanisms of chloroplast biogenesis. Plant Cell Environ 30:350–365

Zagari N, Sandoval-Ibañez O, Sandal N, Su J, Rodriguez-Concepcion M, Stougaard J et al (2017) SNOWY COTYLEDON 2 promotes chloroplast development and has a role in leaf variegation in both Lotus japonicus and Arabidopsis thaliana. Mol Plant 10:721–734

Zechmann B (2019) Ultrastructure of plastids serves as reliable abiotic and biotic stress marker. PLoS One 14:e0214811

Zeiger E, Armond P, Melis A (1981) Fluorescence properties of guard cell chloroplasts evidence for linear electron transport and light-harvesting pigments of photosystems I and II. Plant Physiol 67:17–20

Zemel E, Gepstein S (1985) Immunological evidence for the presence of ribulose bisphosphate carboxylase in guard cell chloroplasts. Plant Physiol 58:586–590

Zhang JH, Zeng JC, Wang XM, Chen SF, Albach DC, Li HQ (2020) A revised classification of leaf variegation types. Flora 272:151703

Zhelyazkova P, Sharma CM, Förstner KU, Liere K, Vogel J, Börner T (2012) The primary transcriptome of barley chloroplasts: numerous noncoding RNAs and the dominating role of the plastid-encoded RNA polymerase. Plant Cell 24:123–136

Zheng M, Liu X, Liang S, Fu S, Qi Y, Zhao J et al (2016) Chloroplast translation initiation factors regulate leaf variegation and development. Plant Physiol 172:1117–1130

Zhou S, Ma K, Mower JP, Liu Y, Zhou R (2024) Leaf variegation caused by plastome structural variation: an example from Dianella tasmanica. Hortic Res 11:uhae009

Zoschke R, Bock R (2018) Chloroplast translation: structural and functional organization, operational control, and regulation. Plant Cell 30:745–770

Role of Abiotic and Biotic Stresses on Chloroplast Biogenesis

7

Anamika Jangra, Jaspreet Kaur, Kishan Kumar, Sudhan S. Kaintura, Sourav Ranjan Mohapatra, Prabhasmita Shatpathy, Ayesha Mohanty, and Kaushik Kumar Panigrahi

Abstract

Chloroplasts, essential organelles for photosynthesis, are also central to various physiological processes critical for growth, development, and stress responses in plants. In addition to contributing to the fixation of atmospheric carbon and producing metabolic energy, chloroplasts also contribute to nitrogen fixation and the synthesis of important biomolecules. Chloroplast biogenesis is a light-exposed process that has evolved to respond to environmental and cellular factors. Chloroplasts are very sensitive to abiotic and biotic stresses and act as both sensors and mediators of plant stress responses. When faced with stress, chloroplasts generate reactive oxygen species and additional signaling molecules to elicit retrograde signaling functions that influence nuclear gene expression to

A. Jangra · J. Kaur · K. Kumar · S. S. Kaintura
Division of Genetics & Tree Improvement, Forest Research Institute, Dehradun, Uttarakhand, India

S. R. Mohapatra (✉)
Department of Forest Biology and Tree Improvement, Odisha University of Agriculture and Technology, Bhubaneswar, Odisha, India
e-mail: srmohapatra.forestry@ouat.ac.in

P. Shatpathy
Department of Plant Physiology, College of Agriculture, Odisha University of Agriculture and Technology, Bhubaneswar, Odisha, India

A. Mohanty
AINP on Soil Biodiversity Biofertilizers, Odisha University of Agriculture and Technology, Bhubaneswar, Odisha, India

K. K. Panigrahi
AICRN on Potential crops, Odisha University of Agriculture and Technology, Bhubaneswar, Odisha, India
e-mail: kkpanigrahi@ouat.ac.in

V. K. Dalal, A. N. Misra (eds.), *Chloroplast Biogenesis and Plastid Interconversions*,
https://doi.org/10.1007/978-981-95-0772-6_7

allow the plant to adapt to stress. In addition to this operation, chloroplasts generate secondary metabolites and phytohormones that contribute to anti-herbivory responses in plants. As compelling and interesting findings in plant chloroplast research continue to emerge, it will be increasingly more difficult to ignore their multi-functional roles in plant stress responses. This review will consider the impact abiotic and biotic stresses have on chloroplast biogenesis, stressing the importance of the chloroplast for plant stress resilience and chloroplasts as potential targets for reinforcing crop durability and adaptability through biotechnologies.

Keywords

Chloroplast biogenesis · Abiotic stress · Biotic stress · Plant resilience · Environmental adaptation

7.1 Introduction

Chloroplasts are vital organelles within plant cells, functioning not only in photosynthesis but also in a myriad of physiological functions essential for plant growth, development, and responses to stress. These organelles, along with other plastids, constitute a closely related group and have overall important functions associated with the relevant metabolism. Chloroplasts are responsible for carbon fixation and energy metabolism, as well as the synthesis of key biomolecules such as essential amino acids, fatty acids, and lipids, all of which are critical for integral membranes and membrane function (Waters and Langdale 2009). Chloroplasts function as key compartments in nitrogen assimilation, specifically converting nitrite into ammonia via nitrite reductase, which is crucial for amino acid biosynthesis (Dobrogojsk et al. 2020).

Chloroplast biogenesis, or the formation of functionally competent chloroplasts, is a highly regulated and light-mediated developmental process that occurs mainly through photomorphogenesis as seedlings emerge from the soil (Pogson et al. 2015). Achieving these complex development transitions entails pigment biosynthesis, protein import, and a coordinated expression of genes that originate from the plastid and nuclear genomes. These processes occur in specialized developmental locations, primarily in shoot apical meristems and leaf primordia, and involve the transition of undifferentiated proplastids into fully competent chloroplasts (Kendrick and Kronenberg 2012). The differentiation of the proplastid and other plastid types and varieties such as chloroplasts, chromoplasts, and amyloplasts depends on environmental signals and intrinsic cellular factors (Gray et al. 2003).

The differentiation of chloroplasts occurs at an intermediate level in darkness, at which point chloroplasts turn into etioplasts. Etioplasts contain a structure, termed prolamellar body (PLB), consisting of protochlorophyllide, the precursor to chlorophyll that is bound by protochlorophyllide oxidoreductase (POR), alongside NADPH, and lipids and carotenoids (Pogson et al. 2015). Once light becomes available, the PLB disassembles, leading to the construction of the thylakoid membranes that will eventually be the site of photosynthesis. This transformation is mediated by photoreceptors such as phytochrome and plant hormones, including brassinosteroids, to trigger the light-based development of the chloroplasts and associated pigments (Cackett et al. 2022). Upon phytochrome activation by light, it migrates to the nucleus and triggers transcription factors involved in the light-mediated synthesis of pigments and chloroplasts (Yoo et al. 2019).

Chloroplasts do more than facilitate photosynthesis or light-mediated construction and activate the developmental pathway; they are also crucial in plant responses to environmental stress (Liu et al. 2018a, b). Plants experience many abiotic stresses associated with being sessile, including drought, extreme temperature, salinity, nutrient deficiency, and biotic stresses (herbivores and pathogens) (Gull et al. 2019). Abiotic and biotic stressors have a detrimental effect on agricultural yields and cause economic losses, which are exacerbated by population growth, climate change, and the loss of arable land. Chloroplasts, as dynamic organelles, play a critical role as early sensors for those stress signals and initiate inherent processes to cope with the stress. Chloroplasts are environmental sensors because their homeostasis is remarkably affected by external events (Gan et al. 2019). Under stress, chloroplasts build pathogenicity-related ROS and other stress-induced signaling molecules as messages back to the nucleus, acting as retrograde signals. Specifically, retrograde signals, as the transition from chloroplasts to regulating signals, occur when chloroplasts initiate a buildup of several molecules: tetrapyrroles, phosphoadenosines, carotenoid oxidation products, and isoprenoid relatives that guide chlorophyll biosynthesis. These retrograde signals under stress conditions induce nuclear gene expression to signal plant cells to the onset of stress conditions (Chi et al. 2015). For instance, to assist the plant adapt to environmental stress, singlet oxygen produced at photosystem II under high light conditions activates specific genes in the nucleus linked to photosynthesis, carbon metabolism, and the processing of plastid mRNA (Sachdev et al. 2021).

The chloroplasts not only are involved in the production of secondary metabolites and phytohormones involved in plant defense (Kachroo et al. 2021), but they also produce and release biologically active compounds under biotic stress to promote resistance to microorganisms and herbivores (Suman et al. 2021). Chloroplast-derived signals also interact and coordinate with other cellular pathways to regulate the gene expression that mediates the response of the plant to a specific stress condition (Farooq et al. 2019). As research continues to advance, the multifunctional capacity of chloroplasts to mediate plant response to stress is becoming increasingly clear that chloroplasts are implicated in many resilience strategies and that governments have multipurpose roles that cross wider metabolic and signaling networks to promote plant survival during stress. Furthermore, the union of chloroplast biogenesis and response to stress signaling paths work to show how plants can finely tune their responses to the environment and how chloroplasts navigate this dynamic system. This review examines how abiotic and biotic stresses influence chloroplast biogenesis, highlighting the evolving roles of chloroplasts in mediating plant resilience and adaptation to environmental challenges. We will summarize recent literature that showcases the potential role of chloroplast-derived metabolites and signaling pathways in contributing to plant defense mechanisms in these circumstances. Additionally, we will consider the consequences of environmental change on chloroplast biology, as this area of research has recently gained much attention. All things considered, the new research confirms our knowledge of the crucial function chloroplasts play in plant resilience and suggests that agricultural durability may be enhanced by biotechnological advancements that target chloroplast biology.

7.2 Molecular Mechanism of Chloroplast Biogenesis

Chloroplast biogenesis is a complex process wherein chloroplasts develop from predecessor proplastids through a light-stimulated integration of cellular events such as gene expression, protein delivery and a range of signaling events. Understanding chloroplast formation requires an understanding of how chloroplasts talk to the nucleus and what internal and environmental signals trigger chloroplast biogenesis (refer to Table 7.1).

7.2.1 Light-Mediated Regulation and Photoreceptor Involvement

Chloroplast biogenesis is heavily influenced by light, which can even regulate the differentiation of proplastids into functional chloroplasts. The light-dependent differentiation process is initiated through the activation of photoreceptors (i.e., PHYA, PHYB and CRY1), which sense light wavelengths that are then transduced to initiate a signaling pathway involving various signaling molecules (Albaqami 2022). While CRY1 is sensitive to blue light, PHYA and PHYB photoreceptors are sensitive to red and far-red light. Through their downstream signaling pathway, the activated photoreceptors effectively control the production of genes required for chloroplast formation upon absorption of light (Bhatla and Lal 2023). For example, the bZIP transcription factor called HY5 (elongated hypocotyl 5) is a crucial transcription factor upregulating light-induced gene expression through binding to G-box elements located in the promoter region of light-induced genes (Hao et al. 2019). HY5 regulates the transcription of genes participating in both chloroplast biogenesis and photosynthesis. Absent light, HY5's activity can be suppressed by COP1 (CONSTITUTIVE PHOTOMORPHOGENIC1), which functions by targeting HY5 for proteasomal degradation through the ubiquitin-proteasome pathway (Li et al. 2011). Photoreceptor pathway targets also include PIFs (phytochrome interacting factors), PIF3 being the best-studied example, which can act as a negative regulation of light-induced gene expression/repression for skotomorphogenesis to occur (Tripathi et al. 2019; Wang et al. 2022). In other words, depending on specific signaling events, photoreceptors activate photoreceptor network signal regulation in chloroplast processes or, more broadly, can activate light processes.

Table 7.1 Proteins and enzymes involved in chloroplast biogenesis: functions, interactions, and regulatory mechanisms

Protein/enzyme	Gene	Function	Location	Interaction partners	Regulatory mechanisms	References
Rubisco	rbcL, rbcS	Catalyzes carbon fixation in the Calvin cycle.	Stroma	Ribulose-1,5-bisphosphate	Activator proteins (e.g., Rubisco activase)	Andersson and Backlund (2008)
Chlorophyll a-b binding protein	CAB1, CAB2	Forms part of light-harvesting complexes, helping absorb light energy.	Thylakoid membranes	Photosystem I and II complexes	Transcriptional regulation by light signals (e.g., HY5)	Härtel et al. (2000)
ATP synthase	atpA, atpB	Synthesizes ATP, essential for energy production during photosynthesis.	Thylakoid membrane, stroma	ADP, phosphate	Proton gradient across the thylakoid membrane	Martins (2009)
Ferredoxin-NADP+ reductase	FNR	Transfers electrons from ferredoxin to NADP+, producing NADPH.	Stroma	Ferredoxin, NADP+	Light-dependent activation by photosystem I	Knaff (2000)
Glyceraldehyde-3-phosphate dehydrogenase	GPDH	Converts 3-phosphoglycerate to glyceraldehyde-3-phosphate in the Calvin cycle.	Stroma	NADP+	Regulation via redox state, feedback inhibition	Zhang et al. (2010)
Chlorophyll synthase	CHLH	Final step in chlorophyll biosynthesis: conversion of protochlorophyllide to chlorophyll.	Stroma	Protochlorophyllide	Regulation by light signals and chlorophyll precursors	Tanaka and Tanaka (2011)
Porphyrinogen oxidase	PPOX	Converts porphyrinogen to a chlorophyll precursor.	Stroma	Protochlorophyllide	Controlled by chlorophyll biosynthesis pathway intermediates	Park et al. (2017)
Light-harvesting complex II (LHCII)	Lhcb1, Lhcb2	Absorbs light energy and transfers it to photosystem II for photosynthesis.	Thylakoid membranes	Photosystem II, chlorophyll	Light intensity regulates expression; influenced by circadian rhythm	Kovacs et al. (2006)

(continued)

Table 7.1 (continued)

Protein/enzyme	Gene	Function	Location	Interaction partners	Regulatory mechanisms	References
Translocon of the outer chloroplast membrane	TOC75, TOC34, TOC159	Mediates protein import into the chloroplast from the cytosol.	Outer membrane of chloroplast	TIC complex, precursor proteins	Regulated by import signals, chaperones like HSP70	Inoue (2015)
Chloroplast division protein FtsZ	FtsZ	Forms a ring at the division site to facilitate chloroplast division.	Inner membrane of chloroplast	MinD, MinE, other division-related proteins	Regulated by dynamin-like proteins (e.g., ARC6)	Vitha et al. (2001)
Pora-ATPase	AtpF	Involved in ATP hydrolysis for maintaining chloroplast envelope homeostasis.	Inner envelope membrane	ATP, ADP	Regulated by ion balance, pH changes across membranes	Liu et al. (2018a, b)
Sigma factor	SigD	Regulates transcription of chloroplast genes, vital for biogenesis.	Stroma	Plastid genome, RNA polymerase	Activated by environmental cues, plastid genome feedback	Zoschke et al. (2013)
Plastidial ribosomal proteins	rpL2, rpS9	Involved in protein synthesis within the chloroplast.	Stroma	Plastid ribosome	Coordinated with plastid translation machinery	Barkan (2011)

7.2.2 Chloroplast Protein Import and Transport Mechanisms

Chloroplast protein import and processing play an indispensable role in chloroplast biogenesis and function. Translocon at the Outer Envelope Membrane (TOC) and Translocon at the Inner Envelope Membrane (TIC) complexes are involved in a closely regulated system that imports proteins generated in the cytosol into chloroplasts. The TOC recognizes and translocates the precursor proteins across the outer membrane, and the TIC allows movement across the inner membrane (Nakai 2015). Upon arrival at the chloroplast, the precursor proteins undergo post-translational modifications that include cleavage of N-terminal transit peptides that are essential for subcellular localization and functionality (Rowland et al. 2015). The ubiquitin-proteasome system (UPS) pathway is also involved in the regulation of chloroplast protein abundance and protein quality control. The RING-type E3 ligase SUPPRESSOR OF PPI1 LOCUS1 (SP1) plays a critical role in facilitating the proteins targeted for degradation at the TOC complex, which helps plastids move more conveniently and ensures less bypassing during chloroplast development (Ling and Jarvis 2015a). Mutants that disrupt these import and processing pathways lead to compromised chloroplasts and impaired plant growth; they highlight the importance of these import and processing processes for maintaining cellular metabolism and health (Padmanabhan and Dinesh-Kumar 2010).

7.2.3 Transcriptional and Hormonal Regulation of Chloroplast Development

Chloroplast biogenesis is influenced not only by light but also by hormones that provide signals to regulate chloroplast development relative to overall plant growth. Light signals are integrated by photoreceptors, including PHYA, PHYB, and CRY1, which promote chloroplast development through a suite of transcription factors, including HY5 (Jing and Lin 2020). Light signal transcription factors regulate genes associated with chloroplast structure, formation, or photosynthesis (Cackett et al. 2022). Hormonal signals also play important functions in chloroplast biogenesis, including signals from gibberellins (GAs) and brassinosteroids (BRs). GAs and BRs have regulatory roles in chloroplast biogenesis through DELLA proteins and GOLDEN2-LIKE1 (GLK1) and GLK2 transcription factors, respectively (Zhang et al. 2024). DELLA proteins are a class of negative regulators of GA signaling, and inhibit chloroplast development by downregulating TOC159 protein levels and thereby inhibiting the accumulation of photosynthetic proteins (Shanmugabalaji et al. 2018). BRs regulate chloroplast biogenesis through GLK1 and GLK2 transcription factors via BRASSINOSTEROID insensitive2 (BIN2). BIN2 acts by phosphorylating GLK1 and GLK2 transcription factors. Loss-of-function mutations of either BIN2 or GLK1/GLK2 lead to abnormal chloroplast development, while gain-of-function mutations can produce differences in BR responses and subsequent changes to chloroplast structure (Zhang et al. 2021; Tachibana et al. 2024). Thus, pathways for hormonal signaling for biogenesis, and light-mediated signaling for chloroplast biogenesis, integrate to ensure that chloroplasts develop under optimal signals from the environment and developmental programming.

7.2.4 Retrograde Signaling Pathways in Chloroplast Biogenesis

Retrograde signaling is an important communication pathway for chloroplasts relaying information to the nucleus. In particular retrograde signaling is crucial for coordinating chloroplast function and development with overall plant growth and environmental factors (Nott et al. 2006). One major retrograde signaling pathway consists of plastid pigments and pigment precursors, which include tetrapyrroles, apocarotenoids, and carotenoids, as retrograde signals, to influence the expression of genes in the nucleus and synchronize the development of chloroplasts with the functional demands of photosynthesis in plants (Cazzonelli et al. 2020). In addition, signals relating to the plastid gene expression (PGE) influence the nuclear gene regulation to adjust gene expression in the nucleus based on the operational status of the chloroplast (Liebers et al. 2022). Reactive oxygen species (ROS), byproducts of photosynthesis, provide retrograde signals to regulate stress responses and adapt gene expression to manage oxidative stress (Banerjee and Roychoudhury 2017). Other retrograde signals involving redox signals, which are associated with the operation of photosynthetic and energy-generating pathways, include plastoquinone and ferredoxin (Hanke et al. 2009). Furthermore, changes in metabolite pools that traverse mediating pathways in chloroplast metabolism act as retrograde signals to the nucleus to regulate gene expression and adapt chloroplasts for optimized performance (Unal et al. 2020). Key players in retrograde signaling pathways include Pentatricopeptide Repeat (PPR) proteins, which regulate plastid gene expression and retrograde signaling. Nuclear-encoded transcription factors, such as ABI4 (a bZIP transcription factor responding to plastid-derived signals to regulate stress-responsive genes) and GUN1 (Genomes Uncoupled 1, which links chloroplast function to nuclear gene expression), are also critical components (León et al. 2013).

Studies focusing on model systems, such as *Arabidopsis thaliana*, have highlighted the role of retrograde signaling in chloroplast biogenesis, demonstrating that mutants in retrograde signaling pathways show altered development and function of chloroplasts. Overall, retrograde signaling integrates chloroplast signals to synchronize nuclear gene expression to support the physiological and developmental facility of the chloroplast, fundamental to supporting chloroplast function and developing responsive mechanisms to environmental change.

7.3 Abiotic Stress Responses in Chloroplast

The major issues of plant growth and physiological functions are generally impacted by common abiotic stresses such as extreme temperature range, salinity and drought (Table 7.2). Retarded plant growth due to drought stress eventually leads to a decline in plant cell division, distorts protein synthesis, and decreases photosynthetic rate

Table 7.2 Overview of chloroplast functions in response to various stressors across species

Species	Stress	Component	Stimulus	Function	References
Arabidopsis thaliana	Drought	NTRC (Thioredoxin reductase)	Reduced photosynthetic efficiency under drought	Maintains redox homeostasis and chloroplast function by detoxifying ROS	Ishiga et al. (2016)
		SAL1-PAP (Retrograde signaling component)	Water deficiency causing oxidative stress	Regulates the expression of nuclear stress-inducible genes via retrograde signaling	Estavillo et al. (2011)
	Salinity	Clp protease	Salinity-triggered oxidative stress	Controls the quality of thylakoid membrane proteins under salt stress	Pulido et al. (2017)
		SP1 (Ubiquitin E3 ligase)	High salt concentrations causing oxidative stress	Protection against salt stress by causing translocon breakdown at the outer chloroplast membrane	Ling and Jarvis (2015b)
	Temperature	Hsp21 (Small heat-shock protein)	High temperatures causing protein denaturation	Binds and stabilizes denatured proteins, preventing aggregation	Bernfur et al. (2017)
	Biotic	MPK3, MPK6 (Mitogen-activated protein kinases)	Pathogen invasion leading to oxidative stress	Modulates chloroplast-generated ROS and regulates immune responses	Su et al. (2018)
Brassica rapa	High light	RH22 (DEAD-box RNA helicase)	Exposure to excessive light	Regulates stress-inducible nuclear genes to mitigate light-induced damage	Nawaz et al. (2018)
Glycine max	Drought	GmDREB2A (Transcription factor)	Drought-induced stress in chloroplasts	Enhances drought tolerance by activating stress-responsive genes	Visioni et al. (2019)
	Salinity	GmNAC109 (Transcription factor)	Salt stress-induced ROS production	Regulates antioxidant enzymes to alleviate oxidative stress	Singh et al. (2019)
Hordeum vulgare	Drought	HvCBF4 (Transcription factor)	Drought-induced osmotic stress	Enhances proline accumulation and drought tolerance	Shen (2013)
	Salinity	HvHKT1;5 (Transport protein)	Sodium ion accumulation under salinity	Maintains ion homeostasis by reducing sodium transport into the shoot	Han et al. (2018)
Musa acuminata	Temperature	SGR (Stay-green protein)	Heat exposure affecting chlorophyll breakdown	Maintains chlorophyll stability during heat stress	Yang et al. (2009)

(continued)

Table 7.2 (continued)

Species	Stress	Component	Stimulus	Function	References
Nicotiana benthamiana	Biotic	PsbQ (Photosystem II protein)	Pathogen infection compromising photosynthetic apparatus	Phosphorylates thylakoid-associated proteins to maintain photosystem II stability	Rodríguez-Herva et al. (2012)
		Rpi-vnt1.1 (NLR protein)	Recognition of pathogen effector proteins	Mediates immune response through chloroplast-associated light-dependent signaling	Gao et al. (2020)
Oryza sativa	Drought	PhyB (Phytochrome B)	Drought-induced chloroplast signaling disruption	Affects chloroplast gene translation and interacts with chloroplast-localized proteins	Yoo et al. (2017)
	High light	CspA, CspB (Cold shock proteins)	High light intensity damaging chloroplasts	Involves in maintaining protein homeostasis under high-light conditions	Guddimalli et al. (2021)
	Temperature	WHY1 (WHIRLY proteins)	Thylakoid membrane damage under cold stress	Regulates chloroplast development and stabilizes the thylakoid membrane	Zhuang et al. (2020a, b)
Solanum lycopersicum	Temperature	CDJ2 (Chloroplast-targeted DnaJ)	Heat stress affecting protein folding	Protects Rubisco activity by assisting in protein refolding during heat stress	Wang et al. (2015)
Solanum tuberosum	Temperature	StHSP90 (Heat shock protein)	High temperatures causing oxidative stress	Enhances antioxidant enzyme activity to reduce heat-induced damage	Zhu et al. (2024)
	Biotic	StNPR1 (Non-expressor of pathogenesis-related genes)	Pathogen-induced oxidative stress	Activates defense-related genes to improve pathogen resistance	Xue et al. (2021)
Triticum aestivum	Temperature	Rca1 (Rubisco activase)	Increased temperatures impairing photosynthetic efficiency	Modulates Rubisco activity to sustain photosynthesis under heat	Kumar et al. (2016)
		Ub2 (Ubiquitin/26S proteasome system)	Stress-induced protein misfolding	Enhances antioxidant capacity by degrading misfolded proteins	Tian et al. (2014)
Zea mays	Temperature	DUA1 (RNA-binding protein)	Low temperature stress reducing photosynthetic activity	Compensates for reduced photosynthesis by regulating cold-responsive genes	Cui et al. (2019)

(Ma et al. 2017). In parallel, shattering ion homeostasis, suppressing seed germination and changing plants' membrane lipid composition via salinity causes prolonged oxidative stress (Feng et al. 2015). Additionally, continuous low and high-temperature exposure can degrade plant cell structures, impede biochemical and physiological processes, slow crop yields, and restrict plant allocation in specific zones (Zhuang et al. 2020a).

Photosynthetic organs are vulnerable to abiotic stress. Whenever plants encounter abiotic stress, these organs also respond to environmental impact as structural and metabolic regulations (Chaves et al. 2009). In unfavorable environmental conditions, chloroplasts activate their metabolic machinery and produce biologically active compounds and phytohormones, which are necessary to protect plants from abiotic stress. The communication network of the chloroplast, as plastid-to-nucleus retrograde signaling, provided extra points to cope with environmental stress (Chan et al. 2016).

The rate of chloroplast malondialdehyde contents, lipid peroxidation, and the enzymatic activity of SOD instantly rise under the influence of abiotic stress (Zhang et al. 2013). Both external and internal agents such as water availability, ion balance, light quantity, leaf age and leaf water potential, temperature and nutrition also affect chloroplast development (Mohanty and Tripathy 2011). The seedlings that are exposed to chill, heat or salt stress suffer from the downregulation of protein clusters and gene expression or post-translational modification of various enzymes that participate in tetrapyrrole metabolism, and these circumstances diminish chlorophyll (Chl) biosynthesis. As shown in Fig. 7.1, environmental stresses, both biotic and abiotic, can disrupt the chloroplast's normal function, thus triggering signals that regulate gene expression and protein behavior. Stress signals activate chloroplast activities to restore homeostasis in chloroplasts. While retrograde transmission from chloroplasts to the nucleus conveys the stress status, biological activity and phytohormones produced by chloroplasts aid in adaptation to shifting environmental conditions.

7.3.1 Response to Drought Stress

Unpredicted climatic conditions result in insufficient rainfall, which ultimately causes drought stress. Drought stress is the major limiting factor affecting plant growth and metabolism and initiates alterations in various physiological processes in plants, such as stunted growth, inefficient photosynthesis, and gross productivity, including impaired chloroplast biogenesis and development (Lawlor and Tezara 2009). A lot of studies have investigated the influence of water scarcity on photosynthesis, especially in fully grown leaves (Massacci et al. 2008). Water stress conditions taper off photosynthesis, and there are two principal channels responsible for this: (1) confined dispersal of CO_2 into the intercellular spaces of leaves due to disturbed stomatal opening and (2) lower metabolic rate. Reducing Chl biosynthesis under water stress led to fewer and irregular gatherings of photosynthetic complexes in thylakoid membranes. The consequence of this activity is metabolic attenuation

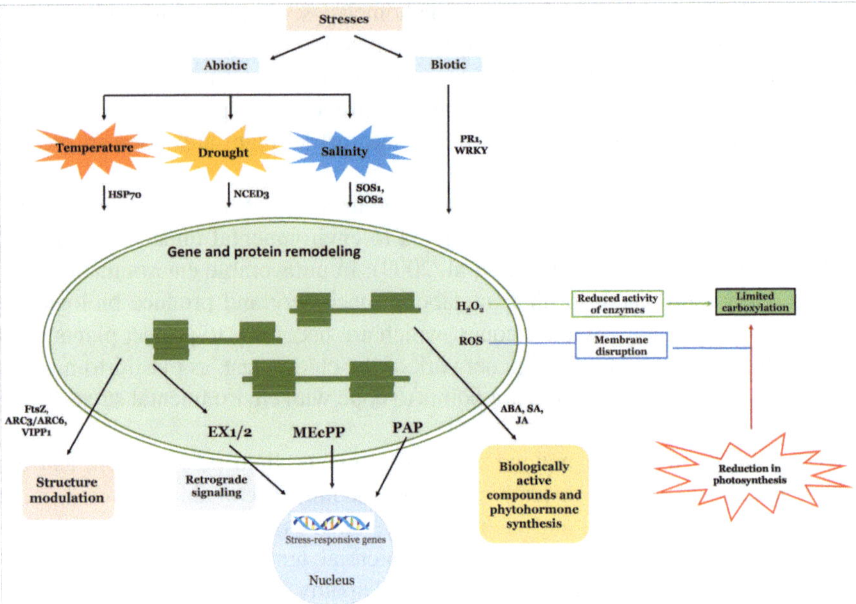

Fig. 7.1 Chloroplast response to environmental stresses

in the development of green leaves, which ultimately leads to inhibited plant growth. Chlorophyll and carotenoids are both necessary pigments for the arrangement, proper folding and insertion of photosynthetic proteins into the membranes of thylakoid.

Chloroplast biogenesis is particularly sensitive to water deficit due to its complex interplay with photosynthetic and metabolic pathways. In water-stressed plants, the depletion in photosynthetic CO_2 fixation leads to an underutilization of NADPH, resulting in an extremely decreased electron transport chain that becomes vulnerable to oxygen attack (Chaves et al. 2009). Most of the time, superoxide (O_2^-) and singlet oxygen are the result of this process, both of which tend to injure the photosynthetic system and other cellular constituents. Furthermore, ROS agents originating under the influence of these conditions can down-regulate the synthesis of tetrapyrrole intermediates, for example, protochlorophyllide (Pchlide), likely due to an imbalance Mg-protoporphyrin monomethyl ester (MPE) cyclase activity (Stenbaek et al. 2008), leading to a decrease in chlorophyll synthesis in drought-stressed plants.

Oxidative damage to the stromal and thylakoid proteins aggravates the disruption in chloroplast formation, affecting the structure and function of the chloroplast. For instance, the oxidation of D1 and D2 proteins, which are essential for preserving electron transport efficiency, upsets the balance of the photosynthetic machinery, especially photosystems I and II, during drought. The supply of ATP for metabolic and biosynthetic processes, including chlorophyll biosynthesis, is jeopardized by the ATP synthase complex, which is essential for energy production and shows decreased activity under water stress. Generally, a lower accumulation of chlorophyll is found in developing seedlings under water stress. The causes behind this lower Chl synthesis are predominantly due to the decreased aggregation of key chlorophyll biosynthetic intermediates, such as Mg-protoporphyrin IX monomethyl ester, glutamate-1-semialdehyde (GSA), protochlorophyllide, and 5-aminolevulinic acid (5ALA). However, the synthesis of 5ALA is diminished under water stress, but the upregulation of the gene expression and plenty of the protein accountable for its synthesis, GSA aminotransferase, rises, emphasizing its vital importance in the greening system under stressful conditions. Various chlorophyll biosynthetic enzymes, for example, Mg-chelatase, 5-aminolevulinic acid dehydratase, protochlorophyllide oxidoreductase, porphobilinogen deaminase, porphyrinogen IX oxidase, and coproporphyrinogen III oxidase, are found in downregulated gene expression and decreased protein quantity, causing a reduction in their biochemical activities in water-scarred seedlings. The primary cause of the vitiated Shibata shift is the down-regulation of protochlorophyllide oxidoreductase. GSA and 5ALA are the foremost intermediates, and these decreases in synthesis may influence the gene expression of subsequent enzymes in the Chl biosynthesis pathway (Dalal and Tripathy 2012). These valuable outcomes are crucial for understanding the molecular mechanisms of fundamental chlorophyll biosynthesis under water stress and can aid in developing stress-resilient plant varieties.

Two-dimensional electrophoresis result of purified chloroplasts evaluate impact of drought on chloroplast proteins, and the determined 26 distinctive chloroplast proteins that respond to water stress. These drought-related proteins were first and foremost involved in amino acid metabolism, photosynthesis and carbon and nitrogen metabolism. Remarkably, most of the proteins related to photosynthesis were downregulated under drought conditions, resulting in reduced photosynthetic efficiency. Rubisco and carbonic anhydrase, these two proteins linked to carbon and nitrogen metabolism, were upregulated among several proteins, which is contradictory to drought-responsive proteins. In cassava, this up-regulation enhanced drought tolerance by improving the carbohydrate conversion transfer rate and defending the plant from oxidative stress.

Some chloroplast proteins also function as messengers, transmitting retrograde signals from chloroplasts to the nucleus in response to drought stress. The study by Li et al. (2020) identified 3936 total leaf proteins and 1087 chloroplast proteins that were distinctively assembled under drought conditions, and they classified them into 16 categories. A particular group of proteins, carbonic anhydrase 1 (CA1), showed a noteworthy reduction in chloroplasts but a significant rise in leaves during drought stress out of the classified protein group. Some plant studies related to CA1 proteins such as moss (*Physcomitrella patens*), *Oryza sativa* and *Arabidopsis thaliana* in Physcomitrella patens protoplasts, consistently demonstrated that CA1 proteins gradually reduced within chloroplasts but accumulated in the cytosol under osmotic stress. These insights reveal that CA1 proteins may be displaced from chloroplasts to the cytosol and assist in response to drought stress as potential signal messengers. Drought-induced damage to chloroplast biogenesis not only limits the plant's photosynthetic capacity but also diminishes its ability to acclimate to other abiotic stresses, highlighting the need for further research into the mechanisms regulating chloroplast biogenesis under drought conditions.

7.3.2 Response to Salt Stress

Similar to drought, another Osmotic stress, salinity, also impacts plants in numerous ways, affecting physiological and biochemical processes including lipid composition, ion balance, malate and nitrogen metabolism, photosynthesis, antioxidative enzyme activity and chloroplast structure (Parida and Das 2005). Several studies have highlighted understanding the effects of salt stress on carbon assimilation, photosynthetic activity, electron transport, pigment composition and the efficiency of photosystems I and II (Koyro 2006). Sustaining the ideal concentrations of potassium and chloride within the chloroplast is crucial for regulating pH, thylakoid stacking, chloroplast volume, electron transport, and overall photosynthetic efficiency (Pottosin and Shabala 2016). However, salt stress disrupts this balance by increasing the concentrations of sodium (Na^+) and chloride (Cl^-) while reducing potassium (K^+) levels inside chloroplasts, resulting in osmotic, ionic, and oxidative stress (Flowers et al. 2015). These disparities within chloroplasts eventually hamper ionic homeostasis and lead to compressed photosynthetic performance, stunted growth, and reduced crop yield.

Regardless of the understanding of these effects, the critical transporters accountable for regulating Na^+, Cl^-, and K^+ concentrations during salt stress remain unclear due to limited studies. More intensive research in this domain is essential to develop a confined approach for improving plant resilience to salinity stress. Most of the physiological and biochemical processes of plants go through the initial platform membranes, and in response to environmental stress factors, including salinity, plants can effortlessly customize their membrane lipid compositions. Various studies demonstrated that salinity can influence significant alterations in the lipid content of plant membranes (Sui et al. 2010). Galactolipids are key components of the photosynthetic membranes in plants, which are the most abundant membranes in nature. Monogalactosyldiacylglycerol (MGD) is a critical galactolipid involved in the formation of photosynthetic membranes, with its synthesis catalyzed by the enzyme monogalactosyldiacylglycerol synthase (MGD synthase). Research has proven that MGD plays an important role in development and plant growth. Defects in chloroplast ultrastructure, disruption of photosynthetic membranes, Pale-green phenotype, complete impairment of photosynthetic capacity and photoautotrophic growth are the result of destruction in MGD activity. These findings reveal that MGD is essential for photosynthetic growth, structural organization of plastidic membranes and chlorophyll formation. The overexpression of the incorporated OsMGD gene in tobacco plants has been found to improve salt tolerance. Monogalactosyldiacylglycerol (MGDG) and digalactosyldiacylglycerol (DGDG) are two galactolipids that play an important role in regulating chloroplast structure and function during salt stress, underlining their importance in the plant's response to salinity.

Research in soybeans by He et al. (2014) noticed the impact of salt stress in the reduction of chloroplast Rubisco levels, a process regulated by the formation of Rubisco-containing bodies (RCBs). During salinity stress, plants acclimatize by modulating the import of protein into chloroplasts to elevate stress tolerance, and the protein that assists via upregulating their transcription is ubiquitin E3 ligase suppressor of PPI1 locus 1 (SP1). SP1 stimulates the deterioration of the translocon in chloroplasts' outer envelope membrane, which subsequently mitigates photosynthesis and reduces the formation of reactive oxygen species (Ling and Jarvis 2016). Among the numerous nucleus-encoded, chloroplast-targeted RNA-binding proteins (RBPs), DEAD-box RNA helicases (RHs) are the gatekeepers of chloroplast gene expression. The downregulation and up-regulation of chloroplast genes depend on them. For example, in *Brassica rapa*, the expression of the chloroplast-targeted DEAD-box RNA helicase BrRH22 is induced by drought stress, high salinity, and abscisic acid treatment. The RNA chaperone functions of BrRH22 help the translation of chloroplast transcripts under salt stress and are significant in regulating abiotic stress (Nawaz et al. 2018). Furthermore, it has been seen that plants expressing BrRH22 ameliorate the greening of cotyledon and enhance growth under ABA treatment, due to lower contents of ABI3, ABI4, and ABI5 to some extent. The insights into BrRH22 and its significant role in modulating chloroplast gene translation and stress responses provide valuable targets for improving chloroplast biogenesis and plant adaptation to abiotic and biotic stresses.

7.3.3 Response to Temperature Stress

Chloroplast differentiation is controlled by a critical environmental factor, light. Exposure to light initiates the biogenesis of the thylakoid during de-etiolation of plant seedlings, escorted by the systematized synthesis and gathering of chlorophylls, proteins, and lipids. At the same time, chloroplast development and function are also regulated by other environmental factors such as temperature. Either high or low temperatures can inhibit crop productivity and growth. Due to their sessile nature, plants must adjust to cope with fluctuating temperatures to sustain normal photosynthesis and chloroplast function (Li et al. 2022). The chloroplast growth and development in juvenile plants are significantly affected by cold stress. Immediate low-temperature occurrence during early seedling development in spring can directly impair chlorophyll biosynthesis (Kusumi and Iba 2014). Various studies explain that plants alter early chloroplast development in response to cold stress. The case of virescent mutants similarly acts as initially it shows low chlorophyll levels in young leaves, but a gradual increment of normal levels is seen when they mature (Yoo et al. 2009). OsV4, V3, St1, TCD9, and TSV are some temperature-sensitive, virescent mutant genes engaged in the regulation of chloroplast development under cold stress (Jiang et al. 2014; Sun et al. 2017).

Zhao et al. (2020) suggest that during the early seedling green stage, low temperatures below 18 °C have been shown to affect chloroplast biogenesis and development in *Oryza sativa*. A rice mutant, white stripe leaf 5 (wsl5), evolves white-striped leaves during early leaf development and becomes albino under cold stress (Liu et al. 2018a, b). In cold stress conditions, genetic and molecular analyses revealed that WSL5 translates a novel chloroplast-targeted pentatricopeptide repeat protein necessary for functional chloroplast development and continuing retrograde signaling from plastids to the nucleus.

The story of high-temperature stress (37 °C) is slightly different from lower temperatures, and it significantly impacts the chloroplast transcriptome in Arabidopsis (*Arabidopsis thaliana*) (Castandet et al. 2016). Although heat stress hinders photosynthesis and chloroplast biogenesis, the exact molecular mechanism regulating chloroplast development under various temperature conditions remains unclear (Zhang et al. 2010). The initial response of heat on chloroplasts includes protein denaturation and accumulation, as well as increased membrane fluidity. Protein denaturation is combated by the aggregation of molecular chaperones, while increased membrane fluidity is balanced through higher biosynthesis of saturated fatty acids, such as digalactosyldiacylglycerol (Rütgers et al. 2017).

The import and modulation of chloroplast protein is processed by Heat shock proteins (HSPs) (Feng et al. 2014). Under heat stress, the expression of chlorophyll biosynthesis-responsive genes is upregulated, and chloroplast biogenesis is nurtured in novel SOC1-like genes and petals overexpressing SOC1 (suppressor of overexpression of CO). MADS-box transcription factors, which are available in many photosynthetic eukaryotes, are also expressed in chloroplast-containing tissues, with multiple homologs observed (Wang et al. 2019). These insights reveal the complex regulatory networks that plants use to modulate chloroplast development and function in response to diverse environmental conditions. Understanding these mechanisms is important for improving crop resilience to temperature extremes.

In angiosperms, plastid genes are predominantly transcribed by two sets of RNA polymerases: plastid-encoded RNA polymerase (PEP) and nuclear-encoded RNA polymerase (NEP) (Börner et al. 2015). A new PEP-associated protein (PAP), WLP2, has been determined through the study of two allelic mutants, wlp2w and wlp2s, which unveil a panicle phenotype and white leaf. These mutants are albino and fatal at high temperatures, demonstrate reduced chlorophyll accumulation, hinder chloroplast ultrastructure, and minimize photosynthetic activity. Research revealed that WLP2, along with its paralog OsFLN2, can physically interact with the thioredoxin OsTRXz, forming a TRX-FLN regulatory module. The transcription of PEP-encoded genes and maintaining redox balance within chloroplasts under heat stress is primarily regulated by the TRX-FLN module (Lv et al. 2017). These results suggest off-center insights into the tissue-specific regulation of chloroplast biogenesis and suggest potential methods for originating green petals. They also align with the extensive understanding of how heat stress impacts chloroplast biogenesis in plants.

Abiotic and biotic stresses both affect chloroplast structure and function. Disrupting various metabolic pathways and giving rise to new ones. Both have a significant effect on overall plant growth and survival. Oftentimes, there can also be an interplay between various types of stresses, further complicating the stress tolerance mechanisms. Thus, understanding both types of stress responses and the links between these responses can lead to a better understanding and development of new strategies to improve overall crop resistance.

7.4 Biotic Stress Responses in Chloroplast

Plants possess a multi-faceted defense mechanism that helps recognize the pathogens and eventually launch a defensive response. The changes in the overall structure and photoelectron transport of chloroplasts experiencing biotic stress (Nanda and Biswal 2008) differ fairly significantly. The chloroplast genome codes for many antimicrobial peptides, which act as a shield against bacterial or viral infections as well as insect herbivory (Jin et al. 2012; Lee et al. 2011; Heil 2009). It is also very probable that retrograde signals from developing plastids are distinct from the ones generated under adverse conditions and might include transcripts, proteins, and many other catalytic biomolecules. The exploration into Plant-microbe relationships demonstrates the crucial role that chloroplasts play in plant survival, primarily acting as metabolic powerhouses of cells as well as mediators of interactions between plants and the environment (Fernandez and Burch-Smith 2019). The chloroplasts are the site of origin of several minor signaling biomolecules, which include the plant hormones like abscisic acid (ABA), jasmonic acid (JA) and salicylic acid (SA) and all essential components that determine how plant-biotic interactions turn out (Lu and Yao 2018). One of the most explored molecules when it comes to plant responses to microbial attack is the lipid hormone JA, along with many other compounds produced from lipids that function as signal molecules during responses to biotic stress in plants (Lim et al. 2017). Experimental evidence in tobacco (*Nicotiana tabacum*) plants shows that during response prompted by *Pseudomonas syringae* pv *phaseolicola,* the Jasmonic Acid and Salicylic acid pathways work synergistically. Similar results were observed in *Arabidopsis thaliana* plants.

One of the absolute key changes that take place as a reaction to pest infestation and pathogen attacks is the manufacture of reactive oxygen species (ROS) by attacked plants. However, as a part of the stress response, secondary sites also become a hub for ROS production, including the cell wall, cell membrane, apoplast and the endoplasmic reticulum. ROS refers to various free radicals such as $O_2{\bullet}^-$ and OH•, as well as some non-radical species like H_2O_2 and $^1O_2$ (Suman et al. 2021). An increased concentration of ROS results in oxidative stress, resulting in chlorophyll damage, nucleic acid and protein degradation, lipid peroxidation and redox imbalance overall (Lv et al. 2017). In the case of *E. carotovora* infection in plants, an exudation of cell wall lysing enzymes brought about the activation of signaling pathways, oxidative burst and host defense mechanism (Palva et al. 1993). As a direct result of this, stromal proteins are transported from the chloroplasts into the cytoplasm. These proteins, including many transcription factors, induce nuclear stress responses and trigger antioxidant genes to protect the plant (Sun et al. 2011).

Likewise, the production of ROS is also a part of the initial multilayered response resulting from a pathogen strike. When the ROS interacts with other plant signaling molecules such as jasmonic acid (JA), salicylic acid (SA), and ethylene (ET), it culminates in the activation of plant defense responses (Torres et al. 2006). As an immediate effect of this, the number of distinct metabolites and proteins increases significantly.

In addition to this, a wide range of volatile organic compounds (VOCs) are produced by chloroplasts. Plants can communicate with nearby plants up to a meter away through systemic signaling that is orchestrated by their volatile organic compounds (VOCs) (Benevenuto et al. 2018). The MEP route is responsible for the production of these volatile organic compounds (VOCs), including terpenoid and carotenoid derivatives as well as fatty acid derivatives, namely alcohols and aldehydes (Bobik and Burch-Smith 2015). The occurrence of disease and predation can cause the production of volatile organic compounds (VOCs), which can activate different defense responses. By protecting neighboring plants as well as sick plants, this defense priming may result in the achievement of population-level resistance (Benevenuto et al. 2018). For example, Birds that typically eat sawfly larvae and adults increased their feeding when sawfly eggs were deposited on Scots pine needles, due to the production of VOC (E)-b-farnesene (Mantyla et al. 2018). Comparably, feeding of *Frankliniella occidentalis* (thrips) on *Phaseolus vulgaris* first increased the activity of antioxidant enzymes in the plant, but eventually, the thrips were able to develop a counter-defense mechanism (Qian et al. 2018). Thus, the synthesis and release of volatile organic molecules from the chloroplasts confer defense against subsequent attacks inside the same plant and inhibit initial attacks in neighboring plants.

Viral infection exhibits a negligible inhibitory effect on the primary light reaction in Malmsey grape vines (Sampol et al. 2003), while the effect on PS II quantum yield of the active reaction center was found to be much less compared to the PS II-PS I electron transport in a toxic bloom-forming alga (*Heterosigma akashiwo*) (Nanda and Biswal 2008; Juneau et al. 2003) and there were no visible changes in the efficiency of PS II (FV/FM) in plum pox virus (PPV) infected peach plants (Hernandez et al. 2004). Studies also show that in TMV-infected tobacco leaves, PS II is affected to a relatively lesser extent (Wilhelmová et al. 2005) and remains completely unaffected in juvenile leaves of *Nicotiana benthamiana* infected with either of the two tobamovirus strains (PMMoV-I and PMMoV-S) (Sajnani et al. 2007). Virosis has also been proven to promote leaf senescence in plants (Phipps and Buis 1973; Kabi et al. 1979; Kar et al. 1985). The chloroplasts are more prone to senescence compared to any other cell organelle (Blswal and Biswal 1988). Due to a change in the composition and capabilities of chloroplasts, necrosis occurs during senescence as a direct result of a sharp decline in photosynthetic activity (Buchanan-Wollaston 1997). Likewise, viral infections impair photosynthesis in plants as a direct result of chlorophyll loss, often coupled with the breakdown of chloroplasts (Zaitlin and Jagendorf 1960; Almasi et al. 2001). Studies into the chloroplast structure have revealed distinct abnormalities in the case of tobacco-TMV (flavum strain) and tobacco-TMV (wild type) infections (Lehto et al. 2003). Additionally, an increase in the number of plastoglobuli was discovered along with a decrease in thylakoid amount in the case of Styrian pumpkin-zucchini yellow mosaic virus (ZYMV) infection (Zechmann et al. 2003). PSII activity and ROS production are negatively affected by viral infections (Rodriguez-Herva et al. 2012). The virulence factor HopI1 produced by *P. syringae* pathogen modifies the thylakoid structures and hinders the increase of salicylic acid (SA) (Jelenska et al. 2007).

Thylakoids are a site for the storage of chlorophyll in plants. It has been reported that the total chlorophyll content of PMV-infected leaves is almost half of that of non-infected leaves, with a significant decrease in the contents of Chl a and Chl b. The total carotenoid content of the leaves also decreased drastically (11.52% of the original). However, as far as the Chl a/b ratio is concerned, it was found to be higher in the infected leaves; in contrast to this, the Car/Chl ratio decreased in infected leaves (Nanda and Biswal 2008). All of this seems to be a result of the pathogen-induced weakening of pigment metabolism. An investigation into the chloroplast illustrates that PTM, a plant homeodomain transcription factor present on the chloroplast envelope, is involved in retrograde signal pathways (Sun et al. 2011). The breakdown of chloroplasts to release antimicrobial proteins can give an easy clarification; however, the retention of antimicrobial peptides within the bounds of the plastid did not support this theory (Lee et al. 2011). Another example is the interaction of the virulence factor protein C4, from the geminivirus Tomato yellow leaf curl virus (TYLCV), with the BAM1 (BARLEY ANY MERISTEM 1) and BAM2 at the plasma membrane, suppressing the cell-to-cell transport of RNA interference (RNAi), an important anti-viral defense apparatus (Rosas-Diaz et al. 2018).

A new pathway was reported by Medina-Puche et al. (2020) C4 shifts its location from the cell membrane to chloroplasts due to the activation of plant defense responses. In *Arabidopsis* plants, Calcium-dependent kinase 16 (CPK16), which is attached to the plasma membrane through N-myristoylation, relocates to the chloroplasts when subjected to flg22 treatment, enhancing the anti-pathogenic reactions and SA marker gene expression (Stael et al. 2011). A very high level of protection from pathogens is conferred to plants through the production of biomass hydrolyzing enzymes within chloroplasts (Verma et al. 2010), demonstrating a mercurial retort activated by proteins expressed within the chloroplasts. In the instance of a tobacco mosaic virus (TMV) attack, a receptor-interacting protein (NRIP1) present within the chloroplast exhibited interaction with the P50 helicase enzyme within the cytoplasm (Caplan et al. 2008). This is found to be the general trend, as a large majority of nucleotide-binding receptors are found to be restricted to the chloroplasts.

As a direct response to biotic stresses, several bioactive peptides are released from their protein precursors (Tavormina et al. 2015). Most of the proteins, including systemins, hydroxyproline-rich systemin (HypSys) and plant elicitor peptides (Peps), act exclusively as damage-associated molecular patterns (DAMPs), encouraging defensive gene expression as well as synthesis of protective secondary metabolites. During pathogen occurrence, ROS signaling also encourages protein transportation from chloroplasts into the cytoplasm, which provides additional sources of antimicrobial proteins as a part of biotic stress responses (Kwon et al.

2013). There may also be a presence of "hidden" peptides in functional proteins that may have comparable properties to DAMPs. One illustration of this is inceptin, a peptide 11–13 aa in length, found in the oral secretions of larvae of *Spodoptera frugipeda*. It was found to encourage the synthesis of volatile compounds, mostly from chloroplasts, that can appeal to the predators of herbivorous insects (Schmelz et al. 2007). Despite various evidence suggesting the involvement of retrograde signaling pathways in the chloroplast responses to biotic stresses, clear-cut relationships have not yet been established, as certain areas like GUN-type signaling, Heme, DHA, etc., have remained relatively opaque (Chan et al. 2016). Although the plant defense against pathogen attack is well understood, the interplay and impact of different signals to generate defense responses against biotic stress still remain somewhat vague.

7.5 Supplementary Effect of Light on Biotic Stress Response in Chloroplasts

The plant defense against pathogens is greatly impacted by the redox state of the chloroplast plastoquinone (PQ) pool. When facing biotic stress, plants distinguish and facilitate light signals through chloroplasts or other photoreceptors. For example, *Arabidopsis* plants regulate the SA levels by red light through the phytochrome-regulated pathway and via blue light through a CRY2/PHOT2-mediated pathway. COP1 is negatively regulated by CRY2 and PHOT2, which consequently regulates the steadiness of the R protein (plant disease resistance proteins), HRT (Hypersensitive Response to Turnip Crinkle Virus), hence affecting the resistance to pathogens. In addition to this, shade environments affect SA-related defense, weaken jasmonic acid (JA)-induced defense, and enhance gibberellin (GA) activity. Calcium signals affecting biotic stress defense are mediated by the calcium-sensing receptor (CAS) present inside the chloroplast. CAS also regulates the buildup of SA inside the cell (Roeber et al. 2021).

Predation of plants by insects can be perceived via the detection of damage-associated molecular patterns (DAMPs) in addition to the herbivore-associated molecular patterns (HAMPs), which comprise the fatty acid-amino couplets, found as part of the oral secretions of maximum lepidopteran larvae (Yoshinaga et al. 2010; Heil 2009). Escobar-Bravo et al. (2019) showed the positive effect of UV light on a compromised tomato mutant resulting from a robust activation of SA-associated defense responses by exposure to UV light after a thrips attack. Additionally, there is also proof that UV-B impacts the attraction of *Arabidopsis* plants to diamondback moths (*Plutella xylostella*), also suggesting that biosynthesis of JA is essential for the exhibition of a defensive reaction (Caputo et al. 2006). UV-B treatment also increased the resistance of *Arabidopsis* to predation by *Spodoptera litura* (Qi et al. 2018). These instances demonstrate that there is probably a proximate relationship between light exposure and the exhibition of chloroplast reactions to herbivores and pathogen outbreaks in plants.

7.6 Conclusion and Future Perspectives

Chloroplasts, traditionally recognized for their role in photosynthesis, are now acknowledged as critical players in plant responses to environmental stresses. Under adverse conditions, the efficiency of photosynthesis is significantly reduced, leading to potential yield losses in crops. However, chloroplasts do more than host photosynthesis; they also serve as sensors of environmental stress, initiating retrograde signaling that modulates nuclear gene expression to help plants adapt and survive.

Recent advances have highlighted the chloroplast's ability to regulate its structure, gene expression, and associated metabolic pathways, including the production of oxylipins, phytohormones, and reactive oxygen species. These adaptations are essential for plant survival under stress, but the mechanisms by which chloroplasts detect stress, trigger appropriate responses, and communicate with the nucleus remain complex and not fully understood. Future research should address several key challenges: determining whether chloroplast responses are direct consequences of environmental stress or active components of stress tolerance; unraveling how external stresses activate chloroplast responses; and exploring the full spectrum of retrograde signaling pathways beyond those involving reactive oxygen species. By answering these questions, we can gain deeper insights into the precise roles of chloroplasts in stress responses, laying the groundwork for developing genetically enhanced crops with improved resilience to environmental challenges.

References

Albaqami MN (2022) Investigating the role of Cryptochrome activity in arabidopsis plant and blue light exposure in Saos-2 cell-mediated mineralization

Almási A, Harsányi A, Gáborjányi R (2001) Photosynthetic alterations of virus infected plants. Acta Phytopathologica et Entomologica Hungarica 36(1–2):15–29

Andersson I, Backlund A (2008) Structure and function of Rubisco. Plant Physiol Biochem 46(3):275–291

Banerjee A, Roychoudhury A (2017) Abiotic stress, generation of reactive oxygen species, and their consequences: an overview. In: Reactive oxygen species in plants: Boon or Bane-revisiting the role of ROS, pp 23–50

Barkan A (2011) Expression of plastid genes: organelle-specific elaborations on a prokaryotic scaffold. Plant Physiol 155(4):1520–1532

Benevenuto RF, Hegland SJ, Töpper JP, Rydgren K, Moe SR, Rodriguez-Saona C, Seldal T (2018) Multiannual effects of induced plant defenses: are defended plants good or bad neighbors? Ecol Evol 8(17):8940–8950

Bernfur K, Rutsdottir G, Emanuelsson C (2017) The chloroplast-localized small heat shock protein Hsp21 associates with the thylakoid membranes in heat-stressed plants. Protein Sci 26(9):1773–1784

Bhatla SC, Lal MA (2023) Light perception and transduction. In: Plant physiology, development and metabolism. Springer Nature Singapore, Singapore, pp 363–390

Blswal UC, Biswal B (1988) Ultrastructural modifications and biochemical changes during senescence of chloroplasts. In: International review of cytology, vol 113. Academic Press, pp 271–321

Bobik K, Burch-Smith TM (2015) Chloroplast signaling within, between and beyond cells. Front Plant Sci 6:781

Börner T, Aleynikova AY, Zubo YO, Kusnetsov VV (2015) Chloroplast RNA polymerases: role in chloroplast biogenesis. Biochim Biophys Acta Bioenerg 1847(9):761–769

Buchanan-Wollaston V (1997) The molecular biology of leaf senescence. J Exp Bot 48(2):181–199

Cackett L, Luginbuehl LH, Schreier TB, Lopez-Juez E, Hibberd JM (2022) Chloroplast development in green plant tissues: the interplay between light, hormone, and transcriptional regulation. New Phytol 233(5):2000–2016

Caplan JL, Mamillapalli P, Burch-Smith TM, Czymmek K, Dinesh-Kumar SP (2008) Chloroplastic protein NRIP1 mediates innate immune receptor recognition of a viral effector. Cell 132:449–462

Caputo C, Rutitzky M, Ballare CL (2006) Solar ultraviolet-B radiation alters the attractiveness of Arabidopsis plants to diamondback moths (*Plutella xylostella* L.): Impacts on oviposition and involvement of the jasmonic acid pathway. Oecologia 149(1):81–90

Castandet B, Hotto AM, Strickler SR, Stern DB (2016) ChloroSeq, an optimized chloroplast RNA-Seq bioinformatic pipeline, reveals remodeling of the organellar transcriptome under heat stress. G3 6(9):2817–2827

Cazzonelli CI, Hou X, Alagoz Y, Rivers J, Dhami N, Lee J et al (2020) A cis-carotene derived apocarotenoid regulates etioplast and chloroplast development. elife 9:e45310

Chan KX, Phua SY, Crisp P, McQuinn R, Pogson BJ (2016) Learning the languages of the chloroplast: retrograde signaling and beyond. Annu Rev Plant Biol 67(1):25–53

Chaves MM, Flexas J, Pinheiro C (2009) Photosynthesis under drought and salt stress: regulation mechanisms from whole plant to cell. Ann Bot 103(4):551–560

Chi W, Feng P, Ma J, Zhang L (2015) Metabolites and chloroplast retrograde signaling. Curr Opin Plant Biol 25:32–38

Cui X, Wang Y, Wu J, Han X, Gu X, Lu T, Zhang Z (2019) The RNA editing factor DUA 1 is crucial to chloroplast development at low temperature in rice. New Phytol 221(2):834–849

Dalal VK, Tripathy BC (2012) Modulation of chlorophyll biosynthesis by water stress in rice seedlings during chloroplast biogenesis. Plant Cell Environ 35(9):1685–1703

Dobrogojski J, Adamiec M, Luciński R (2020) The chloroplast genome: a review. Acta Physiol Plant 42(6):98

Escobar-Bravo R, Chen G, Grosser K, Van Dam NM, Leiss KA, Klinkhamer PGL (2019) Ultraviolet radiation enhances salicylic acid-mediated defense signaling and resistance to *Pseudomonas syringae* DC3000 in a jasmonic acid-deficient tomato mutant. Plant Signal Behav 14(4):e1581560

Estavillo GM, Crisp PA, Pornsiriwong W, Wirtz M, Collinge D, Carrie C et al (2011) Evidence for a SAL1-PAP chloroplast retrograde pathway that functions in drought and high light signaling in Arabidopsis. Plant Cell 23(11):3992–4012

Farooq MA, Niazi AK, Akhtar J, Farooq M, Souri Z, Karimi N, Rengel Z (2019) Acquiring control: The evolution of ROS-Induced oxidative stress and redox signaling pathways in plant stress responses. Plant Physiol Biochem 141:353–369

Feng J, Fan P, Jiang P, Lv S, Chen X, Li Y (2014) Chloroplast-targeted Hsp90 plays essential roles in plastid development and embryogenesis in Arabidopsis possibly linking with VIPP1. Physiol Plant 150(2):292–307

Feng ZT, Deng YQ, Zhang SC, Liang X, Yuan F, Hao JL et al (2015) K+ accumulation in the cytoplasm and nucleus of the salt gland cells of Limonium bicolor accompanies increased rates of salt secretion under NaCl treatment using NanoSIMS. Plant Sci 238:286–296

Fernandez JC, Burch-Smith TM (2019) Chloroplasts as mediators of plant biotic interactions over short and long distances. Curr Opin Plant Biol 50:148–155

Flowers TJ, Colmer TD (2015) Plant salt tolerance: adaptations in halophytes. Ann Bot 115(3):327–331

Gan P, Liu F, Li R, Wang S, Luo J (2019) Chloroplasts—beyond energy capture and carbon fixation: tuning of photosynthesis in response to chilling stress. Int J Mol Sci 20(20):5046

Gao C, Xu H, Huang J, Sun B, Zhang F, Savage Z et al (2020) Pathogen manipulation of chloroplast function triggers a light-dependent immune recognition. Proc Natl Acad Sci 117(17):9613–9620

Gray JC, Sullivan JA, Wang JH, Jerome CA, MacLean D (2003) Coordination of plastid and nuclear gene expression. Philos Trans R Soc Lond B Biol Sci 358(1429):135–145
Guddimalli R, Somanaboina AK, Palle SR, Edupuganti S, Kummari D, Palakolanu SR et al (2021) Overexpression of RNA-binding bacterial chaperones in rice leads to stay-green phenotype, improved yield and tolerance to salt and drought stresses. Physiol Plant 173(4):1351–1368
Gull A, Lone AA, Wani NUI (2019) Biotic and abiotic stresses in plants. In: Abiotic and biotic stress in plants, vol 7, pp 1–9
Han Y, Yin S, Huang L, Wu X, Zeng J, Liu X et al (2018) A sodium transporter HvHKT1; 1 confers salt tolerance in barley via regulating tissue and cell ion homeostasis. Plant Cell Physiol 59(10):1976–1989
Hanke GT, Holtgrefe S, Koenig N, Strodtkoetter I, Voss I, Scheibe R (2009) Use of transgenic plants to uncover strategies for maintenance of redox homeostasis during photosynthesis. Adv Bot Res 52:207–251
Hao X, Zhong Y, Ntzmann HW, Fu X, Yan T, Shen Q et al (2019) Light-induced artemisinin biosynthesis is regulated by the bZIP transcription factor AaHY5 in Artemisia annua. Plant Cell Physiol 60(8):1747–1760
Härtel H, Dörmann P, Benning C (2000) DGD1-independent biosynthesis of extraplastidic galactolipids after phosphate deprivation in Arabidopsis. Proc Natl Acad Sci 97(19):10649–10654
He Y, Yu C, Zhou L, Chen Y, Liu A, Jin J, Jiang D (2014) Rubisco decrease is involved in chloroplast protrusion and Rubisco-containing body formation in soybean (Glycine max.) under salt stress. Plant Physiol Biochem 74:118–124
Heil M (2009) Damaged-self recognition in plant herbivore defence. Trends Plant Sci 14(7):356–363
Hernández JA, Rubio M, Olmos E, Ros-Barceló A, Martínez-Gómez P (2004) Oxidative stress induced by long-term plum pox virus infection in peach (Prunus persica). Physiol Plant 122(4):486–495
Inoue K (2015) Emerging knowledge of the organelle outer membranes–research snapshots and an updated list of the chloroplast outer envelope proteins. Front Plant Sci 6:278
Ishiga Y, Ishiga T, Ikeda Y, Matsuura T, Mysore KS (2016) NADPH-dependent thioredoxin reductase C plays a role in nonhost disease resistance against Pseudomonas syringae pathogens by regulating chloroplast-generated reactive oxygen species. PeerJ 4:e1938
Jelenska J, Yao N, Vinatzer BA, Wright CM, Brodsky JL, Greenberg JT (2007) A J domain virulence effector of *Pseudomonas syringae* remodels host chloroplasts and suppresses defenses. Curr Biol 17(6):499–508
Jiang Q, Mei J, Gong XD, Xu JL, Zhang JH, Teng S et al (2014) Importance of the rice TCD9 encoding α subunit of chaperonin protein 60 (Cpn60α) for the chloroplast development during the early leaf stage. Plant Sci 215:172–179
Jin S, Zhang X, Daniell H (2012) Pinellia ternata agglutinin expression in chloroplasts confers broad spectrum resistance against aphid, whitefly, lepidopteran insects, bacterial and viral pathogens. Plant Biotechnol J 10:313–327
Jing Y, Lin R (2020) Transcriptional regulatory network of the light signaling pathways. New Phytol 227(3):683–697
Juneau P, Lawrence JE, Suttle CA, Harrison PJ (2003) Effects of viral infection on photosynthetic processes in the bloom-forming alga Heterosigma akashiwo. Aquat Microb Ecol 31(1):9–17
Kabi T, Nanda HP, Haldar AM (1979) Chlorophyll degradation and starch metabolism in bean (*Dolichos lablab* L.) leaves infected with yellow bean mosaic virus (YBMV)
Kachroo P, Burch-Smith TM, Grant M (2021) An emerging role for chloroplasts in disease and defense. Annu Rev Phytopathol 59(1):423–445
Kar RK, Nanda HP, Kabi T (1985) Activity of catalase and peroxidase during natural and virus induced papaya leaf ageing. Indian J Plant Physiol (India) 28(2):124
Kendrick RE, Kronenberg GH (eds) (2012) Photomorphogenesis in plants. Springer Science & Business Media
Knaff DB (2000) Oxidation-reduction properties of thioredoxins and thioredoxin-regulated enzymes. Physiol Plant 110(3):309–313

Kovacs L, Damkjær J, Kereïche S, Ilioaia C, Ruban AV, Boekema EJ et al (2006) Lack of the light-harvesting complex CP24 affects the structure and function of the grana membranes of higher plant chloroplasts. Plant Cell 18(11):3106–3120
Koyro HW (2006) Effect of salinity on growth, photosynthesis, water relations and solute composition of the potential cash crop halophyte *Plantago coronopus* (L.). Environ Exp Bot 56(2):136–146
Kumar RR, Goswami S, Singh K, Dubey K, Singh S, Sharma R et al (2016) Identification of putative RuBisCo Activase (TaRca1)—the catalytic chaperone regulating carbon assimilatory pathway in wheat (*Triticum aestivum*) under the heat stress. Front Plant Sci 7:986
Kusumi K, Iba K (2014) Establishment of the chloroplast genetic system in rice during early leaf development and at low temperatures. Front Plant Sci 5:386
Kwon KC, Verma D, Jin S, Singh ND, Daniell H (2013) Release of proteins from intact chloroplasts induced by reactive oxygen species during biotic and abiotic stress. PLoS One 8(6):e67106
Lawlor DW, Tezara W (2009) Causes of decreased photosynthetic rate and metabolic capacity in water-deficient leaf cells: a critical evaluation of mechanisms and integration of processes. Ann Bot 103(4):561–579
Lee SB, Li B, Jin S, Daniell H (2011) Expression and characterization of antimicrobial peptides Retrocyclin-101 and Protegrin-1 in chloroplasts to control viral and bacterial infections. Plant Biotechnol J 9:100–115
Lehto K, Tikkanen M, Hiriart JB, Paakkarinen V, Aro EM (2003) Depletion of the photosystem II core complex in mature tobacco leaves infected by the flavum strain of tobacco mosaic virus. Mol Plant-Microbe Interact 16(12):1135–1144
León P, Gregorio J, Cordoba E (2013) ABI4 and its role in chloroplast retrograde communication. Front Plant Sci 3:304
Li J, Li G, Wang H, Deng XW (2011) Phytochrome signaling mechanisms. The Arabidopsis Book/ American Society of Plant Biologists, 9
Li P, Liu H, Yang H, Pu X, Li C, Huo H, Liu L (2020) Translocation of drought-responsive proteins from the chloroplasts. Cells 9(1):259
Li JY, Yang C, Tian YY, Liu JX (2022) Regulation of chloroplast development and function at adverse temperatures in plants. Plant Cell Physiol 63(5):580–591
Liebers M, Cozzi C, Uecker F, Chambon L, Blanvillain R, Pfannschmidt T (2022) Biogenic signals from plastids and their role in chloroplast development. J Exp Bot 73(21):7105–7125
Lim GH, Singhal R, Kachroo A, Kachroo P (2017) Fatty acid–and lipid-mediated signaling in plant defense. Annu Rev Phytopathol 55(1):505–536
Ling Q, Jarvis P (2015a) Functions of plastid protein import and the ubiquitin–proteasome system in plastid development. Biochim Biophys Acta Bioenerg 1847(9):939–948
Ling Q, Jarvis P (2015b) Regulation of chloroplast protein import by the ubiquitin E3 ligase SP1 is important for stress tolerance in plants. Curr Biol 25(19):2527–2534
Ling Q, Jarvis P (2016) Plant signaling: ubiquitin pulls the trigger on chloroplast degradation. Curr Biol 26(1):R38–R40
Liu X, Lan J, Huang Y, Cao P, Zhou C, Ren Y, Wan J (2018a) WSL5, a pentatricopeptide repeat protein, is essential for chloroplast biogenesis in rice under cold stress. J Exp Bot 69(16):3949–3961
Liu X, Zhou Y, Xiao J, Bao F (2018b) Effects of chilling on the structure, function and development of chloroplasts. Front Plant Sci 9:1715
Lu Y, Yao J (2018) Chloroplasts at the crossroad of photosynthesis, pathogen infection and plant defense. Int J Mol Sci 19(12):3900
Lv Y, Shao G, Qiu J, Jiao G, Sheng Z, Xie L, Hu P (2017) White Leaf and Panicle 2, encoding a PEP-associated protein, is required for chloroplast biogenesis under heat stress in rice. J Exp Bot 68(18):5147–5160
Ma X, Wang G, Zhao W, Yang M, Ma N, Kong F, Meng Q (2017) SlCOR413IM1: a novel cold-regulation gene from tomato, enhances drought stress tolerance in tobacco. J Plant Physiol 216:88–99

Mäntylä E, Kleier S, Lindstedt C, Kipper S, Hilker M (2018) Insectivorous birds are attracted by plant traits induced by insect egg deposition. J Chem Ecol 44:1127–1138

Martins NJF (2009) The role of ATP synthase in photosynthetic flux control (Doctoral dissertation)

Massacci A, Nabiev SM, Pietrosanti L, Nematov SK, Chernikova TN, Thor K, Leipner J (2008) Response of the photosynthetic apparatus of cotton (*Gossypium hirsutum*) to the onset of drought stress under field conditions studied by gas-exchange analysis and chlorophyll fluorescence imaging. Plant Physiol Biochem 46(2):189–195

Medina-Puche L, Tan H, Dogra V, Wu M, Rosas-Diaz T, Wang L et al (2020) A defense pathway linking plasma membrane and chloroplasts and co-opted by pathogens. Cell 182(5):1109–1124

Mohanty S, Tripathy BC (2011) Early and late plastid development in response to chill stress and heat stress in wheat seedlings. Protoplasma 248:725–736

Nakai M (2015) The TIC complex uncovered: the alternative view on the molecular mechanism of protein translocation across the inner envelope membrane of chloroplasts. Biochim Biophys Acta Bioenerg 1847(9):957–967

Nanda RM, Biswal B (2008) Biotic stress induced demolition of thylakoid structure and loss in photoelectron transport of chloroplasts in papaya leaves. Plant Physiol Biochem 46(4):461–468. https://doi.org/10.1016/j.plaphy.2007.12.00

Nawaz G, Lee K, Park SJ, Kim YO, Kang H (2018) A chloroplast-targeted cabbage DEAD-box RNA helicase BrRH22 confers abiotic stress tolerance to transgenic Arabidopsis plants by affecting translation of chloroplast transcripts. Plant Physiol Biochem 127:336–342

Nott A, Jung HS, Koussevitzky S, Chory J (2006) Plastid-to-nucleus retrograde signaling. Annu Rev Plant Biol 57(1):739–759

Padmanabhan MS, Dinesh-Kumar SP (2010) All hands on deck—the role of chloroplasts, endoplasmic reticulum, and the nucleus in driving plant innate immunity. Mol Plant-Microbe Interact 23(11):1368–1380

Palva TK, Holmström KO, Heino P, Palva ET (1993) Induction of plant defense response by exoenzymes of *Erwinia carotovora* ssp. *carotovora*. Mol Plant-Microbe Interact 6:190–196

Parida AK, Das AB (2005) Salt tolerance and salinity effects on plants: a review. Ecotoxicol Environ Saf 60(3):324–349

Park JH, Tran LH, Jung S (2017) Perturbations in the photosynthetic pigment status result in photooxidation-induced crosstalk between carotenoid and porphyrin biosynthetic pathways. Front Plant Sci 8:1992

Phipps J, Buis R (1973) Some oxidizing enzymic activities of the tobacco leaf: a physiological study. III. Modifications linked to virus infection. Physiologie végétale

Pogson BJ, Ganguly D, Albrecht-Borth V (2015) Insights into chloroplast biogenesis and development. Biochim Biophys Acta Bioenerg 1847(9):1017–1024

Pottosin I, Shabala S (2016) Transport across chloroplast membranes: optimizing photosynthesis for adverse environmental conditions. Mol Plant 9(3):356–370

Pulido P, Llamas E, Rodriguez-Concepcion M (2017) Both Hsp70 chaperone and Clp protease plastidial systems are required for protection against oxidative stress. Plant Signal Behav 12(3):e1290039

Qi J, Zhang M, Lu C, Hettenhausen C, Tan Q, Cao G et al (2018) Ultraviolet-B enhances the resistance of multiple plant species to lepidopteran insect herbivory through the jasmonic acid pathway. Sci Rep 8(1):277

Qian L, He S, Liu X, Huang Z, Chen F, Gui F (2018) Effect of elevated CO2 on the interaction between invasive thrips, Frankliniella occidentalis, and its host kidney bean, Phaseolus vulgaris. Pest Manag Sci 74(12):2773–2782

Rodríguez-Herva JJ, González-Melendi P, Cuartas-Lanza R, Antúnez-Lamas M, Río-Alvarez I, Li Z et al (2012) A bacterial cysteine protease effector protein interferes with photosynthesis to suppress plant innate immune responses. Cell Microbiol 14(5):669–681

Roeber VM, Bajaj I, Rohde M, Schmülling T, Cortleven A (2021) Light acts as a stressor and influences abiotic and biotic stress responses in plants. Plant Cell Environ 44(3):645–664

Rosas-Diaz T, Zhang D, Fan P, Wang L, Ding X, Jiang Y et al (2018) A virus-targeted plant receptor-like kinase promotes cell-to-cell spread of RNAi. Proc Natl Acad Sci 115(6):1388–1393

Rowland E, Kim J, Bhuiyan NH, van Wijk KJ (2015) The Arabidopsis chloroplast stromal N-terminome: complexities of amino-terminal protein maturation and stability. Plant Physiol 169(3):1881–1896

Rütgers M, Muranaka LS, Mühlhaus T, Sommer F, Thoms S, Schurig J et al (2017) Substrates of the chloroplast small heat shock proteins 22E/F point to thermolability as a regulative switch for heat acclimation in Chlamydomonas reinhardtii. Plant Mol Biol 95:579–591

Sachdev S, Ansari SA, Ansari MI, Fujita M, Hasanuzzaman M (2021) Abiotic stress and reactive oxygen species: generation, signaling, and defense mechanisms. Antioxidants 10(2):277

Sajnani C, Zurita JL, Roncel M, Ortega JM, Barón M, Ducruet JM (2007) Changes in photosynthetic metabolism induced by tobamovirus infection in Nicotiana benthamiana studied in vivo by thermoluminescence. New Phytol 175(1):120–130

Sampol B, Bota J, Riera D, Medrano H, Flexas J (2003) Analysis of the virus-induced inhibition of photosynthesis in malmsey grapevines. New Phytol 160(2):403–412

Schmelz EA, LeClere S, Carroll MJ, Alborn HT, Teal PE (2007) Cowpea chloroplastic ATP synthase is the source of multiple plant defense elicitors during insect herbivory. Plant Physiol 144(2):793–805

Shanmugabalaji V, Chahtane H, Accossato S, Rahire M, Gouzerh G, Lopez-Molina L, Kessler F (2018) Chloroplast biogenesis controlled by DELLA-TOC159 interaction in early plant development. Curr Biol 28(16):2616–2623

Shen J (2013) Characterization of drought stress regulator CBF/DREB genes in Hordeum vulgare: Expression analysis in ten different barley cultivars (Doctoral dissertation, Bonn, Rheinische Friedrich-Wilhelms-Universität Bonn, Diss., 2013)

Singh A, Kumar A, Yadav S, Singh IK (2019) Reactive oxygen species-mediated signaling during abiotic stress. Plant Gene18:100–173

Stael S, Bayer RG, Mehlmer N, Teige M (2011) Protein N-acylation overrides differing targeting signals. FEBS Lett 585(3):517–522

Stenbaek A, Hansson A, Wulff RP, Hansson M, Dietz KJ, Jensen PE (2008) NADPH-dependent thioredoxin reductase and 2-Cys peroxiredoxins are needed for the protection of Mg–protoporphyrin monomethyl ester cyclase. FEBS Lett 582(18):2773–2778

Su J, Yang L, Zhu Q, Wu H, He Y, Liu Y et al (2018) Active photosynthetic inhibition mediated by MPK3/MPK6 is critical to effector-triggered immunity. PLoS Biol 16(5):e2004122

Sui N, Li M, Li K, Song J, Wang BS (2010) Increase in unsaturated fatty acids in membrane lipids of Suaeda salsa L. enhances protection of photosystem II under high salinity. Photosynthetica 48:623–629

Suman S, Bagal D, Jain D, Singh R, Singh IK, Singh A (2021) Biotic stresses on plants: reactive oxygen species generation and antioxidant mechanism. In: Frontiers in plant-soil interaction. Academic Press, pp 381–411

Sun X, Feng P, Xu X, Guo H, Ma J et al (2011) A chloroplast envelope-bound PHD transcription factor mediates chloroplast signals to the nucleus. Nat Commun 2:477

Sun J, Zheng T, Yu J, Wu T, Wang X, Chen G et al (2017) TSV, a putative plastidic oxidoreductase, protects rice chloroplasts from cold stress during development by interacting with plastidic thioredoxin Z. New Phytol 215(1):240–255

Tachibana R, Abe S, Marugami M, Yamagami A, Akema R, Ohashi T et al (2024) BPG4 regulates chloroplast development and homeostasis by suppressing GLK transcription factors and involving light and brassinosteroid signaling. Nat Commun 15(1):370

Tanaka R, Tanaka A (2011) Chlorophyll cycle regulates the construction and destruction of the light-harvesting complexes. Biochim Biophys Acta Bioenerg 1807(8):968–976

Tavormina P, De Coninck B, Nikonorova N, De Smet I, Cammue BP (2015) The plant peptidome: an expanding repertoire of structural features and biological functions. Plant Cell 27(8):2095–2118

Tian F, Gong J, Zhang J, Feng Y, Wang G, Guo Q, Wang W (2014) Overexpression of monoubiquitin improves photosynthesis in transgenic tobacco plants following high temperature stress. Plant Sci 226:92–100

Torres MA, Jones JD, Dangl JL (2006) Reactive oxygen species signaling in response to pathogens. Plant Physiol 141:373–378
Tripathi S, Hoang QT, Han YJ, Kim JI (2019) Regulation of photomorphogenic development by plant phytochromes. Int J Mol Sci 20(24):6165
Unal D, García-Caparrós P, Kumar V, Dietz KJ (2020) Chloroplast-associated molecular patterns as concept for fine-tuned operational retrograde signalling. Philos Trans R Soc B 375(1801):20190443
Verma D, Kanagaraj A, Jin S, Singh ND, Kolattukudy PE et al (2010) Chloroplast-derived enzyme cocktails hydrolyse lignocellulosic biomass and release fermentable sugars. Plant Biotechnol J 8:332–350
Visioni A, Al-Abdallat A, Elenien JA, Verma RPS, Gyawali S, Baum M (2019) Genomics and molecular breeding for improving tolerance to abiotic stress in barley (Hordeum vulgare L.). In: Genomics assisted breeding of crops for abiotic stress tolerance, vol II, pp 49–68
Vitha S, McAndrew RS, Osteryoung KW (2001) FtsZ ring formation at the chloroplast division site in plants. J Cell Biol 153(1):111–120
Wang G, Kong F, Zhang S, Meng X, Wang Y, Meng Q (2015) A tomato chloroplast-targeted DnaJ protein protects Rubisco activity under heat stress. J Exp Bot 66(11):3027–3040
Wang Z, Shen Y, Yang X, Pan Q, Ma G, Bao M, Ning G (2019) Overexpression of particular MADS-box transcription factors in heat-stressed plants induces chloroplast biogenesis in petals. Plant Cell Environ 42(5):1545–1560
Wang P, Abid MA, Qanmber G, Askari M, Zhou L, Song Y et al (2022) Photomorphogenesis in plants: The central role of phytochrome interacting factors (PIFs). Environ Exp Bot 194:104704
Waters MT, Langdale JA (2009) The making of a chloroplast. EMBO J 28(19):2861–2873
Wilhelmová N, Procházková D, Sindelarova M, Sindelar L (2005) Photosynthesis in leaves of Nicotiana tabacum L. infected with tobacco mosaic virus. Photosynthetica 43:597–602
Xue X, Geng T, Liu H, Yang W, Zhong W, Zhang Z et al (2021) Foliar application of silicon enhances resistance against Phytophthora infestans through the ET/JA-and NPR1-dependent signaling pathways in potato. Front Plant Sci 12:609870
Yang X, Pang X, Xu L, Fang R, Huang X, Guan P et al (2009) Accumulation of soluble sugars in peel at high temperature leads to stay-green ripe banana fruit. J Exp Bot 60(14):4051–4062
Yoo SC, Cho SH, Sugimoto H, Li J, Kusumi K, Koh HJ et al (2009) Rice virescent3 and stripe1 encoding the large and small subunits of ribonucleotide reductase are required for chloroplast biogenesis during early leaf development. Plant Physiol 150(1):388–401
Yoo YH, Nalini Chandran AK, Park JC, Gho YS, Lee SW, An G, Jung KH (2017) OsPhyB-mediating novel regulatory pathway for drought tolerance in rice root identified by a global RNA-Seq transcriptome analysis of rice genes in response to water deficiencies. Front Plant Sci 8:580
Yoo CY, Pasoreck EK, Wang H, Cao J, Blaha GM, Weigel D, Chen M (2019) Phytochrome activates the plastid-encoded RNA polymerase for chloroplast biogenesis via nucleus-to-plastid signaling. Nat Commun 10(1):2629
Yoshinaga N, Alborn HT, Nakanishi T, Suckling DM, Nishida R, Tumlinson JH, Mori N (2010) Fatty acid-amino acid conjugates diversification in lepidopteran caterpillars. J Chem Ecol 36(3):319–325
Zaitlin M, Jagendorf AT (1960) Photosynthetic phosphorylation and Hill reaction activities of chloroplasts isolated from plants infected with tobacco mosaic virus. Virology 12(3):477–486
Zechmann B, Müller M, Zellnig G (2003) Cytological modifications in zucchini yellow mosaic virus (ZYMV)-infected Styrian pumpkin plants. Arch Virol 148:1119–1133
Zhang R, Wise RR, Struck KR, Sharkey TD (2010) Moderate heat stress of Arabidopsis thaliana leaves causes chloroplast swelling and plastoglobule formation. Photosynth Res 105:123–134
Zhang Z, Zhang Q, Wu J, Zheng X, Zheng S, Sun X, Lu T (2013) Gene knockout study reveals that cytosolic ascorbate peroxidase 2 (OsAPX2) plays a critical role in growth and reproduction in rice under drought, salt and cold stresses. PLoS One 8(2):e57472

Zhang D, Tan W, Yang F, Han Q, Deng X, Guo H et al (2021) A BIN2-GLK1 signaling module integrates brassinosteroid and light signaling to repress chloroplast development in the dark. Dev Cell 56(3):310–324

Zhang Q, Mao Y, Zhao Z, Hu X, Hu R, Yin N et al (2024) A Golden2-like transcription factor, BnGLK1a, improves chloroplast development, photosynthesis, and seed weight in rapeseed. J Integr Agric 23(5):1481–1493

Zhao Y, Han Q, Ding C, Huang Y, Liao J, Chen T, Yuan M (2020) Effect of low temperature on chlorophyll biosynthesis and chloroplast biogenesis of rice seedlings during greening. Int J Mol Sci 21(4):1390

Zhu X, Li W, Zhang N, Jin H, Duan H, Chen Z et al (2024) StMAPKK5 responds to heat stress by regulating potato growth, photosynthesis, and antioxidant defenses. Front Plant Sci 15:1392425

Zhuang K, Gao Y, Liu Z, Diao P, Sui N, Meng Q, Kong F (2020a) WHIRLY1 regulates HSP21. 5A expression to promote thermotolerance in tomato. Plant Cell Physiol 61(1):169–177

Zhuang K, Wang J, Jiao B, Chen C, Zhang J, Ma N, Meng Q (2020b) WHIRLY1 maintains leaf photosynthetic capacity in tomato by regulating the expression of RbcS1 under chilling stress. J Exp Bot 71(12):3653–3663

Zoschke R, Qu Y, Zubo YO, Börner T, Schmitz-Linneweber C (2013) Mutation of the pentatricopeptide repeat-SMR protein SVR7 impairs accumulation and translation of chloroplast ATP synthase subunits in Arabidopsis thaliana. J Plant Res 126:403–414

Etioplast to Chloroplast Transition in Higher Plants

8

Akhilesh K. Singh, Santendra K. Soni, and Vijay Kumar Dalal

Abstract

Etioplasts are plastids devoid of chlorophyll. They can be easily seen in the tissue showing scotomorphogenesis, i.e., plant development in the dark. When seedlings grow inside the soil in the dark, scotomorphogenesis takes place and photomorphogenesis is initiated as soon as they come out of the ground in light. Besides this, etioplasts are also found in light-grown-shaded tissues, and also form in dark periods of diurnal circadian rhythms. In monocot leaves, etioplasts can be seen at the leaf base, which is covered with the sheath. A developmental gradient of de-etiolation is created along the length of leaf with the oldest mature chloroplasts at the leaf tip and the younger plastid at the leaf base with the process of de-etiolation. Etioplasts contain lattice-like structures called prolamellar bodies, containing NADPH-POR-Pchlide complexes. These lattices are disintegrated in consonance with the conversion of Pchlide into Chlide upon receiving light and form thylakoids. There are numerous processes that influence this transition. Several biophysical, OMICs, physiological, molecular biology and genetics methods have been used to understand and characterize the process of de-etiolation. This chapter details these processes and captures new developments in the area of plastid de-etiolation. Also, how light affects miRNA expression, synthesis, processing, and function, as well as how it is involved in photomorphogenesis and de-etiolation/greening, will be discussed.

A. K. Singh
Department of Biotechnology, School of Life Sciences, Mahatma Gandhi Central University, Motihari, Bihar, India

S. K. Soni · V. K. Dalal (✉)
Department of Botany, Dayalbagh Educational Institute, Agra, Uttar Pradesh, India
e-mail: dalalvijay@dei.ac.in

V. K. Dalal, A. N. Misra (eds.), *Chloroplast Biogenesis and Plastid Interconversions*,
https://doi.org/10.1007/978-981-95-0772-6_8

Keywords

Chloroplasts · Etioplasts · Prolamellar body · Phototransformation · POR · Pchlide

Abbreviations

AGO1	Argonaute 1
ATP	Adenosine triphosphate
ARR	Arabidopsis Response Regulator
BES1	BRI1-EMS-Suppressor 1
BFA1/3	Biogenesis factors required for ATP synthase 1/3
BRs	Brassinosteroids
BSMV	Barley stripe mosaic virus
BZR1/2	Brassinazole Resistant1/2
CAO	Chlorophyllide a oxygenase
CCA1	Circadian Clock Associated 1
ccr2	Carotenoid chloroplast regulation 2
CET	Cyclic electron transport
CGL71	Conserved in Green Lineage 71
CGL160	Conserved Only in the Green Lineage160
Chl27	Subunit of enzyme Mg-protoporphyrin IX monomethylester cyclase
ChlI	Mg-chelatase subunit I
ChlD	Mg-chelatase subunit D
COP1 TF	Constitutively Photomorphogenic 1 transcription factor
Coprox	Coproporphyrinogen III oxidase
cpd	Constitutive Photomorphogenic Dwarf
cpSRP54	Signal Recognition Particle 54 kDa chloroplastic
CRY	Cryptochrome
CURT1	Curvature thylakoid 1
Cyt b6f	Cytochrome b6f
Cyt b559	Cytochrome b6f subunit 559
D1	PSII reaction center protein D1
DAC	Defective accumulation of cytochrome b6/f complex
DEIP1	DE-Etiolation-Induced Protein 1
DGDG	Digalactosyldiacylglycerol
DXS1	1-deoxy D-xylulose 5-phosphate synthase 1
ELIPs	Early Light-Inducible Proteins
FBA	Fructose 1,6-bisphosphate aldolase
FD	Ferredoxin
FNR	Fd NADP Reductase
FPB1	Facilitator of PsbB biogenesis 1
FQR	Ferredoxin-plastoquinone reductase
GNL	GNC-LIKE
GLK 1/2	GOLDEN2-LIKE 1 and 2
GluTR	Glutamyl tRNA reductase

GNC	Gata-Nitrate-Inducible Carbon Metabolism-Involved transcription factor
GSAT	Glutamate-1 Semialdehyde (GSA) Transferase
HAM	Hairy Meristem
HCF	High chlorophyll fluorescence
HID1	Hidden Treasure 1
HTL	Hypersensitive to light
HY5	Elongated Hypocotyl 5
im	Immutans/IM
JMJ17	Jumonji Domain-Containing Protein17
LDC	Light Dark Cycles
LHCL	Light-harvesting complex
LHY1	Late Elongated Hypocotyl 1
LOM	Lost Meristems
LPOR	Light-dependent NADPH: protochlorophyllide oxidoreductase
LSU	Large subunit of rubisco
LZF1	**L**ight regulated **Z**n-**F**inger 1
MEP	2-C-methyl-D-erythritol 4-phosphate
MGDG	Monogalactosyldiacylglycerol
NDH	NAD(P)H dehydrogenase
NADPH	Nicotinamide adenine dinucleotide phosphate
NTA1	New tiny albino 1
NCP	Nuclear Control of Pep Activity
OEE	Oxygen evolving enhancer
OEC	Oxygen evolving complex
OR	Orange
PAM68	Photosynthesis affected mutant 68
PAP1	Photosynthesis-Associated Protein 1
PBGD	Porphobilinogen dehydratase
PDS	Phytoene desaturase
PetM	Photosynthesis Electron Transport M, Cytochrome b6f complex subunit VII
PG	Phosphatidyl glycerol
PGR5	Proton gradient 5
PGRL	Proton gradient regulation
Phot	Phototropin
Phy	Phytochrome
PIF5.1 TF	Phytochrome Interacting Factor 5.1 transcription factor
PIL15	PIF like 15
PLBs	Prolamellar bodies
ppi2	Plastid protein import2
PPIX	Proto porphyrin IX
PPX-I	Protoporphyrin IX oxidase subunit I
PQ	Plastoquinone
PQH_2	Plastoquinol

PTOX	Plastid terminal oxidase
PSI/II	Photosystem I/II
PsaE	Photosystem I subunit E
PsaH	Photosystem I subunit VI
PsaK	Photosystem I subunit K
Psb16	Photosystem II subunit 16
Psb23	Photosystem II subunit 23
Psb33	Photosystem II subunit 33
PTs	Prothylakoids
PYG7	Ycf37/pale yellow green 7
RCB	Regulator of Chloroplast Biogenesis
RISC	RNA-induced silencing complex
RPOTp	Plastid RNA polymerase
RPS6	Ribosomal protein S6
RuBP	Ribulose bisphosphate
SAM	Shoot Apical Meristem
SCA3	SCABRA3: RPoTp, Rounded and protruded lamina Arabidopsis mutant
SCO2	Snowy Coltyledon 2
SIG	Sigma
SQDG	Sulfoquinovosyldiacylglycerol
SPA2	Suppressor of PHYA-2
SSU	Small subunit of rubisco
TAG	Triglyceride
TCP14	Class I **T**eosinte Branched 1, **C**ycloidea, **P**CF 14 TF
TOR	Target of Rapamycin
TIC/TOC	Translocon at the inner/outer envelope membrane of chloroplasts
UroD	Uroporphyrinogen dehydrogenase
UVR8	UVB-Resistance 8
Ycf3,4,37	Hypothetical chloroplast open reading frames 3, 4 and 37
Y3IP1	Ycf3 interacting protein 1
ZDS	z-carotene desaturase
δ-ALA	Delta-Amino Levulinic Acid

8.1 Introduction

Mostly the aerial vegetative parts of the plants are green due to the chlorophyll molecules present in the chloroplasts. These chloroplasts arise from the proplastids present in the meristematic cells in the shoot apical meristems (Charuvi et al. 2012). Although they are not green, these plastids express several genes of chloroplasts (Dalal et al. 2018). The shoot apical meristem (SAM) in dicots, e.g., tomato and Arabidopsis, is divided into three layers: L1, L2 and L3. L1 consists of the epidermal cell layer that contains fairly developed thylakoids in cells above the SAM. These cells further develop thylakoids in leaf primordia. As the cells in L1 are pushed toward the periphery of the mature leaf as a result of anticlinal division, these thylakoids disintegrate and chloroplasts are converted into leucoplasts except in the stomata. The L2 layer includes a cell layer below the epidermis. L3 is the cell mass lying below L2, in the middle of the SAM. The L2 layer in the central zone contains meristematic cells and proplastids. When these cells divide, they are pushed towards the periphery and later develop into the sub-epidermal mesophyll cells of leaves, and develop chloroplasts during this transition. Similarly, the cells of the L3 layer in the central zone are devoid of thylakoids and develop them when pushed towards the base and periphery (Charuvi et al. 2012; Dalal et al. 2018), giving rise to ground tissue of stem, vasculature and leaf mesophyll cells. The other most significant way to develop chloroplasts is by using etioplasts. Etioplasts develop in the dark or during the dark periods of light-exposed plants. Upon exposure to light, the etioplasts are converted to chloroplasts (Sundqvist and Dahlin 1997). These are green plastids that trap the light energy to perform photosynthesis. Chloroplasts contain the electron transport chain components in thylakoids and the Calvin-Benson cycle inside the stroma to fix the CO_2 in the form of glucose and carbohydrates. Chloroplasts also synthesize fatty acids, amino acids, signal molecules, chemicals and secondary metabolites. Chloroplasts are sometimes, also developed from leucoplasts, chromoplasts, gerontoplasts, etc. (Altamura et al. 2024). They can also convert into other plastid forms, e.g., leucoplasts (elaioplasts, amyloplasts, proteinoplasts), chromoplasts and gerontoplasts.

The formation and functions of chloroplasts have been studied thoroughly and in recent decades several assembly factors that help in chloroplast biogenesis have been discovered, in addition to those discovered earlier. Thylakoid membranes inside chloroplasts contain PSII (approx. 25–20 proteins), PSI (15 proteins), ATP synthase (20 proteins) and NDH complexes (35 proteins) (Laughlin et al. 2020). Generation of these complexes requires assembly factors. Beside the well-known assembly factors that have been nicely reviewed frequent times (Rochaix 2011; Rast et al. 2015), several novel assembly factors have been identified for photosystem II (PSII; e.g. FPB1 (Zhang et al. 2024), HCF243 (Zhang. et al. 2011), SCO2 (Zagari et al. 2017), HCF136 (Plücken et al. 2002), and PAM68 (Armbruster et al. 2010)), photosystem I (PSI e.g., Ycf3, Y3IP1, PYG7, and Ycf4 (Rochaix 2011; Yang et al. 2015)), Cyt b6/f (e.g. DnaJ-like protein HCF222 (Hartings et al. 2017), thioredoxin-like protein HCF164 (Lennartz et al. 2001), DAC (Xiao et al. 2012), NTA1 (Li et al. 2023), PetM (Lan et al. 2021)) and the chloroplast ATP synthase (e.g., BFA1, BFA3 and CGL160 (Zhang et al. 2018a, b). Assembly of NDH subunits (PnsB2 and PnsB3) is mediated by NDH assembly factor, NDH-dependent cyclic electron flow 5 (NDF5) in *Arabidopsis* (Kato et al. 2021). It will be interesting to know how many of these proteins are present in the etioplasts.

When the seedlings grow in the dark and under pressure inside soil, they exhibit skotomorphogenic development that is characterized by yellow color of cotyledons, curved apical hooks, elongated hypocotyl and closed cotyledons during epigeal germination in dicotyledonous plants (Sundqvist and Dahlin 1997). During hypogeal germination, the cotyledons remain in the soil. Chlorophylls are formed in leaves, stem, fruits and seeds developed above the soil surface. The cotyledons of plants showing hypogeal germination show this chlorophyll (Chl) and their breakdown products. Shoot parts of some plants remain in the soil after germination. In their tissues, special plastids can be present even in fully developed green plants, which contain Pchlide or Pchlide esters (Kakuszi et al. 2017). Artificial dark incubation of green plants for several days can also induce etioplast formation (Ikeda 1971). Besides this, the covered parts of some plant parts, e.g., cabbage inner leaves, stems, or leaf sheaths covered by leaf sheaths of outer leaves, also contain etioplasts. A small amount of Chl formation, beneath the soil, has also been reported in the hypocotyl of *Phaseolus vulgaris* through the light piping effect (Kakuszi et al. 2014, 2016). Etioplasts are lens-shaped to oval plant plastids, 1–5 μm in diameter and develop in the dark from proplastids. Before photomorphogenesis, in the dark some plants can grow for 2–15 days, e.g., *Arabidopsis* and wheat (Böddi et al. 1989, 1992; Kis-Petik et al. 1999; Franck et al. 2000) and some for up to 1–2 months, e.g., oak and horse chestnut (Skribanek et al. 2001; Solymosi et al. 2006), using the stored food. However, prolonged dark incubation may also lead to autophagy. Etioplasts contain prolamellar bodies that are unique three-dimensional lattice structures and mostly contain Pchlide-LPOR-NADPH oligomeric complexes. The Chl pathway enzymes are present in etioplasts, but the Chl-synthesis pathway is stopped at the Pchlide step as this enzyme needs light to convert Pchlide to Chlide, so as to continue the Chl biosynthesis (Griffiths et al. 1978; Apel et al. 1980). To understand the structure of etioplasts and follow the transition to chloroplasts, several techniques have been used in the past such as electron tomography, electron microscopy, transcriptomics, proteomics, lipidomics, metabolomics analysis, mutant studies and molecular techniques. Electron tomography technique provides a quantitative understanding of thylakoid biogenesis in both space and time (Daum et al. 2011) and has been widely used to decipher the chloroplast/etioplast structure (Shimoni et al. 2005; Liang et al. 2022). The ratio of MGDG to DGDG changes along with the transformation of PLBs and PTs to thylakoids (Deme' et al. 2014; Mazur et al. 2019). More PSII and DGDG are present in the appressed portion of grana, and more of PSI, Cyt b6/f and MGDG in non-appressed stromal exposed region (Koochak et al. 2019; Wietrzynski et al. 2020).

Etioplasts can be formed very early or late during the dark period of seed germination, depending on the type of plant. In wheat leaves, only 10% of the total etioplasts were transformed after 36 hours of illumination. The rest were either proplastids or amyloplast-like plastids. However, after 48 h of illumination, 68% of the plastids were transitioned into chloroplasts (Younis et al. 1995). The rate of de-etiolation can be enhanced by gibberellic acid and delayed by gabaculine, a tetrapyrrole biosynthesis inhibitor. In the presence of gabaculine, PLB formation was slowed down; Pchlide biosynthesis, including phototransformable Pchlide, and dissociation of POR and Chlide were inhibited (Younis et al. 1995). Gabaculine also caused the alterations in the structure of etioplasts up to 48 h of germination (Younis et al. 1995).

POR enzyme plays a significant role in PLB formation in etioplasts and the phototransformation of etioplasts to chloroplasts. In Angiosperms, light POR (LPOR) is present but in gymnosperms a completely unrelated enzyme, DPOR (Dark POR), performs the Pchlide to Chlide phototransformation. There are two or three isoforms of POR in angiosperms. In rice, there are two and in Arabidopsis there are three isoforms, viz., LPORA, LPORB, and LPORC. *porA* transcript decreases upon exposure of etiolated plants to light (Mapleston and Griffiths 1980; Santel and Apel 1981), *porB* remains the same or increases (Holtorf et al. 1995) and *porC* increases (reviewed in Tripathy and Pattnaik 2012, and in Tripathy and Dalal 2013). In Arabidopsis seedlings, *AtPORA* and *AtPORB* are expressed during the early stages of plant development, i.e., etiolation, while *AtPORB* and *AtPORC* are expressed in mature seedlings and plants (Su et al. 2001). Light has a negative impact on plastid import of PORA protein, and the stability of the imported PORA; therefore, it disappears from illuminated seedlings long before Chl accumulation has reached its maximum. *porB* mRNA concentration reaches maximum levels during the day and a minimum during the night, which compensates for the preferential loss of PORB in the light (Holtorf et al. 1995). POR-degrading proteases also modulate the PORB protein concentration. Etioplasts also contain some of the photosynthetic proteins such as some subunits of PSI and PSII, Cyt b6/f subunits, NDH complex subunits, etc. (Björn et al. 2023). When the seedlings come out of the soil, they show photomorphogenic development, which is characterized by the opening of the cotyledons, slowing of hypocotyl growth, straightening of the apical hook and greening of leaves through chloroplast formation. Turning of etioplasts into chloroplasts is marked by a comparatively fast anatomical process of disintegration of PLB to form thylakoids within 30 min; this is called de-etiolation (Mohr et al. 1983). In this process, LPOR activity is proposed to work as a master switch to initiate photomorphogenesis (Heyes et al. 2021). De-etiolation can occur in both red (620–750 nm) and blue light (450–500 nm) in dicots as well as monocots. Red light also influenced greening in wheat by enhancing the expression of chloroplastic lipoproteins, LHCA1, LHCB3, and rubisco small subunits RbcS1 and RbcS2, mediated through PIF5.1 TF (Zhan et al. 2024). Red light can also influence circadian rhythm through CCA1 and LHY1 (Zhan et al. 2024). Slowly thylakoids are organized into grana, and subsequently, increasing the stacks in the grana. Thylakoids are assembled by the concerted expression of genes involved in lipid, Chl, carotenoids, antioxidant synthesis, assembly factors and photosynthetic complex subunits (Wu 2014). Some authors proposed that there is a gradual formation of thylakoids and an increase in grana stacks to form mature chloroplasts; however, others found that distinct phases are present. Pipitone et al. (2021) determined that de-etiolation is marked by two phases: First phase called "Structural establishment phase" showing the buildup of PSI, PSII and Cyt b6f complexes, endoplasmic reticulum, phospholipids, galactolipids and the second phase "Proliferation phase" characterized by proliferation and enlargement of chloroplasts. The spatial transcriptomics of wheat leaves also showed two phases of plastid development: first phase is marked by plastid proliferation and the build-up of genetic machinery; the second phase is marked by the development of photosynthetic chloroplasts (Loudya et al. 2021). Photomorphogenesis and de-etiolation are controlled by the regulation of gene expression at different levels (Wu et al. 2014). In the dark, phytochrome interacting factors (PIFs), which are basic helix-loop helix domain-containing transcription factors, especially PIF3, repress the photosynthetic genes by binding to their promoters. Their degradation in light removes this repression of photomorphogenesis, leading to chloroplast development (Hernández-Verdeja et al. 2020). Transcriptome analyses have revealed that upon light exposure, up to one-third of Arabidopsis genes are differentially expressed, with 3/5 being upregulated and 2/5 downregulated (Ma et al. 2001; Liu et al. 2012). Photomorphogenesis accompanies the chloroplast differentiation and chlorophyll biosynthesis (Solymosi and Aronsson 2013).

In artificially germinated plants in the dark, the shoot development depends on the amount of nutrients stored in the storage parenchyma of the seed (or grain). However, in nature, plants grown under light also have some parts that are not exposed to light such as inner layers of leaves in the shoot buds, the below-ground portion of epicotyl/hypocotyl in seedlings and stem parts. In such parts the survival or development of etioplasts, proplastids or other modified plastids depends on the food material obtained from the adjacent photosynthesizing and green chlorenchyma tissues when shielded from light for a long time. Plastids in the undersoil hypocotyl region of pea, which are buried up to 5–7 cm inside soil, contain Pchlide and Pchlide ester, which show Emission at 633 nm and 655 nm at 77 K, even in 60-day-old seedlings (Kakuszi et al. 2017). In addition, there are organs that develop in natural light, but their outer tissue layers shade the inner layers (Solymosi et al. 2004b), so that the latter develop in complete darkness, e.g., leaf buds of many trees (Solymosi and Böddi 2006; Solymosi et al. 2006). If these organs are rich in storage tissue (e.g., eggplant pericardium, cabbage and Lettuce heads), the non-photosynthesizing inner layers persist for a long time and contain etioplasts and proplastids. In eggplants, the purple outer layers transmit only around 0.6% of the light. This results in the formation of chloroplasts in the outer 1 cm of mesocarp; however, the deeper layers, where sufficient light does not reach, develop proplastids and etioplasts (Kósa et al. 2024). Another such example is of the pumpkin inner seed coat (Sundqvist and Ryberg 1979), in which etioplasts are present in early developmental stages but later they transform into Pchlide ester storing plastids. Some plants contain microcrystals of Pchlide-pigments; e.g., the inner seed coat of *Cyclanthera explodens* contains protochlorophyll crystals (Sundqvist et al. 1980). PLB and PTs turn into thylakoids when exposed to light. Although the single lamellae form soon after illumination, it takes time to develop stacks in the grana.

8.2 Structure of Etioplasts

Many earlier studies found the detailed structure of etioplasts through Electron microscopy, the presence of pigments and varied Phclide-complexes/forms through spectroscopy (reviewed/mentioned in Solymosi and Böddi 2006; Solymosi and Schoefs 2008; Solymosi and Aronsson 2013; Solymosi et al. 2021; Yoshihara et al. 2024). Etioplasts contain PLBs that are lattice-like structures, as can be seen in Fig. 8.1a. Light-induced phototransformation of Pchilde to Chlide leads to the conformational change in LPOR, resulting in the disintegration of PLBs and can be traced spectroscopically (Dujardin and Sironval 1977; Heyes et al. 2008; Belyaeva

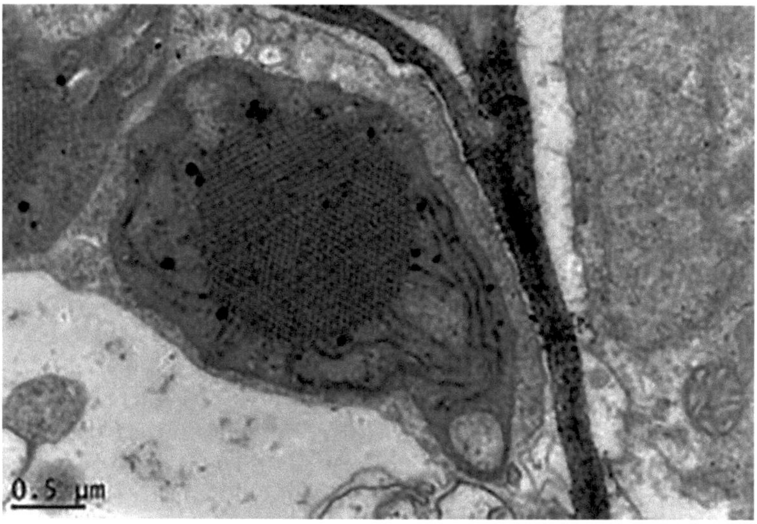

Fig. 8.1 Transmission electron microscope picture illustrating the etioplast strcture in Arabidopsis. Scale bars represent 1200 nm. (Source: Cazzonelli et al. (2020). *A cis—carotene-derived apocarotenoid regulates etioplast and chloroplast development*. Available under CC license 4.0). https://doi.org/10.7554/eLife.45310

and Litvin 2014). With several techniques, especially cryo-tomography, a lot of details of PLBs and PTs formation have been gathered as discussed below.

There are mostly one or two PLBs in the etioplasts but sometimes more (Solymosi and Schoefs 2008). Prothylakoids are the extensions of the PLBs in disc-like structures that look thread-like in 2D electron microscopy images. PLBs contain several chloroplastic proteins but are significantly different from them (Ryberg et al. 1983; Solymosi and Aronsson 2013; Yoshihara et al. 2024). There are several studies that have investigated the structure of PLBs through cyroelectron tomography, molecular biology and microscopy (Solymosi and Schoefs 2008; Lindsten et al. 1993; Shimoni et al. 2005; Liang et al. 2022). PLBs are made up of hollow tubular structures. PLB formation is preceded by the generation of perforated sheets (Lütz 1981). It was proposed that under natural Light Dark Cycles (LDC), chloroplasts of plants do not form PLBs (reviewed in Solymosi and Schoefs 2008). However, a later study by Ikeda (1971) determined that in some plants, PLBs start to appear in chloroplasts as early as ~1 hr. of darkness. These PLBs are more abundant than those found in etiolated seedlings, are smaller in size, and respond faster to the illumination. Later this phenomenon was observed in other plants also (Schoefs and Franck 2008). Upon dark incubation, PLB formation started in 72 h in oat (*Avena sativa)* chloroplasts (Lütz 1981). PLBs grow in size and number with the maturity of the leaf and their structure becomes more regular (Lütz 1981; Liang 2022).

PLBs are comprised of proteins, pigments and lipids, of which POR constitutes 95% protein (Grzyb et al. 2013). In angiosperm etioplasts, paracrystalline PLBs are made up of mostly two types of lattices, i.e., wurtzite type, which is made up of close-packed hexagons forming a hexagonal prism and zinc blende (sphalerite) type, which is made of a cubic close-packed structure (Wójtowicz et al. 1996; Kowalewska et al. 2019). Bean etioplasts show only wurtzite-type PLBs (Solymosi et al. 2021; Kowalewska et al. 2016). Other types of crystal structures are also formed and one PLB may contain more than one type of crystal structure (Gunning 2001). In vitro formation of octamer crystal structures in PLBs was recently reported for cyanobacterial LPOR by Zhang et al. (2021). PLB's structure is influenced by the period of darkness in Arabidopsis (Bykowski et al. 2020). The presence of sucrose in the growing media affected the compactness and size of PLBs. Modulation in the pigment and polyprenol content in the *pif1* and *aba1–1* mutants also changed the nano-morphology of PLBs (Bykowski et al. 2020).

Lipid composition and stoichiometry in etioplasts significantly influence their function and structure. The lipid composition of PLBs, PTs and chloroplasts has been found to be similar (Apostolova and Misra 2014; Selstam and Sandelius 1984; Selstam and Widell-Wigge 1989; Somerville and Browse 1991), although PLBs show a lower presence of unsaturated (18:3) fatty acids (Selstam and Sandelius 1984; Yu et al. 2020). The proportion of DGDG in etioplasts is comparatively higher than that in chloroplasts. Usually the lipid synthesis machinery of etioplasts, although works slowly in the dark, maintains its composition in chloroplasts (Somerville and Browse 1991). The stresses or xenobiotics influence this composition (Di Baccio et al. 2002). In PLBs, glycolipids (monogalactosyldiacylglycerol (MGDG, ~50%) and digalactosyldiacylglycerol (DGDG, ~30%) (Selstam and Sandelius 1984), and sulphoquinovosyl diacylglycerol (SQDG)) are the major lipids. Phospholipids (mainly phosphatidyl choline (PC) and phosphatidyl glycerol (PG)) are around 10–15 mol% of the total amount of polar lipids. Free sterols and sterol esters are present in low amounts (~6 mol%; Ryberg et al. 1983). There are four types of lipids in the thylakoids. Galactolipids MGDG and DGDG are about 80 mol% of total thylakoid membrane lipids (Fujii et al. 2019a), and SQDG and PG are about 20 mol% of total thylakoid membrane lipids (Fujii et al. 2017, 2019b). When SQDGs are reduced, e.g., in *sqd1* mutant of *A. thaliana*, a compensatory increase in PGs has been observed in plants and cyanobacteria (Sato et al. 1996; Endo et al. 2016). Partial reduction of PG by 30% in *A. thaliana pgp1–1* mutant, which accumulated lower Chl, inhibited the Pchlide formation by disrupting the Mg insertion in the tetrapyrrole ring, as well as loosening the lattice structure of PLBs. It also hastens the disintegration of the oligomeric Chlide-LPOR-NADP$^+$ complexes after light exposure and the replacement of NADP+ with NADPH in this oligomeric complex (Solymosi et al. 2007), marked by a shift of fluorescence emission from 690–691 nm to 680 nm (Yoshihara et al. 2024). However, even a complete loss of SDGD did not affect the PLB structure and Pchlide formation (Yoshihara et al. 2024), but it slowed down the disintegration of the oligomeric Chlide-LPOR complexes after light exposure (Yoshihara et al. 2024). Pchlide binding initiates oligomer assembly. Models have been proposed for the involvement of two lid-regions of POR in nucleating higher-order assembly (Zhang et al. 2020). The lid regions also facilitate optimal binding configuration for Pchlide to POR and present the substrate optimally for the photocatalysis and reduction of the C17-C18 double bond (Zhang et al. 2020).

PLBs, although greatly differ from the thylakoid membranes, have similar lipid composition (Fujii et al. 2019a, b). PGs are very important for the thylakoid membranes as their absence severely disrupts photosynthetic complexes in plants as well as in cyanobacteria (Kobayashi et al. 2016). Pchlide-LPOR-NADPH ternary complex formation and oligomerization were disrupted when the MGDG content was decreased by 36% in *A. thaliana*. However, the lattice structure of PLBs was only slightly affected (Fujii et al. 2017). Nevertheless, a higher decrease by about 80% in DGDG was responsible for showing the similar effects on Pchlide-LPOR-NADPH ternary complex organization, but the oligomerization of ternary complexes and Pchlide biosynthesis was not affected in these plants (Fujii et al. 2018). Importance of MGDG in the oligomerization of the ternary complexes was also recognized by Gabruk et al. (2017) in vitro. Tomographic studies by Floris and Kühlbrandt (2021) showed that the inner membrane of etioplast consists of PLBs, covered with LPOR and enriched in plastid ribosomes and PTs, peripheral, disk-shaped membranes enriched in ATP synthase monomers. *A. thaliana PORA* mutants showed reduced size of PLBs and reduced formation of photoactive Phlide. These mutants, when complemented with heterologous expression of cyanobacterial POR, led to the restoration of PLB size, but the photoactive Pchlide concentration was not restored. It showed that the photoactive Pchlide is not necessary for normal-sized PLB formation (Masuda et al. 2009).

8.3 Phototransformation of PLBs and PTs into Thylakoids

With the illumination of plants, PLBs start to disintegrate and form thylakoids that contain various proteins, the green pigment Chl and several carotenoid species. Thus, Chl and carotenoids facilitate the proper conformation and assembly of proteins to bring them into a functional state inside thylakoid membranes (Fig. 8.2).

Aggregates of photoconvertible Phlide–LPOR–NADPH dimeric and oligomeric complexes are present in etioplasts, which have fluorescence emission wavelengths of 637, 644 and the most active aggregate at 655–658 nm (Böddi 1992) and non-photoconvertible Phlide–LPOR–NADPH complexes with emission at 632–633 nm at 77 K. The fluorescence of non-photoactive Pchlide is due to Pchlide-monomers of esterified Pchlide or Protochlorophyll (Lindsten et al. 1988), which dimerizes spontaneously to form $(\text{Phlide–LPOR–NADPH})_2$, having absorption at 638 nm and emission at 645 nm (Ouazzani Chahdi et al. 1998; Lebedev and Timko 1999). These dimers further polymerize to form flash photoactive multimers with absorption at 650 nm and emission at 655–658 nm (Schoefs and Franck 1993; Wiktorsson et al. 1993; Schoefs et al. 2000; Schoefs 2001; Zhang et al. 2019; Gabruk et al. 2015, 2016, 2020; Schneidewind et al. 2019; Nguyen et al. 2021). If the photoreduction is not complete, 640 and 650 nm emitting $NADP^+$-containing Pchlide complexes can also be observed (Böddi et al. 1998; El Hamouri et al. 1981). As per a recent study by Jedynak et al. (2022), the Pchlide 654/Pchlide 633 ratio, which tells about the

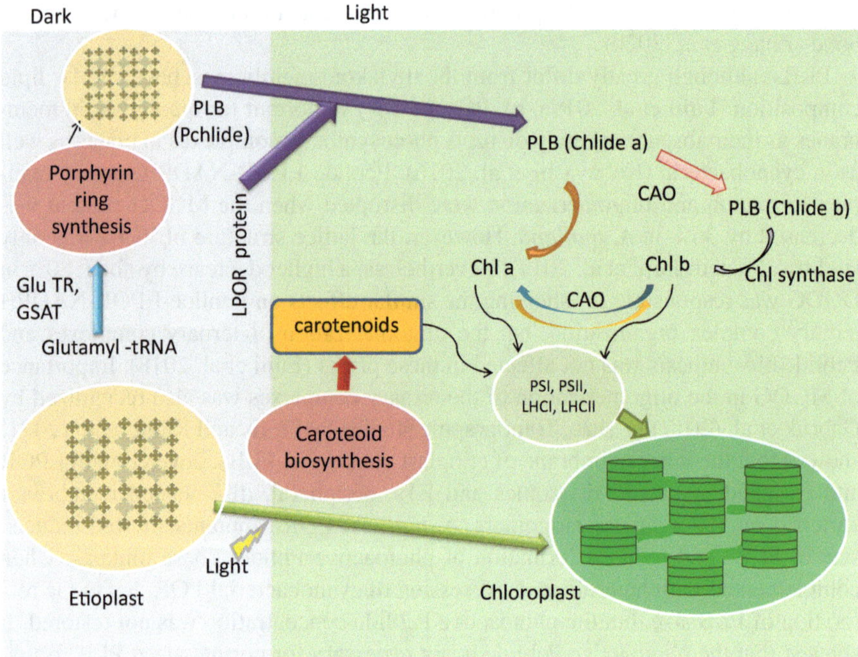

Fig. 8.2 Illustration of the etioplast to chloroplast transition. GluTR Glutamyl tRNA reductase, GSAT Glutamate-1 Semialdehyde Transferase, PLB prolamellar Body, Chl Chlorophyll, PSI/II Photosystem I/II, LHC I/II Light Harvesting Complex I/II, CAO Chlide a Oxygenase

proportion of the photoactive Pchlide, is highest in the youngest seedlings and decreases with age in *Arabidopsis thaliana*.

The 628 and 636 nm emitting forms of Pchlide and/or protochlorophyll, which are called "short-wavelength forms," are present in non-leaf organs such as the epicotyl of etiolated pea (Böddi et al. 1998). These monomers do not give rise to Chlide species (Kósa and Böddi 2012). 628 nm emitting form serves as a Pchlide pool which occupies the site in LPOR NADPH complex when Chlide has exited this complex as the epicotyl is illuminated with continuous light (Böddi et al. 1998; Szenzenstein et al. 2010). Both of these monomeric species are mostly not photoactive. However, the 636 nm emitting form is slightly photoactive (Kósa et al. 2005). This Phclide species can be artificially aggregated into flash-photoactive 644 nm and 655 nm emitting complexes in vitro (Kósa et al. 2006). 636 nm emitting form is considered as an enzyme-bound protochlorophyllide which can be converted into a 678 nm emitting Chlide form with white or 632.8 nm laser light, indicating that this form is an LPOR monomer-Pchlide-NADPH unit (Kósa et al. 2005). Besides these, pea epicotyls also contain protochlorophyll (Pchl) molecules, which also cannot be transformed directly into Chl molecules; rather they serve to capture the sun energy and generate harmful reactive oxygen species (Erdei et al. 2016), i.e., photooxidation products (Hideg et al. 2010).

The ratio and amount of Pchlide forms are highly variable in different types of organs developed in the dark. In older innermost leaves of white cabbage, the short-wavelength Pchlide form(s) are more abundant (Solymosi et al. 2004b). The ratio "flash-photoactive dimer/oligomer ternary complexes to different monomeric Pchlide forms" is much higher in leaves of etiolated seedlings than in the epicotyls of pea (Skribanek et al. 2000).

With the light flash, LPOR–Phlide–NADPH ternary complexes are converted into LPOR–Chlide–$NADP^+$ complexes, i.e., the photoactive Pchlide form is transformed into Chlide form with the absorption at 676 nm and emission at 691 nm at 77 K. This transformation is achieved through many intermediate steps with short lifetimes. When the first hydrogen ion is transferred from NADPH to Pchlide to form semi-reduced Pchlide radical species, it requires light (Lebedev and Timko 1999) and has low fluorescence (Schoefs 2001), but the second hydrogen is transferred spontaneously. Then, in the oligomeric Chlide-LPOR-$NADP^+$ complexes, $NADP^+$ is replaced by NADPH and the oligomeric complexes are dissociated into dimeric forms. One minute after flash, the formation of Chlide–LPOR–NADPH complexes induces a shift in the fluorescence emission from 690–691 nm to 694 nm (Franck et al. 1999). Later, after around 15 min of dark incubation, this emission peak is blue-shifted to 680 nm owing to the release of Chlide and the disintegration of multimeric complexes (Shibata, 1957; Böddi et al. 1990, 1992; Franck 1993; Smeller et al. 2003; Solymosi et al. 2006). Simultaneously, in some of these oligomeric LPOR-Chlide-$NADP^+$ complexes, Chlide is replaced by Pchlide prior to the replacement of $NADP^+$ by NADPH, quickly regenerating photoactive LPOR–Phlide–NADPH ternary complexes (Franck et al. 1999). Chl molecules bind with the protein complexes to form complexes with emission at 684 nm within 2–4 hours (Franck et al. 1993, 1999). Upon further illumination, fully mature Chl-protein complexes form in PSII with emissions at 685 nm and 695 nm and in PSI at 735 nm (Kalaji et al. 2017).

8.4 Factors and Genes Affecting De-etiolation

8.4.1 Factors Affecting De-etiolation

8.4.1.1 Carotenoids

Etioplasts are yellow due to the accumulation of carotenoids, primarily Lutein and Violaxanthin (Rodríguez-Villal6n et al. 2009). Enzymes such as phytoene desaturase (PDS), z-carotene desaturase (ZDS) of the carotenoid pathway are present in etioplast membranes (Cazzonelli and Pogson 2010). The carotenes, besides Chl, are also required for proper greening of seedlings during de-etiolation. In the IMMUTANS/IM (*im*) mutants, greening is affected as the carotenoids do not accumulate to the levels in normal etiolated seedlings. These mutants show a variegated phenotype upon greening. IM encodes for plastid terminal oxidase (PTOX) that is responsible for helping the desaturation reaction of enzymes, phytoene desaturase (PDS) and z-carotene desaturase (ZDS). Phytoene accumulates in these mutants

(Putarjunan et al. 2013). It also oxidizes the plastoquinol (PQH_2) to plastoquinone (PQ) (Kambakam et al. 2016) during the early stages of greening when photosynthetic ETC is not fully functional for oxidizing the PQ pool (Nawrocki et al. 2015). PTOX also acts in one or more redox pathways mediated by NDH (NAD(P)H dehydrogenase) and PGR5 (PROTON GRADIENT REGULATION 5), proteins involved in cyclic electron transport.

Carotenoids also have a significant role in de-etiolation. Accumulation of specific *cis*-carotenes in *Arabidopsis carotenoid chloroplast regulation 2* (*ccr2*) mutants disrupts the etioplast to chloroplast transition due to perturbed retrograde signal (Cazzonelli et al. 2020). *ccr2* mutants did not show PLBs, although they had normal POR protein levels. Some of the factors and genes affecting the structure and function of etioplasts and the greening process are listed in Table 8.1.

8.4.1.2 Autophagy

Autophagy plays an important role during de-etiolation. The absence of autophagy machinery hampers the chloroplast development, which is exacerbated by limited lipid turnover (lipophagy) (Wijerathna-Yapa et al. 2021). Barros et al. (2021) reported that *atg* mutants induce plastoglobule metabolism and lipid oxidation to compensate for reduced lipid catabolism. In a proteomics study of monocot leaves, it was found that the Atp6 mitochondrial protein is most abundant in the basal region (Chotewutmontri et al. 2016), similar to that in the leaves of maize (Majeran et al. 2010). Except for a few, most of the genes of photosynthetic machinery were normally expressed in autophagy (*atg*) mutants (Wijerathna-Yapa et al. 2021). The number of mitochondria decreases during de-etiolation (6 h post-illumination)/photomorphogenesis in cotyledons (Ma et al. 2021) through mitophagy. In the *atg5* mutants, mitophagy and greening are reduced (Wijerathna-Yapa et al. 2021).

8.4.1.3 Tocopherol

Jin et al. (2021) determined that the level of α-tocopherol was significantly upregulated in de-etiolated rice leaves. 1-deoxy D-xylulose 5-phosphate synthase 1 (OsDXS1) enzyme was responsible for the plastidial isoprenoids biosynthesis in de-etiolated rice leaves. Moreover, the genes for the chlorophyll, carotenoid, MEP (2-C-Methyl-D-erythritol 4-phosphate) and tocopherol pathways were coordinately expressed during de-etiolation, and all of them had light-responsive cis-regulatory elements in their promoters (Jin et al. 2021).

8.4.1.4 Sigma (SIG) Factors

SIG factors assist the PEP enzyme complex in identifying the promoter sequences (Schweer et al. 2010) of PhANGs. The expression of SAG12 or SAG13 senescence markers was not observed during etiolation. Reduction in antioxidant molecules and ion leakage resulting from the damage to membranes was not observed during the etiolation (Jedynak et al. 2022). Pchlide accumulation and composition of some xanthophylls was affected in the *lut2* mutant that lacks lutein, a type of xanthophyll (Jedynak et al. 2022).

Table 8.1 Factors and genes affecting etioplast formation and etio-chloroplast transition

S.N.	Factor	Affected pathways	References
Factors affecting etioplasts and etio-chloroplast transition			
1	Carotenoids/IMMUTANS	Greening is affected due to low carotenoid accumulation in normal etiolated seedlings	Putarjunan et al. (2013)
2	Autophagy	In *atg5* mutants, mitophagy and greening are reduced	Wijerathna-Yapa et al. (2021)
3	α-tocopherol	Upregulated in de-etiolated rice leaves; chlorophyll, carotenoid, MEP (2-C-methyl-D-erythritol 4-phosphate) were also increased in expression	Jin et al. (2021)
4	Quality of light	As compared to Red light, Blue light showed faster disintegration of PLBs and de-etiolation in *Arabidopsis* seedlings	Banaś et al. (2024) and Liang et al. (2022)
5	MGDG	In *A. thaliana*, when the MGDG content was decreased by 36%, Pchlide-LPOR-NADPH ternary complex formation and oligomerization was disrupted although the lattice structure of PLBs was affected only slightly	Fujii et al. (2017)
Genes affecting etioplasts and etio-chloroplast transition			
6	Target of rapamycin (TOR)	It phosphorylates the ribosomal protein S6 (RPS6) to boost protein translation in de-etiolating seedlings	Chen et al. (2018)
7	*Sugardependent1*	Delay in de-etiolation due to lower triglyceride mobility in the membranes	Wijerathna-Yapa et al. (2021)
8	*Carotenoid chloroplast regulation 2* (*ccr2*)	Perturbed retrograde signal; absence of PLBs in *ccr2* mutants	Cazzonelli et al. (2020)
9	Early light-inducible proteins (ELIPs)	TCP14 TFs bind to UP1 elements in the promoters of ELIPs and induce their expression	Adamska and Kloppstech (1991)
10	Orange (OR)	OR interacts with TCP14 to repress its action	Sun et al. (2019)
11	Chlorophyllide *a* oxygenase (CAO)	Might have a role in pre-PORA import into the etioplasts	Pollmann et al. (2007), Philippar et al. (2007)
12	LZF1	LZF1 increases the amount of PAP1/MYB75	Chang et al. (2008)
13	*Pale green Arabidopsis 4* (*pga4*)	Encodes for Signal Recognition Particle 54 kDa (cpSRP54) protein, a suppressor of *Var2/FtsH2* mutation, and performs thylakoid quality control	Lei et al. (2023); Shi et al. (2020)
14	CURT1 overexpressing lines	Affects de-etiolation, geometry and density of PLBs	Sandoval-Ibáñez et al. (2021)

(continued)

Table 8.1 (continued)

S.N.	Factor	Affected pathways	References
15	De-etiolation-induced protein 1 (DEIP1)	Interacts with the two cytochrome subunits of the complex, PetA and PetB and assists in the Cyt b_6/f assembly	Sandoval-Ibáñez et al. (2022)
16	SPA2	Promotes seedling etiolation in the dark; In Arabidopsis overexpression of the SPA2 cause hyper-etiolation phenotypes	Banaś et al. (2024)
17	*lut2* mutant	Pchlide accumulation, composition of some xanthophyll affected	Jedynak et al. (2022)
18	*sqd1* mutant	SQDGs are reduced. A compensatory increase in Phosphatidyl Glycerol (PG)	Sato et al. (1996), Endo et al. (2016)
19	*pif1* and *aba1–1* mutants	Changed the nano-morphology of PLBs	Bykowski et al. (2020)
20	*pgp1–1* mutant	Partial reduction of PG and loosened the lattice structure of PLBs; Faster disintegration of the oligomeric Chlide-LPOR-$NADP^+$ complexes after light exposure	Solymosi et al. (2007), Yoshihara et al. (2024)
21	HTL (hypersensitive to light)	Promotes de-etiolation through PHYs and CRYs	Song et al. (2019), Ponnu and Hoecker (2021)
22	Jumonji domain-containing protein17 (JMJ17)	Reduce the expression of $H_3K_4me_2$ and $H_3K_4me_3$ which suppress the expression of photosynthetic genes to influence de-etiolation	Islam et al. (2021)
23	*De-etiolated 2* (*det2*)	Steroid 5α-reductase mutants. Show expression of various PhANGs including Chl biosynthesis genes, under dark conditions	Chory et al. (1991), Kour et al. (2021)
24	*Constitutive photomorphogenic dwarf* (*cpd*), *dwarf4* (*dwf4*), *bri1* and *bin2*	These mutants are related to BZ synthesis (*cpd* and *dwf4*), and signaling (*bri1* and *bin2*); Show expression of PhANGs genes under dark conditions	Li and Jin (2007)

8.4.1.5 Quality of Light

Quality of light also modulates etioplast development. Blue light showed faster disintegration of PLBs and de-etiolation in etiolated *Arabidopsis* seedlings as compared to red light (Banaś et al. 2024). Banaś et al. (2024) suggested that the blue light might enhance the development of PSI, PSII and antennae. The antennae of de-etiolated seedlings under blue light showed more efficiency of heat dissipation as compared to red light (Banaś et al. 2024). Contrarily, red light promoted Chl b and antennae protein synthesis. Besides this, compact spots were observed in PLBs as soon as after 30 min of illumination with red/blue light that spread after longer illumination with white/red/blue light (Banaś et al. 2024; Liang et al. 2022).

8.4.1.6 Nitrogen Monoxide (NO)

The enormous energy demand for the formation of the Pchlide-LPOR-NADPH complex is met through presumably stored food in the seed. The content of Chl, NO and ATP in the chloroplast and cytoplasm increases with the greening (Melo et al. 2016). Using a fluorescence resonance energy transfer-based fluorescent protein sensor (Ateam 1.03-nD/nA), Da et al. (2023) found that during greening, an inhibitor of NO, L-NAME, decreased the Chl and NO content in the chloroplast and cytoplasm of *Arabidopsis* seedlings, and also decreased ATP content, thus suggesting that NO increases the ATP content during de-etiolation (Da et al. 2023).

8.4.1.7 MGDG

Oligomerization of the ternary complexes of Pchlide-LPOR-NADPH is affected due to reduced MGDG content (Gabruk et al. 2017). Reduced MGDG leads to the disruption of Pchlide-LPOR-NADPH ternary complex formation and their oligomerization; however, the PLB lattice was not much affected (Fujii et al. 2017). A higher decrease in MGDG also shows similar effects on Pchlide-LPOR-NADPH (Fujii et al. 2018).

8.4.2 Genes Affecting De-etiolation

8.4.2.1 Target of Rapamycin (TOR)

TOR, an evolutionarily conserved protein kinase, is an important hub gene in mammals and plants, regulating developmental processes. It also regulates the etioplast to chloroplast transition (Rabeh et al. 2024). To establish the photosynthetic machinery in de-etiolating seedlings, it phosphorylates the ribosomal protein S6 (RPS6) to boost protein translation (Chen et al. 2018). Reduction in the TOR activity in *Arabidopsis* through treatment with TOR inhibitors or through mutation in the TOR complex subunit, RAPTOR1B, led to a significant reduction in Pchlide levels in seedlings under darkness (Zhang 2018a, b). However, an increased greening during the de-etiolation, probably due to the saving of food and lesser growth of the seedlings, was observed in such plants (Zhang 2018a, b).

Delay in greening or a defect in de-etiolation was observed in the triacylglycerol lipase mutant *sugardependent1*, which had lower triglyceride (TAG) mobility in the membranes (Wijerathna-Yapa et al. 2021). This was exacerbated by higher carbon resources and minimized by lower carbon resources.

8.4.2.2 Early Light-Inducible Proteins (ELIPs)

ELIPs are co-expressed with greening (Adamska and Kloppstech 1991). They are induced by TCP14 TFs, which directly bind to UP1 elements in their cognate promoter. TCP14 is repressed by interaction with ORANGE (OR, a DnaJ-like zinc finger domain-containing nuclear protein involved in carotenoid biosynthesis) in *Arabidopsis thaliana*. Thus OR suppresses the Chl biosynthesis and thylakoid formation in the germinating etiolated seedlings (Sun et al. 2019).

8.4.2.3 OEP16 and CAO

Outer-envelope solute-channel protein family of 16 kDa (OEP16) and chlorophyllide *a* oxygenase (CAO) had been suggested to play a role in pre-PORA import into the etioplasts by translocation pore formation, and chloroplast biogenesis during de-etiolation (Pollmann et al. 2007). However in a study using *cao* and *oep16* mutants of Arabidopsis, Philippar et al. (2007) found that these proteins are dispensable for de-etiolation. *cao* or *oep16* mutants possessed a normal PLB that contains a protochlorophyllide holochrome. Etioplasts of these double mutants contained PORA protein, suggesting that PORA was imported into the etioplasts (Philippar et al. 2007). Further, Pudelski et al. (2009) used OEP16.1 T-DNA insertion lines to disprove the role of OEP16 in the import of pre-POR into etioplasts and in de-etiolation (Pudelski et al. 2009).

8.4.2.4 De-etiolation-Induced Protein 1 (DEIP1)

Another protein, DEIP1, is essential for photoautotrophic growth. DEIP1 is present in the thylakoid membrane, and its transcript expression is induced similarly to Lhcb2 upon light exposure of etiolated seedlings (Sandoval-Ibáñez et al. 2022). It was found to interact with the two cytochrome subunits of the complex, PetA and PetB and assist in the Cyt b_6/f assembly (Sandoval-Ibáñez et al. 2022).

8.4.2.5 Light-Regulated Zn-Finger 1 (LZF1)

The transition from etioplasts to chloroplasts was also inhibited in *lzf1* mutant that encodes for **L**ight-regulated **Z**n-**F**inger 1 (LZF1; a C2C2-CO B-Box Zn Finger) (Chang et al. 2008). HY5 binds with the promoter of LZF1, which increases the amount of PAP1/MYB75, which in turn enhances the anthocyanin accumulation in Arabidopsis (Chang et al. 2008).

8.4.2.6 Signal Recognition Particle 54 kDa (cpSRP54)

Another mutant of the *Arabidopsis thaliana* gene *pale green Arabidopsis 4* (*pga4*) shows chloroplast development defects during de-etiolation. *pga4* encodes for the cpSRP54 protein that is a suppressor of *Var2/FtsH2* mutation and maintains protein quality control in thylakoid membranes (Lei et al. 2023). Biochemical and genetic evidence showed that in *pga4* and *yellow variegated2* (*var2*) mutants, degradation of LhcB2-GFP to dLhcB2-GFP was reduced, compromising the protein homeostasis in chloroplasts, during de-etiolation (Lei et al. 2023).

8.4.2.7 Curvature Thylakoid 1 (CURT1)

Additionally, a recently identified protein, CURT1 was shown to be involved in the formation of grana margins. Besides this, it also showed the effects on de-etiolation, and on geometry and density of PLBs. *CURT1* mutants showed disturbed formation of PLBs and vesicle fusion. The etio-chloroplast transition showed reduced LPOR accumulation and delay in the photosynthesis initiation in this mutant (Sandoval-Ibáñez et al. 2021). CURT1 overexpressing plants showed excessive PLB membrane bending. Upon illumination, greening and buildup of photosynthetic capacity were normal, but in CURT1 overexpressing lines, the PLBs exhibited lesser disintegration and over-accumulation of LPOR (Sandoval-Ibáñez et al. 2021). *A. andraeanum* cultivar "Sonate" showed a mutation that caused the differential development of plastids (Wang et al. 2018).

8.4.2.8 Suppressor of PHYA-2 (SPA2)

Yet another protein, SPA2, has been reported to promote seedling etiolation in the dark. Overexpression of the SPA2 causes hyper-etiolation phenotypes in Arabidopsis under several light conditions. phyA, phyB, cry1, and COP1 repress protein accumulation or nuclear translocation of SPA2 in response to far-red, red, and blue light treatments (Banaś et al. 2024).

8.4.2.9 NCP and RCB

Dual targeted nuclear control of pep activity (NCP) and regulator of chloroplast biogenesis (RCB), besides helping the Cyt b_6f assembly, assists in the assembly of PEP enzyme complex and in the expression of PhANGs (Yang et al. 2019; Yoo et al. 2019; Hwang et al. 2022).

To further understand the protein profile of the PLBs, several authors used global proteomics, membrane proteomics, temporal proteomics, phosphoproteomics, N-glycosylation and other forms of proteomics studies during de-etiolation (von Zychlinski et al. 2005; Kanervo et al. 2008; Bu et al. 2017; Hamamoto et al. 2012; Armarego-Marriott et al. 2019; Gao et al. 2020).

8.5 Proteomics and Phosphoproteomics of De-etiolation

Proteomics is an important technique for studying the protein repertoire of a cell or an organelle. However, there are very few proteomics studies for etioplasts as compared to many for chloroplasts (von Zychlinski et al. 2005; Blomqvist et al. 2008; Majeran et al. 2010; Plöscher et al. 2011; Hamamoto et al. 2012; Gao et al. 2020). A study by Blomqvist et al. (2008) reported 64 proteins in the proteome of PLBs that were related to photosynthetic light and light reactions, pigment biosynthesis, and protein biosynthesis. ATP synthase is fully mature in PLBs (Plöscher et al. 2011), and the dimeric Cyt b_6f complex only lacks chlorophyll *a* (Reisinger et al. 2008). PSII biogenesis happens at a later stage, and PSI is inserted lastly, only after exposure to the light (Müller and Eichacker 1999). LHCs are not inserted into etioplast membranes, but the supplementation with Zn-pheophytin a/b stably inserts LHCs into etioplast membranes (Kuttkat et al. 1997).

During etiolation, about 998 proteins changed in abundance in maize (Gao et al. 2020). This was about 7% of the total identified proteins, which could be quantitatively estimated, inducing the authors to speculate that the post-transcriptional rather than transcriptional regulation is more important during de-etiolation. However, they observed a higher change (26.6%) in the phosphoproteome during this transition. In the plastid proteome, a lower number of proteins have been identified (Kanervo et al. 2008), e.g., 477 by Kleffmann et al. (2006) and 216 in rice etioplast by von Zychlinski et al. (2005). Etioplasts harbor many other photosynthetic proteins, e.g., Fd NADP Reductase (FNR) (Scheumann et al. 1998), ATP synthase, NDH complex and Cyt b6/f complex (Plöscher et al. 2011). The composition of these complexes has been found to be the same in etioplasts and chloroplasts (Plöscher et al. 2011). N-glycolation, i.e., a posttranslational attachment of N-linked glycans to the N-atom of Asp side chain in the consensus sequence asp-x-ser/thr (where x is any amino acid except pro (Pless and Lennarz 1977)), is an important modification of the proteins that is responsible for plant development, sugar metabolism, signalling, cell wall modification and plant defense (Fitchette et al. 2007). Out of 609 identified N-glycosylated proteins, the abundance of 162 N-glycosylated proteins was changed in maize seedlings during 12 h of de-etiolation (Bu et al. 2017). Hamamoto et al. (2012) observed a well-coordinated expression in the gene and protein levels of Calvin-Benson cycle enzymes in greening rice seedlings, where transcript expression preceded protein expression. Similar coordination was observed for the photorespiration, chloroplast translation machinery and nitrogen assimilation (Hamamoto et al. 2012). However, a significant increase in the levels of respiratory enzymes (glycolysis and TCA cycle) was not reported by Hamamoto et al. (2012).

Using HILIC glycosylation, a process that involves protein enrichment, chemical deglycosylation, HPLC separation and tandem mass spectrometric identification of proteins, Wang et al. (2021) found N-linked glycosylation of 248 unique glycoproteins in wheat seedlings during greening/de-etiolation. Several of them were enzymes such as carboxypeptidase, kinase, peroxidase and phosphatase, and one-fourth of them were related to binding functions. These glycoproteins were involved in the biomacromolecular metabolism, cell wall reconstruction, signal transduction, stress response and endoplasmic reticulum quality control (Wang et al. 2021).

8.6 Timeframe of De-etiolation

Changes in etioplasts start happening immediately after illumination. Shrinkage of PLBs that occur within 10 min of light exposure is one of the first observations made during the photomorphogenesis (Armarego-Marriott et al. 2019). Later the gene expression program sets in (Armarego-Marriott et al. 2019). In tobacco leaves, within 10 min, PLBs are loosened, reduced in size and compactness, and start disintegrating. Some of the PLB proteins are used in forming thylakoids, as detailed in the section “Proteomics of De-etiolation,” although the disintegration of PLB continues for 2 hours. Within the first 30 min, most changes at the ultrastructural level happen (Armarego-Marriott et al. 2019). For 30 min the abundance of mRNAs, responsible for etiolation, is not changed (Armarego-Marriott et al. 2019), although transcription machinery was recently found to respond to light stress within 20 seconds to one min (Suzuki et al. 2015). Transcriptomics results showed that after 30 min, the amount of proteins required for maintenance of etioplasts, e.g., POR, Phy, PIFs, is decreased (Armarego-Marriott et al. 2019). Components of ETC are formed only after 4 hours. PSII starts accumulating earlier and PSI later. However, the PSI accumulation was faster thereafter, and all kinetic parameters of photosynthesis showed sequential PSII and PSI build-up, e.g., Lhcb5, Lhcb6 and PsbS accumulated to high levels that would increase the protective mechanisms such as nonphotochemical quenching. This is useful for plants as the presence of PSII in the absence of PSI would produce oxidants and be very harmful to the greening seedlings. After two hours and up to 4 hours of illumination, a decreased amount of free amino acids, which indicated increased de novo protein synthesis and increased flux through central metabolism, was observed (Armarego-Marriott et al. 2019). On the other hand, in *Arabidopsis* seedlings, the PLBs start turning into thylakoids within one hour upon exposure to light. During de-etiolation in Arabidopsis, the photosynthetic electron transport competency is gained by 14 hours (Pipitone et al. 2021), i.e., the photosynthetic machinery becomes fully functional (PSII fluorescence Fv/Fm reaches around 0.8) and photosynthetic carbon fixation competency soon thereafter. Major structural changes in the photosynthetic machinery occur within 24 hours (Banaś et al. 2024). Other studies reported PSII assembly during the first 16 hours of greening (Schoefs and Franck 1991). The development of chloroplasts along the monocot leaf gradient has been studied in maize and rice leaves through proteomics and transcriptomics, respectively (Li et al. 2010; Pick et al. 2011; Wang et al. 2014a), which hints at two phases of chloroplast greening.

The development of plants that takes place in light is called photomorphogenesis. It includes several processes other than the greening of plants and development of chloroplasts, e.g., slowing of the hypocotyl elongation, opening and flattening of cotyledons in dicots, straightening of the apical hook, etc. The light is an important player in the formation of chloroplasts from etioplasts. Several different receptors are present in plants to perceive light of different wavelengths and the environment around them.

8.7 Photomorphogenesis

8.7.1 Photo-Receptors

The photoreceptors in plants are the red/far-red light (RL/FR) sensitive phytochromes (PHY), the B light sensitive cryptochromes (CRY), phototropins (phot) and the UV-B light sensitive UVR8 (Toman et al. 2024). PHY is present in three clades, i.e., PHYA, PHYB/D/E and PHYC (Mathews 2010). In rice, five PHY A-E genes are present but in Arabidopsis only three (PHY A-C) genes are present. PHYB is the main photoreceptor and it requires abundant light with high RL/FR for its activation. The Pr form of the Phy receptor absorbs RL to convert to the Pfr form, which is the active form. Pfr form converts back to Pr form by absorbing FR light or upon dark incubation. Pr form absorbs RL of 650–670 nm and Pfr form absorbs FR of 705–740 nm (Viczián et al. 2021). Pr form is translocated to the nucleus from the cytoplasm and interacts with TFs, mainly PIFs, to exhibit their response (Viczián et al. 2021). PHYs execute their response during de-etiolation through COP1/SPA/HY5 signaling pathway and through several PIFs (PIF1, PIF3, PIF4, and PIF5) (Song et al. 2019; Ponnu and Hoecker 2021). Wang et al. (2019) discovered five potential ubiquitylation sites, viz., KXG, AXK, AK, KA and TK in maize seedlings during de-etiolation. Further, the authors concluded that most of the ubiquitylated sites were in substrates involved in protein (ribosomes) and DNA (histones) metabolism, and the rest were present in light and hormone signaling (or synthesis proteins) and key photosynthetic enzymes.

An E3 ubiquitin ligase complex, COP1/SPA, inhibits the photomorphogenesis-inducing proteins such as HY5 by marking them for degradation through ubiquitination (Ponnu and Hoecker 2021). HY5 promotes de-etiolation by binding to promoters of many genes responsible for photomorphogenesis, e.g., HTL (hypersensitive to light) (Song et al. 2019; Ponnu and Hoecker 2021). HTL promotes de-etiolation through PHYs and CRYs. Its Arabidopsis mutants *htl-1* and *htl-2* show long hypocotyl phenotype under red, blue, and far-red light and are deficient in Chl formation, cotyledons opening and expansion, and gene expression of Chl a/b binding protein 3 and chalcone synthase (Sun and Ni 2011).

CRYs (cry1 and cry2 in *Arabidopsis thaliana*) also play an important role in de-etiolation. Arabidopsis cry2 is very sensitive to light and gets readily degraded in blue light while cry1 is light stable (Wang et al. 2020). CRY1 also mediates photomorphogenesis through TIC-TOC import apparatus proteins in Arabidopsis. Its deficiency resulted in a lack of some TOC proteins. *Plastid protein import2* (*ppi2*) (lacking TOC159) mutants could not express many photosynthesis-associated nuclear genes upon illumination, although they expressed the photoreceptors to normal levels, implying that CRY1 photoreceptor mediates their effects through TIC/TOC proteins (Fukazawa et al. 2020). De-etiolation is also modulated through blue light photoreceptors, phototropins (phot1 and phot2 in *A. thaliana*), and UV-B light photoreceptor UVR8 and COP1/SPA/HY5 signal pathway (Yadav et al. 2020; Toman et al. 2024). In the dark, a PIF family protein, OsPIL15, represses the growth of etiolated seedlings in rice (Zhou et al. 2014).

8.7.2 Role of Hormones and TFs in De-etiolation

Various phytohormones display significant activities in regulating etioplast development and de-etiolation (Kusnetsov et al. 2020). Etioplast to chloroplast transition is enhanced by the cytokinin plant hormone. It increases the rate of lamellae formation from the PLB. Arabidopsis seedlings, when grown in cytokinin-free medium, showed small etioplasts but larger lens-shaped etioplasts when grown with a cytokinin supplement (Chory et al. 1994). Cytokinin also plays a role in the formation of Pchlide, photoreduction of Pchlide a to Chlide a, as well as the upregulation of many genes involved in Chl biosynthesis during de-etiolation (Cortleven and Schmülling 2015). Chl biosynthesis, light harvesting and greening of cotyledons in *Arabidopsis* is mediated through type B Arabidopsis response regulators (ARRs) (Cortleven et al. 2016). Auxin represses the root greening independent of ARRs. When shoots of Arabidopsis are removed or B-GATA TF is overexpressed, roots show greening, which is mediated through cytokinin, ARR and through well-studied TF LONG HYPOCOTYL5 (HY5) (Kobayashi et al. 2017). Etioplast to chloroplast transition rate was increased in the *ARR1 ARR10 ARR12* triple mutant of Arabidopsis (Cortleven et al. 2016). Similarly, knockout of *ARR1* and *ARR12* increased the Chl biosynthesis in intact roots (Kobayashi et al. 2017) as well as in detached roots. Similar negative regulation of Chl biosynthesis was observed by Balakhonova et al. (2023) through iReenCAM that utilized Chl fluorescence-based quantification of Chl in greening Arabidopsis mutant plants. Balakhonova et al. (2023) demonstrated that during de-etiolation, the cytokinin responding trans-activators, ARR1 and ARR12, and ethylene signaling components, EIN2 and EIN3, inhibited the Chl accumulation. In the detached roots, expression of GNC-LIKE (GNL), a B-type GATA TF, is enhanced, but it can be decreased by Auxin application. ARRs act via up-regulation of nucleus- and plastid-encoded chloroplastic genes. Cytokinin and auxin signaling pathways converge on GNL, which relays the signal from ARR1 and ARR12 to the nucleus (Kobayashi et al. 2017). GLK TFs regulate the cytokinin-dependent effect on the plastome expression through the enhanced transcription of the *SCA3*, a nuclear gene encoding plastid RNA polymerase, RPOTp (Doroshenko et al. 2022).

BRs suppress the photomorphogenesis in the dark through BRASSINAZOLE RESISTANT1 (BZR1) and BRASSINAZOLE-RESISTANT 2 (BZR2)/BRI1-EMS-SUPPRESSOR 1(BES1), which down-regulate the GATA2, GATA4 (Belkhadir and Jaillais 2015), GLK1 and GLK2 TFs (Yu et al. 2011) respectively, all of which are positive regulators of photomorphogenesis. BZR2/BES1 directly binds to the promoter of GLK1, GLK2 to reduce their expression while BZR1 suppresses the expression of GATA2, GATA4 to reduce the expression of PhANGs in darkness. Arabidopsis knockout mutants of *de-etiolated 2* (*det2*), which encodes for the steroid 5α-reductase, an enzyme involved in the BR synthesis (Kour et al. 2021), show photomorphogenesis and dwarfness under darkness, although greening does not occur. *det2* mutants show expression of various PhANGs, including Chl biosynthesis genes under dark conditions (Chory et al. 1991). Several other mutants related to BZ synthesis, e.g., *CONSTITUTIVE PHOTOMORPHOGENIC DWARF* (*cpd*), *dwarf4* (*dwf4*), and mutants involved in BR signaling, *bri1* and *bin2,* exhibited similar characteristics (reviewed in Li and Jin 2007).

8.8 Plant Noncoding RNAs During Photomorphogenesis

Noncoding RNAs (ncRNAs) are mainly comprised of miRNA, siRNA and long ncRNA (lncRNA/lincRNA). miRNA is derived from the single stranded RNA or intron which folds into imperfect loops whereas siRNAs is derived from long, almost perfect complementary double strands of RNA derived from various sources including RNA-dependent RNA polymerase activity, long inverted repeats, convergent transcription, viral or transposon RNAs and that introduced from outside (Tang et al. 2003). miRNA works by degrading the complementary mRNA, thus regulating the transcript abundance in the tissue. They form RNA-induced silencing complex (RISC), which maneuvers the cleavage of mRNA. DCL1, HYL1, AGO1 and other components are present in the miRNA function; therefore, the concentration of these proteins is important for regulating transcript abundance (Tang et al. 2003). Many pathways and processes during photomorphogenesis are regulated by ncRNAs in a light-dependent manner (Sánchez-Retuerta et al. 2018). The concentration of AGO1 changes during de-etiolation. The *ago-1-27* mutant of *Arabidopsis* was not able to de-etiolate during FR Light exposure (Li et al. 2014). PIF4 stabilizes the HYL1 and DCL1 in the dark; in contrast, in red light, they are degraded. HYL1 and DCL1 bind tightly to PIF4 and PHYB. *dcl1* and *hyl1* mutants showed shorter hypocotyls than wild type, suggesting their role in de-etiolation (Sun et al. 2018). During de-etiolation, the expression of most miRNAs did not change except miR163 (Lin et al. 2017). A significant increase in the expression of *PHYA*, *PHYB*, and *PHYC*, the TF genes *PIF4* and *PIF5*, and miRNA genes *miR395, miR408*, and *miR165* was observed in *Eutrema salsugineum* plants when exposed to RL (Pashkovskiy et al. 2021). The highly conserved CCR4-NOT complex is involved in the signaling downstream of PHYA (Schwenk et al. 2021).

HY5 is a bZIP-type TF that regulates the Chl and carotenoid biosynthesis, phytoene synthase (PSY) gene expression, etc. (Gangappa and Botto 2016). In light conditions, HY5 enhances the expression of HEN1, which in turn increases the expression of miR157d and miR319 miRNAs in greening seedlings (Tsai et al. 2014). miR157 itself regulates the amount of HY5 mRNA, thus establishing a feedback loop. The feedback loop is further supported by the *hen1* mutants in which the miR157 miRNA amount is reduced, resulting in uncontrolled higher expression of HY5 mRNA and the protein, and results in a light-hypersensitive phenotype (Tsai et al. 2014). Plants with higher/constitutive expression of miR157 showed reduced HY5 protein and a light-hyposensitive mutant. miR319 also shows a role in photomorphogenesis. Under red light, miR319, miR160, miR167 and miR848, affect hypocotyls elongation in *Arabidopsis* (Sun et al. 2018). Light triggers the miRNA synthesis in de-etiolated seedlings (Choi et al. 2020). The miR172 transcript is downregulated by RL and upregulated by BL in *Arabidopsis* through PHYA, PHYB, and CRY1 and CRY2 (Islam et al. 2023). The results of Park et al. (2021) indicated that FHA2 suppresses miRNA synthesis during de-etiolation. FHA2 protein degrades in the dark, but becomes stable when seedlings are exposed to light. Prolonged darkness induces the FHA2 degradation that deregulates the expression of miRNA upon transfer to light (Park et al. 2021).

In *Arabidopsis*, SCL upregulates the POR protein and Chl biosynthesis. SCL27 directly binds to the promoter of PORC to reduce its expression. SCL is regulated by *HAM (HAIRY MERISTEM)* or *LOST MERISTEMS (LOM)* TFs that are themselves targeted for degradation by miR171. Due to the inactivity of either of these two (i.e., miR171c-resistant SCL27, or *scl6-scl22-scl27* triple mutants) or the overexpression of miR171c, there was a reduction in the POR levels and Chl biosynthesis (Ma et al. 2014). Gibberellin responding DELLA protein interacts with SCL27 to prevent its binding with PORC promoter, thus causing its increased expression and enhanced Chl biosynthesis, in the presence of light (Ma et al. 2014).

Long noncoding (Linc) RNAs or short RNAs (si/miRNA) that regulate the expression of a gene by modulating mainly the transcript level, although Linc functions through other processes. ncRNA RNA Hidden Treasure 1 (HID1) was found to regulate the photomorphogenesis by interacting with PIF3, a bHLH TF. In the *hid1* mutants of *A. thaliana* there was a significant increase in the amount of PIF3 under continuous RL conditions that leads to elongated hypocotyls. HID1 suppresses the transcription of PIF3 by interacting with its first intron-chromatin. *Arabidopsis* mutants can be rescued by the rice homologue of HID1 (Wang et al. 2014b).

8.9 Further Layers of Regulation of De-etiolation

The process of de-etiolation or greening, if misregulated, can be very harmful to the plants, as the Chl intermediates present in the chloroplasts are photodynamic in nature and can form reactive oxygen species if not disposed off immediately after exposure to light. Therefore, etio-chloroplast transition is regulated by several additional layers such as control of gene expression, mRNA processing through non-coding RNAs, alternative splicing, acetylation and other forms of epigenetic controls, etc. These are mentioned below:

8.9.1 Alternative Splicing

An important layer of molecular regulation of de-etiolation is Alternative splicing (AS). AS responds to the stimuli or to the plant development stage through the addition or removal of functional domains, which are present in the exons. AS can respond faster than the transcript modulation; therefore, it is a good strategy to respond fast to any change (Wu et al. 2014). During the de-etiolation, Yan et al. (2020a) observed a huge level of AS in transcripts responding to light. They discovered 41,188 novel transcript isoforms for annotated genes, which brings the 63% of genes in the maize genome showing AS.

8.9.2 Acetylation

Acetylation also occurs on many proteins during de-etiolation. Recently, Yan et al. (2020b) found 462 proteins acetylated at lysine residues during de-etiolation. The acetylated proteins included transcription factors, Chl biosynthesis, photosynthesis light reaction, carbon assimilation, TCA cycle, Glycolysis, histones, lipid, nucleotide and amino acid metabolism proteins (Yan et al. 2020b).

8.9.3 RNA Editing

Editing of RNAs is another aspect of regulating gene expression. Ribosome footprints obtained at the site of RNA editing by Chotewutmontri et al. (2016) showed that the chloroplast translation machinery does not discriminate between the edited and non-edited RNAs. In maize chloroplasts, the editing of start-codon ACG to AUG is required for starting translation at the rpl2 (Chotewutmontri et al. 2016).

8.9.4 Epigenetic Control

Epigenetic control of chromosomal structure by histone methylation that in turn is regulated by JUMONJI DOMAIN-CONTAINING PROTEIN17 (JMJ17) is also found to influence de-etiolation (Islam et al. 2021). Jumanji protein was found to reduce the expression of $H_3K_4me_2$ and $H_3K_4me_3$, which caused the suppression of chloroplastic genes during photomorphogenesis (Islam et al. 2021). JMJ17 was expressed in etiolated seedlings and when transiently expressed in tobacco leaf cells, showed $H_3K_4me_2$ and $H_3K_4me_3$ demethylase activity in vivo (Islam et al. 2021).

Plants are exposed to various types of environmental stresses as a result of their static nature. Several pathways and processes are altered in the plants as a result of stresses (Jain et al. 2022). In addition, to affecting green plants, stresses also influence the etio-chloroplast transition as the seedlings are more susceptible to stresses at the early stages.

8.10 Response to Stresses

Etio-chloroplast transition has been studied under many stresses (Tewari and Tripathy 1998; Le Lay et al. 2000, 2001; Savchenko et al. 2002; Abdelkader et al. 2008; Mohanty and Tripathy 2011; Turan and Tripathy 2015; Tripathy and Dalal 2013; Ounoki et al. 2021, 2023). Leaf segments showed the characteristic swelling of the thylakoid lumen during de-etiolation when subjected to salt stress in the dark. However, after 4 hours of light treatment, the swelling disappeared, and the normal thylakoid arrangement was discernible (Ounoki et al. 2021, 2023). The isosmotic PEG solution did not show the swelling and had only a slight effect on plastid structure and function, which prompted the authors to suggest that ionic, not the osmotic component, is responsible for swollen thylakoids (Ounoki et al. 2023).

8.10.1 Heat Stress

Heat stress caused increased lipid fluidity in the membranes of *Arabidosis* etioplasts that showed higher amounts of Pchlide and carotenoid accumulation in them and relatively higher levels of short-wavelength Pchlide form (Pchlide 630) (Savchenko et al. 2002). Heat stress (HS) leads to a decrease in the size of PLBs and an increase in the size of etioplasts and length of PTs (Savchenko et al. 2002). The amount of POR and the reduction of phlide to chlide didn't change in the PLBs of HS plants, but their membranes showed more protein fluorescence. For the fluorescent amino-acid residues, the hydrophobicity of the microenvironment remained the same

(Savchenko et al. 2002). The pigments in membranes of HS plants were present in the protein-lipid interface (Savchenko et al. 2002).

Heat and chill stress (CS) also slowed down the Shibata shift in wheat seedlings (Mohanty and Tripathy 2011). However, the shift was arrested more in HS than in CS. Salt-stressed wheat (Abdelkader et al. 2008) and rice seedlings (Turan and Tripathy 2015) also showed similar patterns during de-etiolation. Chl accumulation in Gingko also slowed down at 10 °C as compared to 20 °C when illuminated with 15 μmol m^{-2} s^{-1} light (Skribanek et al. 2008).

8.10.2 Heavy Metal Stress

Application of cadmium (Cd) to etiolated plants resulted in a decrease in the 657 nm phototransformable fluorescence emission peak and an increase in the 633 nm non-phototransformable peak, along with the appearance of a new peak at 614.5 nm at 77 K (Böddi et al. 1995). Irradiation resulted in the inhibition of the phototransformation of the 657 nm emission peak, although a small emission peak at 678 nm appeared. This inhibition was removed, and the alterations in the phototransformable/non-phototransformable fluorescence species were reverted when excess NADPH was supplied along with Cd (Böddi et al. 1995). In another work, 15 min incubation with 10^{-2} M Hg^{2+} was found to completely inhibit the activity of the POR enzyme (Plant biology, < 2008). Hg^{2+} also reacted directly with Pchlide to form protopheophorbide (Solymosi et al. 2004a). At even lower Hg^{2+} concentration of 10^{-3} M, the phototransformation was complete and faster, as evident from a faster blue shift of the chlorophyllide emission band, indicating the oxidation of the NADPH molecules (Solymosi et al. 2004a). Lower efficiency of Pchlide to Chlide phototransformation as a result of Cd(II) (most pronounced), Cr(VI) or Fe(III) (lowest) treatment was discovered in the wheat seedlings, due to lesser accumulation of phototransformable Pchlide (655 nm emitting) (Myśliwa-Kurdziel and Strzałka 2005). The Chlide fluorescence peak of metal-treated samples was also blue-shifted (highest for Cd(II), medium for Fe(III) and lowest for Cr(VI)) as compared to untreated samples and depended on the concentration as well as on the time of incubation with metal ions (Myśliwa-Kurdziel and Strzałka 2005).

8.10.3 Pathogen Stress

The pathogen infection is also known to alter the chlorophyll biosynthesis during greening. The amount of POR in etioplasts of leaves of barley infected with Barley stripe mosaic virus (BSMV) was found to be reduced (Harsanyi et al. 2006); as a result, the Chl accumulation was also compromised, giving the characteristic yellow strip appearance. Etioplasts in these seedlings contained 26% less Pchlide, smaller PLBs with irregular membranes but more PTs (Harsányi et al. 2002). The ratio of photoactive (655 nm emitting) to non-photoactive Pchlide forms (645 and 633) was also reduced in these plants (Harsányi et al. 2002). The Shibata shift was also delayed by 40 min (Harsányi et al. 2002).

8.10.4 Drought Stress

When the plants germinate inside the soil and are subjected to water stress during the skotomorphogenic development, their response is important for the survival of seedlings. This determines the germination percentage and seedling establishment for the crops; therefore it is also important from a commercial point of view. There are some studies wherein the stress is provided to the seedlings, and plants are exposed to light after a certain period of stress. Dalal and Tripathy (2012) observed reduced Chl biosynthesis and reduced Chl biosynthesis intermediates, transcripts, proteins and activities of certain Chl biosynthesis pathway enzymes (Fig. 8.3). Biosynthesis of delta-Amino Levulinic Acid (δ-ALA) was reduced; however, out of two responsible enzymes for its synthesis (GluTR and GSAT), protein level of one (GSAT) was increased. Similarly, reduced GSA and/or ALA content had been observed in heat/chill/salt stressed maize, rice, cucumber and Pinus seedlings (Hodgins and Van Huystee 1986; Tewari and Tripathy 1998; Hodgins and Oquist 1989; Turan and Tripathy 2015). Other enzymes of this pathway, e.g., UroD, Cpo, Protoporphyrinogen oxidase, Magnesium chelatase subunit I (CHLI), Geranyl-geranyl reductase and POR, were also reduced due to waster stress (Dalal and Tripathy 2012). When the stress is already provided to the seedlings in the dark, they downregulate the photosynthetic apparatus (since the seedlings generate the photosynthetic apparatus de novo), as per their capacity to utilize the incident light to prevent photo-inhibitory losses. Therefore, the abundance of photosynthetic proteins Cyt b559, Lhcb1, Lhcb2, Cyt f, PSI subunit VI (PsaH), PSI subunit V (PsaK), Lhca1, Lhca4, Oxygen evolving complex 33 (OEC33), OEC23, OEC16 in water stressed seedlings was significantly lower than the control seedlings (Fig. 8.3), although the PsaE subunit showed contrastingly higher abundance suggesting its other roles in stress conditions (Dalal and Tripathy 2018). The thylakoid development, including granal stacking, gets affected due to alterations in these proteins. The development of stacking in the grana during seedling greening under water stress conditions is not as good as during normal greening, as can be seen from Fig. 8.4. Similar results can be seen under chill, heat and salt stresses (Tewari and Tripathy 1998; Turan and Tripathy 2015). The PLBs are disorganized when detached leaves from the dark-grown seedlings are floated on salt solution and transferred to light for four hours (Ounoki et al. 2023). Granum size and photosynthesis significantly decreased when they were grown in light and similarly floated on salt solution in light (Ounoki et al. 2021, 2023).

Shibata shift was delayed by some minutes due to water stress in the rice seedlings, during de-etiolation (Dalal and Tripathy 2012). The ratio of non-phototransformable/phototransformable Pchlide (F632/F657) increased in these

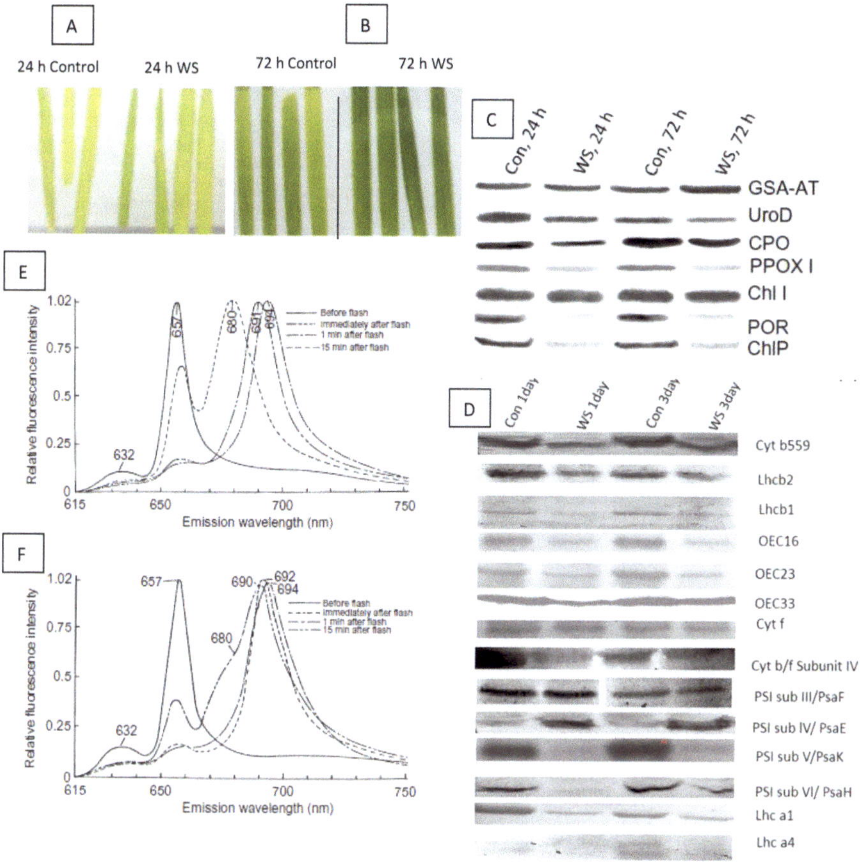

Fig. 8.3 De-etiolation of rice seedlings. (**a**) Rice seedlings after 24 h of greening, (**b**) Rice seedlings after 72 h of greening, (**c**) Immunoblots of Chl biosynthesis enzymes (**d**) Immunoblots of photosynthetic proteins, (**e**) Shibata shift in control and (**f**) water stressed seedlings after 24 h and 72 h of greening/de-etiolation. Water stress (WS) was provided to seedlings by immersing their roots in (50 mM PEG6000 MS nutrient solution) 16 h prior to their transfer to the light of 100 μmol photons m^{-2} s^{-1}; seedlings were maintained for 3 days at 28 °C and at 75% relative. (Source: Dalal and Tripathy 2012, 2018)

seedlings. Le Lay et al. (2000, 2001) did not see any change in the Phlide to Chlide phototransformation in dessicated plucked etiolated wheat leaves after 90 min of light flash, although they observed the formation of a new 692 nm emitting Chlide and higher formation of 676 nm emitting Chlide, so-called "free Chlide." Authors also observed inhibition of the photoactive Pchlide regeneration process (Le Lay et al. 2000, 2001). After rehydration, greening of the dehydrated resurrection plants has also been studied as an equivalent of de-etiolation.

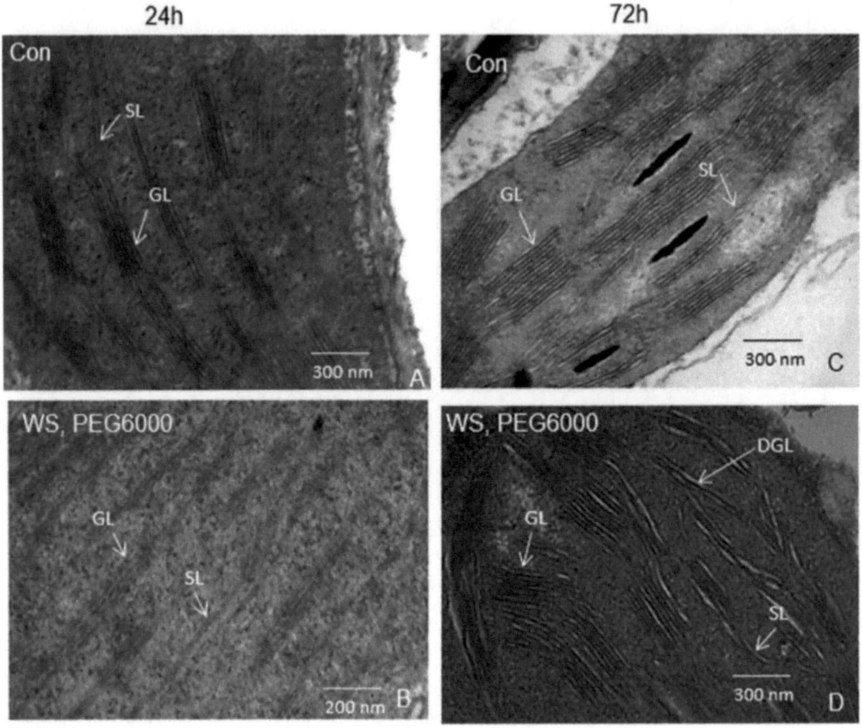

Fig. 8.4 Ultrastructure of chloroplasts. Transmission electron micrograph depicting the ultrastructure of control (Con) and water-stressed (WS) rice seedlings after 24 h and 72 h of greening/de-etiolation. Water stress was provided to seedlings by immersing their roots in (50 mM PEG6000 MS nutrient solution), 16 h prior to their transfer to the light of 100 μmol photons m^{-2} s^{-1}; seedlings were maintained for 3 days at 28 °C and at 75% relative humidity. SL Stromal lamellae, GL: Granal lamellae; DGL: Disintegrated granal lamellae. (Source: Dalal and Tripathy 2018)

8.10.5 Salt Stress

In the etiolated leaf segments of wheat, kept in different salt solutions under darkness for 1.5 hours, the etioplast to chloroplast transition and Chl accumulation were fully inhibited on 600 mM NaCl, $NaNO_3$ or NaCl:KCl (1:1) solutions but there was only slightly slower transition on 300 mM NaCl or $CaCl_2$, 600 mM KNO_3 or KCl solutions indicating that the ionic component is responsible for inhibition of greening in etiolated wheat seedlings instead of osmotic component (Sóti et al. 2023). Swollen prothylakoids were also observed by others under high salt concentration (Abdelkader et al. 2007). Inhibition of greening due to salt stress was also observed in rice seedlings (Turan and Tripathy 2013).

8.10.6 Diethyl Ether

Diethyl ether affects de-etiolation, positioning the plants in skoto-photomorphogenetic state by inhibiting the phytochrome and retrograde signalling, Chl biosynthesis, expression of PhANGs, although photoreduction of Pchlide to Chlide was not affected (Pavlovič et al. 2024).

8.10.7 Greening of Resurrection Plants

When dehydrated, poikilochlorophyllous resurrection plants disintegrate their Chls and thylakoid membranes, turning their chloroplasts into dessicoplasts (Farrant and Moore 2011). Rehydration of these plants such as *Xerophyta humilis* shows similarities with the etioplast–chloroplast transition (Ingle et al. 2008). In the mature region of the leaf, after one week of rehydration, pregranal plastids form that contain PTs and nonphotoactive Pchlide whereas in the basal region of the leaf, etioplasts containing photoactive Pchlide and PLBs are formed after 2 weeks, which transform into normal chloroplasts upon three days of illumination (Solymosi et al. 2013).

8.11 Conclusions and Future Perspectives

Etioplasts are very important organelles of plants. They determine the health of crops when the seedlings germinate and emerge from the soil. It is very important for establishing healthy plants and future yield. Although biophysical studies, evaluating the conversion of Pchlide-Chlide-Chl pigments and POR protein in the process of de-etiolation have been studied since many decades, the role of other proteins regulating de-etiolation is being determined only recently. Recently, there has also been a boost in tomography studies during de-etiolation. The soil is characterized by various abiotic and biotic stresses. The stress during de-etiolation can affect seedling health; therefore, it is very pertinent and of commercial value to study the de-etiolation process under various stresses. Therefore, some recent studies have focused on de-etiolation under various stresses, e.g., heavy metals (Cd, Cr, Hg, etc.), salt, drought, temperature and heat. A recent systems-level exploration pointed out the timeframe during which transcripts, proteins and structural changes take place in the etiolated leaves. An investigation of the metabolomics of etioplasts is further required. In the future, the differences between the translocons of various plastids should also be looked into so as to analyze how plastids (etioplasts) majorly import the NADPH as compared to other metabolites to form an abundance of Pchlide-POR-NADPH aggregates. Moreover, it should also be investigated how, in such a short time, big anatomical changes that lead to the conversion of the PLBs into thylakoids take place and if such a fast-acting mechanism could be transferred to some other process in cells. Besides this, it should be investigated whether chloroplastic proteins are already translated in the cytosol of etioplasts and upon illumination, immediately imported into the transitioning etio-chloroplasts or the sysnthesis of all the proteins which are absent in the etioplasts but present in chloroplasts are upregulated in unison, to maintain the stoichiometry of proteins in the chloroplasts.

References

Abdelkader AF, Aronsson H, Solymosi K et al (2007) High salt stress induces swollen prothylakoids in dark-grown wheat and alters both prolamellar body transformation and reformation after irradiation. J Exp Bot 58:2553–2564. https://doi.org/10.1093/jxb/erm085

Abdelkader AF, Aronsson H, Solymosi K, Böddi B, Sundqvist C (2008) Chlorophyll accumulation, protochlorophyllide formation and prolamellar body conversion are held back in wheat leaves exposed to high salt stress. In: Photosynthesis. Energy from the Sun: 14th International Congress on Photosynthesis. Springer, Netherlands, pp 1133–1136

Adamska I, Kloppstech K (1991) Evidence for an association of the early light-inducible protein (ELIP) of pea with photosystem II. Plant Mol Biol 16:209–223

Altamura MM, Piacentini D, Della Rovere F, Fattorini L, Valletta A, Falasca G (2024) Plastid dynamism integrates development and environment. Plant Physiol Biochem 108813

Apel K, Santel H-J, Redlinger TE, Falk H (1980) The protochlorophyllide holochrome of barley. lsolation and characterization of the NADPH:protochlorophyllide oxidoreductase. Eur J Biochem 111:251–258

Apostolova EL, Misra AN (2014) Alterations in structural organization affect the functional ability of photosynthetic apparatus. Handbook of Plant and Crop Physiology:103–120

Armarego-Marriott T, Kowalewska Ł, Burgos A, Fischer A, Thiele W, Erban A et al (2019) Highly resolved systems biology to dissect the etioplast-to-chloroplast transition in tobacco leaves. Plant Physiol 180(1):654–681

Armbruster U et al (2010) The Arabidopsis thylakoid protein PAM68 is required for efficient D1 biogenesis and photosystem II assembly. Plant Cell 22:3439–3460

Balakhonova et al (2023) iReenCAM: automated imaging system for kinetic analysis of photosynthetic pigment biosynthesis at high spatiotemporal resolution during early deetiolation. Front Plant Sci 14:1093292. https://doi.org/10.3389/fpls.2023.1093292

Banaś AK, Leja K, Zgłobicki P, Jedynak P, Kowalska E, Strzałka W et al (2024) De-etiolation is almost colour blind: the study of photosynthesis awakening under blue and red light. Plant Cell Physiol 65(12):1993–2017

Belkhadir Y, Jaillais Y (2015) The molecular circuitry of brassinosteroid signaling. New Phytol 206(2):522–540

Belyaeva OB, Litvin FF (2014) Mechanisms of Phototransformation of Protochlorophyllide into Chlorophyllide. Biochem Mosc 79(4):337348

Barros JA, Magen S, Lapidot-Cohen T, Rosental L, Brotman Y, Araújo WL, & Avin-Wittenberg T (2021) Autophagy is required for lipid homeostasis during dark-induced senescence. Plant Physiol. 185(4):1542–1558

Björn LO, Shevela D, Govindjee G (2023) What is photosynthesis?—a broader and inclusive view. A closer look at photosynthesis. In: Dalal VK, Misra AN (eds) A closer look at photosynthesis. © 2023 Nova Science Publishers, IncISBN: 979-8-88697-815-5

Blomqvist LA, Ryberg M, Sundqvist C (2008) Proteomic analysis of highly purified prolamellar bodies reveals their significance in chloroplast development. Photosynth Res 96:37–50

Böddi B, Kis-Petik K, Kaposi AD, Fidy J, Sundqvist C (1998) The two spectroscopically different short wavelength protochlorophyllide forms in pea epicotyls are both monomeric. Biochim Biophys Acta (BBA) Bioenerg 1365(3):531–540

Böddi B, Lindsten A, Ryberg M, Sundqvist C (1989) On the aggregational states of protochlorophyllide and its protein complexes in wheat etioplasts. Physiol Plant 76(2):135–143

Böddi B, Lindsten A, Ryberg M, Sundqvist C (1990) Phototransformation of aggregated forms of protochlorophyllide in isolated etioplast inner membranes. Photochem Photobiol 52(1):83–87
Böddi B, Oravecz AR, Lehoczki E (1995) Effect of cadmium on organization and photoreduction of protochlorophyllide in dark-grown leaves and etioplast inner membrane preparations of wheat. Photosynthetica 31(3):411–420
Böddi B, Ryberg M, Sundqvist C (1991) The formation of a short-wavelength chlorophyllide form at partial phototransformation of protochlorophyllide in etioplast inner membranes. Photochem Photobiol 53:667–673
Böddi B, Ryberg M, Sundqvist C (1992) Identification of four universal protochlorophyllide forms in dark-grown leaves by analyses of the 77 K fluorescence emission spectra. J Photochem Photobiol B Biol 12(4):389–401
Bu TT, Shen J, Chao Q, Shen Z, Yan Z, Zheng HY, Wang BC (2017) Dynamic N-glycoproteome analysis of maize seedling leaves during de-etiolation using Concanavalin A lectin affinity chromatography and a nano-LC–MS/MS-based iTRAQ approach. Plant Cell Rep 36:1943–1958
Bykowski M, Mazur R, Buszewicz D, Szach J, Mostowska A, Kowalewska Ł (2020) Spatial nanomorphology of the prolamellar body in etiolated Arabidopsis thaliana plants with disturbed pigment and polyprenol composition. Front Cell Develop Biol 8:586628
Cazzonelli CI, Pogson BJ (2010) Source to sink: regulation of carotenoid biosynthesis in plants. Trends Plant Sci 15:265–274
Cazzonelli CI, Hou X, Alagoz Y, Rivers J, Dhami N, Lee J, Marri S, Pogson BJ (2020) A cis-carotene derived apocarotenoid regulates etioplast and chloroplast development. elife 9:e45310. https://doi.org/10.7554/eLife.45310
Chang CSJ, Li YH, Chen LT, Chen WC, Hsieh WP, Shin J et al (2008) LZF1, a HY5-regulated transcriptional factor, functions in Arabidopsis de-etiolation. Plant J 54(2):205–219
Charuvi D, Kiss V, Nevo R, Shimoni E, Adam Z, Reich Z (2012) Gain and loss of photosynthetic membranes during plastid differentiation in the shoot apex of Arabidopsis. Plant Cell 24(3):1143–1157
Chen GH, Liu MJ, Xiong Y, Sheen J, Wu SH (2018) TOR and RPS6 transmit light signals to enhance protein translation in deetiolating Arabidopsis seedlings. Proc Natl Acad Sci 115(50):12823–12828
Choi SW, Ryu MY, Viczián A, Jung HJ, Kim GM, Arce AL et al (2020) Light triggers the miRNA-biogenetic inconsistency for de-etiolated seedling survivability in Arabidopsis thaliana. Mol Plant 13(3):431–445
Chory J, Nagpal P, Peto CA (1991) Phenotypic and genetic analysis of det2, a new mutant that affects light-regulated seedling development in Arabidopsis. Plant Cell 3(5):445–459
Chory J, Reinecke D, Sim S, Washburn T, Brenner M (1994) A role for cytokinins in de-etiolation in Arabidopsis (det mutants have an altered response to cytokinins). Plant Physiol 104(2):339–347
Chotewutmontri P, Barkan A (2016) Dynamics of chloroplast translation during chloroplast differentiation in maize. PLoS Genet 12(7):e1006106. https://doi.org/10.1371/journal.pgen.1006106
Cortleven A, Marg I, Yamburenko MV, Schlicke H, Hill K, Grimm B et al (2016) Cytokinin regulates the etioplast-chloroplast transition through the two-component signaling system and activation of chloroplast-related genes. Plant Physiol 172(1):464–478
Cortleven A, Schmülling T (2015) Regulation of chloroplast development and function by cytokinin. J Exp Bot 66(16):4999–5013
Da XW, Li L, Wang X, Pang HL, Jia LY, Sun K et al (2023) The accumulation of chlorophyll and changes of ATP during greening of etiolated Arabidopsis seedlings are mediated by nitric oxide. Biol Plant 67:184–188
Dalal VK, Tripathy BC (2012) Modulation of chlorophyll biosynthesis by water stress in rice seedlings during chloroplast biogenesis. Plant Cell Environ 35(9):1685–1703
Dalal V, Dagan S, Friedlander G, Aviv E, Bock R, Charuvi D et al (2018) Transcriptome analysis highlights nuclear control of chloroplast development in the shoot apex. Sci Rep 8(1):8881
Dalal VK, Tripathy BC (2018) Water-stress induced downsizing of light-harvesting antenna complex protects developing rice seedlings from photo-oxidative damage. Sci Rep 8(1):5955

Daum B, Kühlbrandt W (2011) Electron tomography of plant thylakoid membranes. J Exp Bot 62(7):2393–2402

Di Baccio D, Quartacci MF, Vecchia FD, La Rocca N, Rascio N, Navari-Izzo F (2002) Bleaching herbicide effects on plastids of dark-grown plants: lipid composition of etioplasts in amitrole and norflurazon-treated barley leaves. J Exp Bot 53(376):1857–1865

Doroshenko AS, Malyukova AM, Danilova MN, Kuznetsov VV, Kusnetsov VV (2022) Transcription factors of the glrs family are involved in cytokinin-dependent regulation of plastid RNA polymerase SCA3 gene expression during deetiolation of *Arabidopsis thaliana*. In: Doklady Biochemistry and Biophysics, vol 506(1). Pleiades Publishing, Moscow, pp 195–201

Dujardin E, Sironval C (1977) Transitory pigment-protein complexes similar to photosynthesis active centers during protochlorophyll(ide) photoreduction. Plant Sci Lett 10:347–355

El Hamouri B, Brouers M, Sironval C (1981) Pathway from photoinactive P633−628 protochlorophyllide to the P696−682 chlorophyllide in cucumber etioplast suspensions. Plant Sci Lett 21(4):375–379

Endo K, Kobayashi K, Wada H (2016) Sulfoquinovosyldiacylglycerol has an essential role in Thermosynechococcus elongatus BP-1 under phosphate-deficient conditions. Plant Cell Physiol pcw159

Erdei AL, Kósa A, Kovács-Smirová L, Böddi B (2016) Wavelength-dependent photooxidation and photoreduction of protochlorophyllide and protochlorophyll in the innermost leaves of cabbage (*Brassica oleracea* var. *capitata* L.). Photosynth Res 128:73–83

Farrant JM, Moore JP (2011) Programming desiccation-tolerance: from plants to seeds to resurrection plants. Curr Opin Plant Biol 14(3):340–345

Fitchette A-C, Dinh OT, Faye L, Bardor M (2007) Plant proteomics and glycosylation. In: Thiellement H, Zivy M, Damerval C, Me'chin V (eds) Plant proteomics: methods and protocols. Humana Press, Totowa, pp 317–342. https://doi.org/10.1385/1-59745-227-0:317

Floris D, Kühlbrandt W (2021) Molecular landscape of etioplast inner membranes in higher plants. Nat Plants 7(4):514–523

Franck F, Barthélemy X, Strzaeka K (1993) Spectroscopic characterization of protochlorophyllide photoreduction in the greening leaf. Photosynthetica-PRAHA 29:185–185

Grzyb JM, Solymosi K, Strzałka K, Mysliwa-Kurdziel B (2013) Visualization and characterization of prolamellar bodies with atomic force microscopy. J Plant Physiol 170(14):1217–1227

Franck F, Bereza B, Böddi B (1999) Protochlorophyllide-NADP+ and protochlorophyllide-NADPH complexes and their regeneration after flash illumination in leaves and etioplast membranes of dark-grown wheat. Photosynth Res 59:53–61

Franck F, Sperling U, Frick G, Pochert B, van Cleve B, Apel K, Armstrong GA (2000) Regulation of etioplast pigment-protein complexes, inner membrane architecture, and protochlorophyllide a chemical heterogeneity by light-dependent NADPH: protochlorophyllide oxidoreductases A and B. Plant Physiol 124(4):1678–1696

Fujii S, Kobayashi K, Nagata N, Masuda T, Wada H (2018) Digalactosyldiacylglycerol is essential for organization of the membrane structure in etioplasts. Plant Physiol 177(4):1487–1497

Fujii S, Kobayashi K, Nagata N, Masuda T, Wada H (2017) Monogalactosyldiacylglycerol facilitates synthesis of photoactive protochlorophyllide in etioplasts. Plant Physiol 174(4):2183–2198

Fujii S, Nagata N, Masuda T, Wada H, Kobayashi K (2019a) Galactolipids are essential for internal membrane transformation during etioplast-to-chloroplast differentiation. Plant Cell Physiol 60(6):1224–1238

Fujii S, Wada H, Kobayashi K (2019b) Role of galactolipids in plastid differentiation before and after light exposure. Plants 8(10):357

Fukazawa H, Tada A, Richardson LG, Kakizaki T, Uehara S, Ito-Inaba Y, Inaba T (2020) Induction of TOC and TIC genes during photomorphogenesis is mediated primarily by cryptochrome 1 in Arabidopsis. Sci Rep 10(1):20255

Gabruk M, Mysliwa-Kurdziel B (2020) The origin, evolution and diversifcation of multiple isoforms of light-dependent protochlorophyllide oxidoreductase (LPOR): focus on angiosperms. Biochem J 477:2221–2236

Gabruk M, Mysliwa-Kurdziel B, Kruk J (2017) MGDG, PG and SQDG regulate the activity of light-dependent protochlorophyllide oxidoreductase. Biochem J 474(7):1307–1320

Gabruk M, Stecka A, Strzałka W, Kruk J, Strzałka K, Mysliwa-Kurdziel B (2015) Photoactive protochlorophyllide-enzyme complexes reconstituted with PORA, PORB and PORC proteins of A. thaliana: fluorescence and catalytic properties. PLoS One 10:e0116990

Gabruk M, Nowakowska Z, Skupien-Rabian B, Kedracka-Krok S, Mysliwa Kurdziel B, Kruk J (2016) Insight into the oligomeric structure of PORA from *A. thaliana*. Biochim Biophys Acta Proteins Proteom 1864:1757–1764

Gangappa SN, Botto JF (2016) The multifaceted roles of HY5 in plant growth and development. Mol Plant 9(10):1353–1365

Gao ZF, Shen Z, Chao Q, Yan Z, Ge XL, Lu T et al (2020) Large-scale proteomic and phosphoproteomic analyses of maize seedling leaves during de-etiolation. Genom Proteom Bioinform 18(4):397–414

Griffiths WT (1978) Reconstitution of chlorophyll formation by isolated etioplast membranes. Biochem J 174:681–692

Gunning B (2001) Membrane geometry of "open" prolamellar bodies. Protoplasma 215:4–15. https://doi.org/10.1007/BF01280299

Hamamoto K, Aki T, Shigyo M, Sato S, Ishida T, Yano K et al (2012) Proteomic characterization of the greening process in rice seedlings using the MS spectral intensity-based label free method. J Proteome Res 11(1):331–347

Hartings S, Paradies S, Karnuth B, Eisfeld S, Mehsing J, Wolff C et al (2017) The DnaJ-like zinc-finger protein HCF222 is required for thylakoid membrane biogenesis in plants. Plant Physiol 174(3):1807–1824

Harsányi A, Böddi B, Bóka K, Almási A, Gáborjányi R (2002) Abnormal etioplast development in barley seedlings infected with BSMV by seed transmission. Physiol Plant 114(1):149–155

Harsanyi A, Ryberg M, Andersson MX, Boka K, Laszlo L, Botond G et al (2006) Alterations of NADPH: protochlorophyllide oxidoreductase quantity and lipid composition in etiolated barley seedlings infected by Barley stripe mosaic virus (BSMV). Mol Plant Pathol 7(6):533–541

Hernández-Verdeja T, Vuorijoki L, Strand A (2020) Emerging from the darkness: interplay between light and plastid signaling during chloroplast biogenesis. Physiol Plant 169(3):397–406

Heyes DJ, Menon BR, Sakuma M, Scrutton NS (2008) Conformational events during ternary enzyme-substrate complex formation are rate limiting in the catalytic cycle of the light-driven enzyme protochlorophyllide oxidoreductase. Biochemistry 47:10991–10998

Heyes DJ, Zhang S, Taylor A, Johannissen LO, Hardman SJ, Hay S, Scrutton NS (2021) Photocatalysis as the 'master switch' of photomorphogenesis in early plant development. Nat Plants 7(3):268–276

Hideg É, Vitányi B, Kósa A, Solymosi K, Bóka K, Won S et al (2010) Reactive oxygen species from type-I photosensitized reactions contribute to the light-induced wilting of dark-grown pea (*Pisum sativum*) epicotyls. Physiol Plant 138(4):485–492

Hodgins RRW, Öquist G (1989) Porphyrin metabolism in chill-stressed seedlings of Scots pine (*Pinus sylvestris*). Physiol Plant 77(4):620–624

Hodgins RR, Van Huystee RB (1986) Delta-aminolevulinic acid metabolism in chill stressed maize (Zea mays L.). J Plant Physiol 126(2–3):257–268

Holtorf H, Reinbothe S, Reinbothe C, Bereza B, Apel K (1995) Two routes of chlorophyllide synthesis that are differentially regulated by light in barley. Proc Natl Acad Sci USA 92:3254–3258

Hwang Y, Han S, Yoo CY, Hong L, You C, Le BH et al (2022) Anterograde signaling controls plastid transcription via sigma factors separately from nuclear photosynthesis genes. Nat Commun 13(1):7440

Ikeda T (1971) Prolamellar body formation under different light and temperature conditions. Shokubutsugaku Zasshi 84(996):363–375

Ingle RA, Collett H, Cooper K, Takahashi Y, Farrant JM, Illing N (2008) Chloroplast biogenesis during rehydration of the resurrection plant Xerophyta humilis: parallels to the etioplast–chloroplast transition. Plant Cell Environ 31(12):1813–1824

Islam MT, Wang LC, Chen IJ, Lo KL, Lo WS (2021) Arabidopsis JMJ17 promotes cotyledon greening during de-etiolation by repressing genes involved in tetrapyrrole biosynthesis in etiolated seedlings. New Phytol 231(3):1023–1039

Islam W, Waheed A, Idrees A, Rashid J, Zeng F (2023) Role of plant microRNAs and their corresponding pathways in fluctuating light conditions. Biochim Biophys Acta (BBA) Molecul Cell Res 1870(2):119304

Jain P et al (2022) Crop Plants, Abiotic Stress, Reactive Oxygen Species Production, Signaling, and Their Consequences. In: Ansari, S.A., Ansari, M.I., Husen, A. (eds) Augmenting Crop Productivity in Stress Environment. Springer, Singapore. https://doi.org/10.1007/978-981-16-6361-1_7

Jedynak P, Trzebuniak KF, Chowaniec M, Zgłobicki P, Banaś AK, Mysliwa-Kurdziel B (2022) Dynamics of etiolation monitored by seedling morphology, carotenoid composition, antioxidant level, and photoactivity of protochlorophyllide in *Arabidopsis thaliana*. Front Plant Sci 12:772727

Jin X, Baysal C, Drapal M, Sheng Y, Huang X, He W et al (2021) The Coordinated upregulated expression of genes involved in MEP, chlorophyll, carotenoid and tocopherol pathways, mirrored the corresponding metabolite contents in rice leaves during de-etiolation. Plants 10(7):1456

Kakuszi A, Böddi B (2014) Light piping activates chlorophyll biosynthesis in the under-soil hypocotyl section of bean seedlings. J Photochem Photobiol B Biol 140:1–7

Kakuszi A, Sarvari E, Solti A et al (2016) Light piping driven photosynthesis in the soil: low-light adapted active photosynthetic apparatus in the under-soil hypocotyl segments of bean (*Phaseolus vulgaris*). J Photochem Photobiol B Biol 161:422–429. https://doi.org/10.1016/j.jphotobiol.2016.06.009

Kakuszi A, Solymosi K, Böddi B (2017) Transformation of plastids in soil-shaded lowermost hypocotyl segments of bean (*Phaseolus vulgaris*) during a 60-day cultivation period. Physiol Plant 159(4):483–491

Kalaji HM, Schansker G, Brestic M, Bussotti F, Calatayud A, Ferroni L et al (2017) Frequently asked questions about chlorophyll fluorescence, the sequel. Photosynth Res 132:13–66

Kambakam S, Bhattacharjee U, Petrich J, Rodermel S (2016) PTOX mediates novel pathways of electron transport in etioplasts of Arabidopsis. Mol Plant 9(9):1240–1259

Kanervo E, Singh M, Suorsa M, Paakkarinen V, Aro E, Battchikova N et al (2008) Expression of protein complexes and individual proteins upon transition of etioplasts to chloroplasts in pea (*Pisum sativum*). Plant Cell Physiol 49:396–410. https://doi.org/10.1093/pcp/pcn016

Kato Y, Odahara M, Shikanai T (2021) Evolution of an assembly factor-based subunit contributed to a novel NDH-PSI supercomplex formation in chloroplasts. Nat Commun 12(1):3685

Kis-Petik K, Böddi B, Kaposi AD, Fidy J (1999) Protochlorophyllide forms and energy transfer in dark-grown wheat leaves. Studies by conventional and laser excited fluorescence spectroscopy between 10 K–100 K. Photosynth Res 60:87–98

Kleffmann T, Hirsch-Hoffmann M, Gruissem W, Baginsky S (2006) plprot: a comprehensive proteome database for different plastid types. Plant Cell Physiol 47:432–436. https://doi.org/10.1093/pcp/pcj005

Kobayashi K (2016) Role of membrane glycerolipids in photosynthesis, thylakoid biogenesis and chloroplast development. Journal of plant research, 129(4):565–580

Kobayashi K, Ohnishi A, Sasaki D, Fujii S, Iwase A, Sugimoto K et al (2017) Shoot removal induces chloroplast development in roots via cytokinin signaling. Plant Physiol 173(4):2340–2355

Koochak H, Puthiyaveetil S, Mullendore DL, Li M, Kirchhoff H (2019) The structural and functional domains of plant thylakoid membranes. Plant J 97:412–429. https://doi.org/10.1111/tpj.14127

Kósa A, Böddi B (2012) Dominance of a 675 nm chlorophyll (ide) form upon selective 632.8 or 654 nm laser illumination after partial protochlorophyllide phototransformation. Photosynth Res 114:111–120

Kósa A, Hideg É, Bóka K, Solti Á, Böddi B (2024) Light dependent differentiation of outdoors developed purple eggplant (*Solanum melongena* L.) pericarp layers: Leaf chlorenchyma characteristics of the pericarp layers dissected in the dark. Plant Physiol Biochem 207:108394

Kósa A, Márton Z, Böddi B (2005) Fast phototransformation of the 636 nm-emitting protochlorophyllide form in epicotyls of dark-grown pea (*Pisum sativum*). Physiol Plant 124(1):132–142

Kósa A, Márton Z, Solymosi K, Bóka K, & Böddi B (2006) Aggregation of the 636 nm emitting monomeric protochlorophyllide form into flash-photoactive, oligomeric 644 and 655 nm emitting forms in vitro. Biochimica et Biophysica Acta (BBA)-Bioenergetics, 1757(7):811–820

Kour J, Kohli SK, Khanna K, Bakshi P, Sharma P, Singh AD et al (2021) Brassinosteroid signaling, crosstalk and, physiological functions in plants under heavy metal stress. Front Plant Sci 12:608061

Kowalewska Ł, Bykowski M, Mostowska A (2019) Spatial organization of thylakoid network in higher plants. Bot Lett 166:326–343. https://doi.org/10.1080/23818107.2019.1619195

Kowalewska Ł, Mazur R, Suski S, Garstka M, Mostowska A (2016) Three-dimensional visualization of the tubular-lamellar transformation of the internal plastid membrane network during runner bean chloroplast biogenesis. Plant Cell 28(4):875–891

Kusnetsov VV, Doroshenko AS, Kudryakova NV, Danilova MN (2020) Role of phytohormones and light in de-etiolation. Russ J Plant Physiol 67:971–984

Kuttkat A, Edhofer I, Eichacker LA, Paulsen H (1997) Light-harvesting chlorophyll a/b-binding protein stably inserts into etioplast membranes supplemented with Zn-pheophytin a/b. J Biol Chem 272:20451–20455

Lan Y, Chen Q, Kong M, Liu Y, Lyu MJA, Perveen S, Mi H (2021) PetM is essential for the stabilization and function of the cytochrome b6f complex in Arabidopsis. Plant Cell Physiol 62(10):1603–1614

Laughlin TG, Savage DF, & Davies KM (2020) Recent advances on the structure and function of NDH-1: The complex I of oxygenic photosynthesis. Biochimica et Biophysica Acta (BBA)-Bioenergetics, 1861(11):148254

Le Lay P, Böddi B, Kovacevic D, Juneau P, Dewez D, Popovic R (2001) Spectroscopic analysis of desiccation-induced alterations of the chlorophyllide transformation pathway in etiolated barley leaves. Plant Physiol 127(1):202–211

Le Lay P, Eullaffroy P, Juneau P, Popovic R (2000) Evidence of chlorophyll synthesis pathway alteration in desiccated barley leaves. Plant Cell Physiol 41(5):565–570

Lebedev N, Timko MP (1999) Protochlorophyllide oxidoreductase B-catalyzed protochlorophyllide photoreduction in vitro: insight into the mechanism of chlorophyll formation in light-adapted plants. Proc Natl Acad Sci 96(17):9954–9959

Lei Y, Li B, Wang X, Wei J, Wang P, Zhao J et al (2023) Chloroplast SRP54 and FtsH protease coordinate thylakoid membrane-associated proteostasis in Arabidopsis. Plant Physiol 192(3):2318–2335

Lennartz K et al (2001) HCF164 encodes a thioredoxin-like protein involved in the biogenesis of the cytochrome b(6)f complex in Arabidopsis. Plant Cell 13:2539–2551

Li P, Ponnala L, Gandotra N, Wang L, Si Y, Tausta SL et al (2010) The developmental dynamics of the maize leaf transcriptome. Nat Genet 42(12):1060

Li J, Jin H (2007) Regulation of brassinosteroid signaling. Trends Plant Sci 12(1):37–41

Li N, Wong WS, Feng L, Wang C, Wong KS, Zhang N et al (2023) The thylakoid membrane protein NTA1 is an assembly factor of the cytochrome b6f complex essential for chloroplast development in Arabidopsis. Plant Communications 4(1)

Li Y, Varala K, Hudson ME (2014) A survey of the small RNA population during far-red light-induced apical hook opening. Front Plant Sci 5:156

Liang Z, Yeung WT, Ma J, Mai KKK, Liu Z, Chong YLF et al (2022) Electron tomography of prolamellar bodies and their transformation into grana thylakoids in cryofixed Arabidopsis cotyledons. Plant Cell 34(10):3830–3843

Lin MC, Tsai HL, Lim SL, Jeng ST, Wu SH (2017) Unraveling multifaceted contributions of small regulatory RNAs to photomorphogenic development in Arabidopsis. BMC Genomics 18:1–14

Lindsten A, Ryberg M, Sundqvist C (1988) The polypeptide composition of highly purified prolamellar bodies and prothylakoids from wheat (Triticum aestivum) as revealed by silver staining. Physiol Plant 72(1):167–176

Lindsten A, Wiktorsson B, Ryberg M, Sundqvist C (1993) Chlorophyll synthetase activity is relocated from transforming prolamellar bodies to developing thylakoids during irradiation of dark-grown wheat. Physiol Plant 88(1):29–36

Liu MJ, Wu SH, Chen HM, Wu SH (2012) Widespread translational control contributes to the regulation of Arabidopsis photomorphogenesis. Mol Syst Biol 8(1):566

Loudya N, Mishra P, Takahagi K, Uehara-Yamaguchi Y, Inoue K, Bogre L et al (2021) Cellular and transcriptomic analyses reveal two-staged chloroplast biogenesis underpinning photosynthesis build-up in the wheat leaf. Genome Biol 22(1):151

Lütz C (1981) Development and ageing of etioplast structures in dark grown leaves of *Avena sativa* (L.). Protoplasma 108:83–98

Ma L, Li J, Qu L, Hager J, Chen Z, Zhao H, Deng XW (2001) Light control of Arabidopsis development entails coordinated regulation of genome expression and cellular pathways. Plant Cell 13(12):2589–2607

Ma Z, Hu X, Cai W, Huang W, Zhou X, Luo Q et al (2014) Arabidopsis miR171-targeted scarecrow-like proteins bind to GT cis-elements and mediate gibberellin-regulated chlorophyll biosynthesis under light conditions. PLoS Genet 10(8):e1004519

Ma J, Liang Z, Zhao J, Wang P, Ma W, Mai K K, ... & Kang BH (2021) Friendly mediates membrane depolarizationinduced mitophagy in Arabidopsis. Current Biology, 31(9):1931–1944

Majeran W, Friso G, Ponnala L, Connolly B, Huang M, Reidel E et al (2010) Structural and metabolic transitions of C4 leaf development and differentiation defined by microscopy and quantitative proteomics in maize. Plant Cell 22:3509–3542. https://doi.org/10.1105/tpc.110.079764

Masuda S, Ikeda R, Masuda T, Hashimoto H, Tsuchiya T, Kojima H et al (2009) Prolamellar bodies formed by cyanobacterial protochlorophyllide oxidoreductase in Arabidopsis. Plant J 58(6):952–960

Mathews S (2010) Evolutionary studies illuminate the structural-functional model of plant phytochromes. Plant Cell 22(1):4–16

Mazur R, Mostowska A, Szach J, Gieczewska K, Wójtowicz J, Bednarska K et al (2019) Galactolipid deficiency disturbs spatial arrangement of the thylakoid network in Arabidopsis thaliana plants. J Exp Bot 70(18):4689–4704

Melo NK, Bianchetti RE, Lira BS, Oliveira PM, Zuccarelli R, Dias DL et al (2016) Nitric oxide, ethylene, and auxin cross talk mediates greening and plastid development in deetiolating tomato seedlings. Plant Physiol 170(4):2278–2294

Mohanty S, Tripathy BC (2011) Early and late plastid development in response to chill stress and heat stress in wheat seedlings. Protoplasma 248:725–736

Mohr H, Schäfer E (1983) Photoperception and de-etiolation. Philosophical Transactions of the Royal Society of London B, Biological Sciences 303(1116):489–501

Müller B, Eichacker LA (1999) Assembly of the D1 precursor in monomeric photosystem II reaction center precomplexes precedes chlorophyll a-triggered accumulation of reaction center II in barley etioplasts. Plant Cell 11:2365–2377

Myśliwa-Kurdziel B, Strzałka K (2005) Influence of Cd (II), Cr (VI) and Fe (III) on early steps of deetiolation process in wheat: fluorescence spectral changes of protochlorophyllide and newly formed chlorophyllide. Agric Ecosyst Environ 106(2–3):199–207

Nawrocki WJ, Tourasse NJ, Taly A, Rappaport F, Wollman FA (2015) The plastid terminal oxidase: its elusive function points to multiple contributions to plastid physiology. Annu Rev Plant Biol 66:49–74

Nguyen HC, Melo AA, Kruk J, Frost A, Gabruk M (2021) Photocatalytic LPOR forms helical lattices that shape membranes for chlorophyll synthesis. Nat Plants 7(4):437–444

Ouazzani Chahdi AM, Schoefs B, Franck F (1998) Isolation and characterization of photoactive complexes of NADPH: protochlorophyllide oxidoreductase from wheat. Planta 206:673–680

Ounoki R, Ágh F, Hembrom R, Ünnep R, Szögi-Tatár B, Böszörményi A, Solymosi K (2021) Salt stress affects plastid ultrastructure and photosynthetic activity but not the essential oil composition in spearmint (*Mentha spicata* L. var. *crispa* "Moroccan"). *Frontiers*. Plant Sci 12:739467

Ounoki R, Sóti A, Ünnep R, Sipka G, Sárvári É, Garab G, Solymosi K (2023) Etioplasts are more susceptible to salinity stress than chloroplasts and photosynthetically active etio-chloroplasts of wheat (*Triticum aestivum* L.). Physiol Plant 175(6):e14100

Oosawa N, Masuda T, Awai K, Fusada N, Shimada H et al (2000) Identification and light-induced expression of a novel gene of NADPH-protochlorophyllide oxidoreductase isoform in *Arabidopsis thaliana*. FEBS Lett 474:133–136

Park SJ, Choi SW, Kim GM, Møller C, Pai HS, Yang SW (2021) Light-stabilized FHA2 suppresses miRNA biogenesis through interactions with DCL1 and HYL1. Mol Plant 14(4):647–663

Pashkovskiy P, Ryazansky S, Kartashov A, Voloshin R, Khudyakova A, Kosobryukhov AA et al (2021) Effect of red light on photosynthetic acclimation and the gene expression of certain light signalling components involved in the microRNA biogenesis in the extremophile Eutrema salsugineum. J Biotechnol 325:35–42

Pavlovič A, Kopečná M, Hloušková L, Koller J, Hřivňacký M, Ilík P, Bartoš J (2024) Diethyl ether anaesthesia inhibits de-etiolation of barley seedlings by locking them in intermediate skoto-photomorphogenetic state. Physiol Plant 176(1):e14144

Philippar K, Geis T, Ilkavets I, Oster U, Schwenkert S, Meurer J, Soll J (2007) Chloroplast biogenesis: the use of mutants to study the etioplast–chloroplast transition. Proc Natl Acad Sci 104(2):678–683

Pick TR, Bräutigam A, Schlüter U, Denton AK, Colmsee C, Scholz U et al (2011) Systems analysis of a maize leaf developmental gradient redefines the current C4 model and provides candidates for regulation. Plant Cell 23(12):4208–4220. https://doi.org/10.1105/tpc.111.090324

Pipitone R, Eicke S, Pfister B, Glauser G, Falconet D, Uwizeye C et al (2021) A multifaceted analysis reveals two distinct phases of chloroplast biogenesis during de-etiolation in Arabidopsis. elife 10:e62709

Pless DD, Lennarz WJ (1977) Enzymatic Conversion of Proteins to Glycoproteins. Proc Natl Acad Sci USA 74:134–138. https://doi.org/10.1073/pnas.74.1.134

Plöscher M, Reisinger V, Eichacker LA (2011) Proteomic comparison of etioplast and chloroplast protein complexes. J Proteome 74:1256–1265

Plücken H, Müller B, Grohmann D, Westhoff P, Eichacker LA (2002) The HCF136 protein is essential for assembly of the photosystem II reaction center in Arabidopsis thaliana. FEBS Lett 532:85–90

Pollmann S, Springer A, Buhr F, Lahroussi A, Duret A, Bonneville JM, ... & Reinbothe S (2007) Plastid import of NADPH: protochlorophyllide oxidoreductase A is essential for seedling survival. Untersuchung und Charakterisierung des Lichtsammelkomplexes (LHPP) in etiolierten Pflanzen, 129

Ponnu J, Hoecker U (2021) Illuminating the COP1/SPA ubiquitin ligase: fresh insights into its structure and functions during plant photomorphogenesis. Front Plant Sci 12:662793

Pudelski B, Soll J, Philippar K (2009) A search for factors influencing etioplast–chloroplast transition. Proc Natl Acad Sci 106(29):12201–12206

Putarjunan A, Liu X, Nolan T, Yu F, Rodermel S (2013) Understanding chloroplast biogenesis using second-site suppressors of immutans and var2. Photosynth Res 116:437–453

Rabeh K, Oubohssaine M, Hnini M (2024) TOR in plants: Multidimensional regulators of plant growth and signaling pathways. J Plant Physiol 154186:154186

Rast A, Heinz S, Nickelsen J (2015) Biogenesis of thylakoid membranes. Biochimica et Biophysica Acta (BBA)-Bioenergetics 1847(9):821–830

Reisinger V, Hertle AP, Plöscher M, Eichacker LA (2008) Cytochrome b6f is a dimeric protochlorophyll a binding complex in etioplasts. FEBS J 275(5):1018–1024

Rochaix JD (2011) Assembly of the photosynthetic apparatus. Plant Physiol 155(4):1493–1500

Rodríguez-Villalón A, Gas E, Rodríguez-Concepción M (2009) Phytoene synthase activity controls the biosynthesis of carotenoids and the supply of their metabolic precursors in dark-grown Arabidopsis seedlings. Plant J 60:424–435

Ryberg M, Sandelius AS, Selstam E (1983) Lipid composition of prolamellar bodies and prothylakoids of wheat etioplasts. Physiol Plant 57(4):555–560

Sánchez-Retuerta C, Suaréz-López P, Henriques R (2018) Under a new light: regulation of light-dependent pathways by non-coding RNAs. Front Plant Sci 9:962

Sandoval-Ibáñez O, Rolo D, Ghandour R, Hertle AP, Armarego-Marriott T, Sampathkumar A et al (2022) De-etiolation-induced protein 1 (DEIP1) mediates assembly of the cytochrome b6f complex in Arabidopsis. Nat Commun 13(1):4045

Sandoval-Ibáñez O, Sharma A, Bykowski M, Borràs-Gas G, Behrendorff JB, Mellor S et al (2021) Curvature thylakoid 1 proteins modulate prolamellar body morphology and promote organized thylakoid biogenesis in *Arabidopsis thaliana*. Proc Natl Acad Sci 118(42):e2113934118

Mapleston ER, Griffiths WT (1980) Light modulation of the activity of the protochlorophyllide oxidoreductase. Biochem J 189:125–133

Sandelius AS, Selstam E (1984) Localization of galactolipid biosynthesis in etioplasts isolated from dark-grown wheat (*Triticum aestivum* L.). Plant Physiol 76(4):1041–1046

Santel HJ, Apel K (1981) The protochlorophyllide holochrome of barley. The effect of light on the NADPH-protochlorophyllide oxidoreductase. Eur J Biochem 120:95–103

Sato N, Sonoike K, Tsuzuki M, Kawaguchi A (1996) Photosynthetic characteristics of a mutant of Chlamydomonas reinhardtii impaired in fatty acid desaturation in chloroplasts. Biochimica et Biophysica Acta (BBA)-Bioenergetics 1274(3):112–118

Savchenko GE, Klyuchareva EA, Abramchik LM, Serdyuchenko EV (2002) Effect of periodic heat shock on the inner membrane system of etioplasts. Russ J Plant Physiol 49:349–359

Scheumann V, Schoch S, Ru¨diger, W. (1998) Chlorophyll a formation in the chlorophyll b reductase reaction requires reduced ferredoxin. J Biol Chem 273:35102–35108

Schneidewind J, Krause F, Bocola M, Stadler AM, Davari MD, Schwaneberg U et al (2019) Consensus model of a cyanobacterial light-dependent protochlorophyllide oxidoreductase in its pigment-free apo-form and photoactive ternary complex. Communications biology 2(1):351

Schoefs B (2001) The protochlorophyllide–chlorophyllide cycle. Photosynth Res 70:257–271

Schoefs B, Franck F (1991) Photosystem II assembly in 2-day-old bean leaves during the first 16 hrs. of greening. *Comptes Rendus de l'Académie des Sciences*. Série III Sciences de la Vie 313(10)

Schoefs B, Franck F (1993) Photoreduction of protochlorophyllide to chlorophyllide in 2-d-old dark-grown bean (Phaseolus vulgaris cv. Commodore) leaves. Comparison with 10-d-old dark-grown (etiolated) leaves. J Exp Bot 44(6):1053–1057

Schoefs B, Franck F (2008) The photoenzymatic cycle of NADPH: protochlorophyllide oxidoreductase in primary bean leaves (Phaseolus vulgaris) during the first days of photoperiodic growth. Photosynth Res 96:15–26

Schoefs B, Bertrand M, Franck F (2000) Spectroscopic properties of protochlorophyllide analyzed in situ in the course of etiolation and in illuminated leaves. Photochem Photobiol 72(1):85–93

Schweer J (2010) Plant sigma factors come of age: flexible transcription factor network for regulated plastid gene expression. Endocytobiosis & Cell Research, (20)

Schwenk P, Sheerin DJ, Ponnu J, Staudt AM, Lesch KL, Lichtenberg E et al (2021) Uncovering a novel function of the CCR4-NOT complex in phytochrome A-mediated light signalling in plants. elife 10:e63697

Selstam E, Sandelius AS (1984) A comparison between prolamellar bodies and prothylakoid membranes of etioplasts of dark-grown wheat concerning lipid and polypeptide composition. Plant Physiol 76(4):1036–1040

Selstam E, Widell-Wigge A (1989) Hydrophobicity of protochlorophyllide oxidoreductase, characterized by means of Triton X-114 partitioning of isolated etioplast membrane fractions. Physiol Plant 77(3):401–406

Shi Y, He Y, Lv X, Wei Y, Zhang X, Xu X, ... & Wu JL (2020) Chloroplast SRP54s are essential for chloroplast development in rice. Rice, 13(1):54

Shimoni E, Rav-Hon O, Ohad I, Brumfeld V, Reich Z (2005) Three-dimensional organization of higher-plant chloroplast thylakoid membranes revealed by electron tomography. Plant Cell 17(9):2580–2586

Shibata K (1957) Spectroscopic studies on chlorophyll formation in intact leaves. The Journal of Biochemistry, 44(3):147–173

Skribanek A, Apatini D, Inaoka M, Böddi B (2000) Protochlorophyllide and chlorophyll forms in dark-grown stems and stem-related organs. J Photochem Photobiol B Biol 55(2–3):172–177

Skribanek A, Böddi B (2001) Light-and cold-stress effects on the greening process in epicotyls and young stems of red oak (*Quercus rubra*) seedlings. Tree Physiol 21(8):549–554

Skribanek A, Solymosi K, Hideg É, Böddi B (2008) Light and temperature regulation of greening in dark-grown ginkgo (*Ginkgo biloba*). Physiol Plant 134(4):649–659

Smeller L, Solymosi K, Fidy J, Böddi B (2003) Activation parameters of the blue shift (Shibata shift) subsequent to protochlorophyllide phototransformation. Biochimica et Biophysica Acta (BBA)-Proteins and Proteomics 1651(1–2):130–138

Solymosi K, Aronsson H (2013) Etioplasts and their significance in chloroplast biogenesis. In: Plastid development in leaves during growth and senescence. Springer Netherlands, Dordrecht, pp 39–71

Solymosi K, Bóka K, Böddi B (2006) Transient etiolation: protochlorophyll (ide) and chlorophyll forms in differentiating plastids of closed and breaking leaf buds of horse chestnut (Aesculus hippocastanum). Tree Physiol 26(8):1087–1096

Solymosi K, Böddi B (2006) Optical properties of bud scales and protochlorophyll (ide) forms in leaf primordia of closed and opened buds. Tree Physiol 26(8):1075–1085

Solymosi K, Lenti K, Myśliwa-Kurdziel B, Fidy J, Strzałka K, Böddi B (2004a) Hg2+ reacts with different components of the NADPH: protochlorophyllide oxidoreductase macrodomains. Plant Biol 6(03):358–368

Solymosi K, Martinez K, Kristóf Z, Sundqvist C, Böddi B (2004b) Plastid differentiation and chlorophyll biosynthesis in different leaf layers of white cabbage (Brassica oleracea cv. capitata). Physiol Plant 121(3):520–529

Solymosi K, Mysliwa-Kurdziel B (2021) The role of membranes and lipid-protein interactions in the Mg-branch of tetrapyrrole biosynthesis. Front Plant Sci 12:663309

Solymosi K, Schoefs B (2008) Prolamellar body: a unique plastid compartment, which does not only occur in dark-grown leaves. *Plant cell organelles—selected topics*. Research Signpost Trivandrum:151–202

Solymosi K, Smeller L, Ryberg M, Sundqvist C, Fidy J, Böddi B (2007) Molecular rearrangement in POR macrodomains as a reason for the blue shift of chlorophyllide fluorescence observed after phototransformation. Biochimica et Biophysica Acta (BBA)-Biomembranes 1768(6):1650–1658

Solymosi K, Tuba Z, Böddi B (2013) Desiccoplast–etioplast–chloroplast transformation under rehydration of desiccated poikilochlorophyllous *Xerophyta humilis* leaves in the dark and upon subsequent illumination. J Plant Physiol 170(6):583–590

Song J, Cao K, Hao Y, Song S, Su W, Liu H (2019) Hypocotyl elongation is regulated by supplemental blue and red light in cucumber seedling. Gene 707:117–125

Somerville C, & Browse J (1991) Plant lipids: metabolism, mutants, and membranes. Science, 252(5002):80-87

Sóti A, Ounoki R, Kósa A, Mysliwa-Kurdziel B, Sárvári É, Solymosi K (2023) Ionic, not the osmotic component, is responsible for the salinity-induced inhibition of greening in etiolated wheat (*Triticum aestivum* L. cv. Mv Béres) leaves: a comparative study. Planta 258(5):102

Su Q, Frick G, Armstrong G, Apel K (2001) POR C of Arabidopsis thaliana: a third light- and NADPH-dependent protochlorophyllide oxidoreductase that is differentially regulated by light. Plant Mol Biol 47:805–813

Sun T, Zhou F, Huang XQ, Chen WC, Kong MJ, Zhou CF et al (2019) ORANGE represses chloroplast biogenesis in etiolated Arabidopsis cotyledons via interaction with TCP14. Plant Cell 31(12):2996–3014

Sun XD, Ni M (2011) Hyposensitive to light, an alpha/beta fold protein, acts downstream of Elongated Hypocotyl 5 to regulate seedling de-etiolation. Mol Plant 4(1):116–126

Sun Z, Li M, Zhou Y, Guo T, Liu Y, Zhang H, Fang Y (2018) Coordinated regulation of Arabidopsis microRNA biogenesis and red light signaling through Dicer-like 1 and phytochrome-interacting factor 4. PLoS Genet 14(3):e1007247

Sundqvist C, Dahlin C (1997) With chlorophyll pigments from prolamellar bodies to light-harvesting complexes. Physiol Plant 100(4):748–759

Sundqvist C, Ryberg H (1979) Structure of protochlorophyll-containing plastids in the inner seed coat of pumpkin seeds (Cucurbita pepo). Physiol Plant 47(2):124–128

Sundqvist C, Ryberg H, Böddi B, Láng F (1980) Spectral properties of a long-wavelength absorbing form of protochlorophyll in seeds of Cyclanthera explodens. Physiol Plant 48(2):297–301

Suzuki N, Devireddy AR, Inupakutika MA, Baxter A, Miller G, Song L et al (2015) Ultra-fast alterations in mRNA levels uncover multiple players in light stress acclimation in plants. Plant J 84(4):760–772

Szenzenstein A, Kósa A, Solymosi K, Sárvári É, & Böddi B (2010) Preferential regeneration of the NADPH: protochlorophyllide oxidoreductase oligomer complexes in pea epicotyls after bleaching. Physiologia Plantarum, 138(1):102–112

Tang G, Reinhart BJ, Bartel DP, Zamore PD (2003) A biochemical framework for RNA silencing in plants. Genes Dev 17(1):49–63

Tewari AK, Tripathy BC (1998) Temperature-stress-induced impairment of chlorophyll biosynthetic reactions in cucumber and wheat. Plant Physiol 117(3):851–858

Toman E, Käs MD, Taubert B, Moritz M, Qi J, Bolle C (2024) Photosynthetic and plastid performance effects of photoreceptors. J Mitochondria Plastids Endosymbiosis 2(1):2330972

Tripathy BC, Dalal V (2013) Modulation of chlorophyll biosynthesis by environmental cues. In: Biswal B, Krupinska K, Biswal U (eds) Plastid Development in Leaves during Growth and Senescence. Advances in Photosynthesis and Respiration, vol 36. Springer, Dordrecht. https://doi.org/10.1007/978-94-007-5724-0_27

Tripathy BC, Pattanayak GK (2012) Chlorophyll biosynthesis in higher plants. Photosynthesis: plastid biology, energy conversion and carbon assimilation:63–94

Tsai HL, Li YH, Hsieh WP, Lin MC, Ahn JH, Wu SH (2014) HUA ENHANCER1 is involved in posttranscriptional regulation of positive and negative regulators in Arabidopsis photomorphogenesis. Plant Cell 26(7):2858–2872

Turan S, Tipathy BC (2013) Salt and genotype impact on antioxidative enzymes and lipid peroxidation in two rice cultivars during de-etiolation. Protoplasma 250(1):209–222. https://doi.org/10.1007/s00709-012-0395-5

Turan S, Tripathy BC (2015) Salt-stress induced modulation of chlorophyll biosynthesis during de-etiolation of rice seedlings. Physiol Plant 153(3):477–491

Viczián A, Klose C, Hiltbrunner A, Nagy F (2021) Plant Phytochromes: From Structure to Signaling and Beyond. Front Plant Sci 12:811379

von Zychlinski A, Kleffman T, Krishnamurthy N, Sjölander K, Baginsky S, Gruissem W (2005) Proteome analysis of the rice etioplast: Metabolic and regulatory networks and novel protein functions. Mol Cell Proteomics 4:1072–1084. https://doi.org/10.1074/mcp.M500018-MCP200

Wang L, Czedik-Eysenberg A, Mertz RA, Si Y, Tohge T, Nunes-Nesi A et al (2014a) Comparative analyses of C(4) and C(3) photosynthesis in developing leaves of maize and rice. Nat Biotechnol 32(11):1158–1165. https://doi.org/10.1038/nbt.3019

Wang Q, Lin C (2020) Mechanisms of cryptochrome-mediated photoresponses in plants. Annu Rev Plant Biol 71(1):103–129

Wang X, Deng X, Zhu D, Duan W, Zhang J, Yan Y (2021) N-linked glycoproteome analysis reveals central glycosylated proteins involved in wheat early seedling growth. Plant Physiol Biochem 163:327–337

Wang YF, Chao Q, Li Z, Lu TC, Zheng HY, Zhao CF et al (2019) Large-scale identification and time-course quantification of ubiquitylation events during maize seedling de-etiolation. Genomic Proteomic Bioinformatic 17(6):603–622

Wang Y, Fan X, Lin F, He G, Terzaghi W, Zhu D, Deng XW (2014b) Arabidopsis noncoding RNA mediates control of photomorphogenesis by red light. Proc Natl Acad Sci 111(28):10359–10364

Wang Y, Liu S, Tian X, Fu Y, Jiang X, Li Y, Wang G (2018) Influence of light intensity on chloroplast development and pigment accumulation in the wild-type and etiolated mutant plants of *Anthurium andraeanum* 'Sonate'. Plant Signal Behav 13(8):e1482174

Wietrzynski W, Schaffer M, Tegunov D, Albert S, Kanazawa A, Plitzko JM, Baumeister W, Engel BD (2020) Charting the native architecture of Chlamydomonas thylakoid membranes with single-molecule precision. eLife 9:e53740. https://doi.org/10.7554/eLife.53740

Wijerathna-Yapa A, Signorelli S, Fenske R, Ganguly DR, Stroeher E, Li L et al (2021) Autophagy mutants show delayed chloroplast development during de-etiolation in carbon limiting conditions. Plant J 108(2):459–477

Wiktorsson, B., Engdahl, S., Zhong, L. B., Böddi, B., Ryberg, M., & Sundqvist, C. (1993). The effect of cross-linking of the subunits of NADPH-protochlorophyllide oxidoreductase on the aggregational state of protochlorophyllide

Wójtowicz J, Węgrzyn A, Bykowski M, Bukat A, Mostowska A, Kowalewska Ł (1996) 3 Prolamellar body—getting ready for photosynthesis. In: Handbook of photosynthesis. CRC Press, pp 3–23

Wu HP, Su YS, Chen HC, Chen YR, Wu CC, Lin WD, Tu SL (2014) Genome-wide analysis of light-regulated alternative splicing mediated by photoreceptors in Physcomitrella patens. Genome Biol 15:1–18

Wu SH (2014) Gene expression regulation in photomorphogenesis from the perspective of the central dogma. Annu Rev Plant Biol 65(1):311–333

Xiao J et al (2012) DAC is involved in the accumulation of the cytochrome b6/f complex in Arabidopsis. Plant Physiol 160:1911–1922

Yadav A, Singh D, Lingwan M, Yadukrishnan P, Masakapalli SK, Datta S (2020) Light signaling and UV-B-mediated plant growth regulation. J Integr Plant Biol 62(9):1270–1292

Yan Z, Shen Z, Gao ZF, Chao Q, Qian CR, Zheng H, Wang BC (2020b) A comprehensive analysis of the lysine acetylome reveals diverse functions of acetylated proteins during de-etiolation in *Zea mays*. J Plant Physiol 248:153158

Yan Z, Shen Z, Li Z, Chao Q, Kong L, Gao ZF et al (2020a) Genome-wide transcriptome and proteome profiles indicate an active role of alternative splicing during de-etiolation of maize seedlings. Planta 252:1–19

Yang H, Liu J, Wen X, Lu C (2015) Molecular mechanism of photosystem I assembly in oxygenic organisms. Biochimica et Biophysica Acta (BBA)-Bioenergetics 1847(9):838–848

Yang EJ, Yoo CY, Liu J, Wang H, Cao J, Li FW et al (2019) NCP activates chloroplast transcription by controlling phytochrome-dependent dual nuclear and plastidial switches. Nat Commun 10(1):2630

Yoo CY, Pasoreck EK, Wang H, Cao J, Blaha GM, Weigel D, Chen M (2019) Phytochrome activates the plastid-encoded RNA polymerase for chloroplast biogenesis via nucleus-to-plastid signaling. Nat Commun 10(1):2629

Yoshihara A, Kobayashi K, Nagata N, Fujii S, Wada H, Kobayashi K (2024) Anionic lipids facilitate membrane development and protochlorophyllide biosynthesis in etioplasts. Plant Physiol 194(3):1692–1704

Younis S, Ryberg M, Sundqvist C (1995) Plastid development in germinating wheat (*Triticum aestivum*) is enhanced by gibberellic acid and delayed by gabaculine. Physiol Plant 95(3):336–346

Yu CW, Lin YT, Li HM (2020) Increased ratio of galactolipid MGDG: DGDG induces jasmonic acid overproduction and changes chloroplast shape. New Phytol 228(4):1327–1335

Yu X, Li L, Zola J, Aluru M, Ye H, Foudree A et al (2011) A brassinosteroid transcriptional network revealed by genome-wide identification of BESI target genes in Arabidopsis thaliana. Plant J 65(4):634–646

Zagari N, Sandoval-Ibañez O, Sandal N, Su J, Rodríguez-Concepcion M, Stougaard J et al (2017) Snowy Cotyledon 2 promotes chloroplast development and has a role in leaf variegation in both *Lotus japonicus* and *Arabidopsis thaliana*. Mol Plant 10(5):721–734

Zhan W, Cui L, Yang S, Geng J, Shi Y, Chen S et al (2024) Light induces the circadian rhythm and chloroplast development during seedling de-etiolation in maize. Environ Exp Bot 226:105935

Zhang D et al (2011) HCF243 encodes a chloroplast-localized protein involved in the D1 protein stability of the Arabidopsis photosystem II complex. Plant Physiol 157:608–619

Zhang L, Ruan J, Gao F, Xin Q, Che LP, Cai L et al (2024) Thylakoid protein FPB1 synergistically cooperates with PAM68 to promote CP47 biogenesis and Photosystem II assembly. Nat Commun 15(1):3122

Zhang L, Pu H, Duan Z, Li Y, Liu B, Zhang Q et al (2018a) Nucleus-encoded protein BFA1 promotes efficient assembly of the chloroplast ATP synthase coupling factor 1. Plant Cell 30(8):1770–1788

Zhang Y, Zhang Y, McFarlane HE, Obata T, Richter AS, Lohse M et al (2018b) Inhibition of TOR represses nutrient consumption, which improves greening after extended periods of etiolation. Plant Physiol 178(1):101–117

Zhang S, Godwin ARF, Taylor A, Hardman SJO, Jowitt TA, Johannissen LO et al (2021) Dual role of the active site 'lid' regions of protochlorophyllide oxidoreductase in photocatalysis and plant development. FEBS J 288:175–189. https://doi.org/10.1111/febs.15542

Zhang S, Godwin ARF, Taylor A, Hardman SJO, Jowitt TA, Johannissen LO, Hay S, Baldock C, Heyes DJ, Scrutton NS (2020) Dual role of the active site 'lid' regions of protochlorophyllide oxidoreductase in photocatalysis and plant development. FEBS J. https://doi.org/10.1111/febs.15542

Zhang S, Heyes DJ, Feng L, Sun W, Johannissen LO, Liu H et al (2019) Structural basis for enzymatic photocatalysis in chlorophyll biosynthesis. Nature 574(7780):722–725

Zhou J, Liu Q, Zhang F, Wang Y, Zhang S, Cheng H et al (2014) Overexpression of OsPIL15, a phytochrome-interacting factor-like protein gene, represses etiolated seedling growth in rice. J Integr Plant Biol 56(4):373–387

Chromoplast Biogenesis: A Journey from Chloroplasts

9

Nur Ardiyana Rejab, Purabi Mazumdar,
and Najiah Mohd. Sadali

Abstract

Chromoplasts are carotenoid-containing plastids. They primarily form through chloroplast-to-chromoplast transition, involving structural modifications such as thylakoid membrane dismantling, chlorophyll degradation, and the formation of plastoglobules for pigment storage.

At the molecular level, chromoplast differentiation is controlled by complex regulatory networks and dynamic regulation of plastid protein import systems through TOC/TIC complexes. Environmental factors such as light, temperature, and abiotic stressors further regulate this transition, influencing carotenoid biosynthesis and stability.

Chromoplasts contribute to plant defence and stress resilience with carotenoids acting as antioxidants and precursors to signalling molecules like abscisic acid (ABA) and volatile organic compounds (VOCs), which mediate drought resistance, herbivore defence, and pathogen response. The antioxidant properties of carotenoids further protect plant cells from reactive oxygen species (ROS).

Future research on chromoplast differentiation and other plastid functions will expand our understanding of the mechanisms involved in these processes.

N. A. Rejab
Centre for Research in Biotechnology for Agriculture (CEBAR), Universiti Malaya, Kuala Lumpur, Malaysia

Institute of Biological Sciences, Faculty of Science, Universiti Malaya, Kuala Lumpur, Malaysia
e-mail: ardiyana@um.edu.my

P. Mazumdar · N. M. Sadali (✉)
Centre for Research in Biotechnology for Agriculture (CEBAR), Universiti Malaya, Kuala Lumpur, Malaysia
e-mail: purabi@um.edu.my; najiahms@um.edu.my

V. K. Dalal, A. N. Misra (eds.), *Chloroplast Biogenesis and Plastid Interconversions*,
https://doi.org/10.1007/978-981-95-0772-6_9

This knowledge has the potential to support the development of resilient, nutrient-rich crops suited to evolving agricultural and environmental challenges.

Keywords

Chromoplast biogenesis · Chloroplast-to-chromoplast transition · Carotenoid biosynthesis · OR protein · TOC/TIC complexes · Stress adaptation · Plastid differentiation · Metabolic engineering · Crop improvement

9.1 Introduction

9.1.1 Overview of Chromoplast and Other Plastid Types

Plastids are a diverse group of double-membrane organelles. They perform essential metabolic and physiological functions in plant cells. Plastids begin as proplastids, undifferentiated precursor organelles found in meristematic tissues. Based on their roles, developmental stages, and environmental context, plastids differentiate into several specialised types.

Chloroplasts are the most well-known type of plastid. They are essential for photosynthesis. Their structure supports this function. Thylakoids contain chlorophyll, which is used in light-dependent reactions. The stroma is the fluid-filled region where the Calvin cycle happens. The Calvin cycle is a series of biochemical reactions that fix atmospheric carbon dioxide (CO_2) into organic molecules. This cycle has three main phases: carbon fixation, where CO2 is combined with ribulose-1,5-bisphosphate (RuBP) by the enzyme RuBisCO to form 3-phosphoglycerate (3-PGA); reduction, where Adenosine triphosphate (ATP) and Nicotinamide Adenine Dinucleotide Phosphate Hydrogen (NADPH) (Calvin and Benson 1948; Bassham et al. 1954; Raines 2003) crucial to produce carbohydrates, which serve as the primary energy source for plant cells. In addition to their photosynthetic role, chloroplasts are involved in the biosynthesis of a range of essential compounds, including fatty acids, amino acids, and secondary metabolites that are critical for plant defence and stress responses. The development and maintenance of chloroplasts are tightly regulated through a coordinated interplay between nuclear and plastid genomes, ensuring that their function is precisely integrated into the overall cellular metabolism.

Non-photosynthetic plastids, like leucoplasts and chromoplasts, are also important for plant physiology. Leucoplasts are usually colourless. They are involved in the biosynthesis and storage of key macromolecules. Depending on their function, leucoplasts development can be triggered by diverse development processes to generate starch, lipid, and protein-enriched sub-types called amyloplasts (starch storage), elaioplasts (lipid storage), or proteinoplasts (protein storage). Chromoplasts, in contrast, are distinguished by their red, orange, or yellow colouration due to the accumulation of carotenoid pigments. These specialised non-photosynthetic plastids are primarily found in non-photosynthetic tissues where they contribute to nutrient storage and mobilization in various plant organs such as roots, tubers, seeds, and storage of pigments that give bright colours in fruits, flowers, and roots. To accomplish these diverse functions, different types of plastids have different proteomes (Daher et al. 2010; Stensballe et al. 2008; Pinheiro et al. 2013; Wang et al. 2013; Siddique et al. 2006; Barsan et al. 2012), which is achieved through import and degradation of different proteins via the TOC and TIC (translocon at the outer/inner envelope membrane of chloroplasts) machinery, depending on the requirement of the tissues and developmental stages (Chu and Li 2015, 2018; Chu et al. 2020; Rödiger et al. 2021; Ling et al. 2021).

Many non-photosynthetic plastids have unique interconversions. This chapter focuses on the transition of chromoplasts from chloroplasts. Chromoplasts usually develop from existing chloroplasts. However, they can also form from proplastids or leucoplasts, depending on the tissue and developmental stage (Cazzonelli and Pogson 2010; Ruiz-Sola and Rodríguez-Concepción 2012). The transition from chloroplasts to chromoplasts is well-characterised during fruit ripening (Egea et al. 2010, 2011; Spurr and Harris 1968; Bathgate et al. 1985). As chloroplasts change into chromoplasts, the thylakoid membranes are dismantled. This results in a decrease in chlorophyll levels, causing green discolouration. At the same time, carotenoid biosynthesis increases. These pigments are stored in specialised substructures, including plastoglobules, fibrils or crystals. This gives the plant a red, orange, or yellow colour. Although the inner envelope membrane of the plastid stays intact, internal membrane remodelling is essential. It helps accommodate new structures and support carotenoid storage. The expansion of internal membrane compartments contributes to increased plastid volume, accommodating the growing carotenoid-rich structures. As chromoplast development continues, plastoglobules increase in number and size because of carotenoid ester accumulation. In some chromoplasts, like those in carrot roots, carotenoids crystallise within the plastid stroma (Oleszkiewicz et al. 2018; Kim et al. 2010). In tomato fruits, carotenoids accumulate as lycopene crystalloids in membranous sacs, either dissolved or lipid-associated (Klee and Giovannoni 2011; Harris and Spurr 1969). Figure 9.1 describes a schematic diagram of the chloroplast to chromoplast transition.

During the transition from chloroplasts to chromoplasts, plastid density decreases because of chlorophyll degradation. At the same time, chromoplasts enlarge as carotenoids accumulate (Egea et al. 2011; Rosso 1968; Hadjeb et al. 1988). This is their main role in pigment storage and synthesis. Carotenoids are isoprenoid pigments made through the plastidial methylerythritol phosphate (MEP) pathway. Starting from pyruvate and G3P, the MEP pathway creates isoprene units. These include isopentenyl diphosphate (IPP) and dimethylallyl diphosphate (DMAPP), which are used to form geranylgeranyl diphosphate (GGPP), a precursor for carotenoid biosynthesis. IPP and DMAPP are condensed to form GGPP, which is the immediate precursor for carotenoid synthesis. GGPP is converted into phytoene by phytoene synthase (PSY), marking the first step in carotenoid biosynthesis. Phytoene is then desaturated by phytoene desaturase (PDS) and carotene desaturase (ZDS). It

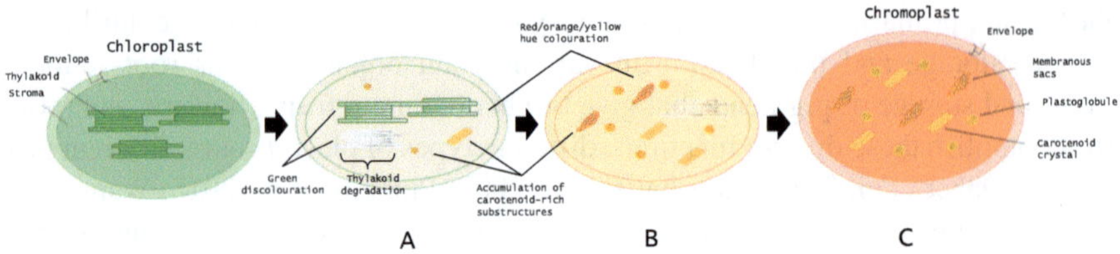

Fig. 9.1 Schematic diagram of chloroplast to chromoplast transition. The transition involves extensive plastid remodelling, leading to carotenoid accumulation and loss of photosynthetic function. The key stages include; (**a**) degradation of chlorophyll and thylakoid membrane (represented by green discolouration), and accumulation of carotenoid-rich substructures including plastoglobules and crystalline deposits (represented by accumulation of red/orange/yellow hue colouration); (**b**) the synthesis of new membrane structures form the inner membrane envelope of the plastid and the appearance of carotenoid-rich membranous sacs, followed by the increase in number and size of plastoglobules; and (**c**) the expansion of internal membrane contributes to increase plastid volume, accommodating the accumulated carotenoids

is also isomerized by lycopene β-cyclase (LCYB) and lycopene ε-cyclase (LCYE) to form various carotenoids, leading to the production of β-Carotene, α-Carotene, and eventually to end zeaxanthin, lutein, and abscisic acid (ABA) (Cazzonelli and Pogson 2010; Ruiz-Sola and Rodríguez-Concepción 2012). Figure 9.2 describes a schematic representation of isoprenoid biosynthesis in plants.

In chloroplasts, carotenoids are essential for light harvesting and photoprotection. They help maintain the stability and function of the photosynthetic apparatus. They absorb light energy to drive photosynthetic electron transport and mitigate oxidative damage caused by excess light and reactive oxygen species (Della Penna and Pogson 2006). During chromoplast differentiation, carotenoid pigments are stored in specialised structures like plastoglobules and fibrils. This compartmentalisation protects carotenoids from oxidative damage, improving their stability, bioavailability, and nutritional value (Ruiz-Sola and Rodríguez-Concepción 2012; Cazzonelli and Pogson 2010). The accumulation of carotenoids in chromoplasts is crucial for the vivid red, orange, and yellow hues characteristic of many fruits and flowers, which serve to attract pollinators and seed dispersers. Chromoplast development is typically associated with fruit ripening and maturation, as well as flower differentiation. In these processes, chromoplasts contribute to the visual and sensory attributes of plant organs, thereby facilitating pollinator attraction and seed dispersal (Ling et al. 2021; Cazzonelli and Pogson 2010). The biosynthesis and accumulation of specific carotenoids are regulated by key enzymes such as phytoene synthase, phytoene desaturase, and lycopene β-cyclase. The activities of these enzymes are tightly controlled by developmental cues and environmental factors (Cazzonelli and Pogson 2010; Ruiz-Sola and Rodríguez-Concepción 2012). The resulting pigment profiles in chromoplasts not only define tissue pigmentation but also have significant implications for human nutrition and health due to their antioxidant properties and roles as precursors to essential vitamins (Cazzonelli and Pogson 2010; Ruiz-Sola and Rodríguez-Concepción 2012).

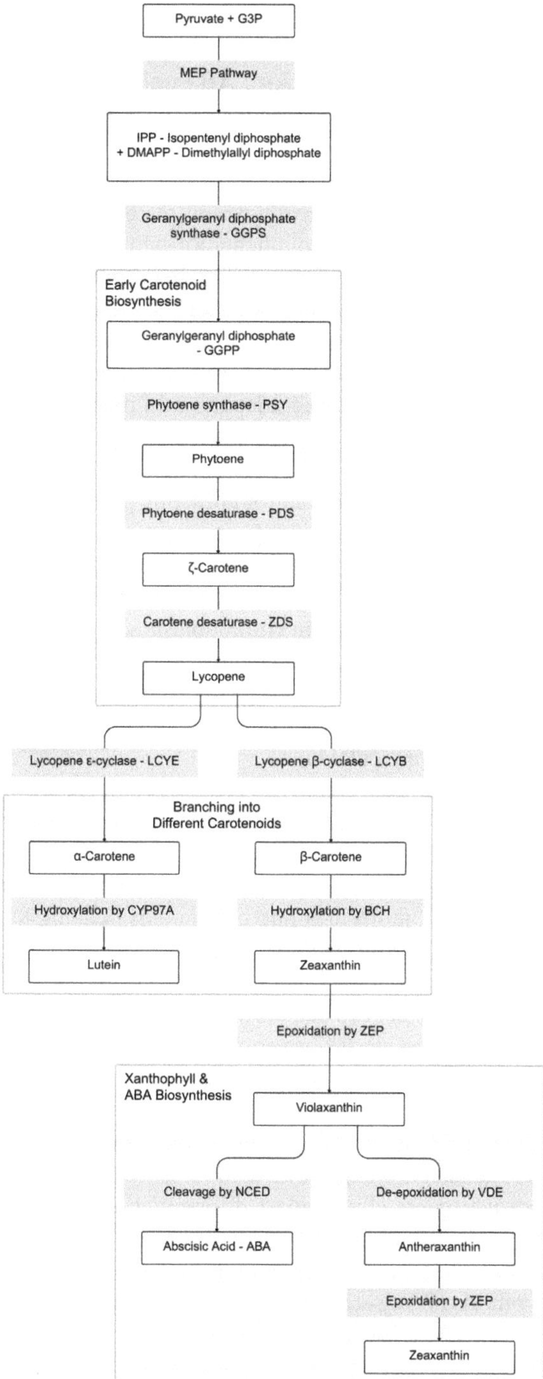

Fig. 9.2 Simplified schematic representation of isoprenoid biosynthesis in plants. MEP (methylerythritol phosphate) pathway in plastid begins with the condensation of G3P (glyceraldehyde-3-phosphate) and pyruvate in equal amounts, leading to the production of β-Carotene, α-Carotene, and eventually leading to end products such as Zeaxanthin, Lutein, and Abscisic Acid (ABA). The key enzymes involved in each step are IPP, DMAPP, GGPS, PSY, PDS, ZDS, LCYB, LCYE, BCH, CYP97A, ZEP, VDE, and NCED

9.1.2 Chromoplast Differentiation: An Evolutionary Strategy for Plant Reproductive Success and Stress Resilience

Chromoplast differentiation is an important evolutionary strategy for plants. It helps improve reproductive success and environmental resilience. Chromoplasts, which accumulate carotenoids, play roles in pigmentation and visual signalling. They also protect plants from oxidative stress and affect plant-animal interactions. Understanding the molecular mechanisms underlying chromoplast differentiation and its evolutionary significance will contribute to knowledge on plant defence mechanisms, herbivore-plant interactions, and stress resilience.

Chromoplast differentiation is a key adaptation in plants, supporting reproductive success and ecological fitness. The transformation of plastids into chromoplasts allows the accumulation of carotenoids, pigments that give vivid colour to flowers and fruits. These pigments are important for attracting pollinators and seed dispersers, which are essential for reproduction. Carotenoids stored in chromoplasts also signal the quality and ripeness of plant tissues. This increases the chances of seed dispersal and supports the ecological interactions that sustain plant populations. Besides their visual role, carotenoids in chromoplasts serve as precursors to important signalling molecules like ABA. ABA regulates seed dormancy and helps plants respond to environmental stresses. Chromoplast differentiation gives plants an evolutionary advantage. It helps them coordinate reproductive and stress-related processes in response to environmental cues.

Plastid differentiation into chromoplasts involves complex changes in molecular and physiological processes. This includes changes in pigment composition, membrane structure, and proteome dynamics. Central to this process is the carotenoid biosynthesis, which is triggered by key enzymes like PSY, LCYB and PDS. These enzymes convert precursor molecules into carotenoids, leading to their accumulation in the chromoplast. The differentiation process also involves major changes in plastid membranes. Thylakoid membranes break down, and new structures form to store carotenoids (Sun et al. 2018; Yuan et al. 2015b; Li and Yuan 2013). Along with these structural changes, the plastid proteome is reprogrammed. Proteins involved in carotenoid biosynthesis and stress responses increase, while those related to photosynthesis decrease (Ling et al. 2021). Chromoplast differentiation is triggered by both external and internal factors. Light, temperature, and hormonal signals, especially ethylene, influence chromoplast formation. Ethylene is important in climacteric fruits, where it triggers chromoplast differentiation during ripening (Müller and Munné-Bosch 2021). Additionally, environmental factors such as temperature also modulate carotenoid accumulation, with lower temperatures typically enhancing carotenoid synthesis (Wang et al. 2022; Ding and Yang 2022; Farneti et al. 2012).

Besides their role in reproduction, chromoplasts help plants resist stress. Carotenoids stored in chromoplasts act as antioxidants, protecting plant cells from oxidative damage caused by stressors like drought, heat, and UV radiation. By scavenging reactive oxygen species (ROS), carotenoids reduce oxidative stress, protect cellular integrity, and help overall plant health during tough conditions (Muñoz and Munné-Bosch 2018; Caverzan et al. 2016; Ding and Yang 2022; Gill and Tuteja

2010; Bouvier et al. 1998). Chromoplasts also interact with plant stress signalling pathways. For example, carotenoids are precursors to ABA, a key regulator for plant responses to abiotic stresses. ABA mediates processes such as stomatal closure, seed dormancy, and stress tolerance, further underscoring the role of chromoplasts in coordinating plant responses to environmental changes (Cazzonelli and Pogson 2010; Liu et al. 2015a). Recent studies suggest that chromoplasts help with heat stress responses. Proteins like Orange (Or) in chromoplasts act as chaperones, stabilising the photosystem II complex under high temperature conditions and improving heat tolerance (Sun et al. 2023). This ability to support photosynthesis under stress further shows the resilience of chromoplasts.

Chromoplasts also help with defence by combining antioxidant activities and regulating signalling pathways that control plant responses to herbivore attacks and other biotic stresses. Carotenoids in chromoplasts are strong antioxidants. They can neutralise ROS created during herbivore feeding or environmental stress. By reducing oxidative damage, carotenoids protect plant cells from the harmful effects of herbivory, helping maintain plant health during stressful periods (Kerchev et al. 2012). The bright colours produced by chromoplasts can also deter herbivores. Brightly coloured tissues may signal that the plant is toxic or less palatable, discouraging herbivores from feeding on it (Su et al. 2022). Although this defence is indirect, it can greatly reduce feeding pressure on plants. Chromoplasts may also affect chemical defence mechanisms by producing volatile organic compounds (VOCs). These VOCs can be released when an herbivore attacks. They help attract natural predators of herbivores or trigger defence responses in nearby plants (Meena et al. 2017). The carotenoids stored in chromoplasts may be involved in the biosynthesis of these VOCs, linking chromoplast differentiation with broader plant defence networks.

Recent advances in genetic engineering have made it possible to manipulate chromoplast differentiation. This can improve crop quality and stress resilience. For example, transgenic tomatoes with higher levels of lycopene β-cyclase and β-carotene hydroxylase show increased carotenoid accumulation, improving nutritional value and antioxidant properties (Dharmapuri et al. 2002; Karniel et al. 2020). Carotenoid accumulation in plants is mainly controlled at the transcriptional level by regulating biosynthetic gene expression. For example, the expression of *PSY-1* and *DXS* is linked to high colour intensity, which strongly influences carotenoid formation in ripening fruit, as seen in tomato, pepper, and chilli pepper (Berry et al. 2019; Lois et al. 2000; Fraser et al. 2000; Huh et al. 2001). Besides transcriptional regulation, carotenoid accumulation is also affected by post-transcriptional mechanisms. This explains the weak link between gene expression patterns and carotenoid content. For example, the OR proteins interact directly with PSY, the rate-limiting enzyme in carotenoid biosynthesis, to regulate its protein level post-transcriptionally. This interaction boosts PSY activity and carotenoid accumulation without changing *PSY* gene expression (Zhou et al. 2015; Yuan et al. 2015a).

Manipulating zeaxanthin, a carotenoid pigment essential for both plant and human health, through transgenic metabolic engineering or traditional genetic breeding has greatly improved food sources with high antioxidant and nutritional value, benefiting human health. In tomatoes, transgenic expression of lycopene β-cyclase and β-carotene hydroxylase led to the accumulation of zeaxanthin and β-cryptoxanthin in fruit (Dharmapuri et al. 2002). The development of "Xantomato" showed that carotenoid biosynthesis in tomatoes can be manipulated without genetic modification. This was done using a quadruple homozygous mutant $B^{Sh}/hp3/gs/hp2^{dg}$, which affects carotenoid biosynthesis, achieving 50% zeaxanthin in total fruit carotenoids (Karniel et al. 2020). Other successful transgenic plants with accumulated zeaxanthin include potatoes and corn (Diretto et al. 2007; Römer et al. 2002; Naqvi et al. 2011; Zhu et al. 2008).

Chromoplast differentiation is a complex evolutionary strategy that improves plant reproductive success, stress resilience, and defence against herbivores. Through carotenoid accumulation and the regulation of metabolic and signalling pathways, chromoplasts contribute to plant fitness and survival across a range of environmental conditions. Advances in genetic engineering and breeding programs targeting chromoplast differentiation promise to improve crop quality, nutritional value, and resilience. This will help ensure agricultural sustainability in the face of global environmental challenges.

9.2 Chromoplast Differentiation

Chromoplasts are specialised plastids that mainly synthesize and store carotenoids. Carotenoids are isoprenoid pigments responsible for the red, orange, and yellow colours in many plant tissues. Chromoplast differentiation from chloroplasts involves major structural and metabolic changes. This includes chlorophyll degradation, the breakdown of thylakoid membranes, and the formation of substructures like plastoglobules, fibrils, or crystals to store carotenoids (Egea et al. 2010, 2011). These substructures allow chromoplasts to store large amounts of carotenoids and protect them from oxidative damage. We will discuss the structural organization of chromoplasts in Sect. 9.2.1.

Chromoplasts are considered the most recently evolved plastid type, adapting to specific ecological and physiological stimuli (Kuntz and Rolland 2012). This adaptation includes major changes in plastid protein content, reprogramming of nuclear gene expression, and regulation by hormones, epigenetic modifications, and post-translational changes. These processes help chromoplasts perform specialised functions related to pigmentation and secondary metabolism (Llorente et al. 2020). We will discuss the complex regulatory framework for chromoplast differentiation in Sect. 9.2.2, which is influenced by various environmental and internal stimuli. We will explain the role of protein import machinery, which is key to chromoplast differentiation, in Sect. 9.3.

Chromoplast differentiation is important during fruit ripening and flower development. It contributes to the visual and sensory qualities of plant organs. The accumulation of specific carotenoids, like lycopene in tomatoes and β-carotene in carrots, gives distinct colouration and improves the nutritional value of these tissues. Carotenoids are precursors to essential vitamins like vitamin A and have strong antioxidant properties that reduce oxidative stress in human cells. Therefore, regulating chromoplast differentiation is important for both plant health and human nutrition. We will provide a more detailed explanation on carotenoid biosynthesis, regulation, and functions in Sect. 9.4.

9.2.1 Structural Diversity and Functional Adaptations of Chromoplasts

Chromoplasts show remarkable structural diversity. Each type is optimised for storing, stabilising, and making carotenoids bioavailable. This diversity is influenced by species, tissue type, and environmental factors, leading to different chromoplast classifications. Chromoplasts can be broadly categorised into crystalline, globular, fibrillary, and membranous forms based on their internal organisation (Simkin et al. 2004; Jeffery et al. 2012). Crystalline chromoplasts, found in tomato and watermelon, contain rigid, crystal-like carotenoid structures. These structures enhance pigment stability by reducing degradation (Bangalore et al. 2008; Harris and Spurr 1969). In contrast, β-carotene-enriched chromoplasts in species like carrot and cauliflower usually form globular or fibrillary structures. These structures store carotenoids efficiently (Frey-Wyssling and Schwegler 1965; Paolillo et al. 2004).

The structure of chromoplasts directly affects their functionality. Chromoplasts with complex structures, like extensive membranous or reticulo-tubular forms, are found in tissues that need high carotenoid levels for pigmentation or physiological functions. These structural changes improve the visual and sensory qualities of plant tissues. They also increase the bioavailability of carotenoids, which are strong dietary antioxidants with health benefits.

Chromoplasts come from chloroplasts, especially during fruit ripening, highlighting their special role in plant development. This change involves major structural and biochemical modifications, including plastid reorganisation and carotenoid accumulation, which give ripened fruits their vivid colour (Harris and Spurr 1969; Bathgate et al. 1985; Egea et al. 2010). During chromoplast differentiation, chloroplasts undergo distinct changes, such as membrane reorganisation. The transition from chloroplasts to chromoplasts involves major changes in plastid structures. This process includes thylakoid breakdown, chlorophyll degradation, loss of the inner envelope membrane, fusion of existing membranes, and the formation of new structures through vesicle budding (Zita et al. 2022; Egea et al. 2011; Llorente et al. 2020; Bréhélin et al. 2007). The transition from chloroplasts to chromoplasts is marked by a significant increase in carotenoid production and storage. This process includes the growth and increase in plastoglobuli, carotenoid crystals formation, and a higher capacity to store β-carotene and other carotenoids (Li and Yuan 2013; Rottet et al. 2015; Egea et al. 2011; Llorente et al. 2020; Austin et al. 2006; Schweiggert et al. 2012).

The change from chloroplast to chromoplast shows how chromoplasts specialise in storing carotenoids, improving pigment stability, and increasing bioavailability to meet the tissue's metabolic needs. These changes enhance the visual appeal, sensory qualities, and nutritional value of plant tissues, making chromoplasts key organelles in both plant physiology and human nutrition. Studies using real-time fluorescence imaging and confocal microscopy have shown how these structural transitions happen over time. Carotenoid fluorescence increases significantly after 10 h, while chlorophyll fluorescence declines gradually, and reaches maximum carotenoid accumulation after 14 h (Egea et al. 2011). The chloroplast-to-chromoplast transition is highly synchronised within individual cells, with over 80% of plastids undergoing the change within 6 h (Egea et al. 2011).

Recent research has given new insights into the molecular mechanisms regulating this transition. It shows that photosynthetic interference acts as a metabolic switch for plastid reprogramming. Phytoene overproduction weakens chloroplast identity, and converting phytoene into downstream carotenoids is necessary for chromoplast differentiation. This change is driven by a complete reprogramming of nuclear gene expression, highlighting the complex coordination between plastid and nuclear regulatory networks. These findings provide strong evidence for the structural and molecular changes that control chromoplast differentiation, helping us better understand fruit ripening and carotenoid biosynthesis (Egea et al. 2010).

The chloroplast-to-chromoplast transition is clear during tomato fruit ripening. Changes in autofluorescence show the gradual degradation of chlorophyll and the formation of carotenoid-rich plastid membranes as explained above (Egea et al. 2010). During leaf senescence, chloroplasts transition into gerontoplasts. These are characterised by chlorophyll degradation without carotenoid biosynthesis, causing the typical yellowing of leaves (Matile 2000; Sadali et al. 2019). Figure 9.3 shows a simplified model of different transformation pathways of chloroplasts into either chromoplasts or gerontoplasts, driven by various environmental and molecular factors.

9.2.2 Regulation of Chromoplast Differentiation

Plastids are versatile and multifunctional organelles in plant cells. They can change between different forms in response to developmental cues, environmental conditions, and physiological needs. This ability to transform highlights their dynamic role in plant metabolism and adaptation. The change between plastid types is not random but a carefully controlled process governed by complex regulatory networks. These networks combine external signals, like light, temperature, and nutrients, with internal signals, including hormones and gene expression patterns. For example, the transition from chloroplasts to chromoplasts, which occurs during fruit ripening and flower development, is controlled by phytohormones like ethylene, abscisic acid, and jasmonic acid (JA). These hormones control genes involved in carotenoid biosynthesis and suppress chlorophyll biosynthesis, which drives the change from green, photosynthetic chloroplasts to brightly coloured chromoplasts (Howitt and Pogson 2006; Toledo-Ortiz et al. 2010). Figure 9.4 shows how different plastid types differentiate and reprogram in response to environmental signals and transcriptional or post-transcriptional regulation.

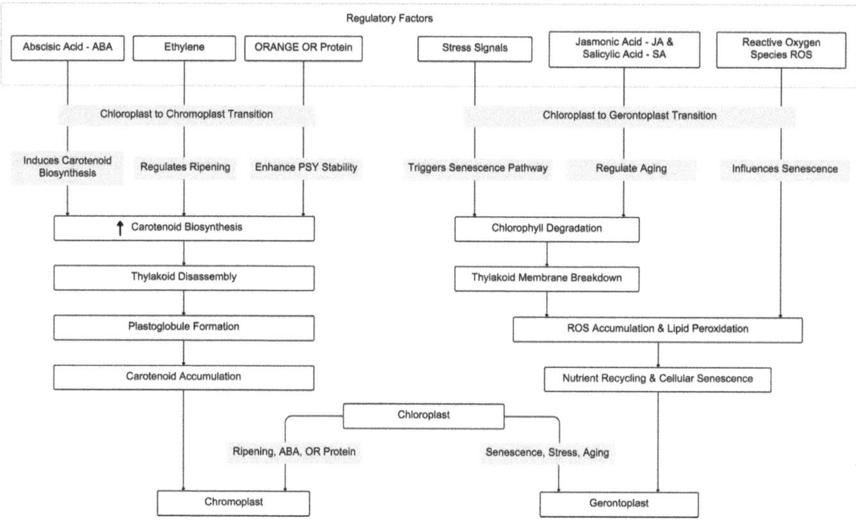

Fig. 9.3 Schematic illustration of the transition of chloroplasts into either chromoplasts or gerontoplasts, triggered by different environmental and molecular factors. The key regulatory factors that could trigger the chloroplasts differentiation into chromoplasts during ripening include absicic acid, ethylene, and OR protein (by enhancing PSY stability). The chromoplast differentiation includes the process of carotenoid biosynthesis, thylakoid disassembly, plastoglobule formation, and carotenoid accumulation. The chloroplasts to gerontoplasts transition during senescence is driven by the presence of stress signals, jasmonic acid, salicylic acid, and reactive oxygen species. The gerontoplast differentiation involves ROS accumulation which can accelerate chlorophyll degradation, lipid peroxidation, photosynthesis reduction, and nutrient recycling due to oxidative damage

Chloroplast to chromoplast differentiation is controlled by both external and internal stimuli, which affect the transition between the two. This process is key to fruit ripening and flower colouration in many plants. Light and temperature are the main external stimuli. Their combined effects can greatly influence fruit colour and nutritional value. Internal stimuli include genetic factors, hormonal regulation, and post-transcriptional and protein regulation, including epigenetic control. The chromoplast differentiation network has been widely studied, with recent research offering new insights into this complex process.

9.2.2.1 External Stimuli

Chromoplast differentiation and function are highly responsive to environmental stimuli, which are crucial for shaping pigment biosynthesis, plastid structure, and carotenoid accumulation. Light and temperature are key environmental factors that strongly affect chromoplast biogenesis. They influence the nutritional and aesthetic qualities of plant tissues, especially in fruits and flowers.

Light as a Modulator of Chromoplast Development

Light is an important environmental signal that controls the expression of plastid and nuclear genes involved in carotenoid biosynthesis. Light quality (wavelength),

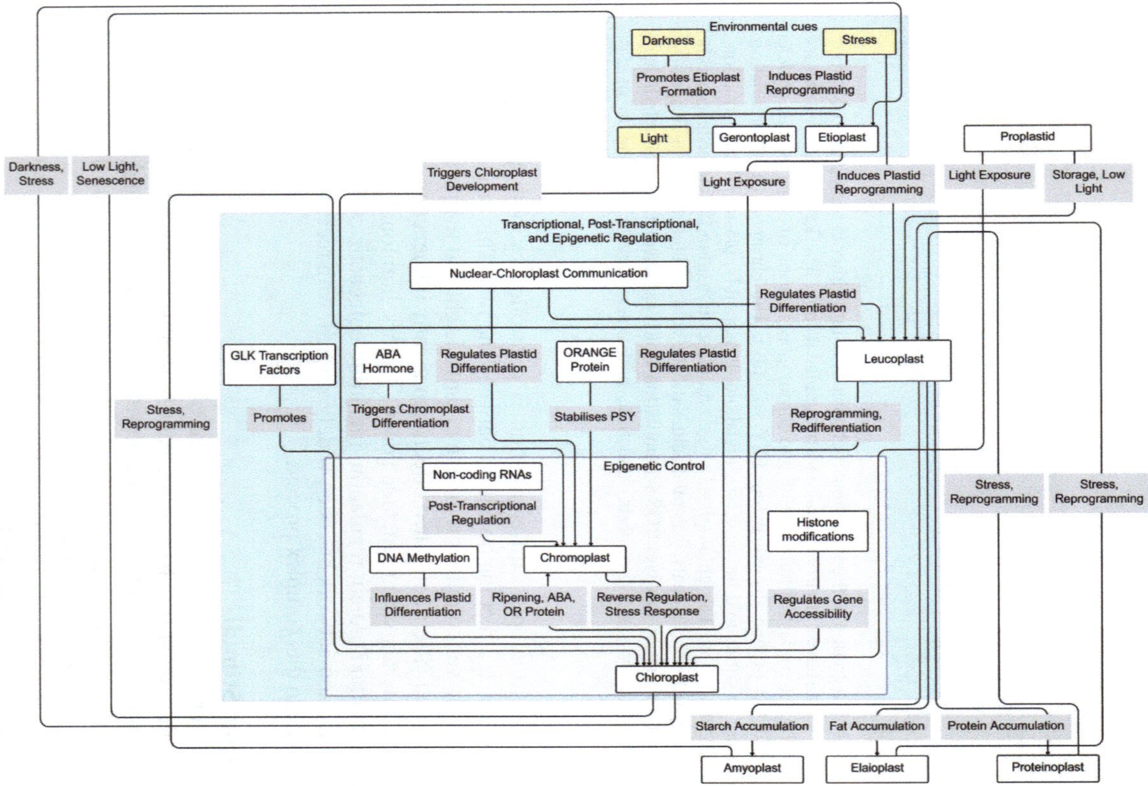

Fig. 9.4 The diagram illustrates the complex process of plastid differentiation in plants. Proplastids are the starting point for various types of plastids, developing into different forms. Chloroplasts develop from proplastids when exposed to light and can further differentiate into chromoplasts through ripening or stimuli from ABA and OR proteins. Chloroplasts may convert to etioplasts under darkness and stress, which can revert back to chloroplasts with light exposure. Leucoplasts emerge from proplastids when stored or in low light conditions and can convert into specialized forms like amyloplasts, elaioplasts, or proteino- … The diagram also illustrates the role of transcriptional regulators and environmental cues, repre-

intensity, and duration (photoperiod) are essential for controlling chromoplast differentiation. Light affects the expression of genes involved in pigment biosynthesis, influencing both the chloroplast-to-chromoplast transition and carotenoid accumulation in chromoplasts. Light presence and intensity can directly impact carotenoid synthesis, contributing to the vibrant colours seen in many ripened fruits like tomatoes, peppers, and papayas.

In high-light environments, carotenoid biosynthetic pathways are upregulated, leading to more chromoplast formation and deeper pigmentation. Light-enhanced carotenoid production is controlled by photoreceptors like phytochromes and cryptochromes. These receptors detect light changes and trigger signalling pathways that activate pigment-related gene expression. For example, in tomatoes and bell peppers, more light exposure leads to higher concentrations of carotenoids like lycopene and β-carotene. This not only improves visual appeal but also boosts the nutritional quality of the fruit due to the antioxidants these compounds provide (Dannehl et al. 2021; Hoffmann et al. 2016).

Under low-light or shaded conditions, chromoplast development may slow down, and chloroplasts could even turn back into etioplasts due to reduced chlorophyll and carotenoid biosynthesis. This reversion helps plants conserve resources in environments with insufficient light for optimal chloroplast and chromoplast function (Cazzonelli and Pogson 2010). By adjusting chromoplast formation based on light availability, plants optimise their pigmentation and energy use, ensuring they are well adapted to their light environment (Bhatla and Lal 2023).

Temperature Effects on Chromoplast Biogenesis

Temperature fluctuations also affect chromoplast development by influencing enzyme stability, carotenoid synthesis, and plastid membrane fluidity. Temperature changes can alter the biochemical environment inside chromoplasts, affecting carotenoid accumulation, stability, and degradation, as well as the structural integrity of chromoplasts. Temperature changes can also influence chromoplast formation by affecting enzyme activity and protein stability within the organelle.

Cold temperatures trigger carotenoid accumulation and change chromoplast structure in cold-tolerant species. This suggests chromoplast differentiation may be an adaptive response to environmental stress (Sun et al. 2018). For example, cooler temperatures increase β-carotene synthesis, which acts as an antioxidant in chromoplasts, improving plant resilience to cold stress (Tomes 1963; Wang et al. 2022). The increased carotenoid content stabilises chromoplast structure and improves cold tolerance, helping with stress protection and stronger pigmentation.

In contrast, high temperatures speed up carotenoid degradation and may damage chromoplast structure. Prolonged heat can disrupt plastid membrane fluidity, reducing the accumulation of heat-sensitive carotenoids like lycopene and β-carotene. This degradation affects the colour of fruits and flowers and lowers their nutritional quality, as carotenoids are important antioxidants with health benefits (Wijesekara and Xu 2024). For species in warm climates, chromoplasts have structural adaptations that help stabilise carotenoids despite high temperatures, keeping pigmentation and nutritional content stable.

The plant's response to temperature changes shows a finely tuned adaptation, where chromoplasts adjust their carotenoid synthesis and structure according to environmental temperature (Ding and Yang 2022). This temperature-sensitive adaptation helps plants optimise pigment retention and colour stability in different climates, improving their ecological fitness and the appeal of fruits and flowers.

Integrated Environmental Responses in Chromoplast Function

The interplay between light and temperature provides a complex regulatory framework for chromoplast differentiation, allowing plants to adapt their pigmentation and carotenoid content to changing environmental conditions. By modulating chromoplast development in response to these stimuli, plants enhance their resilience to stress and optimize resource allocation for both survival and reproductive success.

For example, in fruit-bearing plants, the combined influence of light and temperature can significantly impact fruit colouration and nutritional value. Tomatoes grown in sunny, moderate climates, for instance, develop a rich red colour due to high lycopene accumulation, which is enhanced by adequate sunlight and moderate temperatures. On the other hand, bell peppers exposed to both high light and cooler temperatures may exhibit intensified carotenoid accumulation, contributing to brighter colourations and increased β-carotene levels. This adaptation not only appeals to visual and sensory preferences but also aligns with the plant's strategy to attract seed-dispersing animals.

Light and temperature are powerful factors that modulate chromoplast differentiation, enabling plants to adjust pigment biosynthesis and chromoplast structure. This ability to respond to environmental factors ensures that chromoplasts support plant survival and human nutrition, showing the vital role of chromoplasts in plant adaptability and ecological interactions.

9.2.2.2 Internal Stimuli

Chromoplasts are specialised plastids that are essential for synthesising and storing carotenoid pigments, which contribute to the bright colours in fruits, flowers, and other plant organs. Chromoplast formation usually happens during specific stages, like fruit ripening and flower maturation, when chloroplasts change significantly to support pigmentation, secondary metabolism, and antioxidant protection. This chapter explores the molecular, genetic, hormonal, and epigenetic factors that drive chromoplast differentiation. It looks at how these regulatory mechanisms work together to control chromoplast development and function.

Gene Regulation in Chromoplast Differentiation

Chromoplast differentiation is controlled by a complex network of transcriptional regulators and signalling pathways that coordinate the expression of nuclear and plastid genes. This regulatory network allows chromoplasts to develop metabolic functions for carotenoid biosynthesis, storage, and stabilisation, ensuring efficient pigment accumulation. This contributes to bright colour and better nutritional quality in plant tissues.

The *Or* gene plays a crucial role in chromoplast differentiation, encoding a protein with a DnaJ-like zinc-finger domain that enhances carotenoid accumulation by promoting chromoplast formation in various plant organs, including fruits, roots, and flowers (Lu et al. 2006; Yuan et al. 2015a, b). This gene stabilizes structures that store carotenoids within plastids, thereby facilitating plastid differentiation into chromoplasts. Overexpression of the *Or* gene has been demonstrated to significantly increase carotenoid content in multiple plant species, making it a valuable target for bioengineering aimed at improving crop nutritional quality (Yuan et al. 2015a, b). Its influence extends to diverse plant organs, including flower petals of *Osmanthus fragrans*, where it enhances pigmentation and improves aesthetic appeal (Liu et al. 2024); potato tubers, where it contributes to increased carotenoid accumulation and nutritional enhancement (Bai et al. 2014); ice and sweet potato calli, where its expression is linked to enhanced carotenoid biosynthesis, supporting the development of biofortified staple crops (Lu et al. 2006; Kim et al. 2013); and melons, where it promotes enriched pigment production, thereby enhancing fruit quality and market value (Zhou et al. 2023).

The *PSY* gene family also plays a key role in chromoplast differentiation. PSY catalyses the first step in carotenoid biosynthesis, converting GGPP into phytoene. This reaction is the starting point for producing carotenoids like lycopene, β-carotene, and xanthophylls, which give fruits and flowers their characteristic red, orange, and yellow colours (Fraser et al. 2007). PSY expression is carefully controlled by developmental signals and hormones, ensuring that carotenoid production matches chromoplast formation and is optimised for the plant's needs during stages like ripening.

Other transcription factors, like RIN (Ripening Inhibitor) in tomatoes, help chromoplast differentiation by controlling genes involved in carotenoid biosynthesis, plastid formation, and lipid metabolism. RIN plays a key role in the transition from chloroplasts to chromoplasts. It controls chlorophyll breakdown and the increase in carotenoid biosynthesis, driving the ripening process (Fujisawa et al. 2013).

Hormonal Control of Chromoplast Formation

Phytohormones, particularly ethylene and ABA, play crucial roles in chromoplast differentiation. They regulate the transformation of chloroplasts into chromoplasts, especially during fruit ripening. These hormones orchestrate metabolic shifts necessary for carotenoid accumulation, which enhances pigmentation and fruit appeal (Fig. 9.5).

Ethylene is a key regulator of fruit ripening and chromoplast differentiation in crops like tomatoes, bananas, and peppers. It triggers the expression of carotenoid biosynthetic genes, such as PSY, initiating carotenoid production. Simultaneously, ethylene suppresses photosynthetic genes, including those encoding light-harvesting chlorophyll a/b-binding proteins and Rubisco. This shift redirects plastid metabolism from photosynthesis to carotenoid biosynthesis (Barry and Giovannoni, 2007). As chloroplasts convert into chromoplasts, fruits and flowers develop vivid red and orange hues. These colours attract seed-dispersing animals and enhance market

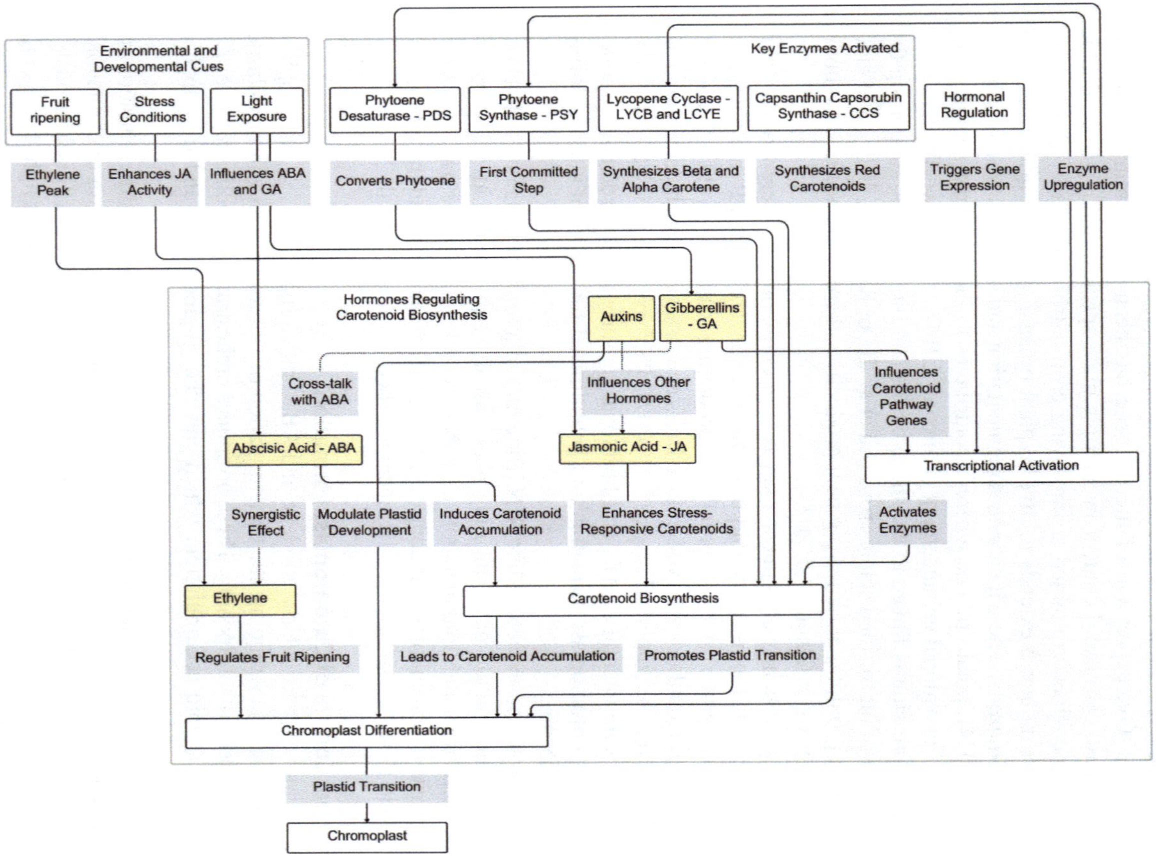

Fig. 9.5 Hormonal regulation of carotenoid biosynthesis and chromoplast differentiation. This diagram illustrates the interactions between different hormones, … collectively contribute to the processes leading to stable chromoplast development

appeal (Cazzonelli and Pogson 2010). Ethylene also promotes thylakoid membrane disassembly and chlorophyll degradation, making carotenoids more visible.

ABA contributes to chromoplast differentiation, especially in response to environmental stress. It interacts with ethylene signalling to fine-tune the timing and extent of chromoplast formation. This coordination balances fruit ripening with stress adaptation, improving fruit quality and shelf life (Sun et al. 2018). In citrus fruits, ABA enhances carotenoid accumulation by modulating gene expression. This process boosts the levels of β-cryptoxanthin and lutein, improving both pigmentation and nutritional content (Della Penna and Pogson 2006; Keawmanee et al. 2023). The complex interplay between ABA and ethylene highlights the sophisticated hormonal control of chromoplast differentiation.

Epigenetic Control in Chromoplast Differentiation

Epigenetic modifications, such as DNA methylation and histone modifications, add another layer of regulation to chromoplast differentiation. These modifications alter chromatin structure and gene expression, influencing carotenoid biosynthesis and plastid development. By fine-tuning gene activity, plants can optimize chromoplast formation in response to developmental and environmental signals. This has potential applications in biofortification, enhancing carotenoid content in crops without altering DNA sequences (Rudolf et al. 2024).

DNA methylation, which adds methyl groups to cytosine residues, often silences gene expression when occurring in promoter regions. In tomatoes, DNA demethylation in the promoters of carotenoid biosynthetic genes correlates with increased gene expression and carotenoid accumulation during ripening (Zhang et al. 2016). For example, demethylation upregulates PSY, facilitating chromoplast differentiation. However, PSY regulation is complex, involving additional factors such as OR protein-mediated post-transcriptional regulation (Zhou et al. 2015).

DNA methylation patterns change dynamically during fruit ripening. The DEMETER-like DNA demethylase SlDML2 plays a crucial role in tomato ripening by actively demethylating ripening-related gene promoters (Liu et al. 2015b; Lang et al. 2017). This process extends beyond carotenoid biosynthesis genes, affecting transcription factors involved in ripening control. Studies suggest a feedback loop between DNA methylation and transcription factor binding, reinforcing regulatory mechanisms in fruit development (Liu et al. 2015b; Zhong et al. 2013; Manning et al. 2006; Chen et al. 2015).

Histone modifications also regulate chromoplast differentiation. Histone acetylation, which relaxes chromatin and enhances transcription, increases around carotenoid biosynthetic genes during ripening. Conversely, histone methylation can either activate or repress gene expression depending on the specific modification (Anwar et al. 2021). In strawberries and oranges, histone modifications influence carotenoid biosynthesis, affecting fruit pigmentation and antioxidant content (Li et al. 2022). Understanding the regulatory role of histone modifications in chromoplast formation may provide novel insights for manipulating chromoplast-related traits in crop breeding.

Although direct evidence remains limited, epigenetic modifications may influence chromoplast differentiation by regulating the expression of genes involved in carotenoid biosynthesis. For instance, epigenetic changes could modulate the activity of transcription factors that regulate the expression of key enzymes such as PSY and PDS. Future research employing clustered regularly interspaced short palindromic repeats (CRISPR)-based epigenome editing or high-throughput sequencing could provide valuable insights into the role of epigenetic modifications in chromoplast differentiation. Investigating epigenetic marks on carotenoid biosynthesis genes during chromoplast development may uncover novel regulatory mechanisms governing this process.

9.3 Protein Dynamics and Import Machinery in Chromoplast

Chromoplasts are specialized plastids that produce and store carotenoid pigments, giving fruits and flowers their vibrant colours and nutritional value. Unlike chloroplasts, which focus on photosynthesis, chromoplasts adapt their functions to accumulate pigments and other metabolites essential for plant growth and appeal. A key aspect of chromoplast differentiation is the import of nuclear-encoded proteins that support chromoplast-specific functions. This section discusses the molecular machinery for protein import, focusing on the TOC (Translocon at the Outer Chloroplast Membrane) and TIC (Translocon at the Inner Chloroplast Membrane) complexes, along with the adaptation of the chromoplast proteome. Understanding these protein dynamics reveals chromoplasts' unique metabolism and suggests ways to enhance fruit quality and nutrition by modifying plastid function.

9.3.1 The TOC/TIC Machinery in Chromoplast Protein Import

Nuclear-encoded precursor proteins enter chromoplasts through TOC/TIC complexes, which transport proteins across the plastid envelope. In chloroplasts, the TOC/TIC system mainly imports photosynthetic proteins. However, during chromoplast differentiation, it reorganizes to meet the needs of a non-photosynthetic plastid. During chloroplast-to-chromoplast conversion, the TOC/TIC system shifts to import proteins essential for carotenoid biosynthesis, lipid metabolism, and chromoplast-specific functions (Ling et al. 2021; Sadali et al. 2019).

The SP1 (SUPPRESSOR OF PPI1 LOCUS1) protein, a RING-type ubiquitin E3 ligase, regulates TOC/TIC reorganization during chromoplast differentiation. SP1 controls plastid protein import by tagging specific TOC components for degradation, shifting the system toward chromoplast-specific proteins (Ling et al., 2012). SP1 selectively degrades TOC isoforms linked to photosynthetic protein import. This restructuring enables chromoplasts to prioritize enzymes for carotenoid biosynthesis and metabolism (Ling et al. 2021; Sadali et al. 2019). SP1 specifically targets and ubiquitinates TOC proteins, marking them for degradation by the 26S proteasome (Jarvis and López-Juez 2013; Ling et al. 2012). This process enables chromoplasts to fine-tune substrate-specific protein import, ensuring that the plastid proteome is optimized for non-photosynthetic functions.

The TOC complex includes Toc159, Toc132, Toc120, Toc75, Toc34, and Toc33. These proteins assemble into stable complexes, and their precise ratio is critical for efficient protein import. Toc75, Toc33, and Toc159 specialize in importing photosynthetic proteins, whereas Toc75, Toc34, and Toc132/120 focus on housekeeping and non-photosynthetic protein import. Plastids adjust their TOC complexes based on tissue type and developmental stage to match changing protein import needs.

The TOC complex at the outer plastid membrane identifies precursor proteins using specific targeting signals. After passing through TOC, proteins enter the intermembrane space. The TIC complex then guides them into the plastid stroma. During stress, SP1 reduces TOC protein levels, limiting plastid protein import (Ling and Jarvis 2015). This reduction depends on SP1, as *sp1* mutants retain normal TOC levels, while SP1 overexpression causes even greater depletion (Ling and Jarvis 2015). The reduction in TOC protein levels has limited the import of photosynthetic apparatus components, attenuating photosynthetic activity and reducing the risk of reactive oxygen species (ROS) overproduction and photo-oxidative damage, improving plant stress tolerance (Ling and Jarvis 2015).

During chromoplast differentiation, the TOC/TIC system adapts to preferentially import enzymes involved in carotenoid biosynthesis, lipid metabolism, and pigment stabilization. Figure 9.6 shows that the remodelling of the protein import machinery occurs during the chloroplast to chromoplast transition mediated by SP1. Recent

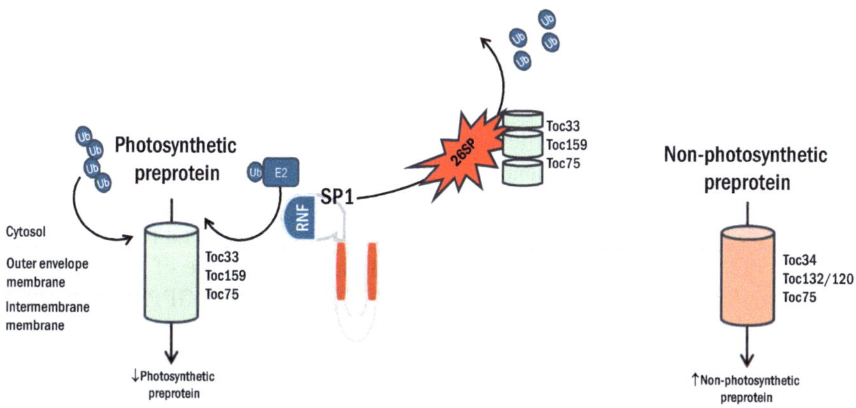

Fig. 9.6 Hypothetical model showing a mechanism for the regulation of the chloroplast-to-chromoplast transition. The remodelling of the protein import machinery occurs during this transition is mediated by the outer membrane E3 ligase, SP1, which targets unwanted TOC components for degradation by the 26S proteasome (26SP) in the cytosol. During chromoplast differentiation, the TOC/TIC system adapts to preferentially import enzymes involved in carotenoid biosynthesis, lipid metabolism, and pigment stabilization. SP1 reduces TOC protein levels to limit photosynthetic preprotein import by targeting and ubiquitinates TOC proteins complex (Toc33, Toc75, Toc159), which specialize in importing photosynthetic proteins and marking them for degradation by the 26S proteasome. RNF, RING-finger domain; Ub, ubiquitin; E2, E2 conjugase

studies suggest that SP1-mediated modifications of TOC/TIC receptors enhance the import efficiency of chromoplast-specific proteins, enabling more rapid chromoplast development and increased carotenoid accumulation, whereby the knockdown of SP1 delays tomato fruit ripening, whereas overexpression of SP1 accelerates ripening (Ling et al., 2021). This selective import mechanism ensures that chromoplasts meet the metabolic demands of pigment production, even as they undergo extensive proteomic and metabolic reprogramming.

9.3.2 Functional Adaptation of the Chromoplast Proteome

The chromoplast proteome changes as plastids transition from chloroplasts to chromoplasts, reflecting their metabolic shift. Chloroplasts contain proteins mainly involved in photosynthesis, carbon fixation, and other light-dependent processes. Chromoplasts, in contrast, contain enzymes and proteins for carotenoid biosynthesis, pigment stabilisation, and protein quality control as key to their function and cellular stability. Proteomic shifts help plants regulate and optimise carotenoid biosynthesis in response to metabolic needs and environmental cues.

A study on *Areca catechu* showed that proteomic changes regulate carotenoid biosynthesis by altering enzyme expression. Fluctuations in PSY, PDS, and LCYB levels directly affected carotenoid production and composition. Notably, reduced light exposure from bagging treatment downregulated carotenoid biosynthesis genes, lowering carotenoid levels (Zheng et al. 2024). These findings highlight how environmental factors shape the proteome, influencing carotenoid accumulation and chromoplast function.

In *Arabidopsis*, the FIBRILLIN6 (FBN6) was identified as a key regulator of carotenoid biosynthesis. FBN6 binds to PSY, a major enzyme in carotenoid biosynthesis, promoting carotenogenesis in bacterial systems and chloroplasts. This interaction reveals a novel protein-level regulation mechanism, offering insights into how protein interactions influence carotenogenesis at the plastid level (Iglesias-Sanchez et al. 2023). Such findings underscore the importance of proteomic interactions in fine-tuning carotenoid biosynthesis beyond transcriptional regulation.

A study on *Nicotiana tabacum* highlighted the essential role of geranylgeranyl diphosphate synthase (GGPPS) in carotenoid biosynthesis. NtGGPPS1 functions as a homodimer in chloroplasts, working with NtPSY1 to enhance carotenoid biosynthesis. An engineered NtGGPPS1 variant increased carotenoid levels, enhanced photosynthesis, and boosted biomass production. These findings indicate that modifying GGPPS could improve plant productivity and carotenoid accumulation (Yan et al. 2024).

A proteomic-metabolomic study on cigar tobacco leaves under shading revealed how environmental factors reshape the proteome and influence carotenoid production. GGPPS was found to be essential for producing various terpenoids, including carotenoids. Shading treatment modified the expression of carotenoid biosynthetic proteins, affecting the production of β-carotene, zeaxanthin, and violaxanthin, key photoprotective compounds (Yan et al., 2024). This study shows how environmental factors drive proteomic changes, optimising carotenoid biosynthesis and plastid function.

Proteomic shifts serve as a key regulatory mechanism in carotenoid biosynthesis and accumulation, influencing enzyme expression, protein-protein interactions, and metabolic adaptation to environmental stimuli. Recent studies on *Areca catechu*, *Arabidopsis*, and *Nicotiana tabacum* as discussed above, demonstrate how modifications in the plastid proteome affect carotenoid production, metabolic efficiency, and photoprotection. Advances in proteomics and metabolomics provide valuable insights into these regulatory mechanisms, paving the way for biotechnological applications aimed at enhancing carotenoid content in crops through genetic engineering and breeding programs.

9.4 Carotenoid Biosynthesis in Chromoplasts

Carotenoids are a diverse group of isoprenoid pigments essential for plant physiology and human nutrition. They produce the red, orange, and yellow colours in many fruits and flowers and play key roles in photoprotection, light harvesting, and signalling. In humans, carotenoids are precursors to essential vitamins and function as antioxidants, supporting vision and reducing the risk of chronic diseases. In plant cells, chromoplasts are the main sites of carotenoid synthesis and storage, undergoing major metabolic and structural changes as they develop from chloroplasts. This section examines the molecular mechanisms of carotenoid biosynthesis in chromoplasts, focusing on key enzymes, regulatory networks, lipid remodelling, and carotenoid storage strategies.

9.4.1 Overview of the Carotenoid Biosynthesis Pathway

Carotenoid biosynthesis is a multi-step pathway starting with the production of GGPP, a 20-carbon compound. It then involves desaturation, cyclisation, and hydroxylation reactions to generate different carotenoids. The first and rate-limiting step is catalysed by PSY, which converts GGPP into phytoene, a colourless carotenoid (Llorente et al. 2020). In bacteria, PSY is known as crtB. Studies show that expressing crtB in the green tissues of *Nicotiana tabacum*, *Nicotiana benthamiana*, and *Arabidopsis thaliana* can trigger chloroplast-to-chromoplast transition (Llorente et al. 2020). This highlights how carotenoid accumulation drives plastid differentiation, linking biosynthesis to plastid structural changes. Carotenoids not only contribute to pigmentation but also regulate plastid protein content, nuclear gene expression, and primary metabolism.

Lycopene is a crucial intermediate in carotenoid biosynthesis, serving as a precursor for various downstream carotenoids. These derivatives are important for plant function and human health, mainly due to their antioxidant and nutritional benefits. Lycopene is converted into downstream carotenoids through two pathways: one producing β-carotene via LCYB and the other generating α-carotene via

LCYE. LCYB converts lycopene into β-carotene, a precursor to vitamin A. β-Carotene also acts as an antioxidant, protecting cells from oxidative stress. LCYE converts lycopene into α-carotene, which, like β-carotene, contributes to vitamin A synthesis and has potent antioxidant effects.

Xanthophylls such as lutein and zeaxanthin, derived from α-carotene, are essential for eye health, protecting the retina from oxidative stress. These carotenoids accumulate in the macula, where they help prevent age-related macular degeneration (AMD) by absorbing high-energy blue light. These xanthophylls play a key role in the xanthophyll cycle, allowing plants to adapt to changing light conditions. They are crucial for photoprotection, allowing plants to release excess light energy as heat, reducing photoinhibition and oxidative stress.

Tomatoes and carrots serve as prominent examples of crops enriched with lycopene and its downstream carotenoids. Tomatoes are an abundant source of lycopene, which is converted into β-carotene in smaller amounts. Lycopene's strong antioxidant properties make it beneficial for human health, particularly in reducing the risk of certain cancers and cardiovascular diseases. Similarly, carrot varieties exhibit variation in carotenoid composition based on their genetic background. Orange carrots primarily accumulate β-carotene, whereas red carrots contain higher levels of lycopene due to differences in lycopene cyclase activity. These natural variations have significant implications for breeding programs aimed at increasing provitamin A content in carrots and other staple crops.

The downstream carotenoids derived from lycopene are integral to both plant physiology and human health. Their biosynthesis involves key enzymatic transformations, leading to compounds with diverse biological functions, including antioxidant defence, photoprotection, and vitamin A biosynthesis. Understanding these biosynthetic pathways and regulatory mechanisms provides opportunities for biotechnological interventions, such as genetic engineering and breeding strategies, to enhance carotenoid content in crops, ultimately improving their nutritional value and health benefits.

9.4.2 Lipid Remodelling in Chromoplast Development

Lipid remodelling is crucial for chromoplast differentiation, as lipids stabilise hydrophobic carotenoids and maintain chromoplast structure. During chromoplast differentiation, galactolipid and triacylglycerol (TAG) synthesis increases, forming lipid droplets in the chromoplast stroma (Suzuki et al. 2015). These lipid droplets store carotenoids, protecting them from oxidation and improving pigment stability.

Plastoglobules, spherical lipid-protein structures in chromoplast membranes, are key to carotenoid storage and stabilisation. Plastoglobule abundance and composition change as chromoplasts develop, marking their transition to a carotenoid-rich state (Austin et al. 2006). Plastoglobules also synthesise tocopherols and other isoprenoid compounds, enhancing chromoplast stability and function. Carotenoid stability in chromoplasts relies on lipid-protein interactions involving plastid lipid-associated proteins (PLAPs) and fibrillin. These proteins bind to carotenoids, forming stable lipid-carotenoid complexes that maintain chromoplast structure and shield carotenoids from light and oxygen damage (Rottet et al. 2015).

Lipid composition significantly impacts carotenoid stability and bioavailability, protecting them from degradation and improving absorption in humans. Lipids protect carotenoids from oxidation by limiting their exposure to oxygen and light. Encapsulating carotenoids in solid lipid nanoparticles (SLNs) enhances their stability by restricting oxidative species diffusion (de Abreu Martins et al. 2021). Fatty acyl composition affects carotenoid stability. Longer-chain fatty acids increase lipid hydrophobicity, improving carotenoid protection and transfer during digestion (Huo et al. 2007).

Lipids are vital for carotenoid absorption in humans. Dietary fats help dissolve carotenoids in mixed micelles, improving their uptake in the intestines. Lipid digestion significantly affects carotenoid bioavailability. Longer acyl-chain triglycerides improve carotene transfer from food, increasing absorption efficiency (Huo et al. 2007). Studies on microalgal paste show that drying and storage conditions affect lipid and carotenoid stability. However, a suitable lipid environment protects carotenoids, preserving them long-term (Ryckebosch et al. 2011). Similarly, in low-fat carrot and tomato-based systems, carotenoids remain stable during long-term storage with minimal colour changes, thanks to lipid protection (Mutsokoti et al. 2017).

9.4.3 Regulation of Enzymes in Carotenoid Biosynthesis

Several key enzymes drive the sequential reactions in the carotenoid biosynthetic pathway (Fig. 9.2), producing specific carotenoid classes that accumulate in chromoplasts. Key enzymes involved include the following:

- Phytoene Synthase (PSY): The first enzyme in the pathway, PSY, converts GGPP into phytoene, initiating carotenoid biosynthesis. PSY expression is tightly regulated by developmental cues, hormones, and environmental factors to ensure synchronisation with chromoplast differentiation (Fraser et al. 2007).
- Phytoene Desaturase (PDS) & ζ-Carotene Desaturase (ZDS): These enzymes introduce double bonds into phytoene, converting it into lycopene, a red carotenoid that acts as a precursor to β-carotene and other derivatives.
- Lycopene β-Cyclase (LCYB) and Lycopene ε-Cyclase (LCYE): These enzymes cyclise lycopene, forming β-carotene and α-carotene, respectively. Their activity balance determines the chromoplast's overall carotenoid composition.
- Capsanthin/Capsorubin Synthase (CCS): CCS produces capsanthin and capsorubin, two keto-carotenoids responsible for the red pigmentation in ripe pepper fruits (Hugueney et al. 1995). Studies on CCS have revealed its role in regulating chromoplast pigmentation and its potential for metabolic engineering (Kuntz et al. 1998).

Chromoplast differentiation requires major metabolic and structural adjustments, particularly in carotenoid biosynthesis regulation. Enzymes like PSY and CCS are central to this process. They regulate carotenoid accumulation, affecting fruit and flower pigmentation and nutritional value.

This step is often upregulated during chromoplast differentiation, allowing carotenoids to accumulate and shape fruit and flower pigmentation (Fraser et al. 2000; Zhou et al. 2015). Studies show that overexpressing PSY boosts carotenoid accumulation. In *Arabidopsis thaliana*, PSY overexpression increased carotenoid levels, confirming its role in regulating carotenoid biosynthesis (Zhou et al. 2015). PSY activity is also post-transcriptionally regulated by OR proteins, which stabilise PSY and enhance its activity without affecting gene expression. This regulation is essential for efficient carotenoid biosynthesis during chromoplast differentiation (Zhou et al. 2015; Yuan et al. 2015b).

Capsanthin/capsorubin synthase is a specialized enzyme involved in the biosynthesis of capsanthin and capsorubin, two distinctive carotenoids that contribute to the red pigmentation of ripe pepper fruits (*Capsicum* spp.). The expression of capsanthin/capsorubin synthase is tightly regulated during pepper fruit ripening, with studies showing a significant upregulation as the fruit matures, coinciding with the accumulation of capsanthin and capsorubin (Huh et al. 2001). This suggests a strong correlation between gene expression patterns and carotenoid accumulation in chromoplasts.

The regulation of carotenoid biosynthesis through transcriptional and post-transcriptional mechanisms directly influences pigment accumulation and fruit ripening. Advances in gene expression analysis and enzymatic activity quantification have deepened our understanding of how chromoplast development is orchestrated, offering insights for biotechnological applications aimed at enhancing carotenoid content in crops.

9.4.4 Hormonal and Developmental Control of Carotenoid Accumulation

Hormones play a key role in regulating carotenoid biosynthesis and chromoplast differentiation by activating transcriptional pathways, enzyme expression, and metabolic transitions. Phytohormones such as ABA, ethylene, gibberellins (GA), auxins, and jasmonic acid (JA) regulate this process by influencing gene expression and enzyme activity, ensuring effective carotenoid accumulation. These hormones work together or independently to regulate plastid transitions, promoting chromoplast differentiation. Figure 9.5 illustrates how phytohormones regulate carotenoid biosynthesis and chromoplast differentiation, highlighting their role in transcriptional activation, enzyme regulation, and plastid transitions.

Hormonal regulation starts with transcriptional activation. ABA boosts carotenoid accumulation, especially during ripening, while ethylene controls colour changes by promoting chromoplast differentiation. GA regulates genes in the carotenoid biosynthetic pathway, while auxins influence plastid development and chromoplast formation. JA is vital in stress responses, increasing the production of protective carotenoids. These pathways upregulate key carotenoid biosynthesis enzymes, ensuring a coordinated metabolic flow towards carotenoid accumulation.

Key enzymes include PSY, which catalyses the first step in carotenoid biosynthesis by converting GGPP into phytoene. PDS then converts phytoene into lycopene, while LCYB and LCYE direct lycopene towards β-carotene and α-carotene synthesis, respectively. Figure 9.2 provides a simplified schematic of key enzymes involved in carotenoid biosynthesis in plants. In some plants, CCS modifies carotenoids to produce capsanthin and capsorubin, which give peppers their red colour. Hormonal signals regulate these enzymes, optimising carotenoid production in response to developmental and environmental cues. Alongside hormonal regulation, environmental and developmental factors also influence carotenoid biosynthesis and chromoplast differentiation. Light affects ABA and GA activity, regulating carotenogenic genes. Stress factors, such as drought or high light, enhance JA activity, increasing protective carotenoid synthesis.

As fruit ripens, ethylene levels increase, promoting chromoplast differentiation and pigment accumulation (Egea et al. 2010). These external cues adjust hormonal responses to align carotenoid biosynthesis with developmental and environmental conditions. As carotenoids accumulate, they drive plastid structural changes, leading to chromoplast differentiation. This process includes thylakoid membrane breakdown, plastoglobule formation for carotenoid storage, and major protein composition changes, shaping chromoplast identity (Egea et al. 2010). Once differentiation is complete, chromoplasts function as stable pigment-storing organelles, supporting fruit colour, nutritional quality, and antioxidative properties. Hormonal interactions continue to regulate chromoplast stability, with synergistic effects between ABA and ethylene reinforcing regulatory mechanisms, while crosstalk between GA and ABA modulates gene expression balance (Sun et al. 2018). Auxins also play a role in adjusting plastid metabolism, highlighting the complexity of hormonal regulation in plastid transitions (Müller and Munné-Bosch 2021).

The intricate interplay between hormonal regulation, environmental factors, and enzyme activity ensures the precise modulation of carotenoid biosynthesis and chromoplast differentiation. Understanding these mechanisms is essential for crop improvement strategies, as manipulating hormonal pathways provides opportunities to enhance carotenoid content, fruit ripening control, and stress resilience (Yuan et al. 2015b). Advances in plant biotechnology and metabolic engineering can leverage this knowledge to optimize carotenoid accumulation for improved nutritional value and commercial appeal in agricultural crops.

9.4.5 Carotenoid Storage and Stability in Chromoplasts

In chromoplasts, carotenoids are stored in specialised structures, including plastoglobules, fibrils, and lipid droplets. Carotenoid stability improves through esterification with fatty acids, increasing their solubility in the lipid-rich chromoplast environment. Xanthophyll acyl-transferase (XAT) and xanthophyll esterase (XES) catalyse esterification, stabilising pigments such as violaxanthin and neoxanthin during ripening (Berry et al. 2019; Watkins et al. 2019). In tomato chromoplasts, XES homologues, such as PYP1 (pale yellow petal 1), esterify violaxanthin and neoxanthin, enhancing their stability and accumulation during ripening (Ariizumi et al. 2014).

Esterification affects both carotenoid solubility and chemical stability. Xanthophylls esterified with unsaturated fatty acids, such as myristic and linoleic acid, are more susceptible to oxidation. In contrast, those with saturated fatty acids, such as lauric and palmitic acid, are more stable (Watkins and Pogson 2020). This differential esterification contributes to the distinct pigment profiles observed across various fruits and vegetables.

9.4.6 Chemical Modifications and Carotenoid Transformation

As chromoplasts differentiate, carotenoids undergo chemical modifications such as oxidation, epoxidation, isomerisation, and polyene chain cleavage. These modifications generate novel carotenoids with distinct structural and functional properties. Epoxy-xanthophylls, such as violaxanthin and antheraxanthin, convert into keto-xanthophylls like capsanthin and capsorubin, producing the deep red colour of ripe peppers (Mohd Hassan et al. 2019; Camara and Brangeon 1981; Kapoor et al. 2022). These modifications occur alongside metabolic changes in lipid composition, creating a hydrophobic environment suitable for carotenoid storage. During chromoplast differentiation, galactolipid levels decline while phospholipid levels rise, indicating plastid membrane adaptation to support carotenoid accumulation (Sun et al. 2007). This restructuring stabilises hydrophobic carotenoids, improving their storage and functionality.

Chemical modifications affect carotenoid colour, stability, and bioavailability. Modifications such as hydroxylation, epoxidation, and glycosylation change carotenoid physicochemical properties and biological functions. Hydroxylation improves carotenoid solubility and antioxidant activity. Lutein and zeaxanthin, hydroxylated derivatives of α-carotene and β-carotene, have higher polarity and stronger antioxidant properties, making them crucial for photoprotection and human health (Demmig-Adams et al. 1996). Epoxidation produces xanthophylls like violaxanthin and neoxanthin, which are vital for the xanthophyll cycle. The xanthophyll cycle is crucial for photoprotection, enabling plants to release excess light as heat and prevent photooxidative damage in photosynthetic tissues (Demmig-Adams et al. 1996). Furthermore, glycosylation involves the attachment of sugar moieties to carotenoids, enhancing their solubility and stability. An example is crocin, a glycosylated derivative of crocetin, which is significantly more soluble and stable than its aglycone form. These properties make crocin valuable for food colouring and pharmaceutical applications (Polyakov et al. 2023).

There are several effects of chemical modifications on plant functional properties. Chemical modifications contribute to pigment diversity in plants. For instance, the conversion of β-carotene into astaxanthin introduces keto groups, shifting the pigment hue from yellow to deep red, a transformation particularly evident in crustaceans and certain microalgae (Maoka 2020). Chemical modification via hydroxylation can enhance antioxidant properties by improving their ability to neutralize ROS. These properties are critical for plant stress responses and contribute to the nutritional value of carotenoid-rich foods. In addition, lipophilic modifications, such as fatty acid esterification, enhance the bioavailability of carotenoids by improving their solubility in dietary lipids and micelle formation during digestion. This enhances intestinal absorption, increasing their nutritional impact (González-Peña et al. 2023).

Chemical modifications of carotenoids are fundamental to their structural diversity and functional versatility, influencing their colour expression, stability, antioxidant potential, and bioavailability. Understanding these modifications provides valuable insights for enhancing the nutritional quality of food crops, improving industrial applications, and optimizing biotechnological strategies for carotenoid production.

9.4.7 Carotenoid Degradation and Metabolism

Carotenoid degradation maintains a balance between biosynthesis and accumulation, regulating pigment levels in chromoplasts and facilitating plastid transitions. Carotenoid degradation is controlled by carotenoid cleavage dioxygenases (CCDs), which break carotenoids at specific double bonds to form apocarotenoids. Apocarotenoids are not just metabolic by-products; they serve vital roles in plant signalling, stress responses, and defence mechanisms.

Carotenoid cleavage dioxygenases (CCDs) are the key enzymes responsible for the conversion of carotenoids into apocarotenoids with diverse physiological functions. 9-cis-epoxycarotenoid dioxygenase (NCED) catalyses the cleavage of 9-cis-violaxanthin and 9-cis-neoxanthin to produce ABA, a hormone crucial for stress responses, seed dormancy, and stomatal regulation. These apocarotenoids act as signalling molecules that regulate plant development and defence responses. The cleavage of β-carotene by CCD1 produces β-ionone, a volatile compound that enhances fruit aroma and floral scent (Auldridge et al. 2006a, b).

Apocarotenoids from carotenoid degradation are essential for plant signalling and defence. ABA is one of the best-studied apocarotenoids, acting as a key signalling molecule in drought responses and seed germination. ABA triggers stomatal closure, reducing water loss in drought conditions, and regulates stress-adaptive gene expression (Cutler et al. 2010). Beyond ABA, apocarotenoids regulate plant defence by modulating pathogen resistance pathways. ABA interacts with SA and JA pathways to fine-tune plant immunity and defence responses (Asselbergh et al. 2008).

Apocarotenoids have been shown to be physiologically important in various plant species. In *Arabidopsis thaliana*, NCED mutants with defective ABA biosynthesis show reduced stress tolerance, underscoring the role of apocarotenoids in stress adaptation (Iuchi et al. 2001). Another key group of apocarotenoids, strigolactones, regulates shoot branching and promotes symbiotic interactions with arbuscular mycorrhizal fungi (Koltai 2015). Strigolactones also influence plant-parasitic nematode interactions, highlighting their dual function in growth regulation and defence.

Apocarotenoids produced through carotenoid degradation serve as essential signalling molecules that regulate stress responses, plant defence, and developmental processes. The understanding of these metabolites extends beyond their role in plastid metabolism to their broader significance in agricultural and biotechnological applications. Investigating these pathways further may offer insights into enhancing crop resilience, improving stress tolerance, and developing sustainable agricultural strategies through targeted metabolic engineering.

9.5 Potential Applications and Future Directions

9.5.1 Potential Biotechnological Applications of Chromoplast Research

Studies on chromoplast differentiation offer significant potential for agriculture and biotechnology. Modifying chromoplast development pathways could improve the nutritional quality, shelf life, and appearance of fruits and vegetables. A deeper understanding of carotenoid biosynthesis, storage, and degradation in chromoplasts has led to new biotechnological opportunities. Regulating key enzymes like PSY and CCS can increase carotenoid content and stability in crops, improving their nutritional value and shelf life. Genetically engineering chromoplasts to produce novel carotenoids could yield new natural colourants and bioactive compounds with health benefits.

The potential to fine-tune carotenoid profiles through genetic and metabolic engineering offers exciting opportunities to develop biofortified crops and functional foods that promote human health. Further research on carotenoid biosynthesis regulation in chromoplasts will lead to new strategies for enhancing crop quality and sustainability.

9.5.2 Current Challenges and Future Directions in Non-photosynthetic Plastid Research

While our understanding of chloroplast biology has greatly advanced, research on non-green plastids, such as chromoplasts, leucoplasts, and elaioplasts, remains comparatively underexplored. The metabolic properties and regulatory networks governing the differentiation and function of these specialised plastids are only beginning to be understood (Kuntz et al. 2024).

Proteomic analyses have shown that each plastid type has a distinct protein profile that reflects its unique function. For example:

- *Chromoplasts* are rich in proteins related to carotenoid biosynthesis and storage (Egea et al. 2011; Sadali et al. 2019).
- *Elaioplasts* exhibit high activity in lipid biosynthesis and storage (Zhu et al. 2018).
- *Amyloplasts*, which specialize in starch storage, have proteomes dominated by enzymes involved in carbohydrate metabolism and biosynthesis (Balmer et al. 2006; Dupont 2008).

Future research efforts should focus on elucidating the complex regulatory networks that control plastid differentiation, interconversion, and function. Advanced techniques such as proteomics, transcriptomics, and metabolomics will be critical in uncovering new aspects of plastid biology. These studies will provide a deeper understanding of the unique roles that non-green plastids play in plant development, stress response, and adaptation, ultimately facilitating the development of crop varieties with enhanced nutritional quality, resilience to stress, and aesthetic appeal.

References

Anwar S, Brenya E, Alagoz Y, Cazzonelli CI (2021) Epigenetic control of carotenogenesis during plant development. Crit Rev Plant Sci 40(1):23–48. https://doi.org/10.1080/07352689.2020.1866829

Ariizumi T, Kishimoto S, Kakami R, Maoka T, Hirakawa H, Suzuki Y, Ozeki Y, Shirasawa K, Bernillon S, Okabe Y, Moing A, Asamizu E, Rothan C, Ohmiya A, Ezura H (2014) Identification of the carotenoid modifying gene 1 as an essential factor in xanthophyll esterification and yellow flower pigmentation in tomato (Solanum lycopersicum). Plant J 79(3):453–465. https://doi.org/10.1111/tpj.12570

Asselbergh B, De Vleesschauwer D, Höfte M (2008) Global switches and fine-tuning—ABA modulates plant pathogen defense. Mol Plant-Microbe Interact 21(6):709–719. https://doi.org/10.1094/mpmi-21-6-0709

Auldridge ME, Block A, Vogel JT, Dabney-Smith C, Mila I, Bouzayen M, Magallanes-Lundback M, Della Penna D, McCarty DR, Klee HJ (2006a) Characterization of three members of the Arabidopsis carotenoid cleavage dioxygenase family demonstrates the divergent roles of this multifunctional enzyme family. Plant J 45(6):982–993. https://doi.org/10.1111/j.1365-313X.2006.02666.x

Auldridge ME, McCarty DR, Klee HJ (2006b) Plant carotenoid cleavage oxygenases and their apocarotenoid products. Curr Opin Plant Biol 9(3):315–321. https://doi.org/10.1016/j.pbi.2006.03.005

Austin JR, Frost E, Vidi P-A, Kessler F, Staehelin LA (2006) Plastoglobules are lipoprotein subcompartments of the chloroplast that are permanently coupled to thylakoid membranes and contain biosynthetic enzymes. Plant Cell 18(7):1693–1703

Bai C, Rivera SM, Medina V, Alves R, Vilaprinyo E, Sorribas A, Canela R, Capell T, Sandmann G, Christou P (2014) An in vitro system for the rapid functional characterization of genes involved in carotenoid biosynthesis and accumulation. Plant J 77(3):464–475

Balmer Y, Vensel WH, DuPont FM, Buchanan BB, Hurkman WJ (2006) Proteome of amyloplasts isolated from developing wheat endosperm presents evidence of broad metabolic capability. J Exp Bot 57(7):1591–1602. https://doi.org/10.1093/jxb/erj156

Bangalore DV, McGlynn WG, Scott DD (2008) Effects of fruit maturity on watermelon ultrastructure and intracellular lycopene distribution. J Food Sci 73(5):S222–S228. https://doi.org/10.1111/j.1750-3841.2008.00778.x

Barry CS, Giovannoni JJ (2007) Ethylene and fruit ripening. J Plant Growth Regul 26(2):143–159. https://doi.org/10.1007/s00344-007-9002-y

Barsan C, Zouine M, Maza E, Bian W, Egea I, Rossignol M, Bouyssie D, Pichereaux C, Purgatto E, Bouzayen M, Latche A, Pech JC (2012) Proteomic analysis of chloroplast-to-chromoplast transition in tomato reveals metabolic shifts coupled with disrupted thylakoid biogenesis machinery and elevated energy-production components. Plant Physiol 160(2):708–725. https://doi.org/10.1104/pp.112.203679

Bassham JA, Benson AA, Kay LD, Harris AZ, Wilson AT, Calvin M (1954) The path of carbon in photosynthesis. XXI. The cyclic regeneration of carbon dioxide Acceptor1. J Am Chem Soc 76(7):1760–1770. https://doi.org/10.1021/ja01636a012

Bathgate B, Purton ME, Grierson D, Goodenough PW (1985) Plastid changes during the conversion of chloroplasts to chromoplasts in ripening tomatoes. Planta 165(2):197–204. https://doi.org/10.1007/BF00395042

Berry HM, Rickett DV, Baxter CJ, Enfissi EM, Fraser PD (2019) Carotenoid biosynthesis and sequestration in red chilli pepper fruit and its impact on colour intensity traits. J Exp Bot 70(10):2637–2650

Bhatla SC, Lal MA (2023) Light perception and transduction. In: Plant physiology, development and metabolism. Springer, Singapore, pp 363–390. https://doi.org/10.1007/978-981-99-5736-1_13

Bouvier F, Backhaus RA, Camara B (1998) Induction and control of chromoplast-specific carotenoid genes by oxidative stress. J Biol Chem 273(46):30651–30659

Bréhélin C, Kessler F, van Wijk KJ (2007) Plastoglobules: versatile lipoprotein particles in plastids. Trends Plant Sci 12(6):260–266. https://doi.org/10.1016/j.tplants.2007.04.003

Calvin M, Benson AA (1948) The path of carbon in photosynthesis. Science 107(2784):476–480. https://doi.org/10.1126/science.107.2784.476

Camara B, Brangeon J (1981) Carotenoid metabolism during chloroplast to chromoplast transformation in Capsicum annuum fruit. Planta 151(4):359–364. https://doi.org/10.1007/BF00393291

Caverzan A, Casassola A, Brammer SP (2016) Antioxidant responses of wheat plants under stress. Genet Mol Biol 39(1):1–6

Cazzonelli CI, Pogson BJ (2010) Source to sink: regulation of carotenoid biosynthesis in plants. Trends Plant Sci 15(5):266–274. https://doi.org/10.1016/j.tplants.2010.02.003

Chen W, Kong J, Qin C, Yu S, Tan J, Chen Y-r, Wu C, Wang H, Shi Y, Li C, Li B, Zhang P, Wang Y, Lai T, Yu Z, Zhang X, Shi N, Wang H, Osman T, Liu Y, Manning K, Jackson S, Rolin D, Zhong S, Seymour GB, Gallusci P, Hong Y (2015) Requirement of CHROMOMETHYLASE3 for somatic inheritance of the spontaneous tomato epimutation colourless non-ripening. Sci Rep 5(1):9192. https://doi.org/10.1038/srep09192

Chu C-C, Li H-m (2015) Protein import into isolated pea root leucoplasts. Front Plant Sci 6:690. https://doi.org/10.3389/fpls.2015.00690

Chu C-C, Li H-m (2018) Developmental regulation of protein import into plastids. Photosynth Res 138(3):327–334. https://doi.org/10.1007/s11120-018-0546-4

Chu C-C, Swamy K, Li H-m (2020) Tissue-specific regulation of plastid protein import via transit-peptide motifs. Plant Cell 32(4):1204–1217. https://doi.org/10.1105/tpc.19.00702

Cutler SR, Rodriguez PL, Finkelstein RR, Abrams SR (2010) Abscisic acid: emergence of a core signaling network. Annu Rev Plant Biol 61:651–679. https://doi.org/10.1146/annurev-arplant-042809-112122

Daher Z, Recorbet G, Valot B, Robert F, Balliau T, Potin S, Schoefs B, Dumas-Gaudot E (2010) Proteomic analysis of Medicago truncatula root plastids. Proteomics 10(11):2123–2137. https://doi.org/10.1002/pmic.200900345

Dannehl D, Schwend T, Veit D, Schmidt U (2021) Increase of yield, lycopene, and lutein content in tomatoes grown under continuous PAR spectrum LED lighting. Front Plant Sci 12:611236. https://doi.org/10.3389/fpls.2021.611236

de Abreu Martins HH, Turmo-Ibarz A, Piccoli RH, Martín-Belloso O, Salvia-Trujillo L (2021) Influence of lipid nanoparticle physical state on β-carotene stability kinetics under different environmental conditions. Food Funct 12(2):840–851

Della Penna D, Pogson BJ (2006) Vitamin synthesis in plants: tocopherols and carotenoids. Annu Rev Plant Biol 57:711–738

Demmig-Adams B, Adams WW III, Barker DH, Logan BA, Bowling DR, Verhoeven AS (1996) Using chlorophyll fluorescence to assess the fraction of absorbed light allocated to thermal dissipation of excess excitation. Physiol Plant 98(2):253–264. https://doi.org/10.1034/j.1399-3054.1996.980206.x

Dharmapuri S, Rosati C, Pallara P, Aquilani R, Bouvier F, Camara B, Giuliano G (2002) Metabolic engineering of xanthophyll content in tomato fruits. FEBS Lett 519(1):30–34. https://doi.org/10.1016/S0014-5793(02)02699-6

Ding Y, Yang S (2022) Surviving and thriving: how plants perceive and respond to temperature stress. Dev Cell 57(8):947–958. https://doi.org/10.1016/j.devcel.2022.03.010

Diretto G, Welsch R, Tavazza R, Mourgues F, Pizzichini D, Beyer P, Giuliano G (2007) Silencing of beta-carotene hydroxylase increases total carotenoid and beta-carotene levels in potato tubers. BMC Plant Biol 7(1):11. https://doi.org/10.1186/1471-2229-7-11

Dupont FM (2008) Metabolic pathways of the wheat (Triticum aestivum) endosperm amyloplast revealed by proteomics. BMC Plant Biol 8(1):39. https://doi.org/10.1186/1471-2229-8-39

Egea I, Barsan C, Bian W, Purgatto E, Latche A, Chervin C, Bouzayen M, Pech JC (2010) Chromoplast differentiation: current status and perspectives. Plant Cell Physiol 51(10):1601–1611. https://doi.org/10.1093/pcp/pcq136

Egea I, Bian W, Barsan C, Jauneau A, Pech JC, Latche A, Li Z, Chervin C (2011) Chloroplast to chromoplast transition in tomato fruit: spectral confocal microscopy analyses of carotenoids and chlorophylls in isolated plastids and time-lapse recording on intact live tissue. Ann Bot 108(2):291–297. https://doi.org/10.1093/aob/mcr140

Farneti B, Schouten RE, Woltering EJ (2012) Low temperature-induced lycopene degradation in red ripe tomato evaluated by remittance spectroscopy. Postharvest Biol Technol 73:22–27. https://doi.org/10.1016/j.postharvbio.2012.05.008

Fraser PD, Enfissi EMA, Halket JM, Truesdale MR, Yu D, Gerrish C, Bramley PM (2007) Manipulation of phytoene levels in tomato fruit: effects on isoprenoids, plastids, and intermediary metabolism. Plant Cell 19(10):3194–3211. https://doi.org/10.1105/tpc.106.049817

Fraser PD, Schuch W, Bramley PM (2000) Phytoene synthase from tomato (Lycopersicon esculentum) chloroplasts—partial purification and biochemical properties. Planta 211(3):361–369. https://doi.org/10.1007/s004250000293

Frey-Wyssling A, Schwegler F (1965) Ultrastructure of the chromoplasts in the carrot root. J Ultrastruct Res 13(5):543–559. https://doi.org/10.1016/S0022-5320(65)90013-4

Fujisawa M, Nakano T, Shima Y, Ito Y (2013) A large-scale identification of direct targets of the tomato MADS box transcription factor ripening inhibitor reveals the regulation of fruit ripening. Plant Cell 25(2):371–386. https://doi.org/10.1105/tpc.112.108118

Gill SS, Tuteja N (2010) Reactive oxygen species and antioxidant machinery in abiotic stress tolerance in crop plants. Plant Physiol Biochem 48(12):909–930

González-Peña MA, Ortega-Regules AE, Anaya de Parrodi C, Lozada-Ramírez JD (2023) Chemistry, occurrence, properties, applications, and encapsulation of carotenoids—a review. Plants (Basel) 12(2):313

Hadjeb N, Gounaris I, Price CA (1988) Chromoplast-specific proteins in Capsicum annuum. Plant Physiol 88(1):42–45

Harris WM, Spurr AR (1969) Chromoplasts of tomato fruits. I. Ultrastructure of low-pigment and high-beta mutants. Carotene analyses. Am J Bot 56(4):369–379. https://doi.org/10.2307/2440812

Hoffmann AM, Noga G, Hunsche M (2016) Alternating high and low intensity of blue light affects PSII photochemistry and raises the contents of carotenoids and anthocyanins in pepper leaves. Plant Growth Regul 79(3):275–285. https://doi.org/10.1007/s10725-015-0132-0

Howitt CA, Pogson BJ (2006) Carotenoid accumulation and function in seeds and non-green tissues. Plant Cell Environ 29(3):435–445

Hugueney P, Badillo A, Chen H-C, Klein A, Hirschberg J, Camara B, Kuntz M (1995) Metabolism of cyclic carotenoids: a model for the alteration of this biosynthetic pathway in Capsicum annuum chromoplasts. Plant J 8(3):417–424. https://doi.org/10.1046/j.1365-313X.1995.08030417.x

Huh JH, Kang BC, Nahm SH, Kim S, Ha KS, Lee MH, Kim BD (2001) A candidate gene approach identified phytoene synthase as the locus for mature fruit color in red pepper (Capsicum spp.). Theor Appl Genet 102(4):524–530. https://doi.org/10.1007/s001220051677

Huo T, Ferruzzi MG, Schwartz SJ, Failla ML (2007) Impact of fatty acyl composition and quantity of triglycerides on bioaccessibility of dietary carotenoids. J Agric Food Chem 55(22):8950–8957

Iglesias-Sanchez A, Navarro-Carcelen J, Morelli L, Rodriguez-Concepcion M (2023) Arabidopsis FIBRILLIN6 influences carotenoid biosynthesis by directly promoting phytoene synthase activity. Plant Physiol 194(3):1662–1673. https://doi.org/10.1093/plphys/kiad613

Iuchi S, Kobayashi M, Taji T, Naramoto M, Seki M, Kato T, Tabata S, Kakubari Y, Yamaguchi-Shinozaki K, Shinozaki K (2001) Regulation of drought tolerance by gene manipulation of 9-cis-epoxycarotenoid dioxygenase, a key enzyme in abscisic acid biosynthesis in Arabidopsis. Plant J 27(4):325–333. https://doi.org/10.1046/j.1365-313x.2001.01096.x

Jarvis P, López-Juez E (2013) Biogenesis and homeostasis of chloroplasts and other plastids. Nat Rev Mol Cell Biol 14(12):787–802

Jeffery J, Holzenburg A, King S (2012) Physical barriers to carotenoid bioaccessibility. Ultrastructure survey of chromoplast and cell wall morphology in nine carotenoid-containing fruits and vegetables. J Sci Food Agric 92(13):2594–2602. https://doi.org/10.1002/jsfa.5767

Kapoor L, Simkin AJ, George Priya Doss C, Siva R (2022) Fruit ripening: dynamics and integrated analysis of carotenoids and anthocyanins. BMC Plant Biol 22(1):27. https://doi.org/10.1186/s12870-021-03411-w

Karniel U, Koch A, Zamir D, Hirschberg J (2020) Development of zeaxanthin-rich tomato fruit through genetic manipulations of carotenoid biosynthesis. Plant Biotechnol J 18(11):2292–2303. https://doi.org/10.1111/pbi.13387

Keawmanee N, Ma G, Zhang L, Kato M (2023) Regulation of chlorophyll and carotenoid metabolism in citrus fruit during maturation and regreening. Rev Agric Sci 11:203–216. https://doi.org/10.7831/ras.11.0_203

Kerchev PI, Fenton B, Foyer CH, Hancock RD (2012) Plant responses to insect herbivory: interactions between photosynthesis, reactive oxygen species and hormonal signalling pathways. Plant Cell Environ 35(2):441–453. https://doi.org/10.1111/j.1365-3040.2011.02399.x

Kim JE, Rensing KH, Douglas CJ, Cheng KM (2010) Chromoplasts ultrastructure and estimated carotene content in root secondary phloem of different carrot varieties. Planta 231(3):549–558. https://doi.org/10.1007/s00425-009-1071-7

Kim SH, Ahn YO, Ahn M-J, Jeong JC, Lee H-S, Kwak S-S (2013) Cloning and characterization of an Orange gene that increases carotenoid accumulation and salt stress tolerance in transgenic sweetpotato cultures. Plant Physiol Biochem 70:445–454

Klee HJ, Giovannoni JJ (2011) Genetics and control of tomato fruit ripening and quality attributes. Annu Rev Genet 45(1):41–59. https://doi.org/10.1146/annurev-genet-110410-132507

Koltai H (2015) Cellular events of strigolactone signalling and their crosstalk with auxin in roots. J Exp Bot 66(16):4855–4861. https://doi.org/10.1093/jxb/erv178

Kuntz M, Dimnet L, Pullara S, Moyet L, Rolland N (2024) The main functions of plastids. In: Maréchal E (ed) Plastids: methods and protocols. Springer US, New York, pp 89–106. https://doi.org/10.1007/978-1-0716-3726-5_5

Kuntz M, Rolland N (2012) Subcellular and sub-organellar proteomics as a complementary tool to study the evolution of the plastid proteome. In: Organelle genetics. Springer, Heidelberg, pp 217–238

Kuntz C, Simkin R, Shipton D, Schuch B (1998) Upregulation of two ripening-related genes from a non-climacteric plant (pepper) in a transgenic climacteric plant (tomato). Plant J 13(3):351–361. https://doi.org/10.1046/j.1365-313X.1998.00032.x

Lang Z, Wang Y, Tang K, Tang D, Datsenka T, Cheng J, Zhang Y, Handa AK, Zhu J-K (2017) Critical roles of DNA demethylation in the activation of ripening-induced genes and inhibition of ripening-repressed genes in tomato fruit. Proc Natl Acad Sci 114(22):E4511–E4519. https://doi.org/10.1073/pnas.1705233114

Li L, Yuan H (2013) Chromoplast biogenesis and carotenoid accumulation. Arch Biochem Biophys 539(2):102–109. https://doi.org/10.1016/j.abb.2013.07.002

Li X, Wang X, Zhang Y, Zhang A, You C-X (2022) Regulation of fleshy fruit ripening: from transcription factors to epigenetic modifications. Hortic Res 9:uhac013. https://doi.org/10.1093/hr/uhac013

Ling Q, Sadali NM, Soufi Z, Zhou Y, Huang B, Zeng Y, Rodriguez-Concepcion M, Jarvis RP (2021) The chloroplast-associated protein degradation pathway controls chromoplast development and fruit ripening in tomato. Nat Plants 7(5):655–666. https://doi.org/10.1038/s41477-021-00916-y

Ling QH, Huang WH, Baldwin A, Jarvis P (2012) Chloroplast biogenesis is regulated by direct action of the ubiquitin-proteasome system. Science 338(6107):655–659. https://doi.org/10.1126/science.1225053

Ling QH, Jarvis P (2015) Regulation of chloroplast protein import by the ubiquitin E3 ligase SP1 is important for stress tolerance in plants. Curr Biol 25(19):2527–2534. https://doi.org/10.1016/j.cub.2015.08.015

Liu LH, Shao ZY, Zhang M, Wang QM (2015a) Regulation of carotenoid metabolism in tomato. Mol Plant 8(1):28–39. https://doi.org/10.1016/j.molp.2014.11.006

Liu R, How-Kit A, Stammitti L, Teyssier E, Rolin D, Mortain-Bertrand A, Halle S, Liu M, Kong J, Wu C, Degraeve-Guibault C, Chapman NH, Maucourt M, Hodgman TC, Tost J, Bouzayen M, Hong Y, Seymour GB, Giovannoni JJ, Gallusci P (2015b) A DEMETER-like DNA demethylase governs tomato fruit ripening. Proc Natl Acad Sci 112(34):10804–10809. https://doi.org/10.1073/pnas.1503362112

Liu X, Qian J, Xu M, Qing H, Yu G, Lv Y, Zhang C, Fu J (2024) ORANGE gene positively regulates chromoplast formation and carotenoid accumulation in Osmanthus fragrans. Sci Hortic 337:113600

Llorente B, Torres-Montilla S, Morelli L, Florez-Sarasa I, Matus JT, Ezquerro M, D'Andrea L, Houhou F, Majer E, Picó B, Cebolla J, Troncoso A, Fernie AR, Daròs J-A, Rodriguez-Concepcion M (2020) Synthetic conversion of leaf chloroplasts into carotenoid-rich plastids reveals mechanistic basis of natural chromoplast development. Proc Natl Acad Sci 117(35):21796–21803. https://doi.org/10.1073/pnas.2004405117

Lois LM, Rodríguez-Concepción M, Gallego F, Campos N, Boronat A (2000) Carotenoid biosynthesis during tomato fruit development: regulatory role of 1-deoxy-D-xylulose 5-phosphate synthase. Plant J 22(6):503–513

Lu S, Van Eck J, Zhou X, Lopez AB, O'Halloran DM, Cosman KM, Conlin BJ, Paolillo DJ, Garvin DF, Vrebalov J (2006) The cauliflower Or gene encodes a DnaJ cysteine-rich domain-containing protein that mediates high levels of β-carotene accumulation. Plant Cell 18(12):3594–3605

Manning K, Tör M, Poole M, Hong Y, Thompson AJ, King GJ, Giovannoni JJ, Seymour GB (2006) A naturally occurring epigenetic mutation in a gene encoding an SBP-box transcription factor inhibits tomato fruit ripening. Nat Genet 38(8):948–952. https://doi.org/10.1038/ng1841

Maoka T (2020) Carotenoids as natural functional pigments. J Nat Med 74(1):1–16. https://doi.org/10.1007/s11418-019-01364-x

Matile P (2000) Biochemistry of Indian summer: physiology of autumnal leaf colouration. Exp Gerontol 35(2):145–158

Meena RK, Jangra S, Wadhwa Z, Wati M, Wati L (2017) Role of plant volatiles in defense and communication. Int J Curr Microbiol App Sci 6:300–313

Mohd Hassan N, Yusof NA, Yahaya AF, Mohd Rozali NN, Othman R (2019) Carotenoids of capsicum fruits: pigment profile and health-promoting functional attributes. Antioxidants (Basel) 8(10):469

Müller M, Munné-Bosch S (2021) Hormonal impact on photosynthesis and photoprotection in plants. Plant Physiol 185(4):1500–1522. https://doi.org/10.1093/plphys/kiaa119

Muñoz P, Munné-Bosch S (2018) Photo-oxidative stress during leaf, flower and fruit development. Plant Physiol 176(2):1004–1014

Mutsokoti L, Panozzo A, Tongonya J, Kebede BT, Van Loey A, Hendrickx M (2017) Carotenoid stability and lipid oxidation during storage of low-fat carrot and tomato based systems. LWT 80:470–478. https://doi.org/10.1016/j.lwt.2017.03.021

Naqvi S, Zhu C, Farre G, Sandmann G, Capell T, Christou P (2011) Synergistic metabolism in hybrid corn indicates bottlenecks in the carotenoid pathway and leads to the accumulation of extraordinary levels of the nutritionally important carotenoid zeaxanthin. Plant Biotechnol J 9(3):384–393. https://doi.org/10.1111/j.1467-7652.2010.00554.x

Oleszkiewicz T, Klimek-Chodacka M, Milewska-Hendel A, Zubko M, Stróż D, Kurczyńska E, Boba A, Szopa J, Baranski R (2018) Unique chromoplast organisation and carotenoid gene expression in carotenoid-rich carrot callus. Planta 248(6):1455–1471. https://doi.org/10.1007/s00425-018-2988-5

Paolillo DJ, Garvin DF, Parthasarathy MV (2004) The chromoplasts of Or mutants of cauliflower (Brassica oleracea L. var. botrytis). Protoplasma 224(3):245–253. https://doi.org/10.1007/s00709-004-0059-1

Pinheiro CB, Shah M, Soares EL, Nogueira FCS, Carvalho PC, Junqueira M, Araújo GDT, Soares AA, Domont GB, Campos FAP (2013) Proteome analysis of plastids from developing seeds of Jatropha curcas L. J Proteome Res 12(11):5137–5145. https://doi.org/10.1021/pr400515b

Polyakov NE, Focsan AL, Gao Y, Kispert LD (2023) The endless world of carotenoids—structural, chemical and biological aspects of some rare carotenoids. Int J Mol Sci 24(12):9885

Raines CA (2003) The Calvin cycle revisited. Photosynth Res 75(1):1–10. https://doi.org/10.1023/A:1022421515027

Rödiger A, Agne B, Dobritzsch D, Helm S, Müller F, Pötzsch N, Baginsky S (2021) Chromoplast differentiation in bell pepper (Capsicum annuum) fruits. Plant J 105(5):1431–1442. https://doi.org/10.1111/tpj.15104

Römer S, Lübeck J, Kauder F, Steiger S, Adomat C, Sandmann G (2002) Genetic engineering of a zeaxanthin-rich potato by antisense inactivation and co-suppression of carotenoid epoxidation. Metab Eng 4(4):263–272. https://doi.org/10.1006/mben.2002.0234

Rosso SW (1968) The ultrastructure of chromoplast development in red tomatoes. J Ultrastruct Res 25(3):307–322

Rottet S, Besagni C, Kessler F (2015) The role of plastoglobules in thylakoid lipid remodelling during plant development. Biochim Biophys Acta 1847(9):889–899. https://doi.org/10.1016/j.bbabio.2015.02.002

Rudolf J, Tomovicova L, Panzarova K, Fajkus J, Hejatko J, Skalak J (2024) Epigenetics and plant hormone dynamics: a functional and methodological perspective. J Exp Bot 75(17):5267–5294. https://doi.org/10.1093/jxb/erae054

Ruiz-Sola MÁ, Rodríguez-Concepción M (2012) Carotenoid biosynthesis in Arabidopsis: a colourful pathway. Arabidopsis Book 10:e0158. https://doi.org/10.1199/tab.0158

Ryckebosch E, Muylaert K, Eeckhout M, Ruyssen T, Foubert I (2011) Influence of drying and storage on lipid and carotenoid stability of the microalga Phaeodactylum tricornutum. J Agric Food Chem 59(20):11063–11069

Sadali NM, Sowden RG, Ling Q, Jarvis RP (2019) Differentiation of chromoplasts and other plastids in plants. Plant Cell Rep 38(7):803–818. https://doi.org/10.1007/s00299-019-02420-2

Schweiggert RM, Mezger D, Schimpf F, Steingass CB, Carle R (2012) Influence of chromoplast morphology on carotenoid bioaccessibility of carrot, mango, papaya, and tomato. Food Chem 135(4):2736–2742. https://doi.org/10.1016/j.foodchem.2012.07.035

Siddique MA, Grossmann J, Gruissem W, Baginsky S (2006) Proteome analysis of bell pepper (Capsicum annuum L.) chromoplasts. Plant Cell Physiol 47(12):1663–1673

Simkin AJ, Laizet Y, Kuntz M (2004) Plastid lipid associated proteins of the plant fibrillin family: structure, localisation, functions and gene expression. Recent Res Dev Biochem 5:307–316

Spurr AR, Harris WM (1968) Ultrastructure of chloroplasts and chromoplasts in Capsicum annuum. I Thylakoid membrane changes during fruit ripening. Am J Bot 55(10):1210–1224. https://doi.org/10.2307/2440743

Stensballe A, Hald S, Bauw G, Blennow A, Welinder KG (2008) The amyloplast proteome of potato tuber. FEBS J 275(8):1723–1741. https://doi.org/10.1111/j.1742-4658.2008.06332.x

Su H, Song Y, Hou Y, Cheng G (2022) A review on resistance to biotic stress in leaf-colored plant. Adv Biosci Biotechnol 13(5):226–241

Sun T, Wang P, Rao S, Zhou X, Wrightstone E, Lu S, Yuan H, Yang Y, Fish T, Thannhauser T, Liu J, Mazourek M, Grimm B, Li L (2023) Co-chaperoning of chlorophyll and carotenoid biosynthesis by ORANGE family proteins in plants. Mol Plant 16(6):1048–1065. https://doi.org/10.1016/j.molp.2023.05.006

Sun T, Xu Z, Wu C-T, Janes M, Prinyawiwatkul W, No HK (2007) Antioxidant activities of different coloured sweet bell peppers (Capsicum annuum L.). J Food Sci 72(2):S98–S102. https://doi.org/10.1111/j.1750-3841.2006.00245.x

Sun T, Yuan H, Cao H, Yazdani M, Tadmor Y, Li L (2018) Carotenoid metabolism in plants: the role of plastids. Mol Plant 11(1):58–74

Suzuki M, Takahashi S, Kondo T, Dohra H, Ito Y, Kiriiwa Y, Hayashi M, Kamiya S, Kato M, Fujiwara M (2015) Plastid proteomic analysis in tomato fruit development. PLoS One 10(9):1–25

Toledo-Ortiz G, Huq E, Rodríguez-Concepción M (2010) Direct regulation of phytoene synthase gene expression and carotenoid biosynthesis by phytochrome-interacting factors. Proc Natl Acad Sci 107(25):11626–11631. https://doi.org/10.1073/pnas.0914428107

Tomes ML (1963) Temperature inhibition of carotene synthesis in tomato. Bot Gaz 124(3):180–185. https://doi.org/10.1086/336189

Wang Y, Zhang C, Xu B, Fu J, Du Y, Fang Q, Dong B, Zhao H (2022) Temperature regulation of carotenoid accumulation in the petals of sweet osmanthus via modulating expression of carotenoid biosynthesis and degradation genes. BMC Genomics 23(1):418. https://doi.org/10.1186/s12864-022-08643-0

Wang YQ, Yang Y, Fei Z, Yuan H, Fish T, Thannhauser TW, Mazourek M, Kochian LV, Wang X, Li L (2013) Proteomic analysis of chromoplasts from six crop species reveals insights into chromoplast function and development. J Exp Bot 64(4):949–961. https://doi.org/10.1093/jxb/ers375

Watkins JL, Li M, McQuinn RP, Chan KX, McFarlane HE, Ermakova M, Furbank RT, Mares D, Dong C, Chalmers KJ, Sharp P, Mather DE, Pogson BJ (2019) A GDSL esterase/lipase catalyses the esterification of lutein in bread wheat. Plant Cell 31(12):3092–3112. https://doi.org/10.1105/tpc.19.00272

Watkins JL, Pogson BJ (2020) Prospects for carotenoid biofortification targeting retention and catabolism. Trends Plant Sci 25(5):501–512. https://doi.org/10.1016/j.tplants.2019.12.021

Wijesekara T, Xu B (2024) A critical review on the stability of natural food pigments and stabilization techniques. Food Res Int 179:114011. https://doi.org/10.1016/j.foodres.2024.114011

Yan T, Cai B, Li F, Guo D, Xia C, Lv H, Lin B, Gao H, Geng Z (2024) Proteomic and metabolomic revealed the effect of shading treatment on cigar tobacco. Front Plant Sci 15:1433575. https://doi.org/10.3389/fpls.2024.1433575

Yuan H, Owsiany K, Sheeja TE, Zhou X, Rodriguez C, Li Y, Welsch R, Chayut N, Yang Y, Thannhauser TW, Parthasarathy MV, Xu Q, Deng X, Fei Z, Schaffer A, Katzir N, Burger J, Tadmor Y, Li L (2015a) A single amino acid substitution in an ORANGE protein promotes carotenoid overaccumulation in Arabidopsis. Plant Physiol 169(1):421–431. https://doi.org/10.1104/pp.15.00971

Yuan H, Zhang J, Nageswaran D, Li L (2015b) Carotenoid metabolism and regulation in horticultural crops. Hortic Res 2:15036. https://doi.org/10.1038/hortres.2015.36

Zhang B, Tieman DM, Jiao C, Xu Y, Chen K, Fei Z, Giovannoni JJ, Klee HJ (2016) Chilling-induced tomato flavor loss is associated with altered volatile synthesis and transient changes in DNA methylation. Proc Natl Acad Sci 113(44):12580–12585. https://doi.org/10.1073/pnas.1613910113

Zheng X, Huang L, Fan B, Peng C, Iqbal A, Zhang Y, Chen H, Ye J, Yang Y (2024) Integrated transcriptomics and metabolomics analyses of the effects of bagging treatment on carotenoid biosynthesis and regulation of Areca catechu L. Front Plant Sci 15:1364945. https://doi.org/10.3389/fpls.2024.1364945

Zhong S, Fei Z, Chen Y-R, Zheng Y, Huang M, Vrebalov J, McQuinn R, Gapper N, Liu B, Xiang J, Shao Y, Giovannoni JJ (2013) Single-base resolution methylomes of tomato fruit development reveal epigenome modifications associated with ripening. Nat Biotechnol 31(2):154–159. https://doi.org/10.1038/nbt.2462

Zhou X, Sun T, Owens L, Yang Y, Fish T, Wrightstone E, Lui A, Yuan H, Chayut N, Burger J (2023) Carotenoid sequestration protein FIBRILLIN participates in CmOR-regulated β-carotene accumulation in melon. Plant Physiol 193(1):643–660

Zhou X, Welsch R, Yang Y, Álvarez D, Riediger M, Yuan H, Fish T, Liu J, Thannhauser TW, Li L (2015) Arabidopsis OR proteins are the major posttranscriptional regulators of phytoene synthase in controlling carotenoid biosynthesis. Proc Natl Acad Sci 112(11):3558–3563. https://doi.org/10.1073/pnas.1420831112

Zhu C, Naqvi S, Breitenbach J, Sandmann G, Christou P, Capell T (2008) Combinatorial genetic transformation generates a library of metabolic phenotypes for the carotenoid pathway in maize. Proc Natl Acad Sci 105(47):18232–18237. https://doi.org/10.1073/pnas.0809737105

Zhu M, Lin J, Ye J, et al (2018) A comprehensive proteomic analysis of elaioplasts from citrus fruits reveals insights into elaioplast biogenesis and function. Horticulture Research, 5, 6. https://doi.org/10.1038/s41438-017-0014-x

Zita W, Bressoud S, Glauser G, Kessler F, Shanmugabalaji V (2022) Chromoplast plastoglobules recruit the carotenoid biosynthetic pathway and contribute to carotenoid accumulation during tomato fruit maturation. PLoS One 17(12):e0277774. https://doi.org/10.1371/journal.pone.0277774

Amyloplasts: Starch-Synthesising Plastids

10

Vijay Kumar Dalal

Abstract

With the goal of achievement of the UN SDGs in the foreseeable future, the scientific community is engaged in increasing food production and improving the quality of food. Amyloplasts are important organelles in this respect. They contain starch; the starch transitions into sugar when fruits mature. Thus, the formation of starch grains and their degradation is important for the yield/quantity of the fruit/food. Another important aspect of amyloplast is the determination of the quality of grain. There are several parameters of starch grains/starch granules that govern the properties of the flour, including viscosity, chalkiness, dough formation, gelatinization etc., responsible for different characteristic during food processing. Therefore understanding the amyloplast formation, development, structure and composition is of utmost importance. The research in the amyloplast field have led to significant commercial gains and are potentially going to increase in the coming decades with the help of advances in the AI, ML, data handling, automation and OMICS technologies. This chapter describes the current knowledge about the amyloplasts.

Keywords

Amyloplasts · Starch Grains · Starch synthesis · Conversion of amyloplasts · Flo · Starch mutants

V. K. Dalal (✉)
Department of Botany, Dayalbagh Educational Institute, Agra, Uttar Pradesh, India
e-mail: dalalvijay@dei.ac.in

V. K. Dalal, A. N. Misra (eds.), *Chloroplast Biogenesis and Plastid Interconversions*,
https://doi.org/10.1007/978-981-95-0772-6_10

10.1 Introduction

Amyloplasts are organelles that contain starch. They are mostly formed of chloroplasts and proplastids (Fig. 10.1) (Liebers et al. 2017; Jarvis and Lopez-Juez 2013). They are found in storage tissues such as fruits, endosperm of seeds, grains, tuberous roots, e.g., cassava, and tuber-stolons (Sagisaka 2008); but they can also be found in non-storage tissues on a temporary basis such as leaves and non-tuberous roots (Choi et al. 2021; Liebers et al. 2017). Starch accumulates in amyloplasts mostly during the maturation period such as in Italian arum (*Arum italicum*) (Bonora et al. 2000) and banana (Solis-Badillo et al. 2020). In the leaves of Arabidopsis and other plants, the starch content in amyloplasts fluctuates on a daily basis/diurnal basis depending on the photosynthetic activity (Fernandez et al. 2017). During the light period, starch formation is high, and this is transported, mostly later during the dark period, to the other portions of the plant. During the dark period, starch is consumed in respiration or transported to the sink. The amyloplasts transition phases for conversion of amyloplast to chloroplast are also found in tubers of sweet potato and in some fruits such as peach palm (*Bactris gasipaes*), and winter squashes (Hempel et al. 2014).

There might be one or many starch grains (SGs) in amyloplasts. SGs can be simple or compound. Compound SGs, found in rice, are 10–20 μm (Matsushima et al. 2013). They are made of the assembly of polyhedral small (3–8 μm in diameter) SGs and look like a turtle shell. Simple SGs are composed of single SGs and found in barley, sorghum, wheat, potato (*Solanum tuberosum*) and maize (Tateoka 1962; Matsushima et al. 2013, 2014). In barley and wheat, simple SGs are of two

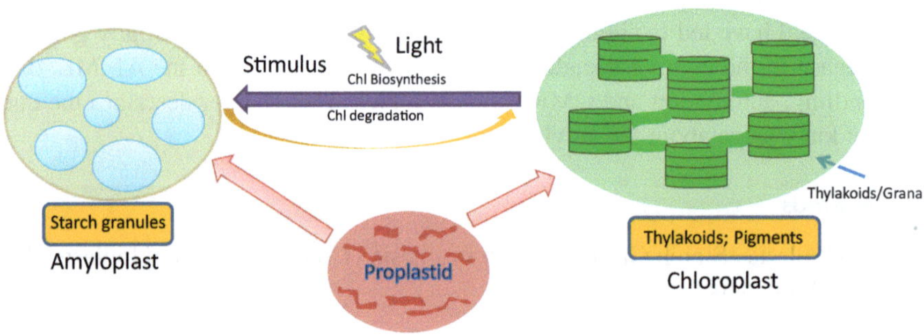

Fig. 10.1 Development of amyloplasts from proplastids and chloroplasts. The major route of amyloplast formation is through the chloroplast transition that involves the degradation of grana and thylakoids, and the formation of plastoglobules and starch grains. Proplastids, found in meristematic cells, are the progenitors of different types of plastids, e.g., chloroplasts, chromoplasts, gerontoplasts and leucoplasts. They contain a very basic structure of lamellae without chlorophyll (Charuvi et al. 2012). Chloroplasts mostly develop from the proplastids in the meristematic cells in the shoot apices. They are present in all green tissues, harbour thylakoid membranes containing pigment-protein complexes and perform photosynthesis. Amyloplasts, in contrast, are mostly found in the storage tissues of plants, calli (in vitro) and roots

types: larger ones of 15–25 μm diameter and smaller ones of less than 10 μm diameter. In Maize and sorghum all SGs are of uniform size, approximately 10 μm in diameter (Matsushima et al. 2013). Multiple compound SGs were reported in ssg6-GFP transgenic lines (Matsushima and Hisano 2019). The SGs constitute a small part of the total plastid during initial kernel development but ~93% at maturity (Briarty et al. 1979). Amyloplasts can be detected in different stages of microsporogenesis, including meiosis and the development of microspores and pollen grains, e.g., in Cannabis (Galán-Ávila et al. 2021). Amyloplasts with single starch grains (SGs) are also present in microspores from the early bicellular stage (Nepi et al. 1996). The number of amyloplasts and their presence in microspores can influence pollen viability and development in barley cultivars (Gajecka et al. 2020). During microspore formation, undifferentiated proplastids divide and differentiate into amyloplasts containing starch. Later these amyloplasts form lobes and divide.

SGs are also categorized according to their size. Although there is a range in the sizes of amyloplasts in barley and wheat, the majority of them can be distinguished by two distinct sizes: size A and B. Size A SGs range from 10 to 25 μm and Size B 1–7 μm (Bechtel et al. 1990). In some crops, even smaller sized SGs are observed, which are denoted Size C. Potatoes are representative organs for amyloplasts. Starch Grains accumulate during the maturation period of amyloplast in banana and Italian arum, and during in vitro callus and endosperm formation in Apple (Sagisaka 2008). In barley and wheat endosperms, one large discoid A-type granule is initiated earlier during endosperm development and smaller-sized B-type spherical SGs are initiated later (Bechtel et al. 1990; Bechtel and Wilson, 2003; Wurtzel et al. 2012; Hanson and Conklin 2020). Starch grains can be present singly, or many SGs can be present in one cell. B types of starch grains are present in stromules, which are the outgrowths of plastids in the form of tubes enclosed by the chloroplast envelope. The presence of a broad range of SGs in fruits/tubers influences the quality of starch through modulation of gelatinisation, viscosity, water absorption and swelling (Chen et al. 2021). These SG characteristics may be changed in some mutants, e.g., ARC5 mutants showed higher starch gelatinizing temperature than the wild type (Yun and Kawagoe 2009).

Many factors such as carotenoids influence amyloplast formation (Wang et al. 2020). Amyloplasts normally accumulate low levels of carotenoids since there are only a few functions for carotenoids in amyloplasts, e.g., ABA synthesis, but some plants/plant varieties may possess higher amounts of carotenoids (Zhou et al. 2011). Among carotenoids, mostly xanthophylls (lutein, violaxanthin and zeaxanthin) are accumulated in amyloplasts (Howitt and Pogson 2006). The accumulation of carotenoids is involved in transforming amyloplasts to amylochromoplasts in sweet potato, thus upregulating the chromoplast biogenesis (Gemenet et al. 2020). This causes a competition in carotenoids and starch synthesis, which is why high carotenoid-containing sweet potato (*Ipomoea batatas*) varieties had low starch content (Drapal et al. 2022). Besides the starch formation, amyloplasts also perform several other functions, e.g., it has been proposed based on the proteomics composition of the kiwi fruit amyloplasts and other grains, several proteins responsible for the fatty acid degradation, chlorophyll formation and degradation, and photosynthetic proteins are found in amyloplasts. Amyloplasts have been found to have a role in nitrogen assimilation also. They contain many enzymes of the GS/GOGAT cycle similar to the root nodules of proplastids (Wise 2007). Amyloplasts are also involved in glucose metabolism, glycolysis, the pentose phosphate pathway, and sulphur metabolism. Moreover, amyloplasts are implicated in diseases, e.g., amyloplasts have been involved in the development of Bitter Pit disease in apple (Qiu et al. 2021).

A good amount of food is coming to the pods from the nearby leaves. Pod photosynthesis is important for the early stages of seed development (Zhang et al. 2017). Formation of amyloplasts in grains inside fruits/pods is a significant factor in determining the yield of plants. In Free Ambient CO_2 Enrichment (FACE) experiments, the yield of grains is determined mainly by starch accumulation (Hasegawa et al. 2013). Normally yield may or may not be directly determined by the starch grain accumulation, which is proposed to be sink-limited (Zi et al. 2018), whereas protein accumulation is proposed as source-limited (Martre et al. 2006; Barneix 2007). For storing protein in the grains, nitrogen supply is important for the type and amount of protein accumulated in the kernels (Bänziger et al. 1994; Rharrabti et al. 2001; Dupont et al. 2006; Martre et al. 2006). There is a limit to the maximum percentage of protein accumulated in the grains of high-yielding varieties (Fowler 2003), with some exceptions. Some conditions stimulate high production of the protein in the grains, e.g., in wheat, supplying nitrogen after anthesis under low temperature and sufficient amount of water induces high grain yield and protein content (Rharrabti et al. 2001; Altenbach et al. 2003; Myers et al. 2011) although there is a negative correlation between the amount of protein and the amount of essential amino acids such as such as methionine, cysteine and lysine accumulated in grain (Andon et al. 2002; Balmer et al. 2006a, b). It is postulated that amyloplasts harbour a regulatory mechanism to respond to the environment for modulating grain filling in terms of starch and protein content (DuPont 2008). The timing of grain filling, its duration and rate determine the amount of protein in the grain. Sucrose transport, from distant leaves to the grains, is not so important for grain filling with starch; rather, the fixed CO_2 in awns and nearby leaves is sufficient to fill the grain. There are some fruits where the photosynthesis activity of the fruit contributes very little to grain filling, e.g., pod (fruit pericarp) photosynthesis rate is only 5% of the leaf in Chickpea (Oury and Godin 2007). The photosynthesis activity of fruit contributes ~75% to grain starch accumulation in *Brassica*. In others, it is more important, e.g., in alfalfa. Photosynthesis is also important for oil production in seeds. Studies in Arabidopsis (Sheoran et al. 1991) and soybean (Cho et al. 2023) reported that seed oil content is increased through the photosynthesis activity of the pod. Lipids present in the starch grains arrive from the degradation of the amyloplast lipid membranes (Tan and Morrison 1979). Another grain trait determined by amyloplasts is kernel hardiness. Little is known about how proteins are imported into amyloplasts in higher plants.

Research indicates that in young amyloplasts of barley and maize, the inner membrane is highly invaginated, forming structures such as tubules, stroma lamellae, and vesicles. Buttrose (1963) proposed that these membrane invaginations could increase surface area, potentially facilitating faster sugar or sugar phosphate uptake into the amyloplast stroma if the inner membrane were the limiting barrier. However, current understanding is that the chloroplast membranes are largely impermeable to neutral sugars and hexose phosphates, suggesting that carbohydrates probably enter amyloplasts primarily as phosphate molecules through the phosphate translocator (Shannon et al. 2009 and the cited literature there). The DNA in the amyloplasts is normally not transcribed or decreases during development, however, the copy number of chloroplast DNA is initially increased and later decreased during wheat amyloplast/endosperm development (Catley et al. 1987).

There has been extensive research in understanding the process of formation of SGs as it is of commercial interest in terms of protein and starch quality parameters of flour that are determinants of food processing, e.g., high protein content helps in the formation of dough. SGs are mostly composed of amylose and amylopectin. Amylose accounts for 0–30% of the total starch content. Its concentration varies widely among species and even among the same species. A protein, *Waxy* (*Wx*), found in the starch granules is a starch-synthesizing enzyme and is responsible for amylose formation. *Wx* transcript was found to be present only in pollen grains, endosperm and embryo sac, inducing authors to propose that it is comprehensively regulated at the transcriptional level. In the *Waxy* wheat, starch was of a more crystalline nature, B-type SGs were higher and formed faster during amyloplast development. Initially amyloplast development was slow in *waxy* wheat, but later it increased along with smaller B-type amyloplasts development (Zhu et al. 2003; Bello et al. 2019).

10.2 Biosynthesis of Starch

Starch synthesis is performed by several enzymes, including sucrose synthase (SUS), which convert sucrose into UDP-glucose that is further converted into ADP-glucose (ADPG) by AGPase. ADP-glucose pyrophosphorylase consists of two subunits: AGPase small subunit encoded by *brittle2* (*bt2 or BT2*) and the large subunit encoded by *shrunken2* (*sh2*). ADPG is transported to amyloplasts where ADPG acts as the activated glucosyl donor for the formation of amylose by granule-bound starch synthase (GBSS, encoded by *waxy* (*wx*)) and for amylopectin by soluble starch synthases (SSs), starch-branching enzyme (BE, encoded by *amylose extender1* (*ae1*)) and starch debranching enzyme (DBE, encoded by *sugary1* (*su1*)) (Hannah 1997; Jeon et al. 2010; Huang et al. 2021; Li et al. 2021) (Fig. 10.2).

Mutations in the GBSS enzyme in the *wx* mutant stop amylose synthesis, resulting in 100% amylopectin, and mutations in the BE enzyme in *ae1* mutants inhibit amylopectin synthesis, resulting in 50% amylose accumulation in the endosperm. The amylose is responsible for starch gelling and firmness, whereas the amylopectin is responsible for the thickening of paste and formation of crystalline starch granules (Whitt et al. 2002). Amylose improves the starch texture, turning it into a slowly digestible carbohydrate (Stinard et al. 1993). Low amylose in starch results in higher paste viscosity of starch (Van Hung et al. 2006). In chloroplasts, starch

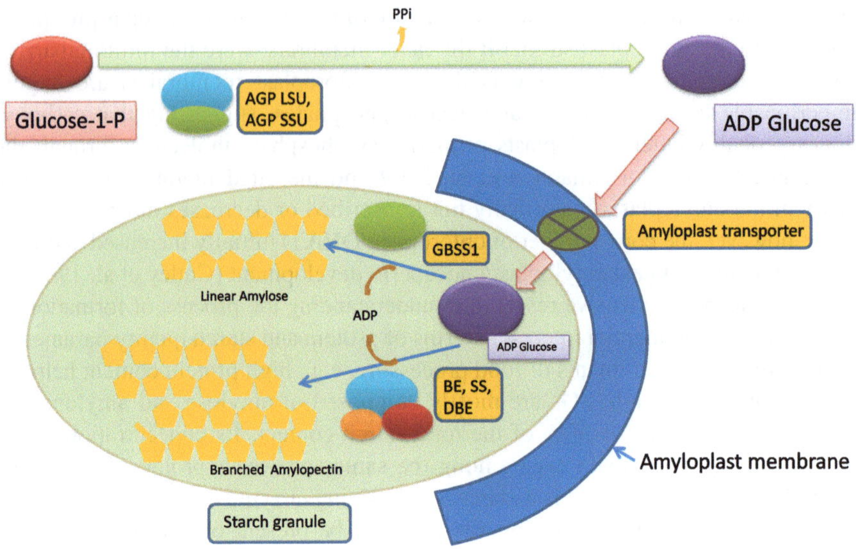

Fig. 10.2 Biosynthesis of starch components amylose and amylopectin in amyloplasts. ADP-glucose (ADPG), ADP-glucose pyrophosphorylase (AGPase), branching enzyme (BE), debranching enzyme (DBE), granule-bound starch synthase (GBSS), large subunit of AGPase (AGP LSU), small subunit of AGPase (AGP SSU), starch synthases (SS) and inorganic phosphate (PPi)

accumulates during light and is consumed during the dark for transport to other tissues. The AGPases are the most abundant enzymes in amyloplasts, accounting for even 50% of the total protein content. Distinct forms of AGPases have been determined outside and inside amyloplasts (Thorbjørnsen et al. 1996). Several starch metabolic enzymes are located in SGs and several outside SGs in amyloplasts in potato tubers (Stensballe et al. 2008).

10.3 Regulation of Amyloplast Formation by Transcription and Other Factors

Transcriptional and translational regulation has been proposed not to act as a dominant factor in amyloplast formation. Ribosomal operon *rrn* and *psbA* genes are transcribed most in amyloplasts through Plastid Encoded Polymerase (PEP; it is a polymerase encoded by the plastid genome that transcribes the genes from the plastid genome) (Brosch et al. 2007). Although many TFs are known to regulate biogenesis of amyloplasts (reviewed in Li et al. 2021, 2024; Liu et al. 2022; Mathura et al. 2024), several of these do so by regulating the formation of starch in amyloplasts e.g. Rice Starch Regulator 1, OsRSR1 (Fu and Xue 2010), *Oryza sativa* Ethylene-responsive element binding protein, clone 89 (OsEBP-89; Zhu et al. 2003), *Oryza sativa* MYC protein (OsBP-5; Zhu et al. 2003), OsbZIP58 (Wang et al. 2013), Nuclear Factor Y, subunit B1 (NF-YB1; Bello et al. 2019) and starch Branching Enzyme IIb (BEIIb; Nagamatsu et al. 2022) (Table 10.1). Proplastid to amyloplast

Table 10.1 TF and genes affecting amyloplasts, starch grains and starch quality

S.N.	Gene name	Function in amyloplast	References
	Genes with a role in amyloplast development		
1	*FLO2*	Gene responsible for the white-core endosperm in *GM645* has not been identified	She et al. (2010), Wu et al. (2015), Bao et al. (2022)
2	*FLO4*	Encodes pyruvate orthophosphate dikinase (PPDK); in *flo4–7* mutants, single starch granules increased, but compound starch granules were decreased, affecting the gelling of starch	Kang et al. (2005), Gao et al. (2023)
3	*FLO5*	Regulates starch synthesis and amyloplast development, especially in the peripheral endosperm of rice;	Fujita et al. (2007), Zhou et al. (2016)
4	*FLO6*	Contains CBM48 domain; interacts with SSIVb and GBSS to regulate starch synthesis in rice; direct starch synthases to the amyloplasts;	Peng et al. (2014), Zhang et al. (2022)
5	*FLO7*	Encodes for a protein of unknown function having a domain of unknown function 1338 (DUF1338); abnormal starch formation in the endosperm of chalky opaque grains in rice, similar to maize opaque2 mutants	Zhang et al. (2016)
	FLO9	Encodes a plant-specific protein homologous to *Arabidopsis* LIKE EARLY STARVATION1 (LESV); OsLESV is involved in starch granule initiation in the endosperm; OsLESV interacts with the starch-binding protein FLO6 to regulate starch biosynthesis and endosperm development	Yan et al. (2024)
6	*FLO10*	Encodes for pentatricopeptide repeat (PPR) protein that plays a role in maintaining endosperm and mitochondrial function; endosperm becomes soft, white, and powdery instead of translucent; smaller starch grains in the starchy endosperm and abnormal cells in the aleurone layer	Wu et al. (2019)
7	*FLO11*	Plastid-localized heat shock protein 70 (OsHsp70cp-2; opaque phenotype in the inner core and the periphery of grains)	Zhu et al. (2018)
8	*FLO15*	Plays an important role in starch synthesis; encodes a protein involved in glyoxalase activity; mutants showed reduced starch content and white endosperm core	You et al. (2019)
9	*FLO16*	Encodes a NAD-dependent cytosolic malate dehydrogenase (CMDH) for conversion of malate to pyruvate; disruption in redox homeostasis, a crucial process for compound starch grain formation and subsequent starch synthesis; mutant shows problems with starch formation, including altered amylose and amylopectin structures, as well as reduced ATP levels	Teng et al. (2019)

(continued)

Table 10.1 (continued)

S.N.	Gene name	Function in amyloplast	References
10	*FLO18*	Involved in the synthesis of galactolipids; Loss-of-function mutations in FLO18 can disrupt galactolipid synthesis, leading to abnormal amyloplast development and chalky endosperm (a condition where the inner part of the grain is not fully developed)	Yu et al. (2018), Chen et al. (2024)
11	*FLO19*	Encodes for class I glutamine amidotransferase; Affects size, shape, and overall function of amyloplasts; altered starch grain morphology; caused damage to the carbon and nitrogen assimilation & metabolism in rice	Lou et al. (2021)
12	FLOURY SHRUNKEN ENDOSPERM1 (FSE1),	Shows floury endosperm phenotype; encodes a phospholipase-like protein; *fse1* mutants showed higher total extraplastidic lipids, phosphatidic acid and total galactolipids in the endosperm plants	Lei et al. (2022), Long et al. (2018), Wang et al. (2021)
13	*Substandard starch grain4 (ssg4)*	Large protein with a domain of unknown function (DUF490), function unknown; floury endosperm due to rounder, loosely packed SG; reduced expression of genes involved in starch and lipid biosynthesis; enlarged SGs; enlargement of SGs in tissues including pollen grains, root caps, and young pericarps	Matsushima et al. (2014), Cai et al. (2021)
14	*Substandard starch grain6 (ssg6)*	Protein homologous to aminotransferase; enlarged SGs; defective galactolipid synthesis,	Matsushima et al. (2016)
15	ARC5	Mutants showed smaller starch granules and enlarged spherical chloroplasts	Yun and Kawagoe (2009)
16	*Enlarged starch grain1 (esg1)*	Starch content decreased; SGs ten-fold larger than the wild-type; putative permease subunit of a bacterial-type ABC (ATP-binding cassette) lipid transporter	Wang et al. (2021)
	Genes with a role in starch synthesis		
17	*OsbZip58*	TF regulates starch synthesis in rice endosperm; mutants exhibited altered starch accumulation in the white belly region and decreased total starch and amylose content; TF directly binds to the promoters of six genes involved in starch-synthesis, *OsAGPL3*, *SBE1*, *Wx*, *OsSSIIa*, *ISA2* and *OsBEIIb*	Wang et al. (2013)
18	*OsRSR1*	Rice Starch Regulator 1; AP2/EREBP family; negatively regulates starch biosynthesis in rice grains; RSR1 mutants exhibit larger grain size and increased rice yield, increased amylose content and altered amylopectin fine structure	Fu and Xue (2010)
19	*OsBP-5*	Rice OsBP-5 and OsEBP-89 proteins act synergistically to modulate the expression of the Wx gene	Zhu et al. (2003)

(continued)

Table 10.1 (continued)

S.N.	Gene name	Function in amyloplast	References
20	*OsEBP-89*	Ethylene-responsive element binding protein, clone 89; Rice OsBP-5 and OsEBP-89 proteins act synergistically to modulate the expression of the Wx gene	Zhu et al. (2003)
21	*NF-YB1*	Nuclear Factor Y, subunit B1; binds to the 'G-box' domain of Wxpromoter and activates Wx transcription; heterotrimer of NF-YC12, NF-YB1 and bHLH144 prevents ubiquitin/26S proteasome-mediated degradation of NF-YB1. Knockdown of these genes caused increased chalkiness; affects sucrose transporter genes, including *OsSUT1*, *OsSUT3*, and *OsSUT4*, to regulate grain filling and chalkiness	Bai et al. (2016), Bello et al. (2019)
22	*OsBT1*	ADP glucose transporter	Cakir et al. (2016)
23	*OspPGM*	Plastidic phosphoglucomutase	Lee et al. (2016)
24	*OsPKpα1*	Pyruvate metabolism-related genes	Cai et al. (2018)
25	Starch branching enzyme IIb (BEIIb)	Enzyme has a crucial role in starch biosynthesis, particularly in the formation of amylopectin; affects starch properties including amylose content, viscosity and gelatinization temperature	Nagamatsu et al. (2022)
26	*OsPFPβ*	Pyruvate metabolism-related genes	Duan et al. (2016)
27	*OsAlaAT1*	Pyruvate metabolism-related genes	Yang et al. (2015)
28	*OsPDIL1–1*	Genes involved in endoplasmic reticulum stress and programmed cell death; *T3612 mutants*; formation of floury endosperm	Han et al. (2012)
29	*OsFSE1*	Genes involved in endoplasmic reticulum stress and programmed cell death	Long et al. (2018)

differentiation is induced by many TFs (Mingo-Castel et al. 1991; Wang et al. 2013; Brosch et al. 2007; Yang et al. 2024). One such TF, StbHLH93, regulates amyloplast development in potato (Yang et al. 2024). The number of amyloplasts differentiated from proplastids was lower in StbHLH93-KD lines, in contrast to a higher number in StbHLH93-OX lines (Yang et al. 2024).

Several other proteins have been found to regulate amylopast formation and starch biosynthesis e.g. ADP glucose transporter *Oryza sativa* BRITTLE1 (OsBT1; shrunken3; Cakir et al. 2016), plastidic phosphoglucomutase (OspPGM; Lee et al. 2016), Small kernel 1 (Li et al. 2024), SUBSTANDARD STARCH GRAIN4 (SSG4), and SSG6 (Matsushima et al. 2014, 2016), pyruvate metabolism-related genes: *Oryza sativa* Pyruvate Kinase plastidic alpha 1 (OsPKpα1; Cai et al. 2018),

Oryza sativa Pyrophosphate: fructose-6-phosphate 1-phosphotransferase β (OsPFPβ; Duan et al. 2016) and *Oryza sativa* Alanine Amino Transferase 1 (OsAlaAT1; Yang et al. 2015), and genes involved in programmed cell death: *Oryza sativa* Floury Shrunken Endosperm 1 (*OsFSE1;* Long et al. 2018) and *Oryza sativa* Protein disulphide isomerase (PDI) like 1–1 (*OsPDIL1–1;* Han et al. 2012). Moreover several *FLO* mutants (floury endosperm) have been identified that affect starch quality and amyloplast structure, e.g., *FLO1* (Long et al. 2018), *FLO2* (She et al. 2010), *FLO6* (Peng et al. 2014), *FLO7* (Zhang et al. 2016), *FLO11* (Zhu et al. 2018) (Table 10.1). Floury endosperm is characterized by a chalky, non-translucent appearance and a tendency to break into powder. Stress in the endoplasmic reticulum also reportedly affects amyloplast development (Kaufman 1999; Han et al. 2012). ER stress is caused by the aggregation of proteins in the ER or the accumulation of misfolded proteins in the ER, which can trigger a cascade of events that disrupt cellular processes, including those involved in amyloplast development (Howell 2013). ER stress can impair the calcium homeostasis, proper folding and trafficking of proteins essential for amyloplast biogenesis, including those involved in starch synthesis and transport (Singh et al. 2021). ER stress can affect calcium homeostasis, which plays a role in signalling pathways that regulate amyloplast development. ER stress activates the unfolded protein response (UPR), a cellular stress signalling response that can increase protein synthesis, remove degraded proteins, and increase the protein folding capacity of cells to affect the development of amyloplasts (Manghwar and Li 2022). ER stress induced by the absence of PDIL1–1 in somatic mutant *T3612* was proposed to be responsible for the formation of floury endosperm (Han et al. 2012). In rice, FLO11 encodes for a plastid-localized heat shock protein 70 (OsHsp70cp-2), associated with TIC and critical for amyloplast development (Zhu et al. 2018). Another mutant *FLO7* shows a floury-white endosperm on the periphery of the grains but not in the inner side, reduced amylose content, and highly disrupted amylopectin in the peripheral white portion of the endosperm. FLO7 encodes for a protein of unknown function having a domain of unknown function 1338 (DUF1338) (Sheng et al. 2015). Its transit peptide is able to transfer GFP into the stroma of amyloplasts. Its expression, although ubiquitous in endosperm, is more specific to the endosperm periphery in rice (Zhang et al. 2016) (Table 10.1).

Rice (*Oryza sativa*) mutants *ssg6* and *ssg4* show an enlargement in the SGs. *Substandard starch grain6* (*ssg6*), which encodes for a protein homologous to aminotransferase, develops large SGs. In the amyloplasts of ssg6, large SGs start forming 3 days after flowering and are found along with small SGs in the same cells. SSG6-GFP protein was localized in the amyloplast membrane near the SGs in rice endosperm, pollen, and pericarp (Matsushima et al. 2016). Another mutant (*c134*) that encodes for *substandard starch grain4* (*ssg4*) protein is involved in starch and lipid metabolism (Cai et al. 2022; Hawkins et al. 2021). This mutant shows reduced amylose content in endosperm and floury endosperm due to loosely packed, rounder SGs. The expression of various genes in starch and lipid biosynthesis was also reduced in these mutants (Cai et al. 2022).

10.4 Lipid Metabolism Defects in Amyloplasts

Lipid metabolism is also known to influence amyloplast development. Many free fatty acids, e.g., lysophosphatidylcholine, lysophospholipids and lysophosphatidylethanolamine are added during the synthesis of amyloplast membranes (Gayral et al. 2019). A rice mutant *floury shrunken endosperm1* (fse1) shows developmental defects in physicochemical properties and yield of starch and shows compound SGs formation (Long et al. 2018). *FSE1* encodes a phospholipase-like protein that is homologous to phosphatidic acid-preferring phospholipase A1 and is expressed ubiquitously in developing seeds and roots. Total extraplastidic lipids, PA and total galactolipid content were increased in the endosperm of *fse1* plants (Long et al. 2018). *ssg4* mutant also exhibits defects in lipid synthesis, e.g., galactosyldiacylglycerol synthesis was reduced, and various genes in its synthesis pathway were reduced, and showed defective floury endosperm (Cai et al. 2022). Another mutant *enlarged starch grain1* (*esg1*), which encodes for a putative permease subunit of a bacterial type ABC lipid transporter (ATP-binding cassette lipid transporter), also decreases starch content in amyloplasts/grains although SGs are enlarged due to a higher number of starch granules per SG. ESG1 protein is located in the chloroplasts and amyloplasts membranes and is involved in galactolipid synthesis, and lipid content is decreased in these mutants (Wang et al. 2021). Lower expression of the OsLTP, decreases the starch grain size and their compactness, protein and fatty acid content, and decreases the chalkiness of the endosperm (Wang et al. 2015).

10.5 DNA Methylation

DNA methylation governs the active or transcribing state of DNA. Therefore, it is important to see the DNA methylation patterns in amyloplasts to determine the active regions of DNA. DNA methylation happens mostly at the CpG sites of DNA, methylation islands in promoter regions or near genes and occasionally at non-CpG sites. CpG methylation occurs by the addition of a methyl group at the fifth carbon of the cytosine nucleotide. In plants, methylation also takes place at CHG and CHH sites. However, there are not many studies examining the methylation pattern in amyloplasts; a higher extent of DNA methylation was observed in chloroplast DNA as compared to amyloplast in pea (Ohta et al. 1991) and rice (Muniandy et al. 2019). It can be found in coding as well as in non-coding regions of plastids (Ngernprasirtsiri et al. 1988, 1990). In amyloplasts, methylation islands are frequently reported (Muniandy et al. 2019). Eleven CHG and two CpG positions were consistently highly methylated in the DNA of young leaf chloroplast and grain amyloplast. Seven of these 13 positions were present in the coding region of plastid genes *ndhB*, 16S ribosomal protein *rrn16s* and large ribosomal protein subunit 2 (*rpl2*) genes, and the remaining six were present in plastid non-coding regions. (Muniandy et al. 2019). *rrn*, *ndhA*, *ndhB* and *psbA* were the most methylated genes (Muniandy et al. 2019) in amyloplasts.

10.6 Proteomics of Amyloplasts

Although there are numerous reports of chloroplast proteomics, there are a few studies that have determined the proteome profile of amyloplasts. One of the reasons is that it is difficult to obtain intact amyloplasts in pure form in large quantities (Mohabir and John 1988). Some investigations compared proteome of amyloplasts with chloroplasts. For example, Balmer et al. (2006a) found that amyloplasts contain many enzymes related to starch metabolism, such as AGPases, and their metabolic activity was high. The proteins related to carbohydrate, nitrogen and sulphur metabolism, cytoskeleton/plastid division, nucleic acid-related reactions, energetics, synthesis of various building blocks, protein-related reactions, signalling, transport, and stress were reported in wheat amyloplasts. 'Miscellaneous function' proteins comprise a big chunk (12%) of identified proteins, opening the possibility for investigation in amyloplasts' functional mechanisms (Balmer et al. 2006a). In another proteomics study, a full-fledged, functional ferredoxin/thioredoxin system was identified from amyloplasts by the same group (Balmer et al. 2006b). In amyloplasts, ferredoxin is reduced by NADPH and not by light as happens in chloroplasts (Balmer et al. 2006b). This study reported 42 potential target proteins of thioredoxin, including BT1 and AGPase (Balmer et al. 2006b). Other authors have reported low metabolic activities in amyloplasts, e.g., Dupont (2008) concluded that amyloplasts do possess some enzymes involved in carbohydrate metabolism, glycolysis, and the pentose phosphate pathway. Transcriptional activity was also reported to be low in potato tuber amyloplasts (Stensballe et al. 2008). In Kiwi fruit amyloplasts, Li et al. (2024) recently reported that the protein import was increased in the amyloplasts; and starch synthesizing/breakdown proteins, ATP synthesizing, Chl breakdown and plastoglobules proteins were increased during ripening. Photosynthesis was reduced in the potato tuber envelope, Rubisco degraded, and AGPase small subunit was profoundly increased during greening and glucose 1-P was imported during amyloplast formation during fruit ripening (Belmont et al. 2022). Ma et al. (2014) used quantitative proteomics and phosphoproteomics to characterize central metabolism during wheat grain development. They found that protein synthesis/assembly/degradation, energy and carbon metabolism, transport, signal transduction, stress/defence and nucleic acid-related processes were involved in amyloplast metabolism in wheat (Ma et al. 2002). In a proteomics investigation of microtubers' amyloplasts obtained from agar agar-grown potatoes and mini tubers' amyloplasts obtained from soil-grown potatoes, soluble proteins were dominated by starch phosphorylase L-1, which was half or 1/3 of the total proteins respectively, and other pathways such as glycolysis, pentose phosphate pathway, amino acid metabolism, starch and sucrose metabolism, and plastid chaperonins (Stensballe et al. 2008). In wheat, DuPont (2008) determined that 135 enzymes of amyloplasts, which could be arranged in 18 pathways, e.g., biosynthesis of carbohydrates, amino acids, lipids, nucleic acids and other proteins, were important for grain filling.

10.7 Tuber Greening

Tuber greening is harmful as per economic considerations due to the accumulation of toxic alkaloids; therefore, it has also been investigated profoundly (Tanios et al. 2018; Zhang et al. 2020). It is a reverse process of amyloplast formation from the chloroplast. Microscopic (Ljubičić et al. 1998) investigations pointed out that Chl fluorescence is emitted from the tuber pericarp, and chloroplasts were found in this region up to 1.5 mm deep (Zhang et al. 2020). A capsule is developed between the plastid membrane and SGs during the greening process (Ljubičić et al. 1998). Nitrogen content and exposure to light increased the potato tuber greening (Zhang et al. 2020). Zhang et al. (2020) reported that the greening involved the formation of grana around amyloplasts, membrane loss and dissolution of SGs. Nitrogen also increased the percentage of small granules in tuber pericarp (Zhang et al. 2020).

10.8 Ca^{2+} Homeostasis

Plastids play an important role in relieving the Ca^{2+} signal by sequestering the Ca^{2+}. Ca^{2+} plays a significant role in amyloplast formation and maintenance. However, Sello et al. (2016) reported no difference in Ca^{2+} signatures in amyloplasts and chloroplasts, which led them to propose that the influx mechanism of Ca^{2+} in both plastids is the same. Amyloplasts sense the environmental stress through Ca^{2+}. If there is a reduction in the Ca^{2+} concentration, the amyloplast membrane starts to degrade (Sello et al. 2016). When membrane degradation is reversed, it leads to the integrity of amyloplasts and the formation of SGs in them (Kumar and Knowles 1993). In some diseases, e.g., in Bitter pit of Apple, Ca^{2+} is decreased in the pulp that causes amyloplast degradation, which in turn increases the carbohydrate concentration in fruit, leading to fruit size enhancement and higher soluble sugar content in fruit. Ca^{2+} homeostasis genes were highly expressed in the pulp of the fruit with bitter pit (Zhou et al. 2009). The Ca^{2+} homeostasis is also related to programmed cell death (PCD). When starch grains accumulate in the wheat pericarp, it leads to the PCD. Contrastingly, when wheat cell suspension is grown under sugar starvation, it leads to starch degradation in amyloplasts (Zhou et al. 2009). Calcium ions are also responsible for the maintenance of the structure of amyloplasts.

10.9 Division of Amyloplasts

Amyloplasts divide by binary fission, multiple fission and stromules. In the growing stromules, which are outgrowths from the chloroplast outer membrane, the amyloplasts divide through binary fission multiple times, giving a beaded appearance. In the stromules, the FtsZ ring helps in the formation of new amyloplasts. Amyloplasts divide in the Arabidopsis integument cells of the ovule by placing the FtsZ ring at multiple places. The large-sized SGs divide by budding. In *minE* mutants, chloroplast division is severely inhibited and is based on the size of the plastids, which leads to a variety of sizes in these mutants; although the process of division is similar to wild type in these mutants. Moreover, *minE* mutants had giant stromules that were stable in structure, leading to assembly of the FtsZ ring and constriction, required for division. SGs could also accumulate in these stromules (Fujiwara et al. 2024). Mutant *ftsZ* also causes the size-proliferation of chloroplasts due to the absence of division; however, it did not lead to proliferation in amyloplasts. In the wheat *PARC6* mutant that was defective in plastid division, giant plastids were reported in both leaves and endosperm (Esch et al. 2023). The size and number of both A- and B-type granules were higher in the mature plastids of these mutants. The A-type granules were aberrant with a lobed surface, and the difference was visible at the early stage of grain development, although the amylose and amylopectin structures were intact in these mutants (Esch et al. 2023). In contrast, *ARC5* mutants had smaller SGs. GFP reporter confirmed the presence of ARC5 and FtsZ2–1 in the division ring of amyloplasts (Yun and Kawagoe 2009). Although *ARC5* mutants had defective chloroplast division, producing enlarged spherical chloroplasts; in amyloplasts, both the loss of function of ARC5 and fusion protein overexpression in the endosperm did not produce spherical amyloplasts. It rather produced fused plastids that were tightly connected or pleomorphic. This underlines the importance of proper stoichiometry of all subunits in the plastid division ring. Starch gelatinizing temperature in these mutants was significantly higher than that of the wild type (Esch et al. 2023).

10.10 Role of Amyloplasts in Gravity Sensing and Geotropism

For many decades, amyloplasts were theorized to have a role in gravi-sensing and gravitropism through their presence in the columella region of the calyptra (Thomson and Whatley 1980). They settle on the lower side towards the gravity vector, in specialized starch sheath cells of shoots, located between leaf traces and cortical cells, and columella cells of the root cap (Ponce et al. 2008; Rioux et al. 2015). This activates the auxin protein PIN2 (PIN FORMED family proteins are polar-efflux carriers that transport the plant hormone auxin from one cell to another across their cell membranes, thus helping in cell elongation, tissue growth and stem bending) to cause uneven growth of tissue that leads to the growth of roots towards gravity (Nakamura et al. 2019). This notion has been challenged due to the recent experiments that determined, through live-cell imaging, that the moment of amyloplast is continuous, random, dynamic and saltatory in the endodermal cells of Arabidopsis stems (Abe et al. 2020). Therefore, amyloplasts may not be working as gravity sensors, but further research is required to confirm this. Another experiment performed in the International Space Station (ISS) (Rioux et al. 2015) found that most amyloplasts were randomly distributed in seedlings and not sedimented on the lower wall.

10.11 Conclusions

Amyloplasts are a type of leucoplasts that synthesize and store starch in them. They store the starch in the form of starch grains (SGs). They are similar to chloroplasts in size. In plants, they are found in storage tissues such as roots, fruits, tubers and seeds. Additionally, they are also found in the root columella and root cap, pollen and young pericarp, etc. There are several genes that play a role in the formation of starch and amyloplasts. Among them, there are *Flo* (Floury endosperm) mutants that change the grain quality and flour characteristics of grains and, in turn, flour texture. Several *Flo* mutants have been discovered, e.g., Flo1, 2, 4, 5, 6, 7, 8, 11, 15, 18 and 19. Moreover, some proteins cause the changes in the size of starch grains, e.g., *substandard starch grain4*, *substandard starch grain6* and *Floury Shrunken Endosperm 1*, which increase the size of SGs. Differences are observed between the DNA methylome and proteome of amyloplasts and chloroplasts. The quality and quantity of the grains are important for commercial purposes; therefore, extensive work has been done on the amyloplast starch synthesis pathway and its degradation pathway. This needs to be further carried out to gain commercial benefits from understanding the amyloplasts.

References

Abe Y, Meguriya K, Matsuzaki T, Sugiyama T, Yoshikawa HY, Morita MT, Toyota M (2020) Micromanipulation of amyloplasts with optical tweezers in Arabidopsis stems. Plant Biotechnol 37(4):405–415

Altenbach SB, DuPont FM, Kothari KM, Chan R, Johnson EL, Lieu D (2003) Temperature, water and fertilizer influence the timing of key events during grain development in a US spring wheat. J Cereal Sci 37(1):9–20

Andon NL, Hollingworth S, Koller A, Greenland AJ, Yates JR III, Haynes PA (2002) Proteomic characterization of wheat amyloplasts using identification of proteins by tandem mass spectrometry. PROTEOMICS 2(9):1156–1168

Bai AN, Lu XD, Li DQ, Liu JX, Liu CM (2016) NF-YB1-regulated expression of sucrose transporters in aleurone facilitates sugar loading to rice endosperm. Cell Res 26(3):384–388

Balmer Y, Vensel WH, DuPont FM, Buchanan BB, Hurkman WJ (2006a) Proteome of amyloplasts isolated from developing wheat endosperm presents evidence of broad metabolic capability. J Exp Bot 57(7):1591–1602

Balmer Y, Vensel WH, Cai N, Manieri W, Schürmann P, Hurkman WJ, Buchanan BB (2006b) A complete ferredoxin/thioredoxin system regulates fundamental processes in amyloplasts. Proc Natl Acad Sci 103(8):2988–2993

Bänziger M, Feil B, Stamp P (1994) Competition between nitrogen accumulation and grain growth for carbohydrates during grain filling of wheat. Crop Sci 34(2):440–446

Barneix AJ (2007) Physiology and biochemistry of source-regulated protein accumulation in the wheat grain. J Plant Physiol 164(5):581–590

Bechtel DB, Wilson JD (2003) Amyloplast formation and starch granule development in hard red winter wheat. Cereal Chem 80(2):175–183

Bechtel DB, Zayas INNA, Kaleikau LORI, Pomeranz Y (1990) Size-distribution of wheat starch granules during endosperm development. Cereal Chem 67(1):59–63

Bello BK, Hou Y, Zhao J, Jiao G, Wu Y, Li Z et al (2019) NF-YB 1-YC 12-bHLH 144 complex directly activates Wx to regulate grain quality in rice (Oryza sativa L.). Plant Biotechnol J 17(7):1222–1235

Belmont R, Bernal L, Padilla-Chacón D, Coello P, Martínez-Barajas E (2022) Starch accumulation in bean fruit pericarp is mediated by the differentiation of chloroplasts into amyloplasts. Plant Sci 316:111163

Bock R (2007) Structure, function, and inheritance of plastid genomes. In: Bock R (ed) Cell and molecular biology of plastids. Springer, Heidelberg, pp 29–63. https://doi.org/10.1007/4735_2007_0223

Bonora A, Pancaldi S, Gualandri R, Fasulo MP (2000) Carotenoid and ultrastructure variations in plastids of Arum italicum Miller fruit during maturation and ripening. J Exp Bot 51(346):873–884. https://doi.org/10.1093/jexbot/51.346.873

Briarty LG, Hughes CE, Evers AD (1979) The developing endosperm of wheat—a stereological analysis. Ann Bot 44(6):641–658

Brosch M, Krause K, Falk J, Krupinska K (2007) Analysis of gene expression in amyloplasts of potato tubers. Planta 227:91–99

Buttrose MS (1963) Ultrastructure of the developing wheat endosperm. Aust J Biol Sci 16(2):305–317

Cai Y, Zhang W, Jin J, Yang X, You X, Yan H et al (2018) OsPKpα1 encodes a plastidic pyruvate kinase that affects starch biosynthesis in the rice endosperm. J Integr Plant Biol 60(11):1097–1118

Cai Y, Chen H, Xiao N, Wu Y, Yu L, Chen Z et al (2022) Substandard starch grain4 may function in amyloplast development by influencing starch and lipid metabolism in rice endosperm. J Plant Physiol 270:153638

Cakir B, Shiraishi S, Tuncel A, Matsusaka H, Satoh R, Singh S et al (2016) Analysis of the rice ADP-glucose transporter (OsBT1) indicates the presence of regulatory processes in the amyloplast stroma that control ADP-glucose flux into starch. Plant Physiol 170(3):1271–1283

Catley MA, Bowman CM, Bayliss MW, Gale MD (1987) The pattern of amyloplast DNA accumulation during wheat endosperm development. Planta 171:416–421

Charuvi D, Kiss V, Nevo R, Shimoni E, Adam Z, Reich Z (2012) Gain and loss of photosynthetic membranes during plastid differentiation in the shoot apex of Arabidopsis. Plant Cell 24(3):1143–1157

Chen J, Hawkins E, Seung D (2021) Towards targeted starch modification in plants. Curr Opin Plant Biol 60:102013

Chen L, Li X, Zheng M, Hu R, Dong J, Zhou L et al (2024) Genes controlling grain chalkiness in rice. Crop J 12:979

Cho YB, Stutz SS, Jones SI, Wang Y, Pelech EA, Ort DR (2023) Impact of pod and seed photosynthesis on seed filling and canopy carbon gain in soybean. Plant Physiol 193(2):966–979

Choi H, Yi T, Ha SH (2021) Diversity of plastid types and their interconversions. Front Plant Sci 12:692024

Drapal M, Gerrish C, Fraser PD (2022) Changes in carbon allocation and subplastidal amyloplast structures of specialised Ipomoea batatas (sweet potato) storage root phenotypes. Phytochemistry 203:113409

Duan E, Wang Y, Liu L, Zhu J, Zhong M, Zhang H et al (2016) Pyrophosphate: fructose-6-phosphate 1-phosphotransferase (PFP) regulates carbon metabolism during grain filling in rice. Plant Cell Rep 35:1321–1331

Dupont FM (2008) Metabolic pathways of the wheat (Triticum aestivum) endosperm amyloplast revealed by proteomics. BMC Plant Biol 8:39

DuPont FM, Hurkman WJ, Vensel WH, Chan R, Lopez ROCIO, Tanaka CK, Altenbach SB (2006) Differential accumulation of sulfur-rich and sulfur-poor wheat flour proteins is affected by temperature and mineral nutrition during grain development. J Cereal Sci 44(1):101–112

Esch L, Ngai QY, Barclay JE, McNelly R, Hayta S, Smedley MA et al (2023) Increasing amyloplast size in wheat endosperm through mutation of PARC6 affects starch granule morphology. New Phytol 240(1):224–241

Fernandez O, Ishihara H, George GM, Mengin V, Flis A, Sumner D et al (2017) Leaf starch turnover occurs in long days and in falling light at the end of the day. Plant Physiol 174(4):2199–2212. https://doi.org/10.1104/pp.17.00601

Fowler DB (2003) Crop nitrogen demand and grain protein concentration of spring and winter wheat. Agron J 95:260–265

Fu FF, Xue HW (2010) Coexpression analysis identifies Rice Starch Regulator1, a rice AP2/EREBP family transcription factor, as a novel rice starch biosynthesis regulator. Plant Physiol 154(2):927–938

Fujiwara MT, Yoshioka Y, Kazama Y, Hirano T, Niwa Y, Moriyama T et al (2024) Principles of amyloplast replication in the ovule integuments of Arabidopsis thaliana. Plant Physiol 196(1):137–152

Gajecka M, Marzec M, Chmielewska B, Jelonek J, Zbieszczyk J, Szarejko I (2020) Plastid differentiation during microgametogenesis determines green plant regeneration in barley microspore culture. Plant Sci 291:110321

Galán-Ávila A, García-Fortea E, Prohens J, Herraiz FJ (2021) Microgametophyte development in Cannabis sativa L. and first androgenesis induction through microspore embryogenesis. Front Plant Sci 12:669424

Gao Z, Li C, Huang F, Xu J, He Y, Dan Y et al (2023) A mutation of cyOsPPDKB affects starch structure and gel consistency in Rice. Agronomy 13(11):2759

Gayral M, Fanuel M, Rogniaux H, Dalgalarrondo M, Elmorjani K, Bakan B, Marion D (2019) The spatiotemporal deposition of lysophosphatidylcholine within starch granules of maize endosperm and its relationships to the expression of genes involved in endoplasmic reticulum–amyloplast lipid trafficking and galactolipid synthesis. Plant Cell Physiol 60(1):139–151. https://doi.org/10.1093/pcp/pcy198

Gemenet DC, da Silva Pereira G, De Boeck B, Wood JC, Mollinari M, Olukolu BA et al (2020) Quantitative trait loci and differential gene expression analyses reveal the genetic basis for negatively associated β-carotene and starch content in hexaploid sweet potato [Ipomoea batatas (L.) Lam.]. Theor Appl Genet 133:23–36

Han X, Wang Y, Liu X, Jiang L, Ren Y, Liu F et al (2012) The failure to express a protein disulphide isomerase-like protein results in a floury endosperm and an endoplasmic reticulum stress response in rice. J Exp Bot 63(1):121–130

Hannah LC (1997) Starch synthesis in the maize seed. In: Cellular and molecular biology of plant seed development, pp 375–405

Hanson MR, Conklin PL (2020) Stromules, functional extensions of plastids within the plant cell. Curr Opin Plant Biol 58:25–32

Hasegawa T, Sakai H, Tokida T, Nakamura H, Zhu C, Usui Y et al (2013) Rice cultivar responses to elevated CO_2 at two free-air CO_2 enrichment (FACE) sites in Japan. Funct Plant Biol 40(2):148–159

Hempel J, Amrehn E, Quesada S, Esquivel P, Jiménez VM, Heller A et al (2014) Lipid-dissolved γ-carotene, β-carotene, and lycopene in globular chromoplasts of peach palm (Bactris gasipaes Kunth) fruits. Planta 240:1037–1050. https://doi.org/10.1007/s00425-014-2121-3

Howell SH (2013) Endoplasmic reticulum stress responses in plants. Annu Rev Plant Biol 64(1):477–499

Howitt CA, Pogson BJ (2006) Carotenoid accumulation and function in seeds and non-green tissues. Plant Cell Environ 29(3):435–445

Huang L, Tan H, Zhang C, Li Q, Liu Q (2021) Starch biosynthesis in cereal endosperms: an updated review over the last decade. Plant Commun 2(5):100237

Jarvis P, Lopez-Juez E (2013) Biogenesis and homeostasis of chloroplasts and other plastids. Nat Rev Mol Cell Biol 14:787–802. https://doi.org/10.1038/nrm3702

Jeon JS, Ryoo N, Hahn TR, Walia H, Nakamura Y (2010) Starch biosynthesis in cereal endosperm. Plant Physiol Biochem 48(6):383–392

Kaufman RJ (1999) Stress signaling from the lumen of the endoplasmic reticulum: coordination of gene transcriptional and translational controls. Genes Dev 13(10):1211–1233

Kumar GM, Knowles NR (1993) Changes in lipid peroxidation and lipolytic and free-radical scavenging enzyme activities during aging and sprouting of potato (Solanum tuberosum) seed-tubers. Plant Physiol 102(1):115–124

Lee SK, Eom JS, Hwang SK, Shin D, An G, Okita TW, Jeon JS (2016) Plastidic phosphoglucomutase and ADP-glucose pyrophosphorylase mutants impair starch synthesis in rice pollen grains and cause male sterility. J Exp Bot 67(18):5557–5569

Li R, Tan Y, Zhang H (2021) Regulators of starch biosynthesis in cereal crops. Molecules 26(23):7092

Li A, Lin J, Zeng Z, Deng Z, Tan J, Chen X et al (2024) The kiwifruit amyloplast proteome (kfALP): A resource to better understand the mechanisms underlying amyloplast biogenesis and differentiation. Plant J 118(2):565–583

Liebers M, Grübler B, Chevalier F, Lerbs-Mache S, Merendino L, Blanvillain R, Pfannschmidt T (2017) Regulatory shifts in plastid transcription play a key role in morphological conversions of plastids during plant development. Front Plant Sci 8:23. https://doi.org/10.3389/fpls.2017.00023

Liu S, Liu T, Wang E, Cheng Y, Liu T, Chen G et al (2022) Dissecting the chloroplast proteome of the potato (Solanum Tuberosum L.) and its comparison with the tuber amyloplast proteome. Plants 11:1915

Ljubičić JM, Wrischer M, Ljubešić N (1998) Formation of the photosynthetic apparatus in plastids during greening of potato microtubers. Plant Physiol Biochem 36(10):747–752

Long W, Wang Y, Zhu S, Jing W, Wang Y, Ren Y et al (2018) FLOURY SHRUNKEN ENDOSPERM1 connects phospholipid metabolism and amyloplast development in rice. Plant Physiol 177(2):698–712

Lou G, Chen P, Zhou H, Li P, Xiong J, Wan S et al (2021) FLOURY ENDOSPERM19 encoding a class I glutamine amidotransferase affects grain quality in rice. Mol Breed 41(5):36

Ma C, Zhou J, Chen G, Bian Y, Lv D, Li X et al (2014) iTRAQ-based quantitative proteome and phosphoprotein characterization reveals the central metabolism changes involved in wheat grain development. BMC Genomics 15:1–20

Manghwar H, Li J (2022) Endoplasmic reticulum stress and unfolded protein response signaling in plants. Int J Mol Sci 23(2):828

Martre P, Jamieson PD, Semenov MA, Zyskowski RF, Porter JR, Triboi E (2006) Modelling protein content and composition in relation to crop nitrogen dynamics for wheat. Eur J Agron 25(2):138–154

Mathura SR, Sutton F, Rouse-Miller J, Bowrin V (2024) The molecular coordination of tuberization: current status and future directions. Curr Opin Plant Biol 82:102655

Matsushima R, Hisano H (2019) Imaging amyloplasts in the developing endosperm of barley and rice. Sci Rep 9(1):3745

Matsushima R, Yamashita J, Kariyama S, Enomoto T, Sakamoto W (2013) A phylogenetic re-evaluation of morphological variations of starch grains among Poaceae species. J Appl Glycosci 60:37–44

Matsushima R, Maekawa M, Kusano M, Kondo H, Fujita N, Kawagoe Y, Sakamoto W (2014) Amyloplast-localized SUBSTANDARD STARCH GRAIN4 protein influences the size of starch grains in rice endosperm. Plant Physiol 164(2):623–636

Matsushima R, Maekawa M, Kusano M, Tomita K, Kondo H, Nishimura H et al (2016) Amyloplast membrane protein SUBSTANDARD STARCH GRAIN6 controls starch grain size in rice endosperm. Plant Physiol 170(3):1445–1459

Mingo-Castel AM, Pelacho AM, De Felipe MR (1991) Amyloplast division in kinetin induced potato tubers. Plant Sci 73(2):211–217

Muniandy K, Tan MH, Song BK, Ayub Q, Rahman S (2019) Comparative sequence and methylation analysis of chloroplast and amyloplast genomes from rice. Plant Mol Biol 100:33–46

Myers AM, James MG, Lin Q, Yi G, Stinard PS, Hennen-Bierwagen TA et al (2011) Maize opaque5 encodes monogalactosyldiacylglycerol synthase and specifically affects galactolipids necessary for amyloplast and chloroplast function. Plant Cell 23:2331–2347

Nagamatsu S, Wada T, Matsushima R, Fujita N, Miura S, Crofts N et al (2022) Mutation in BEIIb mitigates the negative effect of the mutation in ISA1 on grain filling and amyloplast formation in rice. Plant Mol Biol 108(4):497–512
Nakamura M, Nishimura T, Morita MT (2019) Gravity sensing and signal conversion in plant gravitropism. J Exp Bot 70(14):3495–3506
Nepi M, Ciampolini F, Pacini E (1996) Plastid differentiation during Cucurbita pepo (Cucurbitaceae) pollen grain development. Sex Plant Reprod 9:17–24
Ngernprasirtsiri J, Kobayashi H, Akazawa T (1988) DNA methylation as a mechanism of transcriptional regulation in nonphotosynthetic plastids in plant cells. Proc Natl Acad Sci 85(13):4750–4754
Ngernprasirtsiri J, Kobayashi H, Akazawa T (1990) DNA methylation is a determinative element of photosynthesis gene expression in amyloplasts from liquid-cultured cells of sycamore (Acer pseudoplatanus L.). Cell Struct Funct 15(5):285–293
Ohta N, Sato N, Kawano S, Kuroiwa T (1991) Methylation of DNA in the chloroplasts and amyloplasts of the pea, Pisum sativum. Plant Sci 78(1):33–42. https://doi.org/10.1016/0168-9452(91)90159-6
Oury FX, Godin C (2007) Yield and grain protein concentration in bread wheat: how to use the negative relationship between the two characters to identify favourable genotypes? Euphytica 157(1):45–57
Ponce G, Rasgado FA, Cassab GI (2008) Roles of amyloplasts and water deficit in root tropisms. Plant Cell Environ 31(2):205–217
Qiu L, Hu S, Wang Y, Qu H (2021) Accumulation of abnormal amyloplasts in pulp cells induces bitter pit in Malus domestica. Front Plant Sci 12:738726
Rharrabti Y, Elhani S, Martos-Núñez V, García del Moral LF (2001) Protein and lysine content, grain yield, and other technological traits in durum wheat under Mediterranean conditions. J Agric Food Chem 49(8):3802–3807
Rioux D, Lagacé M, Cohen LY, Beaulieu J (2015) Variation in stem morphology and movement of amyloplasts in white spruce grown in the weightless environment of the International Space Station. Life Sci Space Res 4:67–78
Sagisaka S (2008) The proliferation of amyloplasts in meristematic cells of developing stolons of potato and apple callus: progenitors of proplastids. J Plant Physiol 165:1678–1690. https://doi.org/10.1016/j.jplph.2008.02.003
Sello S, Perotto J, Carraretto L, Szabò I, Vothknecht UC, Navazio L (2016) Dissecting stimulus-specific Ca2+ signals in amyloplasts and chloroplasts of Arabidopsis thaliana cell suspension cultures. J Exp Bot 67(13):3965–3974
Shannon JC, Garwood DL, Boyer CD (2009) Genetics and physiology of starch development. In: Starch, pp 23–82. https://doi.org/10.1016/B978-0-12-746275-2.00003-3
She KC, Kusano H, Koizumi K, Yamakawa H, Hakata M, Imamura T et al (2010) A novel factor FLOURY ENDOSPERM2 is involved in regulation of rice grain size and starch quality. Plant Cell 22(10):3280–3294
Sheng ZH, Fang PF, Li SF, Jiao GA, Xie LH, Hu PS et al (2015) Phenotype of rice floury endosperm mutant flo7 and fine mapping of mutated gene. Rice Sci 22(4):162–170
Sheoran IS, Sawhney V, Babbar S, Singh R (1991) In vivo fixation of CO2 by attached pods of Brassica campestris L. Ann Bot 67(4):425–433
Singh MB, Lohani N, Bhalla PL (2021) The role of endoplasmic reticulum stress response in pollen development and heat stress tolerance. Front Plant Sci 12:661062
Solis-Badillo E, Agama-Acevedo E, Tiessen A, Lopez Valenzuela JA, Bello-Perez LA (2020) ADP-glucose pyrophosphorylase is located in the plastid and cytosol in the pulp of tropical banana fruit (Musa acuminata). Plant Foods Hum Nutr 75:76–82. https://doi.org/10.1007/s11130-019-00788-w
Stensballe A, Hald S, Bauw G, Blennow A, Welinder KG (2008) The amyloplast proteome of potato tuber. FEBS J 275(8):1723–1741

Stinard PS, Robertson DS, Schnable PS (1993) Genetic isolation, cloning, and analysis of a mutator-induced, dominant antimorph of the maize amylose extender1 locus. Plant Cell 5:1555–1566

Tan SL, Morrison WR (1979) The distribution of lipids in germ, endosperm, pericarp and tip cap of amylomaize, LG-1 hybrid maize and waxy maize. J Am Oil Chem Soc 56:531–535

Tanios S, Eyles A, Tegg R, Wilson C (2018) Potato tuber greening: a review of predisposing factors, management and future challenges. Am J Potato Res 95:248–257. https://doi.org/10.1007/s12230-018-9648-y

Tateoka T (1962) Starch grains of endosperm in grass systematics. Bot Mag Tokyo 75:377–383

Thorbjørnsen T, Villand P, Denyer K, Olsen OA, Smith AM (1996) Distinct isoforms of ADP glucose pyrophosphorylase occur inside and outside the amyloplasts in barley endosperm. Plant J 10(2):243–250

Van Hung P, Maeda T, Morita N (2006) Waxy and high-amylose wheat starches and flours—characteristics, functionality and application. Trends Food Sci Technol 17:448–456

Wang JC, Xu H, Zhu Y, Liu QQ, Cai XL (2013) OsbZIP58, a basic leucine zipper transcription factor, regulates starch biosynthesis in rice endosperm. J Exp Bot 64(11):3453–3466

Wang X, Zhou W, Lu Z, Ouyang Y, Yao J (2015) A lipid transfer protein, OsLTPL36, is essential for seed development and seed quality in rice. Plant Sci 239:200–208

Wang H, Huang Y, Xiao Q, Huang X, Li C, Gao X et al (2020) Carotenoids modulate kernel texture in maize by influencing amyloplast envelope integrity. Nat Commun 11(1):5346

Wang R, Ren Y, Yan H, Teng X, Zhu X, Wang Y et al (2021) ENLARGED STARCH GRAIN1 affects amyloplast development and starch biosynthesis in rice endosperm. Plant Sci 305:110831

Whitt SR, Wilson LM, Tenaillon MI, Gaut BS, Buckler EST (2002) Genetic diversity and selection in the maize starch pathway. Proc Natl Acad Sci USA 99:12959–12962

Wise RR (2007) The diversity of plastid form and function. In: Wise RR, Hoober JK (eds) The structure and function of plastids, Advances in photosynthesis and respiration, vol 23. Springer, Dordrecht, pp 3–26. https://doi.org/10.1007/978-1-4020-4061-0_1

Wurtzel ET, Cuttriss A, Vallabhaneni R (2012) Maize provitamin a carotenoids, current resources, and future metabolic engineering challenges. Front Plant Sci 3(29):10.3389/fpls.2012.00029

Yan H, Zhang W, Wang Y, Jin J, Xu H, Fu Y et al (2024) Rice LIKE EARLY STARVATION1 cooperates with FLOURY ENDOSPERM6 to modulate starch biosynthesis and endosperm development. Plant Cell 36(5):1892–1912

Yang J, Kim SR, Lee SK, Choi H, Jeon JS, An G (2015) Alanine aminotransferase 1 (OsAlaAT1) plays an essential role in the regulation of starch storage in rice endosperm. Plant Sci 240:79–89

Yang R, Sun Y, Zhu X, Jiao B, Sun S, Chen Y et al (2024) The tuber-specific StbHLH93 gene regulates proplastid-to-amyloplast development during stolon swelling in potato. New Phytol 241(4):1676–1689

Yu H, Yang Y, Chen XY, Lin GX, Sheng JY, Nie JY et al (2018) Comparison of endosperm amyloplast development and degradation in waxy and non-waxy wheat. Cereal Res Commun 46(2):333–343. https://doi.org/10.1556/0806.46.2018.14

Yun MS, Kawagoe Y (2009) Amyloplast division progresses simultaneously at multiple sites in the endosperm of rice. Plant Cell Physiol 50(9):1617–1626

Zhang L, Ren Y, Lu B, Yang C, Feng Z, Liu Z et al (2016) FLOURY ENDOSPERM7 encodes a regulator of starch synthesis and amyloplast development essential for peripheral ENDOSPERM development in rice. J Exp Bot 67(3):633–647

Zhang W, Mao P, Li Y, Wang M, Xia F, Wang H (2017) Assessing of the contributions of pod photosynthesis to carbon acquisition of seed in alfalfa (Medicago sativa L.). Sci Rep 7(1):42026

Zhang H, Zhao Z, Song B, Du P, Liu X (2020) Light-induced ultrastructure changes of amyloplasts and effect of nitrogen fertilization on greening in potato tubers (Solanum tuberosum L.). Postharvest Biol Technol 168:111275

Zhang L, Li N, Zhang J, Zhao L, Qiu J, Wei C (2022) The CBM48 domain-containing protein FLO6 regulates starch synthesis by interacting with SSIVb and GBSS in rice. Plant Mol Biol 108(4):343–361

Zhou Z, Wang L, Li J, Song X, Yang C (2009) Study on programmed cell death and dynamic changes of starch accumulation in pericarp cells of Triticum aestivum L. Protoplasma 236:49–58

Zhou X, McQuinn R, Fei Z, Wolters AMA, Van Eck J, Brown C et al (2011) Regulatory control of high levels of carotenoid accumulation in potato tubers. Plant Cell Environ 34(6):1020–1030

Zhu Y, Cai XL, Wang ZY, Hong MM (2003) An interaction between a MYC protein and an EREBP protein is involved in transcriptional regulation of the rice Wx gene. J Biol Chem 278(48):47803–47811

Zhu X, Teng X, Wang Y, Hao Y, Jing R, Wang Y et al (2018) FLOURY ENDOSPERM11 encoding a plastid heat shock protein 70 is essential for amyloplast development in rice. Plant Sci 277:89–99

Zi Y, Ding J, Song J, Humphreys G, Peng Y, Li C et al (2018) Grain yield, starch content and activities of key enzymes of waxy and non-waxy wheat (Triticum aestivum L.). Sci Rep 8(1):4548

Elaioplast Development: Insights into Lipid Storage Organelle Biogenesis

11

Najiah Mohd Sadali, Mohd Azrai Loh, Zul Ilham, Sitti' Aisyah Mohd Roszelin, Mohd Naqiuddin Husri, and Nur Ardiyana Rejab

Abstract

Elaioplasts are specialized, non-photosynthetic plastids found in certain plant tissues, primarily responsible for synthesizing and storing lipids. As a unique type of leucoplast, elaioplasts play crucial roles in non-photosynthetic organs, such as the tapetum layer of flowers. These organelles are essential for pollen development, providing lipid materials for the pollen coat, which is critical for pollen viability, dispersal, and interactions with pollinators. This chapter explores the structure, function, and metabolic pathways of elaioplasts, emphasizing their roles in lipid metabolism and secondary metabolite production.

Beyond lipid metabolism, elaioplasts are also involved in hormone biosynthesis, such as jasmonates and abscisic acid (ABA), which are essential for regulating plant growth, stress responses, and defence mechanisms. This multifunctional nature places elaioplasts at the intersection of primary and secondary metabo-

N. M. Sadali
Centre for Research in Biotechnology for Agriculture (CEBAR), Universiti Malaya, Kuala Lumpur, Malaysia

M. A. Loh · Z. Ilham · S. A. M. Roszelin
Institute of Biological Sciences, Faculty of Science, Universiti Malaya, Kuala Lumpur, Malaysia

M. N. Husri
Malaysian Palm Oil Board, Kajang, Selangor, Malaysia

N. A. Rejab (✉)
Centre for Research in Biotechnology for Agriculture (CEBAR), Universiti Malaya, Kuala Lumpur, Malaysia

Institute of Biological Sciences, Faculty of Science, Universiti Malaya, Kuala Lumpur, Malaysia
e-mail: ardiyana@um.edu.my

V. K. Dalal, A. N. Misra (eds.), *Chloroplast Biogenesis and Plastid Interconversions*,
https://doi.org/10.1007/978-981-95-0772-6_11

lism, making them key organelles for various physiological and developmental processes in plants.

Overall, this chapter provides an in-depth overview of elaioplast biology and its roles in lipid and secondary metabolite biosynthesis. It offers valuable insights for biotechnological applications aimed at enhancing oil yield and quality in crop plants, with an emphasis on future research to further clarify the factors governing elaioplast development and function for improved agricultural productivity.

The chapter concludes by highlighting future research directions to better understand the genetic and environmental factors regulating elaioplast development and function. A deeper understanding of these regulatory networks could facilitate the manipulation of elaioplast pathways to optimize oil production in agricultural crops, thereby supporting sustainable agriculture and meeting the growing global demand for plant-based oils.

Keywords

Plastid · Non-photosynthetic plastids · Leucoplasts · Elaioplasts · Oilseeds · Tapetal cells

11.1 Introduction

Plastids are a group of double-membrane organelles typically found in photosynthetic organisms. They are distinct from other organelles within plant cells due to their ability to differentiate into various types and perform diverse functions in plant metabolism via the processes of photosynthesis, lipid and amino acid synthesis, nitrogen and sulfur assimilation and the synthesis and storage of secondary metabolites (Jarvis and López-Juez 2013; Wise 2007). Various types of plastids, including etioplasts, chloroplasts, chromoplasts, and elaioplasts, have been characterised, which can interconvert depending on various developmental and environmental conditions (Choi et al. 2021; Jarvis and López-Juez 2013; Sadali et al. 2019). The summary of plastid forms and functions is described in Table 11.1.

Elaioplasts, a specialized type of leucoplast, are essential organelles in plant cells, primarily functioning as reservoirs for oil storage. These small, colourless rounded structures are characterised by their internal accumulation of oil droplets, which define their specialized role in lipid storage. While plastids are typically associated with photosynthesis, elaioplasts exemplify the broader, multifunctional role that these organelles play, contributing to crucial metabolic processes beyond energy capture.

Structurally, elaioplasts are defined by an outer membrane encasing steryl ester-rich globules. These globules are primarily composed of neutral lipids like 24-methylenecholesterol and pollinastanol esterified to various fatty acids. These

Table 11.1 The summary of plastid forms and functions

Plastid type	Function(s)	Distinctive features
Muroplast	Photosynthesis	Walled chloroplast found in laucocystophytic algae; wall similar to that found in prokaryotes
Rhodoplast	Photosynthesis	Red chloroplast of the Rhodophytes; photosynthetically competent at depths up to 268 m
Proplastid		
Germinal	Source of other plastids	Found in egg, meristematic and embryonic cells; source of all other plastids in the plant
Nodule	N_2 assimilation	Site of the glutamate synthase (GOGAT) cycle in nitrogen-fixing root nodules
Etioplast	Transitionary stage	Develops in dark-grown tissue; site of gibberellin synthesis; converts to chloroplast in light
Leucoplast		
Amyloplast	Starch synthesis and storage	Functions in gravi-sensing
Elaioplast	Oil synthesis and storage	Supplies lipids and oils to the exine upon pollen grain maturation
Proteinoplast	Protein synthesis and storage	May not be a functional subcategory; most protein stored in RER-derived seed storage bodies
Chromoplast	Fruit and flower colouration	Rich in carotenoids; used to attract pollinators and seed/fruit-dispersing animals
Gerontoplast	Catabolism	Controls the dismantling of the photosynthetic apparatus during senescence
Gerontoplast	Catabolism	Controls the dismantling of the photosynthetic apparatus during senescence
S-type plastid	Unknown—starch storage?	Found in phloem sieve tubes; relevant in the systematics of many plant taxa
P-type plastid	Unknown—defence?	Found in phloem sieve tubes; systematically relevant; rupture upon damage to sieve tube cell
Kleptoplast	Photosynthesis	Stolen from algal cells by sea slugs; functionally active for lifetime of slug (up to nine months)
Apicoplast	Unknown	Found in parasitic worms that cause malaria; drugs that target apicoplast reduce infectivity
Chloroplast		
Algal	Photosynthesis, etc.	Probably serves many of the same functions as tracheophyte chloroplasts
C_3	Photosynthesis, etc.	Functions in fatty acid, lipid, amino acid and protein synthesis, N and S assimilation
C_4	Photosynthesis, etc.	Dimorphic chloroplasts provide a CO_2-rich, O_2-poor environment for enhanced rubisco activity
Sun/shade	Photosynthesis, etc.	Dimorphic forms develop under different light conditions to optimize photosynthesis
Guard cell	Stomatal functioning	Senses light and CO_2; signals and metabolically drives opening and closing of stomata

Sources: Wise (2007)

globules are stabilized by surface monogalactosyldiacylglycerols (MGDG) and structural proteins (Wu et al. 1999). While plastid lipid-associated proteins (PAPs) are known to play roles in stabilizing plastoglobules in chloroplasts and chromoplasts, their specific involvement in elaioplasts is unclear (Kim et al. 2001). However, structural proteins are important for maintaining the integrity of lipid globules in elaioplasts. For example, fibrillins are structural proteins associated with plastoglobules and could play similar roles in elaioplasts (Choi et al. 2021).

Elaioplasts typically measure around 3–4 μm in diameter and are largely devoid of thylakoids, which distinguishes them from other plastid types (Fig. 11.1). Instead, their interiors are filled with small spherical lipid droplets composed of steryl esters, all enclosed by the structural protein PAP (Hsieh and Huang 2004). This unique composition not only emphasizes the specialized role of elaioplasts in lipid storage but also highlights their significance in supporting the metabolic functions essential for plant vitality.

The oils stored within elaioplasts and other lipid-containing structures in plant cells are synthesized through two distinct biochemical pathways: the prokaryotic and eukaryotic pathways (Fig. 11.2). These pathways are central to lipid production and cellular function, operating in different cellular compartments. The prokaryotic pathway is confined to plastids, where fatty acids are synthesized and incorporated into lipids critical for plastid membrane structure and function. This pathway is

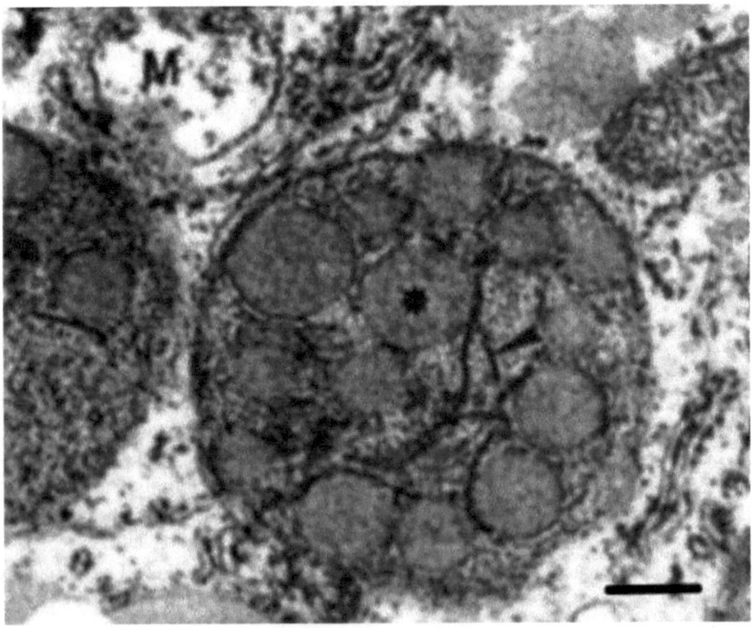

Fig. 11.1 Electron micrograph of an elaioplast in the tapetal cells of *Lilium* prior to pollenkitt release. The stroma shows few thylakoid membranes (indicated by arrowhead) and is densely populated with lipid globules (marked by asterisks). M = mitochondrion (bar = 0.25 μm). (Sources: Clément and Pacini (2001). https://doi.org/10.1007/BF02857849)

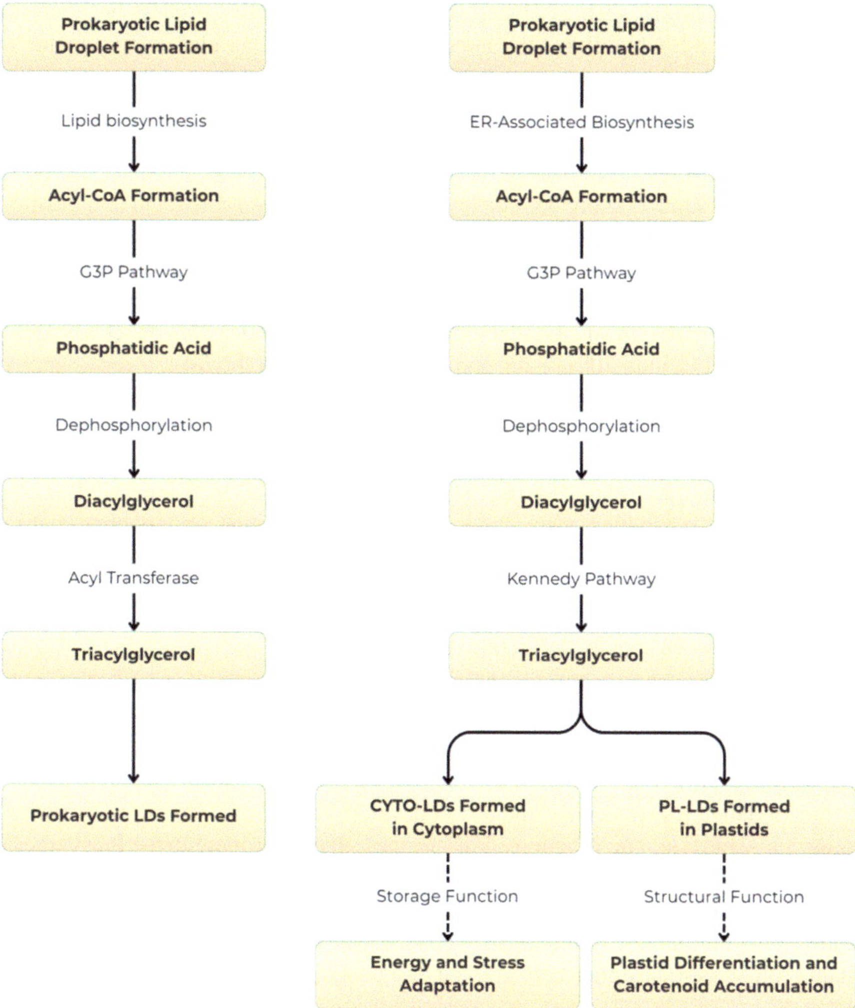

Fig. 11.2 Schematic illustration of lipid droplet formation pathways in prokaryotic and eukaryotic cells. The prokaryotic pathway involves acyl-CoA synthesis followed by the glycerol-3-phosphate (G3P) pathway, culminating in lipid droplet (LD) formation. The eukaryotic pathway occurs via endoplasmic reticulum (ER)-associated biosynthesis, utilizing similar biochemical intermediates, and ultimately branches into cytoplasmic lipid droplets (CYTO-LDs) and plastid lipid droplets (PL-LDs). The diagram also highlights functional roles of LDs, emphasizing their importance in cellular stress adaptation and structural roles in plastid differentiation

especially significant in plastids such as elaioplasts, which depend on these lipids for both their structural composition and their ability to store oils. (Awai et al. 2006; Sato and Awai 2017).

On the other hand, the eukaryotic pathway occurs in the endoplasmic reticulum (ER), where fatty acids produced in the plastids are exported and incorporated into more complex lipids, such as glycerolipids. This cooperative process between the plastids and the ER highlights the interconnectedness of cellular compartments in managing lipid synthesis and distribution. The interaction between these two pathways ensures that plants have a steady supply of lipids not only for structural purposes but also for energy storage in organelles such as elaioplasts (Awai et al. 2006; Negi et al. 2018).

The lipids found in elaioplasts differ significantly in their fatty acid composition and structure, reflecting their classification based on similarities to lipids prevalent in either prokaryotic or eukaryotic cells. Primarily, elaioplasts contain prokaryotic-like triacylglycerols (TAGs) and sterol esters, with monogalactosyldiacylglycerol (MGDG) being the most abundant polar lipid (Hernández-Pinzón et al. 1999; Hernández and Cejudo 2021; Yang and Benning 2018). In the *Brassicaceae* family, TAGs found in lipid droplets within elaioplasts are predominantly composed of sterol esters and MGDG. In contrast, tapetosomes, which are complex organelles originating from the ER, contain lipid droplets enclosed by tapetum-specific oleosins (T-oleosins) and phospholipids (Hsieh and Huang 2004; Yang and Benning 2018). These organelles also store flavonoids and other compounds within ER-derived vesicles.

Elaioplasts have been extensively studied for their crucial role in the pollen development of tapetal cells within anthers (Hsieh and Huang 2004; Quilichini et al. 2014; Ting et al. 1998; Wu et al. 1999). In the structure of an anther, the tapetum is a single-cell layer that encloses the anther cavity, or locule, which contains both elaioplasts and tapetosomes, with a high abundance of plastoglobuli, serving as a reservoir for lipids essential for the formation of the pollen wall (Ting et al. 1998).

Beyond their role in storage and flower development, elaioplasts are also reported to have key biosynthetic functions. For instance, they contribute to the biosynthesis of long-chain fatty acids in developing castor seeds (Eastmond et al. 1997) and terpenes in the outer peel of citrus fruits (Gleizes et al. 1983; Yamasaki and Akimitsu 2007). This multifaceted functionality underscores the importance of elaioplasts in plant metabolism, bridging both storage and biosynthesis within these specialized organelles.

Elaioplasts may also play significant roles during seed germination and stress adaptation. During seed germination, elaioplasts play a crucial role in providing energy reserves. Lipids stored in elaioplasts are mobilised to support seedling growth, similar to how cytoplasmic lipid droplets (LDs) in seeds are mobilised for energy production during germination (Chapman et al. 2012). The mobilization of lipid reserves is essential for seedling establishment before photosynthesis becomes active (Chapman et al. 2012). Additionally, during periods of stress, lipid reserves within elaioplasts become vital for plant survival, highlighting their potential role and adaptation under stress conditions (Michaud and Jouhet 2019).

11.2 Functional Role of Elaioplasts in Plant Physiology and Flower Development

Elaioplasts are specialized plastids primarily associated with lipid storage in non-photosynthetic tissues of plants. Unlike chloroplasts, which are well-recognized for their critical role in photosynthesis, elaioplasts fulfil vital storage and metabolic functions, particularly in reproductive tissues such as flowers and seeds. Their capacity to store energy-dense lipids underscores their significance in plant physiology, particularly concerning flower development, pollen viability, and seed germination.

The processes that govern plastid differentiation and dedifferentiation in plants are multifaceted, influenced by developmental cues, environmental factors, and genetic regulation. Within reproductive tissues, the development of plastids often leads to the formation of elaioplasts, especially during pollen and seed maturation. These plastids are crucial for storing lipids, which can later be mobilised to fuel essential developmental processes.

Within flowers, elaioplasts are predominantly found in tapetal cells, the specialized nutritive cells that surround developing pollen grains. In contrast, non-tapetal cells, such as those in petals and other tissues, may also contain elaioplasts, although they typically serve different roles related to lipid storage and metabolism. The tapetum provides essential lipid materials for the pollen coat, ensuring both the viability and effective dispersal of pollen grains. This differentiation emphasizes the specialized functions of elaioplasts in various cell types and reproductive strategies.

Elaioplasts are integral to the functioning of both reproductive and non-reproductive tissues in plants, particularly in the development of flowers, pollen, and seeds. Their specialized role in storing and mobilizing lipids supports key physiological processes, including pollen viability, seed germination, and early seedling growth. The functional distinctions between elaioplasts in tapetal and non-tapetal cells underscore the versatility of these plastids in meeting the metabolic needs of various plant tissues. By contributing to the production of essential oils and secondary metabolites, elaioplasts also enhance the overall metabolic repertoire of plants. Further exploration of the proteomes and metabolic pathways within elaioplasts will enrich our understanding of their diverse roles in plant physiology and development.

11.2.1 Elaioplasts in Tapetal Cells

In the anther, the tapetum forms a specialized nutritive layer of cells surrounding the anther cavity, or locule, which is essential for pollen development. The behaviour of plastids within the tapetum is highly variable, reflecting species-specific mechanisms and the early stages of microsporogenesis. Generally, elaioplast formation represents the final stage of plastid development in tapetal cells. During this process, proplastids within the tapetum divide and differentiate into elaioplasts, a transition that has been observed in many species (Clément and Pacini 2001). For instance, in the neotropical orchid sub-tribe *Oncidiinae*, tapetal plastids within the floral elaiophore accumulate lipid droplets. Typically, this occurs after starch reserves are depleted, leading to the differentiation of elaioplasts (Pacek et al. 2012; Stpiczyńska et al. 2013).

In contrast, in some anemophilous species, such as *Lolium perenne*, where wind serves as the primary pollination mechanism, the developmental pathway of tapetal proplastids is more complex. Initially, these proplastids differentiate into chromoplasts before transitioning into elaioplasts, which are eventually resorbed alongside other tapetal cell components as part of the normal progression of tapetum degeneration during pollen maturation (Pacini et al. 1992). This sequential differentiation underscores the adaptive plasticity of plastid behaviour within reproductive tissues, reflecting the diverse functional requirements imposed by species-specific reproductive strategies.

11.2.2 Elaioplasts in Non-tapetal Cells

While elaioplasts are commonly associated with tapetal cells, they are also present in a variety of non-tapetal tissues, such as seeds and fruits. Elaioplasts are also present in the cotyledons of oilseeds, where they play a vital role in energy metabolism during germination and early seedling development. The oils stored in these elaioplasts are converted into carbohydrates through gluconeogenesis, providing the growing seedling with essential energy and carbon skeletons (Wijk and Kessler 2017).

In citrus fruits, the edible endocarp stores significant amounts of carotenoids within chromoplasts, which not only contribute to the vibrant red, yellow, and orange hues of the fruit flesh but also serve as important nutrients. In contrast, the pigmented region of the pericarp contains active secretory cells characterised by numerous elaioplasts, which are essential for synthesizing the oils and flavour compounds responsible for the organoleptic properties of citrus fruits (Gleizes et al. 1983; Voo et al. 2012).

Beyond their role in colour and flavour development, elaioplasts are involved in the biosynthesis of terpenes and essential oils, compounds that are widely used in the production of perfumes and flavourings. These oils also serve as natural antiseptic preservatives, enhancing both the safety and shelf life of stored fruits (Patrignani et al. 2015; Tao et al. 2014). Recent proteomic studies on citrus fruit elaioplasts have identified over 600 putative proteins linked to a range of functions, including protein metabolism, lipid transport, and plastidic ATP/ADP transporters. These findings underscore the importance of ATP generation and transport in elaioplast biogenesis and functionality, highlighting their critical role in plant metabolism (Zhu et al. 2018).

11.3 Hormonal and Environmental Cues for Elaioplast Development

11.3.1 Hormonal and Biochemical Regulation in Elaioplasts

Among the thirty most abundant proteins identified in elaioplasts from citrus fruits, lipoxygenase LOX2 (orange1.1 t04376) was notably detected (Zhu et al. 2018). Lipoxygenases (LOX) in plants are known for their involvement in various physiological processes, particularly in the production of jasmonic acid, a hormone essential for stress responses and developmental regulation (Wasternack and Hause 2013). The detection of LOX2 in citrus elaioplasts suggests a potential role in modulating jasmonic-responsive gene expression, indicating that hormonal regulation might play a significant part in elaioplast development. This discovery also raises the possibility that LOX activity in citrus elaioplasts contributes to the formation of volatile compounds, which are important for the aroma and flavour of citrus fruits (Zhang et al. 2017). Furthermore, the involvement of LOX-derived volatiles in the outer peel supports the idea that elaioplasts are integral to the biochemical pathways that influence fruit maturation and sensory attributes.

The identification of proteins associated with the methylerythritol phosphate (MEP) pathway and terpene biosynthesis in citrus elaioplasts highlights their pivotal function in hormone metabolism and the production of secondary metabolites as illustrated in Fig. 11.3. Notably, the complete set of enzymes involved in the MEP pathway, including 1-deoxy-D-xylulose-5-phosphate synthase (DXS), 1-deoxy-D-xylulose-5-phosphate reductoisomerase (DXR), 4-diphosphocytidyl-2C-methyl-D-erythritol kinase (CMK), 2-C-methyl-D-erythritol 4-phosphate cytidylyltransferase (MCT), 2-C-methyl-D-erythritol 2,4-cyclodiphosphate synthase (MDS), 4-hydroxy-3-methylbut-2-enyl diphosphate synthase (HDS), and 4-hydroxy-3-methylbut-2-enyl diphosphate reductase (HDR), has been detected in both elaioplasts and chromoplasts of citrus fruits. This discovery strongly suggests that the MEP pathway is an essential biochemical feature of elaioplasts, playing a significant role in the biosynthesis of terpenes, tocopherols, and carotenoids (Zhu et al. 2018).

The abundance of MEP pathway-associated proteins in elaioplasts emphasizes the critical function of these organelles in terpene biosynthesis, which contributes to the production of essential oils and other secondary metabolites that influence the aroma, flavour, and nutritional value of citrus fruits (Zhu et al. 2018). This proteomic analysis provides a comprehensive understanding of elaioplasts, offering valuable insights into their biogenesis and metabolic roles, particularly in terpene metabolism. Understanding these processes marks a key step in deciphering the broader functions of elaioplasts within plant metabolic networks.

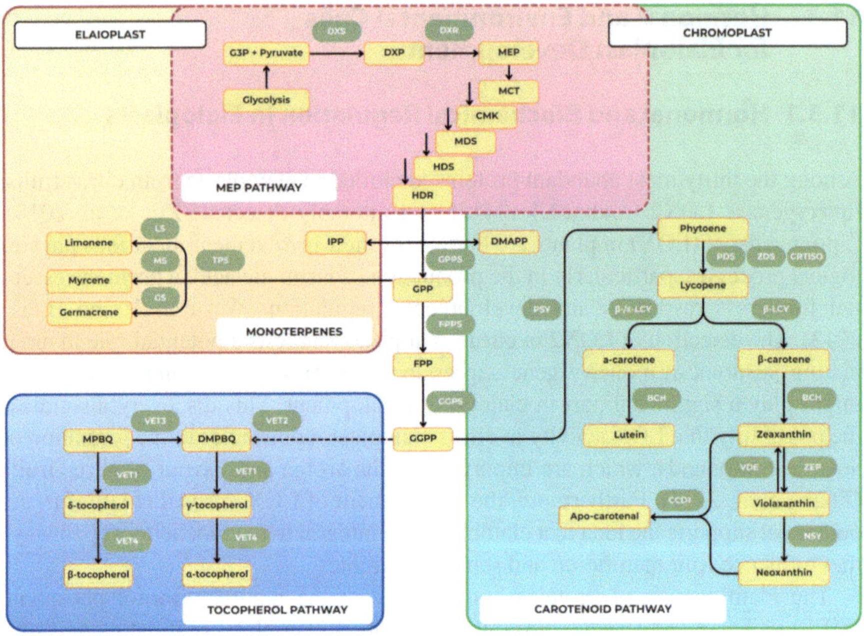

Fig. 11.3 Proteins associated with the methylerythritol phosphate (MEP) pathway and terpene biosynthesis in citrus elaioplasts. The presence of all key MEP pathway enzymes—including 1-deoxy-D-xylulose-5-phosphate synthase (DXS), 1-deoxy-D-xylulose-5-phosphate reductoisomerase (DXR), 2-C-methyl-D-erythritol 4-phosphate cytidylyltransferase (MCT), 4-(cytidine 5′-diphospho)-2-C-methyl-D-erythritol kinase (CMK), 2-C-methyl-D-erythritol 2,4-cyclodiphosphate synthase (MDS), 4-hydroxy-3-methylbut-2-enyl diphosphate synthase (HDS), and 4-hydroxy-3-methylbut-2-enyl diphosphate reductase (HDR)—demonstrates the essential role of elaioplasts in hormone metabolism and the production of secondary metabolites such as monoterpenes, tocopherols, and carotenoids. The monoterpene biosynthesis pathway is facilitated by enzymes such as limonene synthase (LS), myrcene synthase (MS), and germacrene synthase (GS), which convert geranyl diphosphate (GPP) into various monoterpenes. The carotenoid biosynthesis pathway in chromoplasts involves enzymes such as phytoene synthase (PSY), phytoene desaturase (PDS), ζ-carotene desaturase (ZDS), carotenoid isomerase (CRTISO), beta-lycopene cyclase (β-LCY), beta/epsilon-lycopene cyclase (β–/ε-LCY), beta-carotene hydroxylase (BCH), zeaxanthin epoxidase (ZEP), violaxanthin de-epoxidase (VDE), neoxanthin synthase (NSY), and carotenoid cleavage dioxygenase 1 (CCD1), which play essential roles in producing carotenoids such as α-carotene, β-carotene, lutein, zeaxanthin, violaxanthin, and neoxanthin. The tocopherol biosynthesis pathway, which is crucial for vitamin E production, involves key intermediates such as 2-methyl-6-phytylbenzoquinol (MPBQ) and 2,3-dimethyl-6-phytyl-1,4-benzoquinol (DMPBQ), along with the activity of tocopherol biosynthesis enzymes VET1, VET2, VET3, and VET4, which contribute to the synthesis of α-tocopherol, β-tocopherol, γ-tocopherol, and δ-tocopherol. The diagram highlights the integration of these pathways within elaioplasts and chromoplasts, emphasizing their role in plastidial isoprenoid metabolism

11.3.2 The Dynamics of Elaioplasts and Plastoglobuli in Response to Environmental Triggers

Plastoglobuli are plastid-associated lipoprotein particles, characterised by a membrane lipid monolayer embedded with specific proteins and filled with various hydrophobic molecules. They contain specialized proteomes and metabolomes and are found across multiple plastid types, including chloroplasts, chromoplasts, and elaioplasts. The dynamic behaviour of plastoglobuli, both in terms of size and shape, highlights their critical roles in plastid development, particularly during transitions such as thylakoid degradation, and in response to environmental triggers. These structures can rapidly expand or contract during plastid developmental transitions or in reaction to external conditions, such as stress or nutrient availability (Wijk and Kessler 2017).

In tapetal cells, elaioplasts are enriched with plastoglobuli, which store sterol esters necessary for pollen coat formation. Proteomic and metabolomic analyses of plastoglobuli isolated from elaioplasts have revealed a diverse molecular composition, including low levels of galactolipids, moderate levels of TAGs, and high concentrations of sterol esters (Wijk and Kessler 2017). Lipid bodies within elaioplasts accumulate hydrophobic molecules, including sterol esters, which serve as a critical component of the pollen coat. The sterol head groups and acyl chains of sterol esters found in both the pollen coat and plastid lipid bodies are remarkably similar in composition, suggesting that plastid lipid bodies are the primary source of sterol esters in pollen coat formation (Hernández-Pinzón et al. 1999; Wu et al. 1997).

Despite the recognized importance of plastoglobuli in elaioplast function, the proteomes of plastoglobules within elaioplasts in seeds or tapetal cells remain largely uncharacterised. A detailed proteomic analysis would provide critical insights into their metabolic functions, offering a more comprehensive understanding of the pathways involved in elaioplast activity. Currently, the proteomes of plastoglobuli in these non-photosynthetic plastids remain largely unexplored, leaving gaps in our knowledge of their complete metabolic potential. Furthermore, plastoglobuli respond to various external factors, including nutrient availability and environmental stress, such as drought. These factors significantly influence plastid differentiation and lipid storage, further highlighting the role of plastoglobuli in mediating plastid responses to environmental conditions.

11.4 Elaioplasts as the Specialized Plastids for Lipid Storage

Elaioplasts represent a unique class of plastids integral to the synthesis and storage of lipids in plant tissues. Distinguished by their prominent plastoglobules, which are significantly larger than those found in other plastid types, such as chromoplasts, elaioplasts serve essential roles in plant metabolism (Zhu et al. 2018). While it is understood that plastoglobules in chromoplasts arise from the disintegration of the thylakoid membrane during plastid differentiation, the precise origin of plastoglobules within elaioplasts remains a subject of ongoing investigation.

Several hypotheses have emerged to explain the formation of plastoglobules in elaioplasts. One prevailing theory suggests that these structures may originate from the inner envelope membrane of the plastid itself, analogous to the formation of cytoplasmic lipid droplets (CYTO-LDs) from the outer membrane of the ER (Lundquist et al. 2020). Alternatively, elaioplasts may derive from precursor plastids, such as chromoplasts or amyloplasts, through a series of differentiation events that reflect the dynamic nature of plastid development (Choi et al. 2021, 2022). Despite these insights, experimental evidence supporting these hypotheses is still lacking, indicating a need for further research to elucidate the origins and developmental pathways of elaioplast plastoglobules.

The significance of lipid droplets in cellular homeostasis cannot be overstated; they serve as critical reservoirs for neutral lipids and other hydrophobic compounds. Both CYTO-LDs and plastid lipid droplets (PL-LDs) play vital roles across eukaryotic organisms, with plastoglobules being particularly pivotal for lipid metabolism and storage within plastids. The dynamic nature of these lipid droplets, characterised by their ability to store and mobilize lipids in response to cellular demands, underscores their importance in facilitating essential cellular functions and enabling stress responses.

11.4.1 Lipid Types and Their Functions

Elaioplasts represent a specialized class of plastids primarily tasked with the synthesis and storage of lipids in plants. One of the hallmark features of elaioplasts is the presence of plastoglobuli—lipid-filled structures located within the plastid stroma, serving as crucial sites for lipid accumulation (Hernández and Cejudo 2021; Wijk and Kessler 2017). These plastids have been extensively studied, particularly in the context of pollen formation, where they are predominantly located within the tapetal cells lining the anther (Clément and Pacini 2001). Additionally, elaioplasts play a significant role in the peels of citrus fruits, where they contribute to the accumulation of aromatic compounds (Zhu et al. 2018).

In contrast to animal cells, where fatty acid synthesis occurs primarily in the cytosol, approximately 95% of fatty acids in plants are synthesized within plastids. These fatty acids are either metabolised directly into TAGs or exported to the ER for further modification (Choi et al. 2022; Hölzl and Dörmann 2019). Most research on fatty acid synthesis has centred on chloroplasts due to their abundance and well-established roles as the most studied plastid type. The metabolic reactions occurring within chloroplasts are intricately associated with the thylakoid membranes, which house photosynthetic complexes and biochemical pathways that include fatty acid and lipid synthesis (Hölzl and Dörmann 2019).

The process of fatty acid synthesis in chloroplasts initiates with the conversion of acetyl-CoA into malonyl-CoA, a reaction catalysed by acetyl-CoA carboxylase (ACCase), as illustrated in Fig. 11.4. Following this, malonyl-CoA is transferred to an acyl carrier protein (ACP) via malonyl-CoA: ACP malonyl-transferase (MCMT). The first elongation reaction entails the condensation of malonyl-ACP with

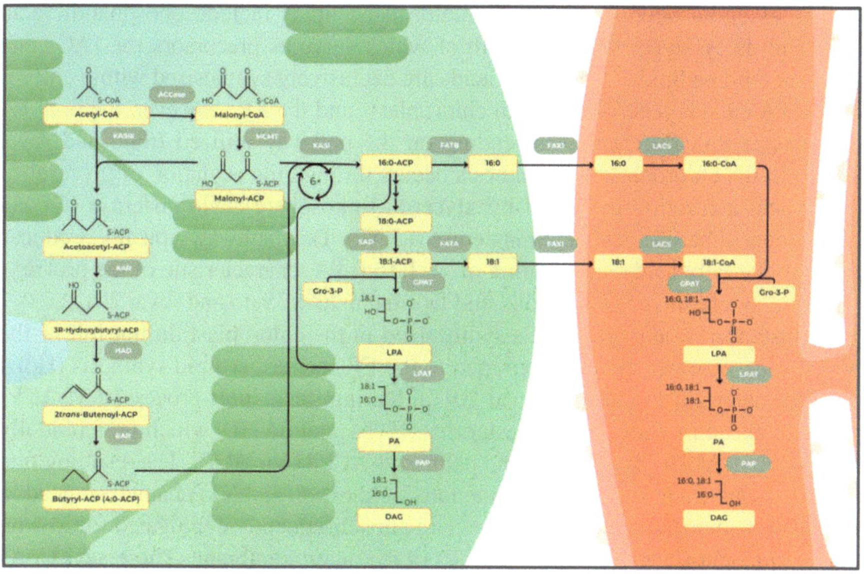

Fig. 11.4 Schematic representation of the fatty acid synthesis pathway in chloroplasts and its extension in the endoplasmic reticulum (ER). The process begins with the conversion of acetyl-CoA to malonyl-CoA by acetyl-CoA carboxylase (ACCase). Malonyl-ACP is subsequently elongated through series of condensation, reduction and dehydration reactions catalised by ketoacyl-ACP synthases (KASs), ultimately producing palmitoyl-ACP (16:0-ACP), and stearoyl-ACP (18:0-ACP). Stearoyl_ACP desaturase (SAD) converts 18:0-ACP into oleoyl-ACP ($18{:}1^{\Delta 9}$-ACP). In this pathway, 16:0 (palmitic acid) is a saturated fatty acid with 16 carbon atoms, 18:0 (stearic acid) is a saturated fatty acid with 18 carbon atoms, and 18:1 (oleic acid) is a monounsaturated fatty acid with 18 carbon atoms and one double bond. Key enzymes involved are indicated at each step: MCMT (Malonyl-CoA:ACP malonyltransacylase), FAT (Fatty acyl-ACP thioesterase), KAR (Ketoacyl-ACP reductase), HAD (Hydroxyacyl-ACP dehydratase), LACS (Long-chain acyl-CoA synthetase), GPAT (Glycerol-3-phosphate acyltransferase), LPAT (Lysophosphatidic acid acyltransferase) and PAP (Phosphatidic acid phosphatase). Other proteins and intermediates include FAX1 (Fatty acid export 1 protein), Gro-3-P (Glycerol-3-phosphate), LPA (Lysophosphatidic acid), PA (Phosphatidic acid), DAG (Diacylglycerol)

acetyl-CoA, facilitated by 3-ketoacyl-ACP synthase III (KASIII), culminating in the formation of acetoacetyl-ACP. This intermediate subsequently undergoes a series of reductions mediated by three essential enzymes—3-ketoacyl-ACP reductase (KAR), hydroxyacyl-ACP dehydratase (HAD), and enoyl-ACP reductase (ENR)—to produce butyryl-ACP (4:0-ACP) (Hölzl and Dörmann 2019). The resulting 4-carbon substrate is subjected to six additional condensation reactions with malonyl-ACP via KASI, ultimately yielding palmitoyl-ACP (16:0-ACP). A fraction of the palmitoyl-ACP is further processed by KASII and stearoyl-ACP-Δ9-desaturase (SAD), leading to the formation of stearoyl-ACP (18:0-ACP) and oleoyl-ACP ($18{:}1^{\Delta 9}$-ACP), which are the primary products of chloroplast fatty acid synthesis (Hernández and Cejudo 2021; Hölzl and Dörmann 2019).

The essential building blocks for chloroplast lipids include phosphatidic acid (PA) and diacylglycerol (DAG), both of which serve as precursors for TAGs and other membrane lipids. While fatty acids are exclusively synthesised within chloroplasts, PA can be generated in both chloroplasts and the ER; however, the final PA products exhibit differences depending on the pathway utilized (Sato and Awai 2017). In the chloroplast, PA synthesis occurs through the acylation of glycerol-3-phosphate, a reaction catalysed by glycerol-3-phosphate acyltransferase (GPAT) and lysophosphatidic acid acyltransferase (LPAT). This pathway typically produces PA with the fatty acid configuration of 18:1,16:0-PA, characteristic of the prokaryotic lipid synthesis pathway (Hölzl and Dörmann 2019; Sato and Awai 2017).

In instances where acyl-ACPs synthesized in the chloroplast do not follow the prokaryotic pathway, they are exported to the ER for further lipid synthesis (Hölzl and Dörmann 2019; Sato and Awai 2017). During this export process, acyl-ACPs are hydrolysed by fatty acid thioesterases (FATA and FATB), which preferentially target oleoyl-ACP ($18{:}1^{\Delta 9}$-ACP) and palmitoyl-ACP (16:0-ACP). The resulting free fatty acids are transported through the fatty acid export (FAX1) channel situated in the inner membrane of the chloroplast and are subsequently conjugated to CoA by long-chain acyl-CoA synthetase (LAC9) in the outer membrane. These acyl-CoAs are then transferred to the ER via ATP-binding cassette transporter subfamily A9 (ABCA9) (Choi et al. 2022; Li et al. 2015; Schnurr et al. 2002). Within the ER-localized pathway, PA is synthesized through a similar process, involving the acylation of glycerol-3-phosphate by GPAT and LPAT2, yielding PA configurations of 16:0,18:1-PA and 18:1,18:1-PA that differ from those produced in chloroplasts (Hölzl and Dörmann 2019; Sato and Awai 2017).

11.4.2 Formation of Lipid Droplets in Plastids

Elaioplasts are specialized plastids responsible for lipid accumulation, particularly in oilseeds and the tapetal tissue surrounding developing pollen grains. Unlike other plastid types, elaioplasts are unlikely to originate directly from precursor chloroplasts but may instead arise from proplastid division or through the transformation of chromoplasts and amyloplasts (Clément and Pacini 2001; Lundquist et al. 2020). The defining feature of elaioplasts is the presence of plastidial lipid droplets (LDs), known as PL-LDs, which are critical for lipid storage and metabolism (Lundquist et al. 2020). While the exact mechanism of PL-LD formation remains unclear, these lipid droplets may originate from a distinct membrane system within the plastid, potentially the inner envelope membrane, or they may persist through multiple plastid divisions and differentiation states (Lundquist et al. 2020). However, empirical evidence remains limited, leaving the precise origins of non-chloroplast PL-LDs unclear.

Beyond their presence in elaioplasts, PL-LDs are also found in chloroplasts and chromoplasts, where they function as major lipid reservoirs. They are structurally distinct from CYTO-LDs and prokaryotic lipid droplets (PRO-LDs). PL-LDs are enclosed by a galactolipid monolayer, which is characteristic of plastid membranes.

While most plastids contain spherical PL-LDs, the PL-LDs in chromoplasts feature elongated fibrils derived from plastoglobules. These plastoglobules are dynamic structures, as studies indicate that they can transform into carotenoid fibrils, demonstrating their structural plasticity during plastid differentiation (Berry et al. 2019).

Structurally, all LDs consist of a neutral lipid core, primarily composed of TAGs and esters, enclosed within an amphipathic monolayer (Zhao et al. 2022). However, the composition of this monolayer varies depending on the LD type. Unlike typical organelle bilayers, where lipids form two distinct layers, LDs are surrounded by a single monolayer of amphipathic lipids, with acyl chains extending into the lipid core and polar headgroups facing the aqueous cytoplasmic environment (Lundquist et al. 2020; Chorlay and Thiam 2020; Giménez-Andrés et al. 2021). Among plastid lipid droplets, plastoglobules in chloroplasts are particularly well studied. These LDs not only accumulate TAGs and prenyl-lipids but also play crucial roles in lipid metabolism, storage, and stress adaptation. Their presence in all photosynthetic eukaryotes highlights their fundamental role in plastid function, particularly in response to environmental stressors and developmental transitions.

11.4.3 Lipid Droplets and Their Role in Cellular Homeostasis

LDs are essential for maintaining cellular homeostasis by serving as reservoirs for excess fatty acids and other lipids, which are essential for metabolism, signalling and stress response. They prevent the toxic accumulation of free fatty acids and their derivatives, safeguarding cellular membranes and signalling pathways (Xu et al. 2020). LDs are classified into three types: CYTO-LDs, PRO-LDs, and PL-LDs. PRO-LDs are exclusive to prokaryotic organisms, whereas CYTO-LDs and PL-LDs are found in eukaryotic species. CYTO-LDs, commonly referred to as oil bodies, originated from the ER and are dispersed throughout the cytoplasm, whereas PL-LDs are localized within plastids, where they play important roles in plastidial lipid metabolism (Lundquist et al. 2020). The sizes of LDs vary significantly depending on their origin. CYTO-LDs can be large, with diameters ranging from 1000 nm in yeast to 250,000 nm in human adipocytes, while PL-LDs, such as plastoglobules, are typically smaller, ranging from 50 to 800 nm in diameter (Lundquist et al. 2020). Despite these differences, both types of LDs serve a common function as storage depots for neutral lipids, including TAGs and esters, as well as for other hydrophobic compounds such as carotenoids (Hölzl and Dörmann 2019).

In plant cells, CYTO-LDs and PL-LDs are often located near their respective biosynthetic sites, the ER and the thylakoid membrane, supporting the hypothesis that these organelles play a role in their formation (Fig. 11.5). CYTO-LDs, which are associated with the ER, are surrounded by a monolayer of phospholipids, whereas PL-LDs, found in plastids, are enveloped by a monolayer of galactolipids (Choi et al. 2022; Lundquist et al. 2020). Plastoglobules, in particular, are hypothesized to form through a blistering process at highly curved regions of the thylakoid membrane, where they serve as dynamic lipid reservoirs (Hölzl and Dörmann 2019; Kirchhoff 2019). Plastoglobules are thought to play an important role in regulating

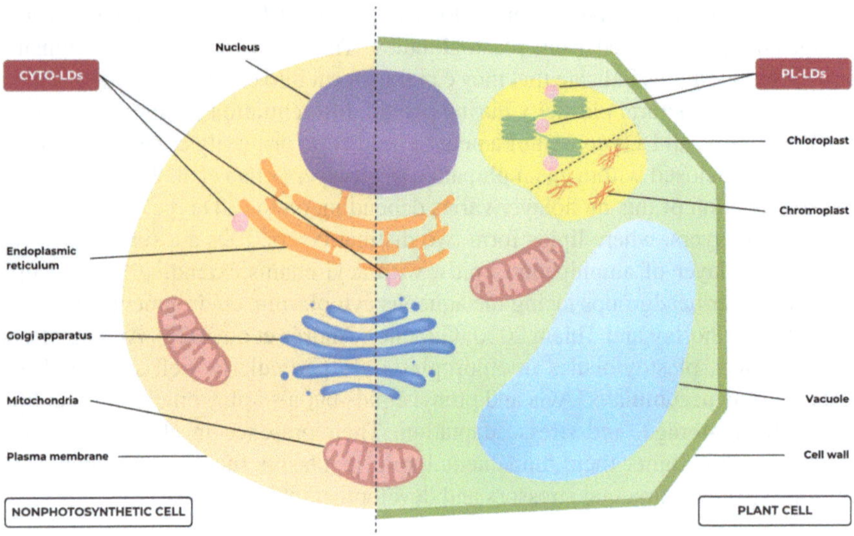

Fig. 11.5 Cellular localization of lipid droplet classes in eukaryotes. The illustration depicts a generic chimeric photosynthetic/ non-photosynthetic eukaryotic cell (outlined by a dashed line) and a plastid represented as a chloroplast/chromoplast chimera. Cytoplasmic lipid droplets (CYTO-LDs) are present in all eukaryotic cells, whereas plastid lipid droplets (PL-LDs) are exclusive to photosynthetic cells

lipid availability within the thylakoid membrane, replenishing lipids when necessary, or storing excess lipids synthesized by the chloroplast (Kirchhoff 2019).

In chloroplasts, plastoglobules function in lipid homeostasis and thylakoid membrane remodelling, whereas in chromoplasts, they contribute to carotenoid accumulation, with evidence suggesting that plastoglobules transform into elongated carotenoid fibrils, reflecting their structural plasticity (Berry et al. 2019). Plastoglobules are also abundant in plastids involved in stress responses, such as etioplasts and gerontoplasts. During stress adaptation and leaf senescence, plastoglobules accumulate large quantities of fatty acid phytyl esters (FA-PE), which are formed from the esterification of free fatty acids and phytol, a byproduct of chlorophyll degradation (Xu et al. 2020). This esterification process serves a protective function, preventing the accumulation of free phytol and fatty acids, which possess detergent-like properties that can disrupt cellular membranes (Gutbrod et al. 2021; Hölzl and Dörmann 2019).

11.5 Elaioplast Biogenesis and Functional Conservation

Proplastids, small undifferentiated plastids, are present in meristematic and reproductive tissues. Their specific roles include organelle transmission between generations—predominantly via female gametes, though occasionally through pollen—and ensuring plastid inheritance across all cells of the organism. In vegetative cells, plastid differentiation typically follows a unidirectional pathway, whereas in reproductive cells, plastid interconversion is frequently observed (Clément and Pacini 2001). Elaioplasts, like other specialized plastids, are largely derived from proplastids through a process of differentiation. All plastids, including elaioplasts, are enclosed by a double-membrane structure termed the envelope (Jarvis and López-Juez 2013).

Although the biogenesis of chloroplasts has been extensively studied, the biogenesis of elaioplasts remains less well characterised. Recent comprehensive proteomic analysis of elaioplasts from citrus fruits by Zhu et al. (2018) identified 655 putative plastid proteins, providing insights into elaioplast biogenesis and function. The identified putative plastid proteins were classified into 32 functional categories based on sequence homology, utilizing both in silico analysis and manual curation (Zhu et al. 2018). MapMan-based functional classification revealed that approximately 50% of these proteins fell into six major categories, including protein metabolism, transport, lipid metabolism, and ATP generation and transport (Zhu et al. 2018). Consistent with prior findings in non-photosynthetic plastids, a substantial number of proteins responsible for ATP production and reducing power were identified, highlighting the conservation of these pathways across plastid types (Wang et al. 2013). The abundance of biosynthetic proteins suggests that elaioplasts possess active metabolic functions, contributing to the production of reducing power, energy, and precursors essential for metabolite biosynthesis.

The biogenesis and homeostasis of plastids are tightly regulated processes, synchronized with the development of cells and organs to ensure that the appropriate number and type of plastids are present during each stage of plant growth. Plastid development is coordinated by the plastid genetic system in conjunction with nucleocytosolic factors. The import and sorting of nucleus-encoded proteins, along with plastid division, are essential for plastid homeostasis. Moreover, these processes are regulated by the ubiquitin–proteasome system, which integrates plastid biogenesis into the broader developmental programs of the plant.

11.5.1 Regulation of Elaioplast Biogenesis

Plastid biogenesis is a highly regulated process, coordinated with both environmental stimuli and endogenous signals. A key aspect of this regulation involves the translocation of proteins from the cytosol into the plastid. Most plastid proteins are encoded by nuclear genes, synthesized in the cytosol as precursors with N-terminal transit peptides, which serve as targeting signals. These precursor proteins are imported into plastids via the translocon complexes located at the outer and inner envelopes of plastids: the TOC (Translocon at the Outer Chloroplast membrane) and TIC (Translocon at the Inner Chloroplast membrane) complexes (Jarvis 2008; Nellaepalli et al. 2023).

The TOC apparatus comprises three major components: TOC159, TOC34, and TOC75, where the numbers reflect their molecular weight in kilodaltons. TOC159 and TOC34 function as receptor proteins, with GTPase domains that regulate the recognition of precursor proteins, while TOC75 forms the β-barrel channel through which preproteins are translocated (Kessler and Schnell 2002; Schnell et al. 1994; Smith et al. 2002; Sun et al. 2002; Wallas et al. 2003). The complex exists in various forms, containing different isoforms of import receptors. At the inner envelope, the TIC complex mediates translocation through the inner membrane, a process that requires significant energy, assisted by molecular chaperones (Theg et al. 1989). Although several components of the TIC complex—TIC20, TIC21, TIC22, TIC32, TIC40, TIC55, TIC56, TIC62, TIC100, TIC110, and TIC214—have been identified, the precise roles of many remain unclear (Jarvis 2008; Nellaepalli et al. 2023).

The import of nuclear-encoded precursor proteins via TOC/TIC complexes is fundamental for the biogenesis of plastids, playing a crucial role not only in chloroplast development (Ling et al. 2012) but also in the formation of chromoplasts (Egea et al. 2010; Ling et al. 2021). Proteomic studies on citrus elaioplasts have identified a significant proportion of plastid proteins—approximately 16.9%—involved in the protein import machinery, including TOC33, TOC75, TOC159 and TIC62, TIC55, TIC40, TIC22, and TIC110. Additionally, several ribosomal proteins and caseinolytic peptidases (Clp proteases), associated with translation and protein homeostasis, have also been detected (Zhu et al. 2018).

While the precise regulatory mechanisms governing elaioplast biogenesis are yet to be fully elucidated, the presence of numerous proteins associated with import, ribosomal assembly, and proteolysis in citrus elaioplasts strongly suggests the involvement of the plastid protein import machinery and the establishment of functional translation and homeostasis systems during elaioplast development.

11.5.2 Functional Conservation of Elaioplasts and Other Plastid Types

Plastids represent a diverse and crucial group of organelles, fundamental to plant cell function and development. While chloroplasts are the most well-known plastids due to their role in photosynthesis and the accumulation of chlorophyll, plastids exist in various forms, each specialized for different biochemical processes. Chloroplasts are unique among plastid types as they are the only ones capable of producing energy through photosynthesis. However, consistent with findings from non-photosynthetic plastids, a significant number of proteins involved in ATP generation and reducing power are conserved across various plastid types (Wang et al. 2013). This observation extends to elaioplasts, where photosynthetic proteins have been identified, as previously noted in other non-photosynthetic plastids such as chromoplasts (Barsan et al. 2010; Ling et al. 2021) and amyloplasts (Balmer et al. 2006).

The functions of plastids, however, extend far beyond photosynthesis. They are metabolically versatile, contributing to the synthesis of essential compounds such as amino acids, fatty acids, starch, chlorophylls, carotenoids, and signalling molecules. Plastids also play pivotal roles in the assimilation of nitrogen and sulfur (Wise 2007; Sakamoto et al. 2008). A comparative proteomic analysis of citrus chromoplasts and elaioplasts has highlighted the conservation of lipid biosynthesis and homeostasis mechanisms in these plastid types (Zhu et al. 2018). This is further supported by metabolic studies on tomato fruit chromoplasts, where radiolabelled precursors revealed insights into carotenoid and lipid biosynthesis, demonstrating a high level of functional conservation (Angaman et al. 2012).

The diversity of plastid functions underscores their critical role in maintaining cellular homeostasis, as well as their ability to perceive and respond to environmental and developmental signals. This is especially relevant in the context of reactive oxygen species (ROS), which are byproducts of photosynthesis and can serve as signalling molecules but may also cause cellular damage if not properly regulated (Asada 2006; Shapiguzov et al. 2012). In elaioplasts, proteins involved in redox regulation and stress response form one of the major functional classes. These include components of the ascorbate-glutathione (AsA-GSH) cycle and other antioxidant systems, such as glutathione peroxidases, peroxiredoxins (PRXs), superoxide dismutases (SODs), and thioredoxins (Zhu et al. 2018).

Furthermore, elaioplasts contain several key enzymes involved in ROS scavenging, including those catalysing reactions in the AsA-GSH cycle, such as ascorbate peroxidases (APXs), as well as the monodehydroascorbate reductase system (Zhu et al. 2018). The presence of these enzymes suggests that elaioplasts possess an active redox system, which plays a critical role in managing oxidative stress during fruit maturation and development. This redox activity mirrors the antioxidant machinery found in chloroplasts, highlighting a conserved mechanism that allows plastids to adapt to environmental challenges and developmental cues.

11.6 Practical Applications: Enhancing Lipid Storage in Plants

11.6.1 Agricultural Significance

Plants are known to synthesize significant quantities of lipids in seeds or fruit mesocarps, which serve as storage reserves of carbon and energy for seed germination and development. Lipids are the preferred energy storage molecules due to their higher energy density compared to carbohydrates and proteins. When extracted from oil crops such as oil palm, soybeans, and sunflower, these lipids are referred to as vegetable oils. In contrast to mineral oils like petroleum, vegetable oils are primarily used as edible oils and are considered a renewable resource (Zafar et al. 2019).

Vegetable oils mainly consist of TAGs, which are composed of three fatty acids linked to a glycerol backbone. These fatty acids, primarily C16 and C18 chains, are synthesized within plastids. The most common fatty acids found in vegetable oils include palmitic acid (16:0), stearic acid (18:0), oleic acid ($18{:}1^{\Delta 9}$), linoleic acid ($18{:}2^{\Delta 9,12}$), and α-linolenic acid ($18{:}3^{\Delta 9,12,15}$) (Hölzl and Dörmann 2019; Msanne et al. 2020). TAGs are typically stored in the cells of seeds or fruits and are extracted using either mechanical methods, such as expeller pressing, or chemical processes like hexane solvent extraction.

Vegetable oil is a critical global commodity, with annual consumption reaching around 200 million metric tons, driven by both food and industrial demands (Msanne et al. 2020). Over the past two decades, the production and consumption of vegetable oils have grown rapidly, with oils such as palm, soybean, and sunflower witnessing at least a twofold increase in production (Giannini et al. 2022; Pilorgé 2020). During this same period, global vegetable oil consumption has also doubled, despite a global population increase of only 25%. This disproportionate rise in demand is largely attributed to the growing use of vegetable oils in non-food applications, such as biodiesel, lubricants, and cosmetics, particularly in developing nations like China and India, alongside the European Union's increasing demand for these renewable resources (Giannini et al. 2022; Parveez et al. 2022; Pilorgé 2020).

11.6.2 Improving the Quality and Yield of Edible Oils

Edible oils have been a staple of the human diet for millennia, contributing approximately 20% of the daily caloric intake in developed countries (Msanne et al. 2020). Over the past fifty years, the global demand for edible oils has surged, with annual consumption rising from around 40 million metric tons to nearly 200 million metric tons (Pilorgé 2020). This substantial increase is driven not only by the growing global population but also by the diversification of edible oils into non-food sectors, including biofuels, cosmetics, and industrial lubricants. As the biodiesel, cosmetics, and lubricant industries expand, their encroachment into the edible oil supply chain is contributing to the intensifying demand for this crucial commodity. Current projections suggest that by the middle of this century, the global demand for vegetable oils could at least double, if not exceed, current levels. Global oilseed production has risen steadily over the past decade, increasing by approximately 20% and reaching around 650 million metric tons in 2023 (Fig. 11.6). This growth has been largely driven by increased soybean, rapeseed, and sunflower production, particularly in South America, North America, and Europe. Soybeans remain the dominant oilseed, accounting for over 60% of total production, followed by rapeseed and sunflower. However, factors such as climate variability and biotic stress led to fluctuations in production levels. To meet this rising demand, two primary strategies are available: expanding agricultural land to increase the area under oil crop cultivation or improving the yield and quality of oil produced per unit area through advances in agronomic practices and breeding techniques. Table 11.2 highlights the overview of some successful studies in improving oil contents. However, the expansion of agricultural land often leads to deforestation, habitat loss, and decreased biodiversity, which has prompted conservation groups and environmental activists to oppose this approach (Tang and Al Qahtani 2020). Given these environmental concerns, the focus has increasingly shifted toward enhancing oil crop yields by improving the genetic and phenotypic traits of plants through selective breeding and biotechnological innovations (Bhat 2010; Zafar et al. 2019).

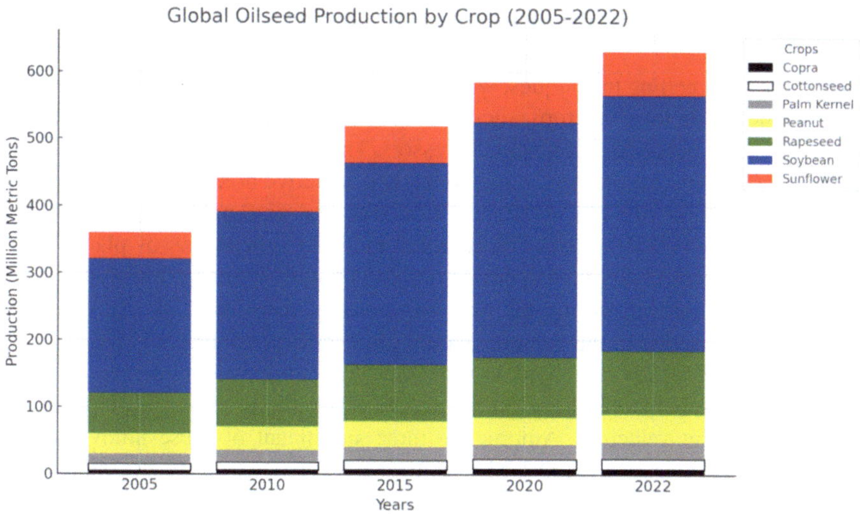

Fig. 11.6 This figure illustrates the trends in global oilseed production from 2005 to 2022, highlighting the contributions of major oilseed crops, including soybeans, rapeseed, sunflower seed, palm kernel, and others. The data reflect the steady increase in production over time, with soybeans remaining the dominant oilseed, followed by rapeseed and sunflower seed. Source: USDA 2025

Table 11.2 Overview of studies exploring three key strategies to enhance TAG biosynthesis: increasing the amount of available substrate, channelling the substrate towards TAG production, and improving the efficiency of TAG accumulation within lipid storage compartments

Protein	Function	Gene origin	Reference
ABCA9	Facilitates the transport of fatty acid and acyl-CoA from the cytoplasm to the ER.	Arabidopsis	Kim et al. (2013)
FAX1	Facilitates the export of fatty acid and acyl-CoA from the chloroplast to the cytoplasm.	Arabidopsis	Tian et al. (2018)
FAX1 ABCA9	Works in tandem for the export of fatty acids and acyl-CoA from the chloroplast to the cytoplasm, and from the cytoplasm to the ER.	Camelina	Cai et al. (2021)
GDPH GPAT LPAT DGAT	Multiple enzymes serving in the Kennedy pathway for the synthesis of TAG in plants.	Rapeseed and yeast	Liu et al. (2015)
LPAT2	Facilitates the synthesis PA from Lysophosphatidic acid (LPA)	Peanuts	Chen et al. (2015)
DGAT1	Facilitate the synthesis of TAG from diacylglycerol (DAG)	Arabidopsis	Savadi et al. (2016)
Seipin1	Facilitates the assembly of CYTO-LDs from the ER.	Arabidopsis	Lunn et al. (2018)

11.6.3 Conventional Breeding and Its Limitations

Conventional breeding techniques, which have been employed for centuries, aim to improve desirable traits in crops, such as enhanced oil content, larger fruit size, and increased resistance to environmental stressors. These techniques rely on the natural genetic variability within plant populations and selective breeding to develop new varieties with improved characteristics. However, conventional breeding is often a slow and unpredictable process. It requires multiple generations of plants to be grown and assessed before the desired traits can be reliably incorporated into a stable cultivar. Moreover, during the breeding process, there is a risk that desirable traits may be lost due to the complex nature of genetic inheritance.

In recent years, advancements in molecular biology have complemented traditional breeding methods, offering tools to accelerate the process and improve the precision of trait selection. Among the most significant of these innovations is marker-assisted selection (MAS). MAS allows breeders to screen plant genotypes at the seedling stage, selecting individuals with desirable traits without waiting for the plants to mature (Fujino et al. 2019). This method saves both time and resources, as large populations can be screened quickly, enabling breeders to focus their efforts on the most promising candidates for future breeding cycles (Burt and Mitchell Fetch 2018; Yue et al. 2021).

11.6.4 Genetic Engineering and Gene Editing: Expanding the Possibilities

As conventional breeding approaches begin to reach their limits, more precise and targeted methods, such as genetic engineering and gene editing, are being explored. These molecular techniques offer the potential to introduce new traits directly into the genome, bypassing the slow and often imprecise process of traditional breeding (Gao 2021). Gene editing technologies, such as CRISPR-Cas9, enable precise modifications to specific genes, allowing for the enhancement of traits like oil content, resistance to pests, or tolerance to environmental stress. The advantage of gene editing over conventional methods is its speed and accuracy. Additionally, gene editing opens the possibility of incorporating beneficial traits from unrelated species, such as bacteria or fungi, into oil crops—an approach that is impossible with traditional breeding.

The primary goal of improving oil crop genetics is to enhance the yield and quality of oil produced. Crops like rapeseed, sunflower, and soybean are the main sources of commercial vegetable oils. Over the years, genetic engineering has made significant strides in improving oil yield by manipulating the metabolic pathways involved in lipid synthesis, particularly triacylglycerol (TAG) production. Three key strategies are commonly employed in these studies: the "push," "pull," and "pack and protect" approaches (Lunn et al. 2018; Vanhercke et al. 2013; Zafar et al. 2019).

Push: Redirecting the metabolic flux of fatty acids towards TAG synthesis by increasing the availability of fatty acid precursors.

Pull: Enhancing the enzymatic machinery responsible for TAG assembly.
Pack and Protect: Improving the storage capacity of TAGs within the cell and ensuring their retention in storage compartments, such as LDs.

11.6.5 Manipulating Fatty Acid Flux to Enhance Oil Yield

In plants, fatty acid synthesis primarily occurs in plastids, which are either used locally or exported to the ER for TAG assembly. Oil bodies, also known as LDs, are the major storage sites for TAGs and are significantly larger than plastoglobules, particularly in oil-rich tissues like seeds and fruit mesocarp (Lundquist et al. 2020).

Studies have demonstrated that manipulating key enzymes involved in the transport of fatty acids from plastids to the ER can significantly increase oil yield. Two enzymes, FAX1 and ABCA9, play crucial roles in this process. FAX1 facilitates the export of fatty acids from plastids, while ABCA9 imports these fatty acids into the ER for TAG synthesis. Overexpression of these transporters has been shown to boost oil content in model plants such as Arabidopsis thaliana and emerging oil crops like Camelina sativa. For instance, the overexpression of ABCA9 in Arabidopsis led to a 40% increase in seed oil content, while FAX1 overexpression resulted in a 21% increase (Kim et al. 2013; Tian et al. 2018). Moreover, simultaneous overexpression of both FAX1 and ABCA9 in Camelina increased seed oil content by 13%, demonstrating the additive effects of these transporters when expressed together (Cai et al. 2021).

11.6.6 The Kennedy Pathway: A Key to Enhancing TAG Synthesis

The Kennedy pathway is a fundamental biochemical route for TAG synthesis to determine oil yield and lipid composition in plants. Schematic representation of the Kennedy pathway for TAG biosynthesis is described in Fig. 11.7. The pathway begins with the reduction of glycerol-3-phosphate (G3P), which serves as the backbone for TAG synthesis. Initially, glycerol-3-phosphate acyltransferase (GPAT) catalyses the acylation of G3P, forming lysophosphatidic acid (LPA). Lysophosphatidic acid acyltransferase (LPAT) then acylates LPA at the sn-2 position to produce phosphatidic acid (PA). Phosphatidic acid phosphatase (PAP) subsequently converts phosphatidic acid (PA) into diacylglycerol (DAG). The final, rate-limiting step is catalysed by diacylglycerol acyltransferase (DGAT), which adds an acyl-CoA to DAG to produce TAG. This DGAT-mediated reaction is widely recognized as the rate-limiting step and thus a primary target for enhancing oil accumulation through metabolic engineering (Chen et al. 2015; Liu et al. 2015).

Recent biotechnology efforts have targeted key Kennedy pathway enzymes to enhance oil yield, aiming to improve agricultural productivity and sustainable biofuel production. For example, the overexpression of yeast-derived *ScLPAT* in transgenic tobacco resulted in an 8% increase in seed oil content compared to wild-type plants, while simultaneous overexpression of multiple genes in the pathway

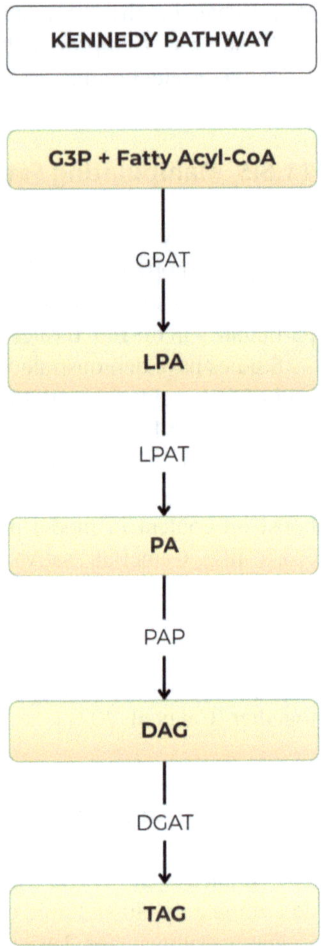

Fig. 11.7 Schematic representation of the Kennedy pathway for triacylglycerol (TAG) biosynthesis. The pathway initiates with glycerol-3-phosphate acyltransferase (GPAT) catalyzing the acylation of glycerol-3-phosphate (G3P) to form lysophosphatidic acid (LPA). Lysophosphatidic acid acyltransferase (LPAT) adds an acyl group to LPA, producing phosphatidic acid (PA). Phosphatidic acid phosphatase (PAP) then dephosphorylates PA, forming diacylglycerol (DAG). In the final, rate-limiting step, diacylglycerol acyltransferase (DGAT) converts DAG into TAG

[*glycerol-3-phosphate dehydrogenases (GPDH), GPAT, LPAT*, and *DGAT*)] led to a 14.46% increase in oil content (Liu et al. 2015). These findings underscore the potential of targeting multiple genes within the TAG biosynthesis pathway to achieve significant improvements in oil yield.

11.6.7 "Pack and Protect" Strategy: Enhancing Storage and Retention of TAGs

The "Pack and Protect" strategy is a genetic engineering approach designed to enhance the storage capacity and retention of TAGs within plant cells by optimising LD formation and safeguarding stored lipids from degradation (Sagun et al. 2023).

Following their synthesis, TAGs are typically stored within cytoplasmic lipid droplets. These LDs are stabilized and structurally supported by key proteins such as oleosin and seipin, which facilitate their formation, budding from the endoplasmic reticulum (ER), and long-term stability.

One notable example of a successful "pack and protect" strategy involved the overexpression of the Arabidopsis seipin1 protein (AtSei1) in transgenic Arabidopsis expressing the *castor fatty acid hydroxylase-12* (*RcFAH12*) gene (Lunn et al. 2018). *RcFAH12* transgenic plants typically exhibit reduced oil content due to inefficient incorporation of hydroxy fatty acids into TAGs. However, the introduction of *AtSei1* led to considerable improvements, including increased seed oil content, formation of larger lipid droplets indicative of enhanced storage capacity, and an overall increase in seed yield (Lunn et al. 2018).

This strategy offers multiple benefits for agricultural productivity, crop improvement and industrial application. By optimising LD formation and protection, plants demonstrate increased oil yield, which improves the economic and nutritional value of seeds for both biofuel production and food industries. Additionally, this strategy can positively influence oil quality by potentially altering fatty acid composition profiles, thus increasing nutritional or industrial applicability. Furthermore, protecting LDs from degradation helps maintain cellular homeostasis under stress conditions, preserving energy reserves and enhancing overall plant resilience (Sagun et al. 2023). Figure 11.8 illustrates the metabolic engineering strategies to enhance LD accumulation and stability in crops, focusing on "Pack and Protect" strategy.

11.7 Conclusion and Future Research Directions

Elaioplasts are specialized lipid storage organelles that play critical roles in plant development and metabolism. This chapter has highlighted the unique structural and functional features of elaioplasts, emphasizing their importance in lipid accumulation, pollen development, and the synthesis of secondary metabolites. Despite recent advancements, our understanding of elaioplast biogenesis and regulation remains limited.

Future research should prioritize elucidating the genetic and environmental factors that influence elaioplast differentiation and lipid storage. Specifically, identifying key regulators of elaioplast development and understanding their interactions with other plastid types are crucial areas for further study. Advanced molecular techniques such as transcriptomic and proteomic analyses, coupled with genetic engineering, can offer new insights into these processes.

Expanding our knowledge of elaioplast biology has significant biotechnological implications. Enhancing lipid storage capacity through targeted manipulation of elaioplast pathways could improve oil yield in agricultural crops, offering a sustainable solution to meet global demands for plant-derived oils. Further research should also explore the potential of engineering elaioplasts for increased production of high-value secondary metabolites, thereby broadening their application in agriculture and industry.

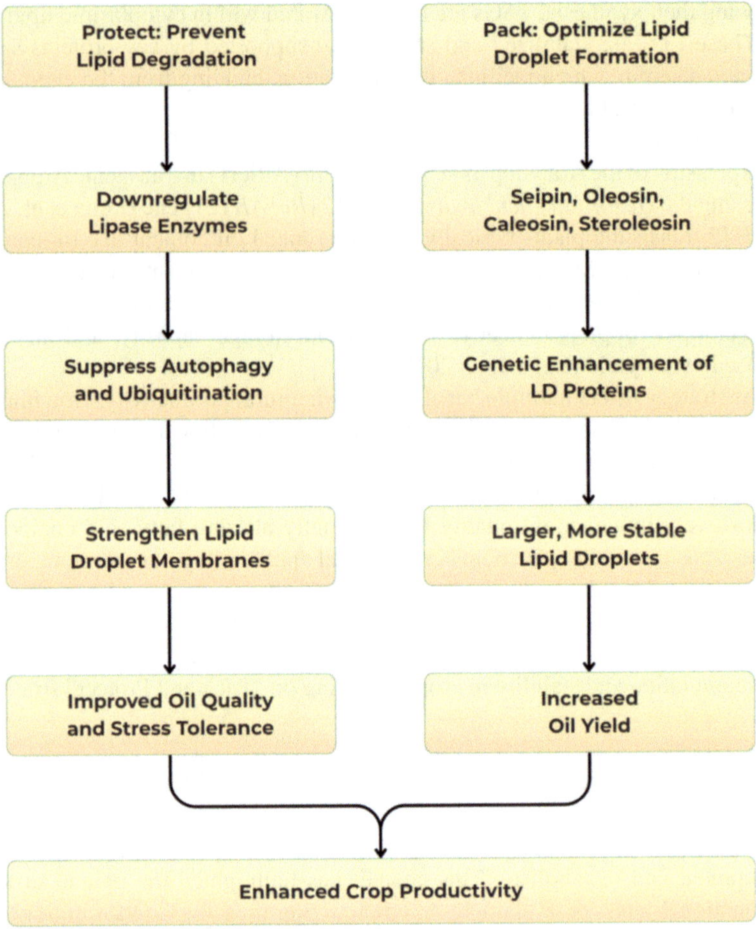

Fig. 11.8 Illustration of metabolic engineering strategies to enhance lipid droplet (LD) accumulation and stability in crops, focusing on two main approaches: "Protect"—protecting lipids from degradation (left) and "Pack"—optimizing lipid droplet formation (right). Protection involves downregulating lipase activity, suppressing autophagy and ubiquitination pathways, and reinforcing lipid droplet membranes, ultimately improving oil quality and plant stress tolerance. Optimization includes genetic enhancement of key lipid droplet-associated proteins such as Seipin, Oleosin, Caleosin, and Steroleosin, leading to larger and more stable lipid droplets, increased oil yield, and enhanced crop productivity

While significant progress has been made in understanding elaioplasts, many fundamental questions remain. Addressing these gaps will advance our knowledge of plastid biology and open new avenues for improving crop yield, quality, and resilience through innovative biotechnological strategies.

References

Angaman DM, Petrizzo R, Hernández-Gras F, Romero-Segura C, Pateraki I, Busquets M, Boronat A (2012) Precursor uptake assays and metabolic analyses in isolated tomato fruit chromoplasts. Plant Methods 8:1–10

Asada K (2006) Production and scavenging of reactive oxygen species in chloroplasts and their functions. Plant Physiol 141(2):391–396

Awai K, Xu C, Lu B, Benning C (2006) Lipid trafficking between the endoplasmic reticulum and the chloroplast. Biochem Soc Trans 34(3):395–398

Balmer Y, Vensel WH, DuPont FM, Buchanan BB, Hurkman WJ (2006) Proteome of amyloplasts isolated from developing wheat endosperm presents evidence of broad metabolic capability. J Exp Bot 57(7):1591–1602

Barsan C, Sanchez-Bel P, Rombaldi C, Egea I, Rossignol M, Kuntz M, Zouine M, Latche A, Bouzayen M, Pech JC (2010) Characteristics of the tomato chromoplast revealed by proteomic analysis. J Exp Bot 61(9):2413–2431

Berry HM, Rickett DV, Baxter CJ, Enfissi EM, Fraser PD (2019) Carotenoid biosynthesis and sequestration in red chilli pepper fruit and its impact on colour intensity traits. J Exp Bot 70(10):2637–2650

Bhat S (2010) Transgenics for increasing productivity of crops. J Plant Biochem Biotechnol 19:1–7

Burt A, Mitchell Fetch J (2018) Marker-assisted breeding of hexaploid spring wheat in the Canadian prairies. Can J Plant Sci 99(2):111–127

Cai G, Wang G, Kim S-C, Li J, Zhou Y, Wang X (2021) Increased expression of fatty acid and ABC transporters enhances seed oil production in camelina. Biotechnol Biofuels 14:1–11

Chapman KD, Dyer JM, Mullen RT (2012) Biogenesis and functions of lipid droplets in plants: thematic review series: lipid droplet synthesis and metabolism: from yeast to man. J Lipid Res 53(2):215–226. https://doi.org/10.1194/jlr.R021436

Chen S, Lei Y, Xu X, Huang J, Jiang H, Wang J, Cheng Z, Zhang J, Song Y, Liao B (2015) The peanut (*Arachis hypogaea L.*) gene *AhLPAT2* increases the lipid content of transgenic Arabidopsis seeds. PLoS One 10(8):e0136170

Choi H, Yi T, Ha SH (2021) Diversity of plastid types and their interconversions. Front Plant Sci 12

Choi YJ, Zaikova K, Yeom S-J, Kim Y-S, Lee DW (2022) Biogenesis and lipase-mediated mobilization of lipid droplets in plants. Plants 11(9):1243

Chorlay A, Thiam AR (2020) Neutral lipids regulate amphipathic helix affinity for model lipid droplets. J Cell Biol 219(4). https://doi.org/10.1083/jcb.201907099

Clément C, Pacini E (2001) Anther plastids in angiosperms. Bot Rev 67:54–73

Eastmond PJ, Dennis DT, Rawsthorne S (1997) Evidence that a malate/inorganic phosphate exchange translocator imports carbon across the leucoplast envelope for fatty acid synthesis in developing castor seed endosperm. Plant Physiol 114(3):851–856

Egea I, Barsan C, Bian W, Purgatto E, Latche A, Chervin C, Bouzayen M, Pech JC (2010) Chromoplast differentiation: current status and perspectives. Plant Cell Physiol 51(10):1601–1611

Fujino K, Hirayama Y, Kaji R (2019) Marker-assisted selection in rice breeding programs in Hokkaido. Breed Sci 69(3):383–392

Gao C (2021) Genome engineering for crop improvement and future agriculture. Cell 184(6):1621–1635

Giannini V, Maucieri C, Vamerali T, Zanin G, Schiavon S, Pettenella DM, Bona S, Borin M (2022) Sunflower: from Cortuso's description (1585) to current agronomy, uses and perspectives. Agriculture 12(12):1978

Giménez-Andrés M, Emeršič T, Antoine-Bally S, D'Ambrosio JM, Antonny B, Derganc J, Čopič A (2021) Exceptional stability of a perilipin on lipid droplets depends on its polar residues, suggesting multimeric assembly. eLife 10:e61401. https://doi.org/10.7554/eLife.61401

Gleizes M, Pauly G, Carde J-P, Marpeau A, Bernard-Dagan C (1983) Monoterpene hydrocarbon biosynthesis by isolated leucoplasts of *Citrofortunella mitis*. Planta 159(4):373–381

Gutbrod P, Yang W, Grujicic GV, Peisker H, Gutbrod K, Du LF, Dörmann P (2021) Phytol derived from chlorophyll hydrolysis in plants is metabolised via phytenal. J Biol Chem 296:100530

Hernández ML, Cejudo FJ (2021) Chloroplast lipids metabolism and function. A redox perspective. Front Plant Sci 12:712022

Hernández-Pinzón I, Ross JH, Barnes KA, Damant AP, Murphy DJ (1999) Composition and role of tapetal lipid bodies in the biogenesis of the pollen coat of *Brassica napus*. Planta 208:588–598

Hölzl G, Dörmann P (2019) Chloroplast lipids and their biosynthesis. Annu Rev Plant Biol 70(1):51–81

Hsieh K, Huang AH (2004) Endoplasmic reticulum, oleosins, and oils in seeds and tapetum cells. Plant Physiol 136(3):3427–3434

Jarvis P (2008) Targeting of nucleus-encoded proteins to chloroplasts in plants. New Phytol 179(2):257–285

Jarvis P, López-Juez E (2013) Biogenesis and homeostasis of chloroplasts and other plastids. Nat Rev Mol Cell Biol 14(12):787–802

Kessler F, Schnell DJ (2002) A GTPase gate for protein import into chloroplasts. Nat Struct Mol Biol 9(2):81

Kim HU, Wu SSH, Ratnayake C, Huang AHC (2001) Brassica rapa has three genes that encode proteins associated with different neutral lipids in plastids of specific tissues. Plant Physiol 126(1):330–341. https://doi.org/10.1104/pp.126.1.330

Kim S, Yamaoka Y, Ono H, Kim H, Shim D, Maeshima M, Martinoia E, Cahoon EB, Nishida I, Lee Y (2013) *AtABCA9* transporter supplies fatty acids for lipid synthesis to the endoplasmic reticulum. Proc Natl Acad Sci 110(2):773–778

Kirchhoff H (2019) Chloroplast ultrastructure in plants. New Phytol 223(2):565–574

Li N, Gügel IL, Giavalisco P, Zeisler V, Schreiber L, Soll J, Philippar K (2015) *FAX1*, a novel membrane protein mediating plastid fatty acid export. PLoS Biol 13(2):e1002053

Ling QH, Huang WH, Baldwin A, Jarvis P (2012) Chloroplast biogenesis is regulated by direct action of the ubiquitin-proteasome system. Science 338(6107):655–659

Ling Q, Sadali NM, Soufi Z, Zhou Y, Huang B, Zeng Y, Rodriguez-Concepcion M, Jarvis RP (2021) The chloroplast-associated protein degradation pathway controls chromoplast development and fruit ripening in tomato. Nat Plants 7(5):655–666

Liu F, Xia Y, Wu L, Fu D, Hayward A, Luo J, Yan X, Xiong X, Fu P, Wu G (2015) Enhanced seed oil content by overexpressing genes related to triacylglyceride synthesis. Gene 557(2):163–171

Lundquist PK, Shivaiah K-K, Espinoza-Corral R (2020) Lipid droplets throughout the evolutionary tree. Prog Lipid Res 78:101029

Lunn D, Wallis JG, Browse J (2018) Overexpression of *Seipin1* increases oil in hydroxy fatty acid-accumulating seeds. Plant Cell Physiol 59(1):205–214

Michaud M, Jouhet J (2019) Lipid trafficking at membrane contact sites during plant development and stress response. Front Plant Sci 10. https://doi.org/10.3389/fpls.2019.00002

Msanne J, Kim H, Cahoon EB (2020) Biotechnology tools and applications for development of oilseed crops with healthy vegetable oils. Biochimie 178:4–14

Negi J, Munemasa S, Song B, Tadakuma R, Fujita M, Azoulay-Shemer T, Engineer CB, Kusumi K, Nishida I, Schroeder JI (2018) Eukaryotic lipid metabolic pathway is essential for functional chloroplasts and CO_2 and light responses in Arabidopsis guard cells. Proc Natl Acad Sci 115(36):9038–9043

Nellaepalli S, Lau AS, Jarvis RP (2023) Chloroplast protein translocation pathways and ubiquitin-dependent regulation at a glance. J Cell Sci 136(18)

Pacek A, Stpiczyńska M, Davies KL, Szymczak G (2012) Floral elaiophore structure in four representatives of the *Ornithocephalus* clade (Orchidaceae: Oncidiinae). Ann Bot 110(4):809–820

Pacini E, Taylor PE, Singh MB, Knox RB (1992) Development of plastids in pollen and tapetum of rye-grass, *Lolium perenne* L. Ann Bot 70(2):179–188

Parveez GKA, Kamil NN, Zawawi NZ, Ong-Abdullah M, Rasuddin R, Loh SK, Selvaduray KR, Hoong SS, Idris Z (2022) Oil palm economic performance in Malaysia and R&D progress in 2021. J Oil Palm Res 34(2):185–218

Patrignani F, Siroli L, Serrazanetti DI, Gardini F, Lanciotti R (2015) Innovative strategies based on the use of essential oils and their components to improve safety, shelf-life and quality of minimally processed fruits and vegetables. Trends Food Sci Technol 46(2):311–319

Pilorgé E (2020) Sunflower in the global vegetable oil system: situation, specificities and perspectives. Oilseeds Fats Crops Lipids 27:34

Quilichini TD, Douglas CJ, Samuels AL (2014) New views of tapetum ultrastructure and pollen exine development in *Arabidopsis thaliana*. Ann Bot 114(6):1189–1201

Sadali NM, Sowden RG, Ling Q, Jarvis RP (2019) Differentiation of chromoplasts and other plastids in plants. Plant Cell Rep 38(7):803–818

Sagun JV, Yadav UP, Alonso AP (2023) Progress in understanding and improving oil content and quality in seeds. Front Plant Sci 14. https://doi.org/10.3389/fpls.2023.1116894

Sakamoto W, Miyagishima SY, & Jarvis P (2008) Chloroplast biogenesis: control of plastid development, protein import, division and inheritance. The Arabidopsis Book, 6, e0110. https://doi.org/10.1199/tab.0110

Sato N, Awai K (2017) "Prokaryotic pathway" is not prokaryotic: noncyanobacterial origin of the chloroplast lipid biosynthetic pathway revealed by comprehensive phylogenomic analysis. Genome Biol Evol 9(11):3162–3178

Savadi S, Naresh V, Kumar V, Bhat SR (2016) Seed-specific overexpression of *Arabidopsis DGAT1* in Indian mustard (*Brassica juncea*) increases seed oil content and seed weight. Botany 94(3):177–184

Schnell DJ, Kessler F, Blobel G (1994) Isolation of components of the chloroplast protein import machinery. Science 266(5187):1007–1012

Schnurr JA, Shockey JM, de Boer G-J, Browse JA (2002) Fatty acid export from the chloroplast. Molecular characterization of a major plastidial acyl-coenzyme A synthetase from Arabidopsis. Plant Physiol 129(4):1700–1709

Shapiguzov A, Vainonen J, Wrzaczek M, Kangasjärvi J (2012) ROS-talk–how the apoplast, the chloroplast, and the nucleus get the message through. Front Plant Sci 3(292):1–9

Smith MD, Hiltbrunner A, Kessler F, Schnell DJ (2002) The targeting of the *atToc159* preprotein receptor to the chloroplast outer membrane is mediated by its GTPase domain and is regulated by GTP. J Cell Biol 159(5):833–843

Stpiczyńska M, Davies KL, Pacek-Bieniek A, Kamińska M (2013) Comparative anatomy of the floral elaiophore in representatives of the newly re-circumscribed *Gomesa* and *Oncidium* clades (Orchidaceae: Oncidiinae). Ann Bot 112(5):839–854

Sun Y-J, Forouhar F, Li H-M, Tu S-L, Yeh Y-H, Kao S, Shr H-L, Chou C-C, Chen C, Hsiao C-D (2002) Crystal structure of pea Toc34, a novel GTPase of the chloroplast protein translocon. Nat Struct Biol 9(2):95–100

Tang KHD, Al Qahtani HM (2020) Sustainability of oil palm plantations in Malaysia. Environ Dev Sustain 22(6):4999–5023

Tao N, Jia L, Zhou H (2014) Anti-fungal activity of *Citrus reticulata* Blanco essential oil against *Penicillium italicum* and *Penicillium digitatum*. Food Chem 153:265–271

Theg SM, Bauerle C, Olsen L, Selman B, Keegstra K (1989) Internal ATP is the only energy requirement for the translocation of precursor proteins across chloroplastic membranes. J Biol Chem 264(12):6730–6736

Tian Y, Lv X, Xie G, Zhang J, Xu Y, Chen F (2018) Seed-specific overexpression of *AtFAX1* increases seed oil content in Arabidopsis. Biochem Biophys Res Commun 500(2):370–375

Ting JT, Wu SS, Ratnayake C, Huang AH (1998) Constituents of the tapetosomes and elaioplasts in *Brassica campestris* tapetum and their degradation and retention during microsporogenesis. Plant J 16(5):541–551

USDA (2025) World oilseeds production. Retrieved 16 March 2025, from, https://www.sopa.org/world-oilseeds-production/

Vanhercke T, El Tahchy A, Shrestha P, Zhou X-R, Singh SP, Petrie JR (2013) Synergistic effect of WRI1 and DGAT1 coexpression on triacylglycerol biosynthesis in plants. FEBS Lett 587(4):364–369

Voo SS, Grimes HD, Lange BM (2012) Assessing the biosynthetic capabilities of secretory glands in *Citrus* peel. Plant Physiol 159(1):81–94

Wallas TR, Smith MD, Sanchez-Nieto S, Schnell DJ (2003) The roles of Toc34 and Toc75 in targeting the Toc159 preprotein receptor to chloroplasts. J Biol Chem 278(45):44289–44297

Wang YQ, Yang Y, Fei Z, Yuan H, Fish T, Thannhauser TW, Mazourek M, Kochian LV, Wang X, Li L (2013) Proteomic analysis of chromoplasts from six crop species reveals insights into chromoplast function and development. J Exp Bot 64(4):949–961

Wasternack C, Hause B (2013) Jasmonates: biosynthesis, perception, signal transduction and action in plant stress response, growth and development. An update to the 2007 review in Annals of Botany. Ann Bot 111(6):1021–1058

Wijk KJV, Kessler F (2017) Plastoglobuli: plastid microcompartments with integrated functions in metabolism, plastid developmental transitions, and environmental adaptation. Annu Rev Plant Biol 68(1):253–289

Wise RR (2007) The diversity of plastid form and function. In: Wise RR, Kenneth Hoober J (eds) The structure and function of plastids, vol 23. Springer, Dordrecht, pp 3–26

Wu SSH, Platt KA, Ratnayake C, Wang T-W, Ting JTL, Huang AHC (1997) Isolation and characterization of neutral-lipid-containing organelles and globuli-filled plastids from *Brassica napus* tapetum. Proc Natl Acad Sci 94(23):12711–12716

Wu SS, Moreau RA, Whitaker BD, Huang AH (1999) Steryl esters in the elaioplasts of the tapetum in developing *Brassica* anthers and their recovery on the pollen surface. Lipids 34(5):517–523

Xu C, Fan J, Shanklin J (2020) Metabolic and functional connections between cytoplasmic and chloroplast triacylglycerol storage. Prog Lipid Res 80:101069

Yamasaki Y, Akimitsu K (2007) In situ localization of gene transcriptions for monoterpene synthesis in irregular parenchymic cells surrounding the secretory cavities in rough lemon (*Citrus jambhiri*). J Plant Physiol 164(11):1436–1448

Yang Y, Benning C (2018) Functions of triacylglycerols during plant development and stress. Curr Opin Biotechnol 49:191–198

Yue GH, Ye BQ, Lee M (2021) Molecular approaches for improving oil palm for oil. Mol Breed 41:1–17

Zafar S, Li Y-L, Li N-N, Zhu K-M, Tan X-L (2019) Recent advances in enhancement of oil content in oilseed crops. J Biotechnol 301:35–44

Zhang H, Xie Y, Liu C, Chen S, Hu S, Xie Z, Deng X, Xu J (2017) Comprehensive comparative analysis of volatile compounds in citrus fruits of different species. Food Chem 230:316–326

Zhao Q, Xu Y, Liu Y (2022) Soybean oil bodies: a review on composition, properties, food applications, and future research aspects. Food Hydrocoll 124:107296

Zhu M, Lin J, Ye J, Wang R, Yang C, Gong J, Liu Y, Deng C, Liu P, Chen C, Cheng Y, Deng X, Zeng Y (2018) A comprehensive proteomic analysis of elaioplasts from citrus fruits reveals insights into elaioplast biogenesis and function. Hortic Res 5(1):6

Chloroplast to Gerontoplast: Degradation of the Photosynthetic Machinery

12

Jitendra Kumar Pandey and Gunjan Dubey

Abstract

Chloroplasts are semi-autonomous organelles. They are the key player in the life cycle of plants from the young seedling stage to the senescing plant stage. This organelle is also accountable for the independent establishment of young seedlings as it is responsible for the photosynthetic activity. During senescence and in response to most stresses, it is the first organelle to undergo degradation. Carbon, nitrogen and micronutrients stored in the chloroplast are remobilized to the sink and play a key role either in the growth or productivity of the plant. Under stressed conditions, chloroplasts are responsible for the survival of the plant. Chloroplast degradation comprises both intra- and extraplastidial degradation of different components. Intraplastidial degradation includes alternation and breakdown of the thylakoid membrane system, degradation of various constituents of the chloroplast such as pigments, proteins, starch, and nucleic acid, along with an increase in the size and number of plastoglobules. Alternation and disruption of the chloroplast envelope also occurs. Extraplastidial degradation basically occurs by autophagy, which includes macro and microautophagy. Autophagy can be vesicle-mediated and/or ubiquitin-mediated. The role of chaperones is not yet reported in chlorophagy (selective degradation of chloroplast materials). In this chapter, various processes taking place in the chloroplast at the time of senescence are described.

J. K. Pandey (✉)
Department of Botany, Government Post Graduate College, Tikamgarh, India

Maharaja Chhatrasal Bundelkhand University, Chhatarpur, India

G. Dubey
Department of Botany, University of Rajasthan, Jaipur, Rajasthan, India

V. K. Dalal, A. N. Misra (eds.), *Chloroplast Biogenesis and Plastid Interconversions*,
https://doi.org/10.1007/978-981-95-0772-6_12

Keywords

Chloroplast · Gerontoplast · Senescence · Intraplastidial degradation · Extraplastidial degradation

12.1 Introduction

Chloroplasts are the receptors of solar energy in higher plants on Earth. This organelle converts light energy into chemical energy and fixes CO_2 into carbohydrates. Chloroplasts are also involved in a variety of metabolic processes such as assimilation of nitrogen and sulphur as well as synthesis of diverse compounds of biological importance including starch, membrane lipids, amino acids, fatty acids, purine and pyrimidine bases, hormones, isoprenoids, phenylpropanoids, and tetrapyrroles (Armbruster and Strand 2020; Cejudo et al. 2021; Domínguez and Cejudo 2021). Massive amounts of proteins, lipids, and nucleic acids are present in the chloroplast. On the basis of the abovementioned facts, the chloroplast may be considered as a central hub during leaf development, and it is necessary to remobilize these nutrients to the sink during senescence. Chloroplasts are semi-autonomous, double membrane organelles containing stroma and the thylakoid membrane system (Kirchhoff 2019). In the biosphere, chloroplasts contain the most abundant pigment, lipid, soluble protein, and membrane proteins such as chlorophylls, monogalactosyldiacylglycerol (MGDG), Rubisco, and light-harvesting complex II (LHCII), respectively (Kirchhoff 2019).

With the initiation of senescence and during stressed conditions, the chloroplast is the first organelle of plants that undergoes degradation concomitant with the visible changes in the colour of the leaf, typically turning from green to yellow or sometimes red. This colour change is due to the degradation of chlorophylls and the revealing of other pigments like carotenoids (yellow/orange) and anthocyanins (red) (Feild et al. 2001; Pandey and Biswal 2024). It was documented much earlier that chloroplasts are degraded ahead of other cellular components (Noodén 1988). Chloroplast turnover plays a critical role in plastid transition and mobilization of nutrients such as carbon and nitrogen (Zhuang and Jiang 2019). Chloroplast turnover is an important source of nitrogen to plants during natural senescence and also especially under stress condition which limits the plant metabolism. Chloroplasts are the main source of nitrogen in leaves (representing up to 70–80% of leaf nitrogen), which can be remobilized (Makino and Osmond 1991; Mae 2004; Krupinska 2007; Wada and Ishida 2013). Rubisco alone comprises 20–30% of the total nitrogen present in the leaf (Wada and Ishida 2013). The amount of terrestrial Rubisco alone has been estimated at about 0.7 Gt (Gt = 1×10^{15} g) (Bar-On and Milo 2019). About 10 billion tons of Rubisco is degraded every year. Turnover of 1.2 billion tons of chlorophylls occurs every year during senescence (Hendry et al. 1987; Otegui 2018). Chloroplast proteins are extensively degraded during the senescence of leaves, and the peptides and amino acids released are exported to other parts of the plants (Mae 2004). During senescence and under various stresses, excess reactive oxygen species (ROS) are produced, which possibly play an important role in the chloroplast degradation, and this helps in nutrient remobilization and plant nutrient economy (Biswal and Pandey 2018).

In this chapter we have sub-categorized the chloroplast degradation into two subsections: intraplastidial and extraplastidial degradation. *Intraplastidial degradation* involves the degradation events that occur inside the chloroplast, and *extraplastidial degradation* involves the vesicle-mediated and ubiquitin-mediated sequestration of chloroplast materials and even whole chloroplasts for their delivery into the lytic vacuole. Extraplastidial degradation is also categorized as *macroautophagy* and *microautophagy*.

12.2 Intraplastidial Degradation: Chloroplast to Gerontoplast Transition

Degrading chloroplast was named as "gerontoplast" by Peter Sitte (1977). Dismantling of the chloroplast is the first event in the senescing leaf cells (Noodén 1988). Components of each chloroplast are progressively broken down as can be seen in a majority of cases where the number of chloroplasts per cell does not seem to decline during senescence; however, some studies have recorded a decrease in the number of plastids per cell (reviewed in Krupinska 2007). The chloroplast loses its volume and changes its shape even before any morphological change happens in the nucleus and mitochondria. During the transformation of chloroplasts into gerontoplasts, breakdown and loss of the thylakoid membrane network, degradation of photosynthetic proteins, and accumulation of numerous plastoglobules occur simultaneously (Kato and Sakamoto 2013). Most obvious changes occur during the chloroplast-to-gerontoplast transition, which include a decrease in the tubular structure and surface area of thylakoids with an increase in the size and number of plastoglobules. Plastoglobules are 100–200 nm spherical organelles, covering less than 3% of the chloroplast cross-section area under non-stress conditions, which increase tremendously under senescence and stress conditions (Arzac et al. 2022).

Ultimately, the organization of the thylakoid membrane changes and eventually thylakoids break down. These changes are accompanied by a reduction in size and shape of chloroplasts (from ellipsoid to rounded shape), and generation of vacuolar space and/or dilated thylakoids in the stroma of gerontoplasts (Mulisch and Krupinska 2013). Ultrastructural changes during chloroplast degradation can be basically summarised in three stages: alternation and breakdown of the thylakoid membrane system, increase in the number and size of plastoglobules, and alternation and disruption of the plastid envelope. Along with these structural changes, cleavage of various constituents of the chloroplast such as pigments, proteins, lipids, and starch occurs continuously (Mulisch and Krupinska 2013; Domínguez and Cejudo 2021).

12.2.1 Alternation in Thylakoid Membrane System

The thylakoid membrane system of higher plants is differentiated into stacked grana thylakoids and unstacked stroma thylakoids. The stacking of grana thylakoids is mainly due to the photosystem (PS) II components, which include major light-harvesting complexes (LHCII), minor light-harvesting complexes (Lhcb4, b5, and b6), as well as core complexes. Along with the PS II components, cations such as Mg^{2+}, Na^{+} and K^{+} play an important role in grana stacking (Wan et al. 2014). The sequence and extent of degradation of both types of thylakoid systems are variable in different plant species and depend on both external and internal conditions. Early degradation of stroma thylakoids is more common than that of granal thylakoids. A decreased number of grana stacks, grana stacks without stroma thylakoids and even only single long thylakoids are observed in different plant species. Generally thylakoid degradation starts with the unstacking of grana thylakoids followed by the formation of elongated and loose lamellae.

Simultaneous swelling and degradation of all thylakoids is also reported. Most of the chlorophylls present in the chloroplast are attached to the light-harvesting complex II (LHCII); thus, the decrease in the chlorophyll content and even specific changes in the chlorophyll *a*/*b* are indicators of change in the thylakoid membrane system. Grana unstacking, flattening of thylakoids, swelling of intrathylakoid space, and degradation of stroma lamellae are the characteristic features of thylakoid disorganisation (Biswal and Biswal 1988; reviewed in Lichtenthaler 2013; Mulisch and Krupinska 2013; Domínguez and Cejudo 2021). Cup-shaped stacked membranes, thylakoid coiling and formation of dilations at the thylakoid ends are reported as anomalous breakdown of the thylakoid membrane system (Wrischer et al. 2009).

12.2.2 Plastoglobules

The characteristic feature of gerontoplast is the occurrence of a large number of lipoprotein particles called plastoglobules. In mature chloroplasts, the size of plastoglobules ranges from 80 to 120 nm, and in gerontoplasts it ranges from 200 to 300 nm. Plastoglobules of extreme size, i.e., 5 μm are also reported (Lichtenthaler and Weinert 1970). Small plastoglobules are spherical in shape with a smooth surface, whereas large plastoglobules are wrinkled in shape (Tuquet and Newman 1980). Plastoglobules are lipid monolayer-surrounded lipid droplets that remain attached to the thylakoid membrane. The plastoglobules function as both lipid biosynthesis and storage sub-compartments of thylakoid membranes. Plastoglobules work as an extra-thylakoidal storage site for excess isoprenoid lipids, such as α-tocopherol and plasotoquinone-9. The lipid molecules contained in the plastoglobule core (carotenoids, plastoquinone, and tocopherol) are in a dynamic equilibrium with those located in the thylakoid membranes (Austin et al. 2006; Lichtenthaler 2013; Coulon et al. 2024). The size, number, shape and density of plastoglobules are variable in different plant species and also depend on internal factors such as senescence, fruit development, and external factors such as high light intensity, high salinity, oxidative stress, drought, chilling, nitrate starvation, viral infection, and ozone etc. (Besagni and Kessler 2013; Domínguez and Cejudo 2021). Upon exposure to abiotic stress and also during senescence, the size and/or number of plastoglobules increases in the leaves due to the lipid accumulation in their hydrophobic core (Besagni and Kessler 2013; Coulon et al. 2024). Austin et al. (2006) reported that during oxidative stress and senescence, plastoglobules form grape-like clusters that are attached to each other and remain continuous with the thylakoid membranes by extensions of the half-lipid bilayer. Arzac et al. (2022) concluded that plastoglobules are ubiquitous under non-stress conditions and their morphology, which is closely related to their composition, changes differently depending on the specific stress that the plant undergoes.

According to Lichtenthaler and Sprey (1966), there are two types of plastoglobules:

(i) The first type of plastoglobules is present in annual plants, where they increase in size rather than in number.
(ii) The second type of plastoglobules is present in perennial plants, where they increase in number, rather than an obvious increase in size (Lichtenthaler 1968; Lichtenthaler 2013; reviewed in Mulisch and Krupinska 2013).

However, in some plants both an increase in number and size of plastoglobules is reported (Tuquet and Newman 1980).

12.2.3 Alteration of Chloroplast Envelope and Activities in the Vicinity of Chloroplast

Alterations in the chloroplast envelope, including perforation, broken envelope and stromule formation, are usually exhibited during senescence (Ishida et al. 2008; Springer et al. 2016). Peripheral vesiculation of the tubular apparatus is reported in gerontoplasts much earlier (Harris 1978; Tuquet and Newman 1980). During senescence and upon exposure to stress conditions, stromules become more abundant than during normal developmental stages (Brunkard et al. 2015; Caplan et al. 2015). Amoeboid morphology, cytoplasmic invaginations, membrane-bound extrusions, and generation of plastidial stromules are reported in virus-induced senescence (Zhao et al. 2016). The intense activity of the Golgi complex is reported in the vicinity of the chloroplast envelope during senescence. Physical contact between the ER and the chloroplast increases as the outer membrane of the chloroplast gets damaged during senescence. The endoplasmic reticulum might deliver hydrolase enzymes involved in chloroplast degradation during this stage. Activity of such hydrolases might be responsible for the empty spaces detected in gerontoplasts (reviewed in Mulisch and Krupinska 2013).

12.2.4 Breakdown of Pigment Protein Complexes

Photosystem I, photosystem II, cytochrome b_6f complex, and LHCs are the multiple pigment protein complexes present in thylakoid membranes. They constitute more than 30% of the chloroplast protein. The structure of thylakoid membranes changes dramatically during leaf senescence. During the chloroplast degradation process, massive accumulation of plastoglobuli occurs with the progressive loss of grana stacking and disappearance of thylakoid membranes (Kato and Sakamoto 2013). The mutual interaction between PSII-LHCII supercomplexes facing each other is a prerequisite for grana stacking. During senescence, LHCII migrates from the stacked grana thylakoid to the margin region of the grana thylakoid and then to stroma lamellae (Domínguez and Cejudo 2021). In senescing leaves, PSI-LHCII megacomplexes are formed. This is possibly the reason behind the unstacking and flattening of the grana structure. This dismantling of photochemical machinery and linear electron transport disruption are main events during senescence (Prakash et al. 2001, 2003; Schwarz et al. 2018; Albanese et al. 2020; Domínguez and Cejudo 2021). The pigment and protein degradation in pigment-protein complexes is summarised in the following section.

12.2.5 Degradation of Macromolecules of Chloroplast

12.2.5.1 Photosynthetic Pigments Degradation

The primary function of leaf senescence is recycling of the nitrogen, where the degradation of protein complexes is the primary activity during leaf senescence. Chlorophylls should also be degraded to avoid uncontrolled photochemistry during leaf senescence. Due to the degradation of chlorophylls, the first visible symptom of senescence is the colour change of leaves from green to yellow or sometimes red (Feild et al. 2001; Pandey et al. 2017b). Chlorophyll *b* exclusively exists in Lhcb proteins whereas chlorophyll *a* is found in both Lhca proteins and Core complexes (Green and Durnford 1996). As chlorophylls are the essential components of all chlorophyll-protein complexes, degradation of chlorophylls is considered as the major symptom of the disorganisation of the photosynthetic apparatus. Degradation of chlorophylls starts in the chloroplast, then it involves the cytoplasm and its final destination is the lytic vacuoles (Biswal 2005). The different steps of degradation are summarised in Fig. 12.1. The chlorophyll degradation pathway is commonly known as *pheophorbide a oxygenase (PAO)/phyllobilin pathway* on the basis of the name of the enzyme PAO, which is responsible for the opening of the pigment macrocyclic ring. The primary steps of chlorophyll degradation are Mg-dechelation and dephytylation, which happen before the cleavage of the porphyrin ring by PAO (Kuai et al. 2018). These steps are subject to a lot of regulations (reviewed in Hu et al. 2021). The first step of chlorophyll *a* degradation is the conversion of chlorophyll *a* to pheophytin. From the centre of chlorophyll *a*, Mg^{2+} is removed by magnesium dechelatase (Matsuda et al. 2016; Shimoda et al. 2016; Biswal and Pandey 2016). Pheophytin *a* is converted into pheophorbide *a* in the next step;

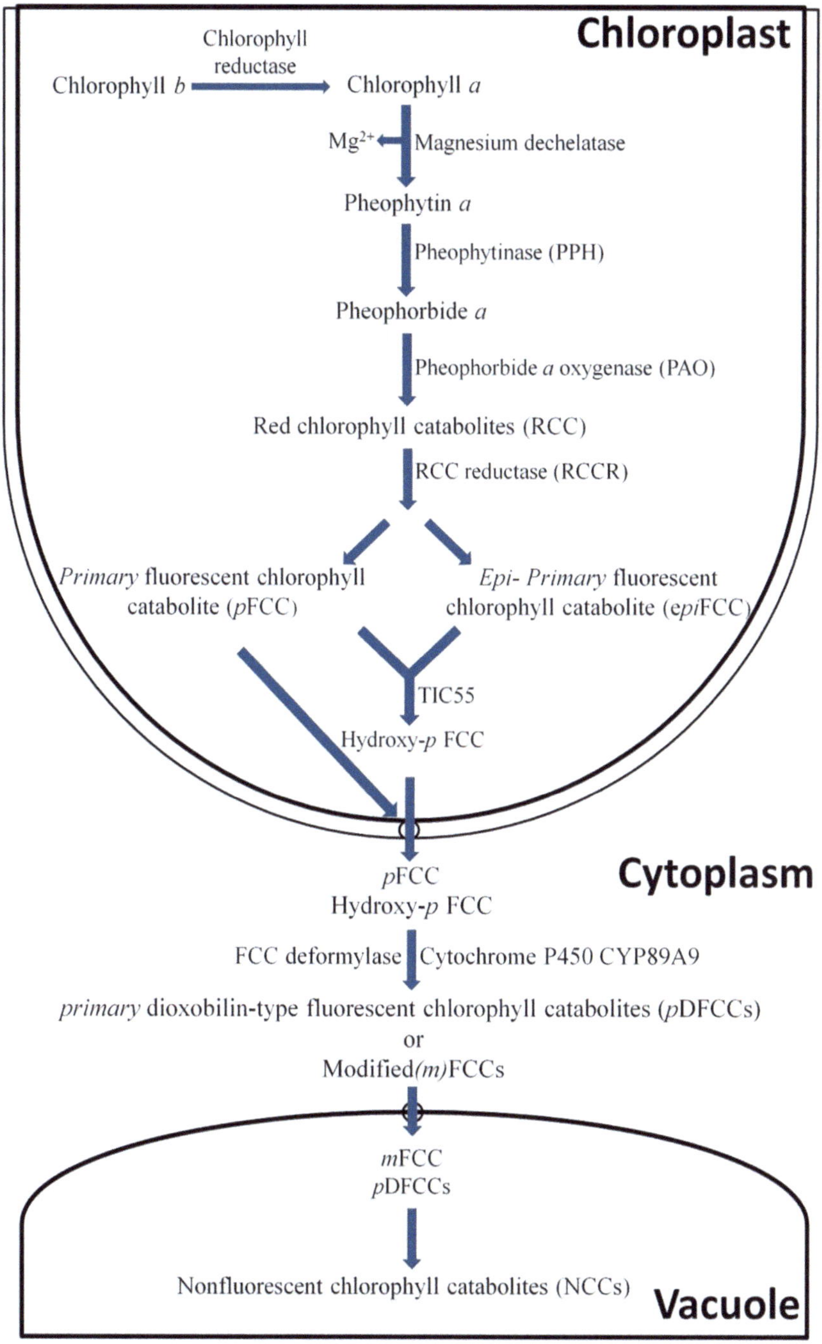

Fig. 12.1 Pheophorbide *a* oxygenase (PAO)/phyllobilin pathway of chlorophyll degradation showing several intermediate fluorescent and non-fluorescent chlorophyll catabolites, the enzyme involved in different intermediate steps, and their location

pheophytin-dephytylating enzyme pheophytinase (PPH) is responsible for this conversion. After the action of PAO, pheophorbide *a* is converted into red-coloured catabolite (RCC), which is immediately reduced by RCC reductase (RCCR) into *primary* fluorescent chlorophyll catabolite (*p*FCC). Two C16 stereoisomers, *p*FCC and *epimeric*-FCC (*epi*-FCC), are formed depending on the plant species, e.g., in gerontoplasts of *Capsicum annuum* (Moser and Matile 1997). Hydroxylation of *p*FCC by TIC55 enzyme, belonging to a family of Fd-dependent Rieske-type monooxygenases, into hydroxy-*p*FCC is also reported (Hauenstein et al. 2016; Kuai et al. 2018). By unidentified transporters, *p*FCC and hydroxy-*p*FCC are exported into the cytoplasm. FCC deformylase, cytochrome P450 CYP89A9, oxidatively deformylates the *p*FCC and hydroxy-*p*FCC at the C5-formyl group to produce *primary* dioxobilin-type fluorescent chlorophyll catabolites (*p*DFCCs) in *Arabidopsis* (Christ et al. 2013; Süssenbacher et al. 2015). This modification can be escaped and some other types of modification are also reported (reviewed in Hu et al. 2021). FCCs and DFCCs are finally imported into the vacuole. In the vacuole, FCCs and DFCCs are non-enzymatically isomerized to formyloxobilin-type non-fluorescent chlorophyll catabolites (NCCs) and dioxobilin-type non-fluorescent chlorophyll catabolites (DNCCs), respectively (Müller et al. 2011; Christ et al. 2013; Kräutler 2016; Hu et al. 2021).

Generally chlorophyll *a* is degraded faster than chlorophyll *b* (Mulisch and Krupinska 2013). Prior to degradation, chlorophyll b/chlorophyllide *b* is converted into chlorophyll a/chlorophyllide *a* by the chlorophyll cycle in a two-step reaction catalysed by chlorophyll *b* reductases (Kuai et al. 2018). Carotenoids remain more or less stable during senescence than chlorophylls. Not much is known about the carotenoids degradation; however, biosynthesis of multiple apocarotenoids and abscisic acid (ABA) is correlated with carotenoids degradation, and these might be using the products of carotenoids degradation as their precursors. Most probably plastoglobules have a major role in the carotenoids degradation as carotenoid cleavage dioxygenase 4 (CCD4) appears as a major component of these structures (Rottet et al. 2016).

12.2.5.2 Proteins

The rapid decrease in the protein content during the leaf senescence supported the idea of protein degradation inside the chloroplast (Kato and Sakamoto 2013; Pandey et al. 2017a). Degradation of the non-photosynthetic stromal proteins, such as those involved in sulphur and nitrogen assimilation, and the stromal proteins involved in the carbon reduction cycle starts early. Degradation of carbon reduction cycle enzymes such as Rubisco, glyceraldehyde 3-phosphate dehydrogenase and ribulose 5-phospahte kinase proceeds roughly parallel at similar rates (Krupinska 2007). In the stroma and thylakoid membranes, a number of proteases are localized; however, the hydrolytic activity against Rubisco is only confirmed for CND41, which is a 41 kD chloroplast nucleoid DNA-binding protease (Kato et al. 2004, 2005; Wada and Ishida 2013).

Cytochrome b_6/f complex is lost before photosystem I and PSII, as well as ATP synthase. Due to the early degradation of b_6/f complex, the connection between PSII and PSI becomes a limiting step in the photochemical reactions of photosynthesis (Holloway et al. 1983; Humbeck et al. 1996; Guiamét et al. 2002). Regarding PSI/PSII, there are both types of reports: firstly, PSII and PSI broken down in a similar time course, and secondly, PSI is lost before PSII (Miersch et al. 2000; Tang et al. 2005). Most possibly, the degradation of PSII core proteins occurs in a coordinated manner, whereas the breakdown of PSII antenna proteins may be uncoupled from the degradation of PSII reaction centres (Hidema et al. 1991, 1992; Miersch et al. 2000; Lu et al. 2001; Guiamét et al. 2002; Costa et al. 2013).

Plastid-localized proteases are involved in the degradation of chloroplast proteins. A large number of proteases, with plastid targeting peptides, are revealed by genomic approaches. Numbers of these proteases, particularly chloroplast proteases of the Clp-, DegP-, FtsH- and Lon-families are upregulated during the senescence in various plant species (Adam and Clarke 2002; Bhalerao et al. 2003; Buchanan-Wollaston et al. 2003; Andersson et al. 2004; Guo et al. 2004; Sakamoto 2006; Gregersen and Holm 2007; Kato and Sakamoto 2010). Proteases of the Clp and Lon families are known to be involved in protein degradation in the stroma. Proteases of the Deg and FtsH families are considered to participate in protein degradation in thylakoid membranes and are also responsible for the unstacking of grana thylakoids (Kato and Sakamoto 2010; Kato and Sakamoto 2013). An ATP-dependent serine protease (Lon protease) and an ATP-independent serine protease (SppA protease) are reported to be closely associated with the stromal side of the thylakoid membrane (Lensch et al. 2001; Ostersetzer et al. 2007). During senescence, D1 protein of photosystem II is possibly degraded by the proteases of the FtsH and DegP families. Proteases of these families are involved in the degradation of D1 protein during the PSII repair cycle following the photoinhibition (Zaltsman et al. 2005; Kato et al. 2009). Deg proteases (Deg1, De2, Deg5, Deg7 and Deg8) are ATP-independent serine-endoproteases. Deg2 and Deg7 are peripherally attached to the stromal side whereas Deg1, Deg5, and Deg8 are localized at the luminal side of thylakoid membranes. Photodamaged D1 proteins are cleaved by Deg2 and Deg7 between helices D and E of the stromal DE loop, whereas the luminal side CD loop, linking helices C and D of the D1 protein, is cleaved by Deg1, Deg5, and Deg8 proteases (Yoshioka-Nishimura 2016). ATP-dependent metalloprotease FtsH is responsible for further degradation of fragments of D1 protein generated by the action of Deg proteases (Kato et al. 2012; Yoshioka-Nishimura 2016). During nitrogen starvation conditions, cytochrome b_6/f complex would be a putative substrate for Clp protease in *Chlamydomonas reinhardtii* (Majeran et al. 2000). There is evidence for degradation of photosystem II proteins within the chloroplast but there is no clear-cut evidence for breakdown of Rubisco inside the chloroplast during the senescence (Reviewed in Costa et al. 2013).

12.2.5.3 Lipids

Lipid degradation is one of the most critical events of chloroplast degradation as the chloroplast is a membrane-bound structure and it contains a thylakoid membrane network inside. The disruption of the chloroplast membrane is induced by 13-lipoxygenase, which catalyses the dioxygenation of unsaturated fatty acids. This disruption plays an important role in the release of stromal components (Springer et al. 2016). Thylakoid membrane is made up of 4 unique classes of lipids: monogalactosyldiacylglycerol (MGDG), digalactosyldiacylglycerol (DGDG), sulfoquinovosyl diacylglycerol (SQDG) and phosphatidylglycerol (PG) that constitute nearly 40%, 32%, 15% and 13% respectively (van Eerden et al. 2015). During leaf senescence, degradation of DGDG is catalysed by alkaline α-galactosidase (Lee et al. 2009). The degradation of MGDG and phosphatidylglycerol (PG) is performed by plastid glycerolipid A1 lipases, PLIP2 and PLIP3, respectively (Wang et al. 2018). The hydrolysis of phosphatidylcholine to phosphatidic acid is catalysed by phospholipase Dδ (Jia et al. 2013; Jia and Li 2015).

12.2.5.4 Starch

Amylase is responsible for the intraplastidial degradation of photosynthesis-related transitory starch at night (Wang et al. 2013; Domínguez and Cejudo 2021). Extraplastidial degradation of starch is discussed in the upcoming Sect. 12.3.1.2.

12.3 Extraplastidial Degradation: Chlorophagy or Chloroplast Autophagy

After the early degradation of the chloroplast and mobilization of nutrients, the chloroplasts converted into shrunken gerontoplasts. The decrease in the number of gerontoplasts in the cytoplasm or whole chloroplast degradation involves the vesicle-dependent degradation and Ubiquitin-mediated degradation. All the mechanisms of bulk chloroplast degradation are not fully elucidated (Wada and Ishida 2013; Lei et al. 2023). Various mechanisms involved in the extraplastidial degradation of the chloroplast are summarised in Fig. 12.2.

Autophagy is a universal process found in all eukaryotes, including plants. In autophagy, cellular cargoes and even whole organelles are transported to the vacuoles for their further degradation or recycling. Autophagy is generally categorised into two types:

(a) Macroautophagy: when autophagosomes (endoplasmic reticulum-derived double membrane vesicles) engulf the substrates to be degraded;
(b) Microautophagy: when cytoplasmic substrates are directly taken up by the vacuoles.

Many core AuTophaGy (ATG) components and receptors are involved in the activation and progression of autophagy (Wang et al. 2023). The first step of macroautophagy is the initiation of phagophore formation, which eventually gives rise to autophagosomes— unique double-membrane structures that sequester cytoplasmic material. These autophagosomes deliver the cargoes into the vacuoles. The autophagy-related protein 8 (ATG8) and vacuolar receptor ATG8-interacting protein 1 (ATI1) play a critical role in macrochlorophagy (Zhuang and Jiang 2019). The ATG8 protein gets covalently conjugated to the membrane lipid

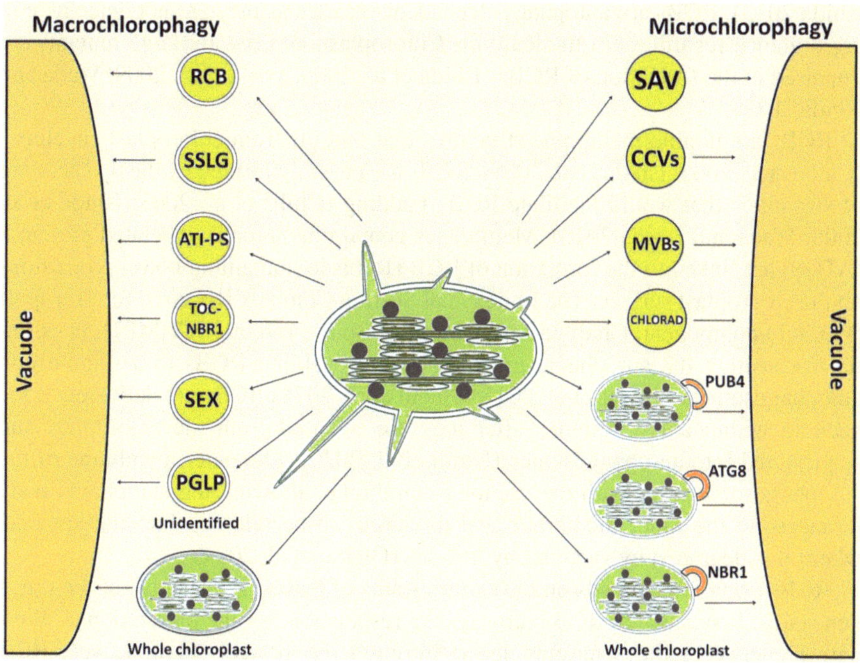

Fig. 12.2 Different pathways involved in extraplastidial degradation of chloroplast during senescence and under various stresses. Different components of chloroplast and even whole chloroplast may be degraded by both macrochlorophagy and microchlorophagy. The final degradation events occur in lytic vacuole

phosphatidylethanolamine of autophagosomes. The ATG8-interacting motif (AIM) of ATI1interacts with ATG8 to complete the autophagy process (Bu et al. 2020; Liu et al. 2021).

12.3.1 Macroautophagy or Macrochlorophagy: Autophagy-Dependent Vesicles

12.3.1.1 Rubisco-Containing Bodies (RCBs)

The emergence of vesicles from the chloroplast envelope was first observed by Guiamét et al. (1999). Park et al. (1999) reported the localization of some chloroplastic proteins, including Rubisco, in the vacuole. Chiba et al. (2003) first identified Rubisco-containing bodies (RCBs) as stromal protein-containing vesicular bodies that appeared both in the cytosol and vacuole of senescing wheat leaves.

RCBs are a class of small vesicular bodies (0.4–1.2 μM in diameter in wheat) containing only envelope components and stromal fraction that include Rubisco and glutamine synthetase II but no thylakoid proteins (Chiba et al. 2003; Ishida et al. 2008; Spitzer et al. 2015; Izumi et al. 2023). With the start of Rubisco loss, the number of RCBs increased in the senescing cells (Chiba et al. 2003; Wada and

Ishida 2013). RCBs are autophagy-dependent vesicles as they are not detectable in *atg* mutants, regardless of the leaf age. Chloroplasts of *atg4* and *atg5* mutants are impaired in the formation of RCBs (Ishida et al. 2008; Wada et al. 2009; Wada and Ishida 2013).

RCBs are double membrane structures that bud off from chloroplast envelopes (Chiba et al. 2003; Ishida et al. 2008). RCBs are possibly formed by the tip shedding of stromules that would facilitate RCBs budding (Chiba et al. 2003; Ishida et al. 2008; Wada and Ishida 2013). Membranes containing autophagy-related protein 8 (ATG8) are linked to the formation of RCB (Rubisco-containing body) protrusions on the chloroplast surface. The Endosomal Sorting Complex Required for Transport (ESCRT) component, Charged Multivesicular Body Protein 1 (CHMP1), is essential for sealing the autophagic membranes and enabling RCBs to assemble into autophagosomes (Spitzer et al. 2015; Otegui et al. 2024).RCBs are delivered to the vacuole within a few minutes after they are released from the chloroplast and sequestered into autophagosomes (Izumi et al. 2023). The outer membrane of the autophagosome fuses with the vacuolar membrane, allowing the autophagic body (comprising the inner membrane and its cargo) to be released into the vacuole, where it is degraded by vacuolar hydrolases (Otegui et al. 2024).

RCB formation depends on the carbon status of leaves, irrespective of the nitrogen status. It seems that RCB pathways are responsible for the production of alternative energy in the senescing leaves through the proteolysis in leaves. RCB formation is also important under stress conditions and for plastid homeostasis (Izumi et al. 2010; Wada and Ishida 2013; Spitzer et al. 2015).

12.3.1.2 Small Starch-Like Granule Bodies (SSGL)

The photosynthetic starch accumulated in the plastid during the day is degraded at night by intraplastidial as well as extraplastidial degradation by piecemeal macroautophagy, which involves SSLG bodies. SSLGs are a class of small vesicular bodies (less than 0.5 μM in diameter). In leaves kept in the dark, stromules extending from the chloroplast envelope, marked by ATG8, bud off SSGLs containing small starch granules labelled with Granule-Bound Starch Synthase I (GBSSI). The large starch granules (1–2 μm) are broken down into small granules (<1 μm) before their transfer to SSGL. These SSGLs are delivered to the vacuoles for further degradation of starch in canonical autophagy-dependent processes (Wang et al. 2013; Otegui et al. 2024).

12.3.1.3 ATI1 (ATG8-Interacting Protein 1)-Plastid-Associated Bodies (ATI-PS)

The name ATI1-PS is given on the basis of the role of autophagy receptor ATI1 (ATG-8 binding transmembrane protein—a membrane-spanning protein) in the formation of these autophagic vesicles (Michaeli et al. 2014). It is of ~50 to 100 nm diameter, contains stromal, thylakoid, and envelope proteins (Avin-Wittenberg et al. 2012; Honig et al. 2012). In carbon-starved senescing cells, ATI1 enters chloroplasts and interacts with chloroplastic proteins localized to stroma, thylakoid, plastoglobuli, and envelope, forming intraplastidic ATI1-positive foci. ATI-PS bodies are only identified in the chloroplast envelope; they are not associated with the stromules. They are extruded from the chloroplast envelope independently of the core ATG machinery, but delivered to the vacuoles for further degradation of proteins in canonical autophagy-dependent processes that depend on the interaction with ATG8 (Michaeli et al. 2014; Wan and Ling 2022; Otegui et al. 2024). ATI-PS bodies are also involved in plant stress tolerance (Michaeli et al. 2014).

12.3.1.4 Neighbour of Brca1 Gene (NBR1)-Mediated Macroautophagy of TOC

Under stressed conditions, such as UV exposure and high temperature, the TOC159 and possibly other translocon subunits of the chloroplast outer membrane (TOC) are primarily ubiquitinated through K63-linked ubiquitination. This ubiquitination enables the binding to the ubiquitin-binding receptor NBR1 and triggers NBR1 recruitment to the surface of chloroplasts (Wan et al. 2023; Saeed et al. 2023). NBR1interacted with ATG8and linked the ubiquitinated TOC cargo to autophagosomal membranes, which was eventually delivered to the vacuole (Wan et al. 2023; Otegui et al. 2024). The ubiquitin-binding NBR1 autophagy receptor has an important role in microautophagy of even whole chloroplasts (Lee et al. 2023).

12.3.1.5 Starch Excess-Associated Chloroplast Autophagy (SEX Chlorophagy): Macrochlorophagy with Unclear Mechanism

Wang et al. (2015) reported a macrophagic degradation mechanism of chloroplast, independent of ATG6-, ATG5- and ATG7-dependent macroautophagy pathways. The autophagosome formation was suppressed, and plastidial starch degradation was impaired by disrupting microtubules via silencing tubulin genes or treating them with a microtubule depolymerising agent. These disorganised chloroplasts are cleared by an autophagy-related pathway in which chloroplasts are selectively transported into the tonoplasts.

12.3.1.6 Unidentified Stroma-Localized Macrochlorophagy Receptor

Ni et al. (2022) reported that under nutrient starvation conditions, 2-phosphoglycolate phosphatase (PGLP) interacts with ATG8 and relocates from the stroma to chloroplast protrusions. The receptor and the exact mechanism of transport are largely unknown (Ni et al. 2022; Otegui et al. 2024).

12.3.2 Microautophagy or Microchlorophagy: Autophagosome-Independent Vesicles

Microautophagy is generally triggered in adverse environmental conditions, as a mechanism evolved for the removal of damaged and defective components of the cell. Microautophagy generally occurs in two ways: the damaged chloroplast gets coated with ATG8 and engulfed by the vacuole; however, no receptors have been identified yet for this canonical autophagy machinery (Izumi et al. 2017). In the second pathway, ubiquitinated chloroplasts undergo piecemeal autophagy by non-canonical autophagy machinery. The microchlorophagy pathway involves NBR1 association with ubiquitinated chloroplasts (Izumi et al. 2017; Otegui et al. 2024). Lee et al. (2023) hypothesized that the cytoplasmic ubiquitination machinery enters the chloroplast as the chloroplasts had lost their membrane integrity. Once thylakoid and stroma proteins are ubiquitinated, they are recognized by NBR1 for autophagic clearance.

12.3.2.1 Senescence–Associated Vacuoles (SAVs)

Senescence-associated vacuoles (SAVs) are identified and named as SAVs by Guiamét and co-authors (Guiamét et al. 1999; Otegui et al. 2005; reviewed in Wada and Ishida 2013). SAVs are a class of small (0.5–0.8 μM in diameter), acidic, lytic vacuoles involved in the natural or stress-induced degradation of the chloroplasts. They develop in mesophyll and guard cells that contain chloroplasts. SAVs are about 0.8 pH units more acidic than the central vacuole and are characterized by high peptidase activity, particularly of cysteine-type proteases (Otegui et al. 2005; Costa et al. 2013). SAG12, which is strictly a senescence-associated cysteine protease and used as the marker for the progression of senescence, is also localized to SAVs (Otegui et al. 2005). SAVs are bound by a single membrane and contain at least one common marker with the central vacuole, i.e., vacuolar H^+ pyrophosphatase (Otegui et al. 2005).

Chloroplast proteins are transported to the SAVs, which contain chloroplast stroma (Rubisco and glutamine synthetase II) proteins, and PSI chlorophyll binding proteins as well as chlorophyll *a* and *b,* but no PSII proteins are detected in SAVs (Martínez et al. 2008; Costa et al. 2013; Wada and Ishida 2013; Gomez et al. 2019).

The origin, as well as the selection and export of chloroplastic proteins to the SAVs, is not very clear. The events taking place in the chloroplast and in the secretory pathway are both hypothesized to be involved in the origin of SAVs. Most probably protuberances being cut off from the chloroplast envelope or stromules are shed off, and the resultant vesicles may later fuse with vesicles coming from the Golgi apparatus and endoplasmic reticulum, containing proteases, hydrolytic enzymes, etc. SAVs eventually fuse with the tonoplast and discharge their contents in the central vacuole (reviewed in Costa et al. 2013).

12.3.2.2 Chloroplast Vesiculation-Containing Vesicles (CCVs)

During developmental senescence and under abiotic stress conditions, chloroplast vesiculation (CV) protein expression is induced in *Arabidopsis*. CV consists of two components, a transit peptide and a transmembrane domain. It interacted with thylakoids and the inner envelope of the plastid after its import into the chloroplast (Otegui et al. 2024). It induces the unstacking of granal thylakoids and other structural changes inside the chloroplast. CV protein's interaction with clathrin, along with the involvement of the dynamin-related protein DRP1A, induces deformation and budding of clathrin-coated vesicles from the inner chloroplast envelope membrane toward the cytosol (Otegui et al. 2024). The CV-positive buds are released into the cytosol and delivered to vacuoles for further degradation. These CV-containing vesicles enclose proteins from the stroma, inner envelopes and thylakoids such as PSII components. Delayed senescence is reported in CV-silenced plants (Wang and Blumwald 2014; Pan et al. 2023; Otegui et al. 2024).

12.3.2.3 Ubiquitin-Mediated Degradation Pathways

Multi-vesicular Bodies (MVBs)

MVBs are specialized endosomes whose content is also degraded in the vacuole. The endosomal sorting complexes required for transport (ESCRT) play an important role in the vacuolar degradation of MVBs. Zhang et al. (2022) reported the role of MVBs in the degradation of chloroplast-destined proteins under heat stress to enhance the thermotolerance of rice plants.

Plant U-Box (PUB4)-Mediated Chloroplast Degradation

Ubiquitination has a very critical role in the chloroplast biogenesis and functions. The role of ubiquitination in chloroplast degradation is still evolving (Wan and Ling 2022). A cytosol-located PLANT U-BOX (PUB4) E3 ubiquitin ligase mediates the ubiquitination of chloroplast proteins under singlet oxygen ($^1O_2$) stress (Woodson et al. 2015). Ubiquitination triggers the chloroplast degradation independent of the autophagic pathway. The ubiquitinated damaged chloroplast could directly fuse with the vacuole (Woodson et al. 2015).

Chloroplast-Associated Protein Degradation (CHLORAD)

Chloroplast-localized SUPPRESOR OF PPI1 LOCUS 1 (SP1), another E3 ubiquitin ligase, is located on the outer envelope membrane (OEM) of the chloroplast. Ubiquitination of TOC proteins is induced by SP1, which are ultimately degraded by the UPS (ubiquitin-proteasome system) (Ling et al. 2019). This pathway involves SUPPRESOR OF PPI1 LOCUS 2 (SP2) and cytosol-located CDC48 proteins; these two proteins form a complex with SP1 at the surface of the chloroplast. The SP2 and CDC48 proteins cooperate for retro-translocation of ubiquitinated proteins from the chloroplast outer envelope for their subsequent degradation in the vacuole (Ling et al. 2019).

12.3.3 Autophagy of Whole Chloroplast

A decrease in the number of chloroplasts is reported during the late phase of natural senescence, dark-induced senescence and senescence induced by nutrient starvation (Wittenbach et al. 1982; Ono et al. 1995; Inada et al. 1998). Electron microscopy and other techniques were used to show that the whole chloroplast can be engulfed in the central vacuole (Minamikawa et al. 2001; Mulisch and Krupinska 2013; Izumi et al. 2017). Wada et al. (2009) showed that engulfment of the whole chloroplast in Arabidopsis is autophagy dependent, as in the double *atg* mutant (*atg4a4b-1*), the number of chloroplasts remains stable and no chloroplasts were found in the vacuole (reviewed in Wada and Ishida 2013; Kato and Sakamoto 2013). The entire chloroplast can be degraded by both macro and microautophagy; however, the chloroplast autophagy is most probably the microautophagy as the size of even shrunken chloroplasts is more than the cargo of macroautophagy, but the exact mechanism of autophagy is still speculative (Wada and Ishida 2013; Otegui et al. 2024). The exposure to high light induces structural damage to the chloroplast envelope. The photodamaged chloroplasts get partially coated with ATG8 and engulfed by the vacuole (Izumi et al. 2017). Nakamura et al. (2018) also reported microautophagic degradation of chloroplasts under high visible light stress. The autophagy-deficient mutants accumulated abnormally swollen chloroplasts under high visible light conditions. GFP-ATG8, a-markers of autophagosomes, partially decorated the membrane of swollen chloroplasts, which were then directly engulfed by the vacuole. Lee et al. (2023) reported the microautophagy-type direct engulfment of the chloroplast into the central vacuole under high light stress. Under intense light, the photodamaged chloroplasts associated with NBR1, both at the surface and in the interior, independently of ATG7. NBR1-coated chloroplasts are directly engulfed into the central vacuole (Lee et al. 2023).

The damaged, abnormally shaped and defective chloroplasts are also degraded by the chloroplast autophagy as the quality control mechanism of cells, throughout the lifespan of leaves and also during the senescence (Niwa et al. 2004; Stettler et al. 2009).

12.4 Conclusions

Undoubtedly, the chloroplast is one of the most important components of a plant cell. Its thylakoid membrane system is the centre of photosynthetic light reaction, and the stroma of dark reaction. Several other key life reactions occur in the stroma that contains starch granules, numerous proteins and metabolites. Biogenesis and degradation of chloroplasts are highly organised, very well programmed, extremely complex, and genetically controlled. Over the last 20 years, there has been tremendous growth in the understanding of the mechanism of bulk chloroplast degradation. However, it is not yet fully elucidated. There are several knowledge gaps in our understanding of chloroplast degradation, e.g., the molecular mechanism of degradation of Rubisco, which comprises 20–30% of total leaf nitrogen, is still largely unclear. Two extraplastidial vesicles, RCVs (macrochlorophagy pathway) and SAVs (microchlorophagy pathway), are involved in the autophagic degradation of Rubisco. RCBs are involved in the vesicle-mediated extraplastidial macroautophagy of Rubisco, but the mechanism of RCB formation from chloroplasts is not yet clearly understood. The molecular mechanisms for the RBC's content selection and whether any specific receptors are involved in this process are currently unknown. SAVs are involved in the vesicle-mediated extraplastidial microautophagy of Rubisco, but the mechanism of SAV formation is not very clear. The mechanism of sorting of stromal and PSI proteins for transport to SAVs is largely unknown. The final fate of SAVs is also not clearly understood. The coordination between multiple degradation pathways is largely unknown. Lipid degradation is one of the key aspects of chloroplast degradation. It has been identified that during senescence, the expression of up to 68 putative lipases increases (Troncoso-Ponce et al. 2013). The role of these lipases during senescence remains to be elucidated. In contrast to macroautophagy, the molecular mechanism of microautophagy is still largely unknown. The molecular mechanism involved in the clustering and release of chloroplast cargo and whole chloroplasts through ATI1- and/or NBR1-mediated macrochlorophagy is largely unknown. Moreover, in what way chloroplasts are recognized for degradation is still unclear and needs further investigation.

References

Adam Z, Clarke AK (2002) Cutting edge of chloroplast proteolysis. Trends Plant Sci 7:451–456

Albanese P, Tamara S, Saracco G, Scheltema RA, Pagliano C (2020) How paired PSII-LHCII supercomplexes mediate the stacking of plant thylakoid membranes unveiled by structural mass-spectrometry. Nat Commun 11:1361

Andersson A, Keskistalo J, Sjödin A, Bhalerao R, Sterky F, Wissel K, Tandre K, Aspeborg H, Moyle R, Ohmiya Y, Bhalerao R (2004) A transcriptional timetable of autumn senescence. Genome Biol 5:R24

Armbruster U, Strand DD (2020) Regulation of chloroplast primary metabolism. Photosynth Res 145:1–3

Arzac MI, Fernández-Marín B, García-Plazaola JI (2022) More than just lipid balls: quantitative analysis of plastoglobule attributes and their stress-related responses. Planta 255:62

Austin JR, Frost E, Vidi PA, Kessler F, Staehelin LA (2006) Plastoglobules are lipoprotein subcompartments of the chloroplast that are permanently coupled to thylakoid membranes and contain biosynthetic enzymes. Plant Cell 18:1693–1703

Avin-Wittenberg T, Michaeli S, Honig A, Galili G (2012) ATI1, a newly identified atg8-interacting protein, binds two different Atg8 homologs. Plant Signal Behav 7:685–687

Bar-On YM, Milo R (2019) The global mass and average rate of rubisco. Proc Natl Acad Sci USA 10:4738–4743

Besagni C, Kessler F (2013) A mechanism implicating plastoglobules in thylakoid disassembly during senescence and nitrogen starvation. Planta 237:463–470

Bhalerao R, Keskitalo J, Sterky F, Erlandsson R, Björkbacka H, JonssonBirve S, Karlsson J, Gardeström P, Gustafsson P, Lundeberg J, Jansson S (2003) Gene expression in autumn leaves. Plant Physiol 131:430–442

Biswal B (2005) Formation and demolition of chloroplast during leaf ontogeny. In: Pessarakli M (ed) Handbook of photosynthesis, 2nd edn. CRC Press, Boca Raton, pp 109–122

Biswal B, Biswal UC (1988) Ultrastructural modifications and biochemical changes during senescence of chloroplasts. Int Rev Cytol 113:271–321

Biswal B, Pandey JK (2016) Development of chloroplast: biogenesis, senescence, and regulations. In: Pessarakli M (ed) Handbook of photosynthesis, 3rd edn. CRC Press, Boca Raton, pp 77–93

Biswal B, Pandey JK (2018) Loss of photosynthesis signals a metabolic reprogramming to sustain sugar homeostasis during senescence of green leaves: role of cell wall hydrolases. Photosynthetica 56:404–410

Brunkard JO, Runkel AM, Zambryski PC (2015) Chloroplasts extend stromules independently and in response to internal redox signals. Proc Natl Acad Sci USA 112:10044–10049

Bu F, Yang M, Guo X, Huang W, Chen L (2020) Multiple functions of ATG8 family proteins in plant autophagy. Front Cell Dev Biol 8:466

Buchanan-Wollaston V, Earl S, Harrison E, Mathas E, Navadpour S, Page T, Pink D (2003) The molecular analysis of senescence – a genomics approach. Plant Biotechnol J 1:3–22

Caplan JL, Kumar AS, Park E, Padmanabhan MS, Hoban K, Modla S, Czymmek K, Dinesh-Kumar SP (2015) Chloroplast stromules function during innate immunity. Dev Cell 34:45–57
Cejudo FJ, Gonzalez MC, Perez-Ruiz JM (2021) Redox regulation of chloroplast metabolism. Plant Physiol 186:9–21
Chiba A, Ishida H, Nishizawa NK, Makino A, Mae T (2003) Exclusion of ribulose-1,5-bisphosphate carboxylase/oxygenase from chloroplasts by specific bodies in naturally senescing leaves of wheat. Plant Cell Physiol 44:914–921
Christ B, Süssenbacher I, Moser S, Bichsel N, Egert A, Müller T, Kräutler B, Hörtensteiner S (2013) Cytochrome P450 CYP89A9 is involved in the formation of major chlorophyll catabolites during leaf senescence in *Arabidopsis*. Plant Cell 25:1868–1880
Costa ML, Martínez DE, Gomez FM, Carrión CA, Guiamet JJ (2013) Chloroplast protein degradation: involvement of senescence-associated vacuoles. In: Biswal B, Krupinska K, Biswal U (eds) Plastid development in leaves during growth and senescence, Advances in photosynthesis and respiration, vol 36. Springer, Dordrecht, pp 417–433
Coulon D, Nacir H, Bahammou D, Jouhet J, Bessoule JJ, Fouillen L, Bréhélin C (2024) Respective role of plastoglobules and lipid droplets in leaf neutral lipid accumulation during senescence and nitrogen deprivation. J Exp Bot erae301
Domínguez F, Cejudo FJ (2021) Chloroplast dismantling in leaf senescence. J Exp Bot 72:5905–5918
Feild TS, Lee DW, Holbrook NM (2001) Why leaves turn red in autumn. The role of anthocyanins in senescing leaves of red-osier dogwood. Plant Physiol 127:566–574
Gomez FM, Carrión CA, Costa ML, Desel C, Kieselbach T, Funk C, Krupinska K, Guiamet J (2019) Extra-plastidial degradation of chlorophyll and photosystem I in tobacco leaves involving 'senescence-associated vacuoles'. Plant J99:465–477
Green BR, Durnford DG (1996) Chlorophyll-carotenoid proteins of oxygenic photosynthesis. Annu Rev Plant Physiol 47:685–714
Gregersen PL, Holm PB (2007) Transcriptome analysis of senescence in the flag leaf of wheat (*TriticumaestivumL.*). Plant Biotechnol J 5:192–206
Guiamét JJ, Pichersky E, Noodén LD (1999) Mass exodus from senescing soybean chloroplasts. Plant Cell Physiol 40:986–992
Guiamét JJ, Tyystjärvi E, Tyystjärvi T, John I, Kairavuo M, Pichersky P, Noodén LD (2002) Photoinhibition and loss of photosystem II reaction center proteins during senescence of soybean leaves. Enhancement of photoinhibition by the "stay-green" mutation *cytG*. Physiol Plant 115:468–478
Guo Y, Cai Z, Gan S (2004) Transcriptome of *Arabidopsis* leaf senescence. Plant Cell Environ 27:521–549
Harris JB (1978) Development of a tubular apparatus in chloroplasts of aging *Cyphomandra* leaves. Cytobios 21:151–164
Hauenstein M, Christ B, Das A, Aubry S, Hörtensteiner S (2016) A role for TIC55 as a hydroxylase of phyllobilins, the products of chlorophyll breakdown during plant senescence. Plant Cell 28:2510–2527
Hendry GAF, Houghton JD, Brown SB (1987) The degradation of chlorophyll – a biological enigma. New Phytol 107:255–302
Hidema J, Makino A, Mae T, Ojima K (1991) Photosynthetic characteristics of rice leaves aged under different irradiances from full expansion through senescence. Plant Physiol 97:1287–1293
Hidema J, Makino A, Kurita Y, Mae T, Ojima K (1992) Changes in the levels of chlorophyll and light harvesting chlorophyll *a/b* protein of PSII in rice leaves aged under different irradiances from full expansion through senescence. Plant Cell Physiol 33:1209–1214
Holloway PJ, Maclean DJ, Scott KJ (1983) Rate limiting steps of electron transport in chloroplasts during ontogeny and senescence of barley. Plant Physiol 72:795–801
Honig A, Avin-Wittenberg T, Ufaz S, Galili G (2012) A new type of compartment, defined by plant-specific Atg8-interacting proteins, is induced upon exposure of *Arabidopsis* plants to carbon starvation. Plant Cell 24:288–303

Hu X, Gu T, Khan I, Zada A, Jia T (2021) Research progress in the interconversion, turnover and degradation of chlorophyll. Cells 10:3134

Humbeck K, Quast S, Krupinska K (1996) Functional and molecular changes in the photosynthetic apparatus during senescence of flag leaves from field-grown barley plants. Plant Cell Environ 19:337–344

Inada N, Sakai A, Kuroiwa H, Kuroiwa T (1998) Three dimensional analysis of the senescence program in rice (*Oryza sativa* L.) coleoptiles – investigations of tissues and cells by fluorescence microscopy. Planta 205:153–164

Ishida H, Yoshimoto K, Izumi M, Reisen D, Yano Y, Makino A, Ohsumi Y, Hanson MR, Mae T (2008) Mobilization of rubisco and stroma-localized fluorescent proteins of chloroplasts to the vacuole by an *ATG* gene-dependent autophagic process. Plant Physiol 148:142–155

Izumi M, Wada S, Makino A, Ishida H (2010) The autophagic degradation of chloroplasts via rubisco-containing bodies is specifically linked to leaf carbon status but not nitrogen status in *Arabidopsis*. Plant Physiol 2010:1196–1209

Izumi M, Ishida H, Nakamura S, Hidema J (2017) Entire photodamaged chloroplasts are transported to the central vacuole by autophagy. Plant Cell 29:377–394

Izumi M, Nakamura S, Otomo K, Ishida H, Hidema J, Nemoto T, Hagihara S (2023) Autophagosome development and chloroplast segmentation occur synchronously for piecemeal degradation of chloroplasts. eLife 12

Jia Y, Li W (2015) Characterisation of lipid changes in ethylene-promoted senescence and its retardation by suppression of phospholipase Dδ in *Arabidopsis* leaves. Front Plant Sci 6:1045

Jia Y, Tao F, Li W (2013) Lipid profiling demonstrates that suppressing *Arabidopsis* phospholipase Dδ retards ABA-promoted leaf senescence by attenuating lipid degradation. PLoS One 8:e65687

Kato Y, Sakamoto W (2010) New insight into the type and function of proteases in plastids. In: Jeon KW (ed) International review of cell and molecular biology, vol 161. Burlington Academic Press, Burlington, pp 185–218

Kato Y, Sakamoto W (2013) Plastid protein degradation during leaf development and senescence: role of proteases and chaperones. In: Biswal B, Krupinska K, Biswal U (eds) Plastid development in leaves during growth and senescence, Advances in photosynthesis and respiration, vol 36. Springer, Dordrecht, pp 453–477

Kato Y, Murakami S, Yamamoto Y, Chatani H, Kondo Y, Nakano T, Yokota A, Sato F (2004) The DNA-binding protease, CND41, and the degradation of ribulose-1,5-bisphosphate carboxylase/oxygenase in senescent leaves of tobacco. Planta 220:97–104

Kato Y, Yamamoto Y, Murakami S, Sato F (2005) Posttranslational regulation of CND41 protease activity in senescent tobacco leaves. Planta 222:643–651

Kato Y, Miura E, Ido K, Ifuku K, Sakamoto W (2009) The variegated mutants lacking chloroplastic FtsHs are defective in D1 degradation and accumulate reactive oxygen species. Plant Physiol 151:1790–1801

Kato Y, Sun X, Zhang L, Sakamoto W (2012) Cooperative D1 degradation in the photosystem II repair mediated by chloroplastic proteases in Arabidopsis. Plant Physiol 159:1428–1439

Kirchhoff H (2019) Chloroplast ultrastructure in plants. New Phytol 223:565–574

Kräutler B (2016) Breakdown of chlorophyll in higher plants-phyllobilins as abundant, yet hardly visible signs of ripening, senescence, and cell death. Angew Chem Int Edit 55:4882–4907

Krupinska K (2007) Fate and activities of plastids during leaf senescence. In: Wise RR, Hoober JK (eds) The structure and function of plastids. Springer, Dordrecht, pp 433–449

Kuai B, Chen J, Hörtensteiner S (2018) The biochemistry and molecular biology of chlorophyll breakdown. J Exp Bot 69:751–767

Lee RH, Hsu JH, Huang HJ, Lo SF, Grace Chen SC (2009) Alkaline α-galactosidase degrades thylakoid membranes in the chloroplast during leaf senescence in rice. New Phytol 184:596–606

Lee HN, Chacko JV, Gonzalez Solís A, Chen K-E, Barros JAS, Signorelli S, Millar AH, Vierstra RD, Eliceiri KW, Otegui MS (2023) The autophagy receptor NBR1 directs the clearance of photodamaged chloroplasts. Elife 12:e86030

Lei P, Yu F, Liu X (2023) Recent advances in cellular degradation and nuclear control of leaf senescence. J Exp Bot 74:5472–5486

Lensch M, Herrmann RG, Sokolenko A (2001) Identification and characterization of SppA, a novel light-inducible chloroplast protease complex associated with thylakoid membranes. J Biol Chem 276:33645–33651

Lichtenthaler HK (1968) Plastoglobules and the fine structure of plastids. Endeavour 27:144–149

Lichtenthaler HK (2013) Plastoglobuli, thylakoids, chloroplast structure and development of plastids. In: Biswal B, Krupinska K, Biswal U (eds) Plastid development in leaves during growth and senescence, Advances in photosynthesis and respiration, vol 36. Springer, Dordrecht, pp 337–361

Lichtenthaler HK, Sprey B (1966) Ultrastructural changes in chloroplasts of detached parsley leaves. Z Naturforsch 21b:690–697

Lichtenthaler HK, Weinert H (1970) Die Beziehungenzwischen Lipochinonbiosynthese und Plastoglobulibildung in den Chloroplasten von *Ficus elastic* Roxb. Zeitschriftfür Naturforschung 25b:619–623

Ling Q, Broad W, Trösch R, Töpel M, DemiralSert T, Lymperopoulos P, Baldwin A, Jarvis RP (2019) Ubiquitin-dependent chloroplast-associated protein degradation in plants. Science 363:eaav4467

Liu W, Liu Z, Mo Z, Guo S, Liu Y, Xie Q (2021) ATG8-interacting motif: evolution and function in selective autophagy of targeting biological processes. Front Plant Sci 12:783881

Lu C, Lu Q, Zhang J, Kuang T (2001) Characterization of photosynthetic pigment composition, photosystem II photochemistry and thermal energy dissipation during leaf senescence of wheat plants grown in the field. J Exp Bot 52:1805–1810

Mae T (2004) Leaf senescence and nitrogen metabolism. In: Noodén LD (ed) Plant cell death processes. Academic, San Diego, pp 157–168

Majeran W, Wollman FA, Vallon O (2000) Evidence for a role of ClpP in the degradation of the chloroplast cytochrome b_6f complex. Plant Cell 12:137–150

Makino A, Osmond B (1991) Effects of nitrogen nutrition on nitrogen partitioning between chloroplasts and mitochondria in pea and wheat. Plant Physiol 96:355–362

Martínez DE, Costa ML, Guiamet JJ (2008) Senescence-associated degradation of chloroplast proteins inside and outside the organelle. Plant Biol 10:15–22

Matsuda K, Shimoda Y, Tanaka A, Ito H (2016) Chlorophyll a is a favorable substrate for Chlamydomonas Mg-dechelatase encoded by STAY-GREEN. Plant Physiol Biochem 109:365–373

Michaeli S, Honig A, Levanony H, Peled-Zehavi H, Galili G (2014) *Arabidopsis* ATG8-INTERACTING PROTEIN1 is involved in autophagy- dependent vesicular trafficking of plastid proteins to the vacuole. Plant Cell 26:4084–4101

Miersch I, Heise J, Zelmer I, Humbeck K (2000) Differential degradation of the photosynthetic apparatus during leaf senescence in barley (*Hordeumvulgare*L.). Plant Biol 2:618–623

Minamikawa T, Toyooka K, Okamoto T, Hara-Nishimura I, Nishimura M (2001) Degradation of ribulose-bisphosphate carboxylase by vacuolar enzymes of senescing French bean leaves: immunocytochemical and ultrastructural observations. Protoplasma 218:144–153

Moser D, Matile P (1997) Chlorophyll breakdown in ripening fruits of *Capsicum annuum*. J Plant Physiol 150:759–761

Mulisch M, Krupinska K (2013) Ultrastructural analyses of senescence associated dismantling of chloroplasts revisited. In: Biswal B, Krupinska K, Biswal U (eds) Plastid development in leaves during growth and senescence, Advances in photosynthesis and respiration, vol 36. Springer, Dordrecht, pp 307–335

Müller T, Rafelsberger M, Vergeiner C, Kräutler B (2011) A dioxobilane as product of a divergent path of chlorophyll breakdown in Norway maple. Angew Chem Int Edit 50:10724–10727

Nakamura S, Hidema J, Sakamoto W, Ishida H, Izumi M (2018) Selective elimination of membrane-damaged chloroplasts via microautophagy. Plant Physiol 177:1007–1026

Ni J, Li Y, Xiang Y, Yang X, Jia L, Yue J, Wang H (2022) Autophagic degradation of the chloroplastic2-phosphoglycolate phosphatase TaPGLP1 in wheat. Plant Cell Rep 41:473–487

Niwa Y, Kato T, Tabata S, Seki M, Kobayashi M, Shinozaki K, Moriyasu Y (2004) Disposal of chloroplasts with abnormal function into the vacuole in *Arabidopsis thaliana* cotyledon cells. Protoplasma 223:229–232

Noodén LD (1988) The phenomena of senescence and aging. In: Noodén LD, Leopold AC (eds) Senescence and aging in plants. Academic, San Diego, pp 1–50

Ono K, Hashimoto H, Katoh S (1995) Changes in the number and size of chloroplasts during senescence of primary leaves of wheat grown under different conditions. Plant Cell Physiol 36:9–17

Ostersetzer O, Kato Y, Adam Z, Sakamoto W (2007) Multiple intracellular locations of Lon protease in *Arabidopsis*: evidence for the localization of AtLon4 to chloroplasts. Plant Cell Physiol 48:881–885

Otegui MS (2018) Vacuolar degradation of chloroplast components: autophagy and beyond. J Exp Bot 69:741–750

Otegui M, Noh Y-S, Martínez D, Vila-Petroff M, Staehelin A, Amasino R, Guiamet JJ (2005) Senescence-associated vacuoles with intense proteolytic activity develop in senescing leaves of *Arabidopsis* and soybean. Plant J 41:831–844

Otegui MS, Steelheart C, Ma W, Ma J, Kang BH, De Medina Hernandez VS, Dagdas Y, Gao C, Goto-Yamada S, Oikawa K, Nishimura M (2024) Vacuolar degradation of plant organelles. Plant Cell 36:3036–3056

Pan T, Liu Y, Hu X, Li P, Lin C, Tang Y, Tang W, Liu Y, Guo L, Kim C, Fang J (2023) Stress-induced endocytosis from chloroplast inner envelope membrane is mediated by CHLOROPLAST VESICULATION but inhibited by GAPC. Cell Rep 42:113208

Pandey JK, Biswal B (2024) Sugar homeostasis in leaves during senescence and abiotic stress: role of cell wall polysaccharides. Biologia 79:2347–2357

Pandey JK, Dash SK, Biswal B (2017a) Loss in photosynthesis during senescence is accompanied by an increase in the activity of β-galactosidase in leaves of *Arabidopsis thaliana*: modulation of the enzyme activity by water stress. Protoplasma 254:1651–1659

Pandey JK, Dash SK, Biswal B (2017b) Nitrogen-deficiency-induced loss in photosynthesis and modulation of β-galactosidase activity during senescence of Arabidopsis leaves. Acta Physiol Plant 39:75

Park H, Eggink LL, Roberson RW, Hoober JK (1999) Transfer of proteins from the chloroplast to vacuoles in *Chlamydomonasreinhardtii* (chlorophyta): a pathway for degradation. J Phycol 35:528–538

Prakash JS, Baig MA, Mohanty P (2001) Senescence induced structural reorganization of thylakoid membranes in *Cucumissativus* cotyledons; LHC II involvement in reorganization of thylakoid membranes. Photosynth Res 68:153–161

Prakash JS, Baig MA, Bhagwat AS, Mohanty P (2003) Characterisation of senescence-induced changes in light harvesting complex II and photosystem I complex of thylakoids of *Cucumissativus*cotyledons: age induced association of LHCII with photosystem I. J Plant Physiol 160:175–184

Rottet S, Devillers J, Glauser G, Douet V, Besagni C, Kessler F (2016) Identification of plastoglobules as a site of carotenoid cleavage. Front Plant Sci 7:1855

Saeed B, Deligne F, Brillada C, Dünser K, Ditengou FA, Turek I, Allahham A, Grujic N, Dagdas Y, Ott T, Kleine-Vehn J (2023) K63-linked ubiquitin chains are a global signal for endocytosis and contribute to selective autophagy in plants. Curr Biol 33:1337–1345

Sakamoto W (2006) Protein degradation machineries in plastids. Annu Rev Plant Biol 57:599–621

Schwarz EM, Tietz S, Froehlich JE (2018) Photosystem I-LHCII megacomplexes respond to high light and aging in plants. Photosynth Res 136:107–124

Shimoda Y, Ito H, Tanaka A (2016) *Arabidopsis* STAY-GREEN, Mendel's green cotyledon gene, encodes magnesium-dechelatase. Plant Cell 28:2147–2160

Sitte P (1977) Chromoplasten – bunte Objekte der modernen Zellbiologie. BIUZ 7:65–74

Spitzer C, Li F, Buono R, Roschzttardtz H, Chung T, Zhang M, Osteryoung KW, Vierstra RD, Otegui MS (2015) The endosomal protein CHARGED MULTIVESICULAR BODY PROTEIN1 regulates the autophagic turnover of plastids in *Arabidopsis*. Plant Cell 27:391–402

Springer A, Kang CH, Rustgi S, von Wettstein D, Reinbothe C, Pollmann S, Reinbothe S (2016) Programmed chloroplast destruction during leaf senescence involves 13-lipoxygenase (13-LOX). Proc Natl Acad Sci USA 113:3383–3388

Stettler M, Eicke S, Mettler T, Messerli G, Hörtensteiner S, Zeeman SC (2009) Blocking the metabolism of starch breakdown products in *Arabidopsis* leaves triggers chloroplast degradation. Mol Plant 2:1233–1246

Süssenbacher I, Hörtensteiner S, Kräutler B (2015) A dioxobilin-type fluorescent chlorophyll catabolite as a transient early intermediate of the dioxobilin-branch of chlorophyll breakdown in *Arabidopsis thaliana*. Angew Chem Int Edit 54:13777–13781

Tang Y, Wen X, Lu C (2005) Differential changes in degradation of chlorophyll-protein complexes of photosystem I and photosystem II during flag leaf senescence of rice. Plant Physiol Biochem 43:193–201

Troncoso-Ponce MA, Cao X, Yang Z, Ohlrogge JB (2013) Lipid turnover during senescence. Plant Sci 205–206:13–19

Tuquet C, Newman DW (1980) Aging and regreening in soybean cotyledons. 1. Ultrastructural changes in plastids and plastoglobuli. Cytobios 29:43–59

van Eerden FJ, de Jong DH, de Vries AH, Wassenaar TA, Marrink SJ (2015) Characterization of thylakoid lipid membranes from cyanobacteria and higher plants by molecular dynamics simulations. BBA-Biomembranes 1848:1319–1330

Wada S, Ishida H (2013) Autophagy of chloroplasts during leaf senescence. In: Biswal B, Krupinska K, Biswal U (eds) Plastid development in leaves during growth and senescence, Advances in photosynthesis and respiration, vol 36. Springer, Dordrecht, pp 435–451

Wada S, Ishida H, Izumi M, Yoshimoto K, Ohsumi Y, Mae T, Makino A (2009) Autophagy plays a role in chloroplast degradation during senescence in individually darkened leaves. Plant Physiol 149:885–893

Wan C, Ling Q (2022) Functions of autophagy in chloroplast protein degradation and homeostasis. Front Plant Sci 13:993215

Wan T, Li M, Zhao X, Zhang J, Liu Z, Chang W (2014) Crystal structure of a multilayer packed major light-harvesting complex: implications for grana stacking in higher plants. Mol Plant 7:916–919

Wan C, Zhang H, Cheng H, Sowden RG, Cai W, Jarvis RP, Ling Q (2023) Selective autophagy regulates chloroplast protein import and promotes plant stress tolerance. EMBO J 42:e112534

Wang S, Blumwald E (2014) Stress-induced chloroplast degradation in *Arabidopsis* is regulated via a process independent of autophagy and senescence-associated vacuoles. Plant Cell 26:4875–4888

Wang Y, Yu B, Zhao J, Guo J, Li Y, Han S, Huang L, Du Y, Hong Y, Tang D, Liu Y (2013) Autophagy contributes to leaf starch degradation. Plant Cell 25:1383–1399

Wang Y, Zheng X, Yu B, Han S, Guo J, Tang H, Yu AYL, Deng H, Hong Y, Liu Y (2015) Disruption of microtubules in plants suppresses macroautophagy and triggers starch excess-associated chloroplast autophagy. Autophagy 11:2259–2274

Wang K, Guo Q, Froehlich JE, Hersh HL, Zienkiewicz A, Howe GA, Benning C (2018) Two abscisic acid-responsive plastid lipase genes involved in jasmonic acid biosynthesis in *Arabidopsis thaliana*. Plant Cell 30:1006–1022

Wang J, Zhang Q, Bao Y, Bassham DC (2023) Autophagic degradation of membrane-bound organelles in plants. Biosci Rep 43:BSR20221204

Wittenbach VA, Lin W, Hebert RR (1982) Vacuolar localization of proteases and degradation of chloroplasts in mesophyll protoplasts from senescing primary wheat leaves. Plant Physiol 69:98–102

Woodson JD, Joens MS, Sinson AB, Gilkerson J, Salom PA, Weigel D, Fitzpatrick JA, Chory J (2015) Ubiquitin facilitates a quality-control pathway that removes damaged chloroplasts. Science 350:450–454

Wrischer M, Prebeg T, Magnus V, Ljubesic N (2009) Unusual thylakoid structures appearing during degradation of the photosynthetic apparatus in chloroplasts. Acta Bot Croat 68:1–9

Yoshioka-Nishimura M (2016) Close relationships between the PSII repair cycle and thylakoid membrane dynamics. Plant Cell Physiol 57:1115–1122

Zaltsman A, Ori N, Adam Z (2005) Two types of FtsH subunits are required for chloroplast biogenesis and Photosystem II repair in *Arabidopsis*. Plant Cell 17:2782–2790

Zhang H, Zhou JF, Kan Y, Shan JX, Ye WW, Dong NQ, Guo T, Xiang YH, Yang YB, Li YC, Zhao HY (2022) A genetic module at one locus in rice protects chloroplasts to enhance thermotolerance. Science 376:1293–1300

Zhao J, Zhang X, Hong Y, Liu Y (2016) Chloroplast in plant-virus interaction. Front Microbiol 7:1565

Zhuang X, Jiang L (2019) Chloroplast degradation: multiple routes into the vacuole. Front Plant Sci 10:359

MIX
Papier aus verantwortungsvollen Quellen
Paper from responsible sources
FSC® C105338

If you have any concerns about our products,
you can contact us on
ProductSafety@springernature.com

In case Publisher is established outside the EU,
the EU authorized representative is:
Springer Nature Customer Service Center GmbH
Europaplatz 3, 69115 Heidelberg, Germany

Printed by Libri Plureos GmbH
in Hamburg, Germany